Nonparametric Regression Methods for Longitudinal Data Analysis

WILEY SERIES IN PROBABILITY AND STATISTICS

Established by WALTER A. SHEWHART and SAMUEL S. WILKS

A complete list of the titles in this series appears at the end of this volume.

Nonparametric Regression Methods for Longitudinal Data Analysis

HULIN WU

University of Rochester
Dept. of Biostatistics and Computer Biology
Rochester, New York

JIN-TING ZHANG

National University of Singapore
Dept. of Biostatistics and Applied Probability
Singapore

WILEY-INTERSCIENCE

A JOHN WILEY & SONS, INC., PUBLICATION

Published by John Wiley & Sons, Inc., Hoboken, New Jersey.
Published simultaneously in Canada.

Library of Congress Cataloging-in-Publication Data is available.

ISBN-13 978-0-471-48350-2
ISBN-10 0-471-48350-8

Printed in the United States of America.

10 9 8 7 6 5 4 3 2 1

To Chuan-Chuan, Isabella, and Gabriella
To Yan and Tian-Hui
To Our Parents and Teachers

Preface

Nonparametric regression methods for longitudinal data analysis have been a popular statistical research topic since the late 1990s. The needs of longitudinal data analysis from biomedical research and other scientific areas along with the recognition of the limitation of parametric models in practical data analysis have driven the development of more innovative nonparametric regression methods. Because of the flexibility in the form of regression models, nonparametric modeling approaches can play an important role in exploring longitudinal data, just as they have done for independent cross-sectional data analysis. Mixed-effects models are powerful tools for longitudinal data analysis. Linear mixed-effects models, nonlinear mixed-effects models and generalized linear mixed-effects models have been well developed to model longitudinal data, in particular, for modeling the correlations and within-subject/between-subject variations of longitudinal data. The purpose of this book is to survey the nonparametric regression techniques for longitudinal data analysis which are widely scattered throughout the literature, and more importantly, to systematically investigate the incorporation of mixed-effects modeling techniques into various nonparametric regression models.

The focus of this book is on modeling ideas and inference methodologies, although we also present some theoretical results for the justification of the proposed methods. The data analysis examples from biomedical research are used to illustrate the methodologies throughout the book. We regard the application of the statistical modeling technologies to practical scientific problems as important. In this book, we mainly concentrate on the major nonparametric regression and smoothing methods including local polynomial, regression spline, smoothing spline and penalized spline approaches. Linear and nonlinear mixed-effects models are incorporated in these smoothing methods to deal with continuous longitudinal data, and generalized linear and additive mixed-effects models are coupled with these nonparametric modeling techniques to handle discrete longitudinal data. Nonparametric models as well as semiparametric and time varying coefficient models are carefully investigated.

Chapter 1 provides a brief overview of the book chapters, and in particular, presents data examples from biomedical research studies which have motivated the use of nonparametric regression analysis approaches. Chapters 2 and 3 review mixed-effects models and nonparametric regression methods, the two important building blocks of the proposed modeling techniques. Chapters 4~7 present the core contents of this book with each chapter covering one of the four major nonparametric regression methods including local polynomial, regression spline, smoothing spline and penalized spline. Chapters 8 and 9 extend the modeling techniques in Chapters 4~7 to semiparametric and time varying coefficient models for longitudinal data analysis. The last chapter, Chapter 10, covers discrete longitudinal data modeling and analysis.

Most of the contents of this book should be comprehensible to readers with some basic statistical training. Advanced mathematics and technical skills are not necessary for understanding the key modeling ideas and for applying the analysis methods to practical data analysis. The materials in Chapters 1~7 can be used in a lower or medium level graduate course in statistics or biostatistics. Chapters 8~10 can be used in a higher level graduate course or as reference materials for those who intend to do research in this area.

We have tried our best to acknowledge the work of many investigators who have contributed to the development of the models and methodologies for nonparametric regression analysis of longitudinal data. However, it is beyond the scope of this project to prepare an exhaustive review of the vast literature in this active research field and we regret any oversight or omissions of particular authors or publications.

We would like to express our sincere thanks to Ms. Jeanne Holden-Wiltse for helping us with polishing and editing the manuscript. We are grateful to Ms. Susanne Steitz and Mr. Steve Quigley at John Wiley & Sons, Inc. who have made great efforts in coordinating the editing, review, and finally the publishing of this book. We would like to thank our colleagues, collaborators and friends, Zongwu Cai, Raymond Carroll, Jianqing Fan, Kai-Tai Fang, Hua Liang, James S. Marron, Yanqing Sun, Yuedong Wang, and Chunming Zhang for their fruitful collaborations and valuable inspirations. Thanks also go to Ollivier Hyrien, Hua Liang, Sally Thurston, and Naisyin Wang for their review and comments on some chapters of the book. We thank our families and loved ones who provided strong support and encouragement during the writing process of this book. We are grateful to our teachers and academic mentors, Fred W. Huffer, Jinhuai Zhang, Jianqing Fan, Kai-Tai Fang and James S. Marron, for guiding us to the beauty of statistical research. J.-T. Zhang also would like to acknowledge Professors Zhidong Bai, Louis H. Y. Chen, Kwok Pui Choi and Anthony Y. C. Kuk for their support and encouragement.

Wu's research was partially supported by grants from the National Institute of Allergy and Infectious Diseases, the National Institutes of Health (NIH). Zhang's research was partially supported by the National University of Singapore Academic

Research grant R-155-000-038-112. The book was written with partial support from the Department of Biostatistics and Computational Biology, University of Rochester, where the second author was a Visiting Professor.

HULIN WU AND JIN-TING ZHANG

University of Rochester
Department of Biostatistics and Computational Biology
Rochester, NY, USA
and
National University of Singapore
Department of Statistics and Applied Probability
Singapore

Contents

Guide to Notation

We use lowercase letters (e.g., a, x, and α) to denote scalar quantities, either fixed or random. Occasionally, we also use uppercase letters (e.g., X, Y) to denote random variables. Lowercase bold letters (e.g., $\mathbf{x}$ and $\mathbf{y}$) will be used for vectors and uppercase bold letters (e.g., $\mathbf{A}$ and $\mathbf{Y}$) will be used for matrices. Any vector is assumed to be a column vector. The transposes of a vector $\mathbf{x}$ and a matrix $\mathbf{X}$ are denoted as $\mathbf{x}^T$ and $\mathbf{X}^T$ respectively. Thus, a row vector is denoted as $\mathbf{x}^T$.

We use diag($\mathbf{a}$) to denote a diagonal matrix whose diagonal entries are the entries of $\mathbf{a}$, and use diag($\mathbf{A}_1, \cdots, \mathbf{A}_n$) to denote a block diagonal matrix. We use $\mathbf{A} \otimes \mathbf{B}$ to denote the Kronecker product, $(a_{ij}\mathbf{B})$, of two matrices $\mathbf{A}$ and $\mathbf{B}$.

The symbol "$\equiv$" means "equal by definition". The L_2-norm of a vector $\mathbf{x}$ is denoted as $\|\mathbf{x}\| \equiv \sqrt{\mathbf{x}^T\mathbf{x}}$. For a function of a scalar x, $f^{(r)}(x) \equiv d^r f(x)/dx^r$ denotes the r-th derivative of $f(x)$. The estimator of $f^{(r)}(x)$ is denoted as $\hat{f}^{(r)}(x)$.

For a longitudinal data set, n denotes the number of subjects, n_i denotes the number of measurements for the i-th subject, and t_{ij} denotes the design time point for the j-th measurement of the i-th subject. The response value, the fixed-effects and random-effects covariate vectors at time t_{ij} are often denoted as y_{ij}, $\mathbf{x}_{ij}$ and $\mathbf{z}_{ij}$, respectively. We use $\mathbf{y}_i = [y_{i1}, \cdots, y_{in_i}]^T$, $\mathbf{X}_i = [\mathbf{x}_{i1}, \cdots, \mathbf{x}_{in_i}]^T$ and $\mathbf{Z}_i = [\mathbf{z}_{i1}, \cdots, \mathbf{z}_{in_i}]^T$ to denote the response vector, the fixed-effects and random-effects design matrices for the i-th subject, and use $\mathbf{y} = [\mathbf{y}_1^T, \cdots, \mathbf{y}_n^T]^T$, $\mathbf{X} = [\mathbf{X}_1^T, \cdots, \mathbf{X}_n^T]^T$ and $\mathbf{Z} =$ diag($\mathbf{Z}_1, \cdots, \mathbf{Z}_n$) to denote the response vector, the fixed-effects and random-effects design matrices for the whole data set. We often use $\boldsymbol{\alpha}, \boldsymbol{\beta}$ or $\boldsymbol{\alpha}(t), \boldsymbol{\beta}(t)$ to denote the fixed-effects or fixed-effects functions, and use $\mathbf{a}_i$, $\mathbf{b}_i$ or $v_i(t)$, $\mathbf{v}_i(t)$ to denote the random-effects or random-effects functions. For the whole longitudinal data set, $\mathbf{b}$ often means $[\mathbf{b}_1^T, \cdots, \mathbf{b}_n^T]^T$.

Acronyms

AIC	Akaike Information Criterion
ASE	Average Squared Error
BIC	Bayesian Information Criterion
CSS	Cubic Smoothing Spline
CV	Cross-Validation
df	Degree of Freedom
GCV	Generalized Cross-Validation
GEE	Generalized Estimating Equation
GLME	Generalized Linear Mixed-Effects
GNPM	Generalized Nonparametric Population Mean
GNPME	Generalized Nonparametric Mixed-Effects
GSPM	Generalized Semiparametric Population Mean
GSAME	Generalized Semiparametric Additive Mixed-Effects
LME	Linear Mixed-Effects
Loglik	Log-likelihood
LPK	Local Polynomial Kernel
LPK-GEE	Local Polynomial Kernel GEE

LPME	Local Polynomial Mixed-Effects
MSE	Mean Squared Error
NLME	Nonlinear Mixed-Effects
NPM	Nonparametric Population Mean
NPME	Nonparametric Mixed-Effects
PCV	"Leave-One-Point-Out" Cross-Validation
SCV	"Leave-One-Subject-Out" Cross-Validation
SPM	Semiparametric Population Mean
SPME	Semiparametric Mixed-Effects
TVC	Time-Varying Coefficient

1

Introduction

Longitudinal data such as repeated measurements taken on each of a number of subjects over time arise frequently from many biomedical and clinical studies as well as from other scientific areas. Updated surveys on longitudinal data analysis can be found in Demidenko (2004) and Diggle et al. (2002), among others. Parametric mixed-effects models are a powerful tool for modeling the relationship between a response variable and covariates in longitudinal studies. Linear mixed-effects (LME) models and nonlinear mixed-effects (NLME) models are the two most popular examples. Several books have been published to summarize the achievements in these areas (Jones 1993, Davidian and Giltinan 1995, Vonesh and Chinchilli 1996, Pinheiro and Bates 2000, Verbeke and Molenberghs 2000, Diggle et al. 2002, and Demidenko 2004, among others). However, for many applications, parametric models may be too restrictive or limited, and sometimes unavailable at least for preliminary data analyses. To overcome this difficulty, nonparametric regression techniques have been developed for longitudinal data analysis in recent years. This book intends to survey the existing methods and introduce newly developed techniques that combine mixed-effects modeling ideas and nonparametric regression techniques for longitudinal data analysis.

1.1 MOTIVATING LONGITUDINAL DATA EXAMPLES

In longitudinal studies, data from individuals are collected repeatedly over time whereas cross-sectional studies only obtain one data point from each individual subject (i.e., a single time point per subject). Therefore, the key difference between longitudinal and cross-sectional data is that longitudinal data are usually correlated within a subject and independent between subjects, while cross-sectional data are often independent.

A challenge for longitudinal data analysis is how to account for within-subject correlations. LME and NLME models are powerful tools for handling such a problem when proper parametric models are available to relate a longitudinal response variable to its covariates. Many real-life data examples have been presented in the literature employing LME and NLME modeling techniques (Jones 1993, Davidian and Giltinan 1995, Vonesh and Chinchilli 1996, Pinheiro and Bates 2000, Verbeke and Molenberghs 2000, Diggle et al. 2002, and Demidenko 2004, among others). However, for many other practical data examples, proper parametric models may not exist or are difficult to find. Such examples from AIDS clinical trials and other biomedical studies will be presented and used throughout this book for illustration purposes. In these examples, LME and NLME models are no longer applicable, and nonparametric mixed-effects (NPME) modeling techniques, which are the focuses of this book, are a natural choice at least at the initial stage of exploratory analyses. Although the longitudinal data examples in this book are from biomedical and clinical studies, the proposed methodologies in this book are also applicable to panel data or clustered data from other scientific fields. All the data sets and the corresponding analysis computer codes in this book are freely accessible at the website: *http://www.urmc.rochester.edu/smd/biostat/people/faculty/WuSite/publications.htm.*

1.1.1 Progesterone Data

The progesterone data were collected in a study of early pregnancy loss conducted by the Institute for Toxicology and Environmental Health at the Reproductive Epidemiology Section of the California Department of Health Services, Berkeley, USA. Figures 1.1 and 1.2 show levels of urinary metabolite progesterone over the course of the women's menstrual cycles (days). The observations came from patients with healthy reproductive function enrolled in an artificial insemination clinic where insemination attempts were well-timed for each menstrual cycle. The data had been aligned by the day of ovulation (Day 0), determined by serum luteinizing hormone, and truncated at each end to present curves of equal length. Measurements were recorded once per day per cycle from 8 days before the day of ovulation and until 15 days after the ovulation. A woman may have one or several cycles. The length of the observation period is 24 days. Some measurements from some subjects were missing due to various reasons. The data set consists of two groups: the conceptive progesterone curves (22 menstrual cycles) and the nonconceptive progesterone curves (69 menstrual cycles). For more details about this data set, see Yen and Jaffe (1991), Brumback and Rice (1998), and Fan and Zhang (2000), among others.

Figure 1.1 (a) presents a spaghetti plot for the 22 raw conceptive progesterone curves. Dots indicate the level of progesterone observed in each cycle, and are connected with straight line segments. The problem of missing values is not serious here because each cycle curve has at least 17 out of 24 measurements. Overall, the raw curves present a similar pattern: before the ovulation day (Day 0), the raw curves

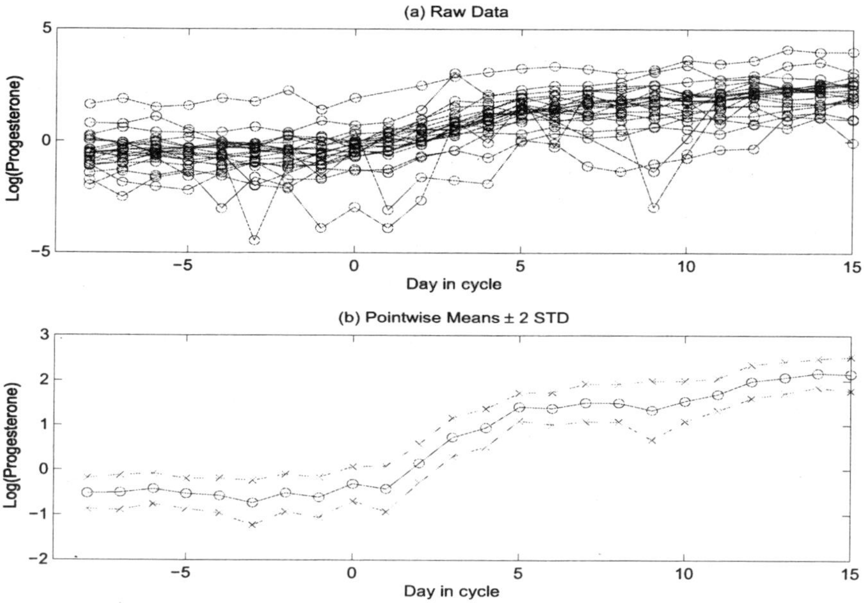

Fig. 1.1 The conceptive progesterone data.

are quite flat, but after the ovulation day, they generally move upward. However, it is easy to see that within a cycle curve, the measurements vary around some underlying curve which appears to be smooth, and for different cycles, the underlying smooth curves are different from each other. Figure 1.1 (b) presents the pointwise means (dot-dashed curve) with 95% pointwise standard deviation (SD) band (cross-dashed curves). They were obtained in a simple way: at each distinct design time point t, the mean and standard deviation were computed using the cross-sectional data at t. It can be seen that the pointwise mean curve is rather smooth, although it is not difficult to discover that there is still some noise appeared in the pointwise mean curve.

Figure 1.2 (a) presents a spaghetti plot for the 69 raw nonconceptive progesterone curves. Compared to the conceptive progesterone curves, these curves behave quite similarly before the day of ovulation, but generally show a different trend after the ovulation day. It is easy to see that, like the conceptive progesterone curves, the underlying individual cycles of the nonconceptive progesterone curves appear to be smooth, and so is their underlying mean curve. A naive estimate of the underlying mean curve is the pointwise mean curve, shown as dot-dashed curve in Figure 1.2 (b). The 95% pointwise SD band (cross-dashed curves) provides a rough estimate for the accuracy of the naive estimate.

The progesterone data have been used for illustrations of nonparametric regression methods by several authors. For example, Fan and Zhang (2000) used them to illustrate their two-step method for estimating the underlying mean function for longitudinal data or functional data, Brumback and Rice (1998) used them to illus-

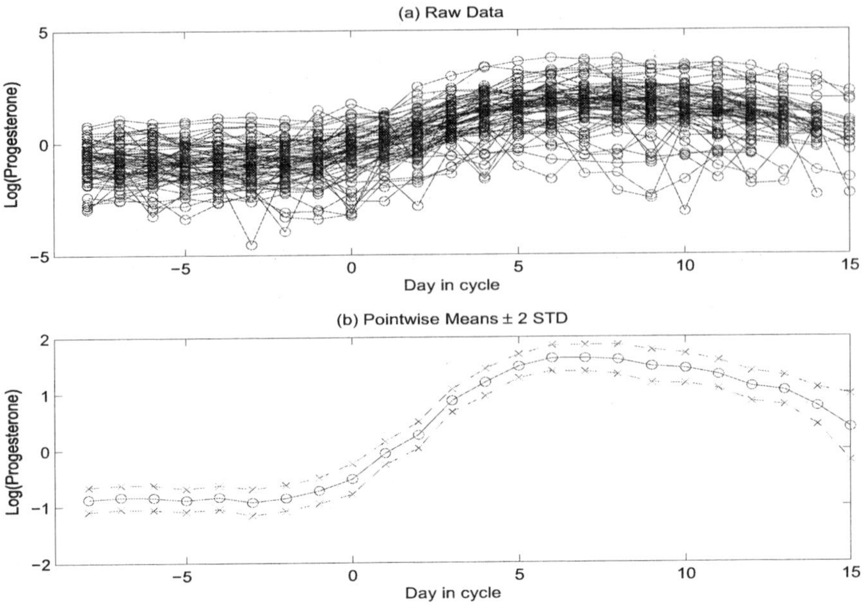

Fig. 1.2 The nonconceptive progesterone data.

trate a smoothing spline mixed-effects modeling technique for estimating both mean and individual functions, while Wu and Zhang (2002a) used them to illustrate a local polynomial mixed-effects modeling approach.

1.1.2 ACTG 388 Data

The ACTG 388 data were collected in an AIDS clinical trial study conducted by the AIDS Clinical Trials Group (ACTG). This study randomized 517 HIV-1 infected patients to three antiviral treatment arms. The data from one treatment arm will be used for illustration of the methodologies proposed in this book. This treatment arm includes 166 patients treated with highly active antiretroviral therapy (HAART) for 120 weeks during which CD4 cell counts were monitored at baseline and at weeks 4, 8, and every 8 weeks thereafter (up to 120 weeks). However, each individual patient might not exactly follow the designed schedule for measurements, and missing clinical visits for CD4 cell measurements frequently occurred. CD4 cell count is an important marker for assessing immunologic response of an antiviral regimen. Of interest are CD4 cell count trajectories over the treatment period for individual patients and for the whole treatment arm. More details about this study and scientific findings can be found in Fischl et al. (2003) and Park and Wu (2005).

The CD4 cell count data from the 166 patients during 120 weeks of treatment are plotted in Figure 1.3 (a). From this spaghetti plot, it is difficult to capture any useful information. It can be seen that the individual CD4 cell counts are quite noisy over time. We usually expect that the CD4 cell counts would increase if the antiviral treatment was effective. But from this plot, it is not easy to see any patterns among the individual patients' CD4 counts. Before a parametric model is found to fit this data set, we would have to assume that these individual curves are smooth but corrupted with noise.

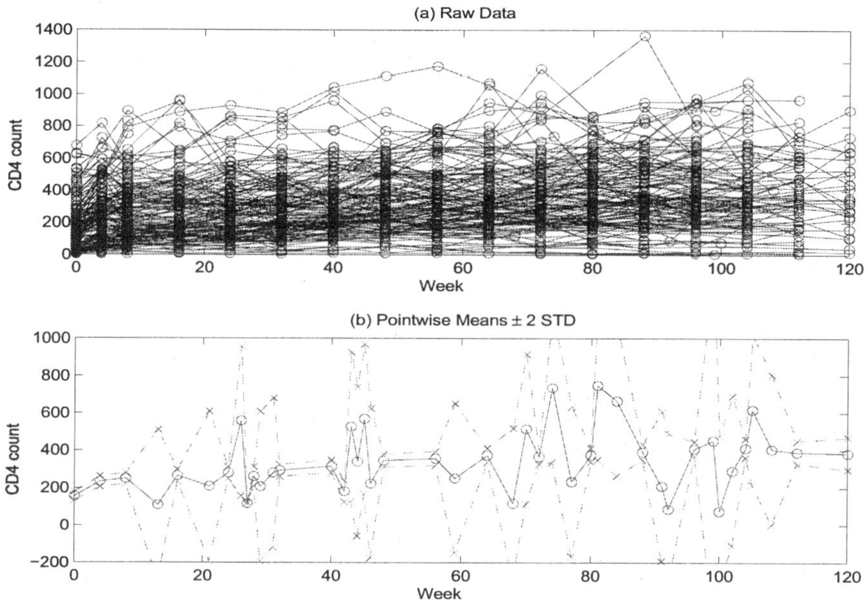

Fig. 1.3 The ACTG 388 data.

Figure 1.3 (b) presents the simple pointwise means (solid curve with dots) of the CD4 counts and their 95% pointwise SD band (cross-dashed curves). This jiggly connected pointwise mean function shows an upward trend, but it is not smooth, although the underlying mean function appears to be smooth. Moreover, the pointwise SDs are not always computable, because at some design time points (e.g., the third design time point from the right end), only a single cross-sectional data point is available. In this case, the pointwise mean is just the cross-sectional measurement itself and the pointwise SD is 0, which is not a proper measure for the accuracy of the pointwise mean. In the plot, we replaced this 0 standard deviation by the estimated standard deviation $\hat{\sigma}$ of the measurement errors, computed using all the residuals. However, this only partially solves the problem.

Without assuming parametric models for the mean and individual curves for the ACTG 388 data, nonparametric modeling techniques are then necessarily involved to handle the aforementioned problems. An example is provided by Park and Wu (2005), where they employed a kernel-based mixed-effects modeling approach.

1.1.3 MACS Data

Human immune-deficiency virus (HIV) destroys CD4 cells (T-lymphocytes, a vital component of the immune system) so that the number or percentage of CD4 cells in the blood of a patient will reduce after the subject is infected with HIV. The CD4 cell level is one of the important biomarkers to evaluate the disease progression of HIV infected subjects. To use the CD4 marker effectively in studies of new antiviral therapies or for monitoring the health status of individual subjects, it is important to build statistical models for CD4 cell count or percentage. For CD4 cell count, Lange et al. (1992) proposed Bayesian models while Zeger and Diggle (1994) employed a semiparametric model, fitted by a backfitting algorithm. For further related references, see Lange et al. (1992).

A subset of HIV monitoring data from the Multi-center AIDS Cohort Study (MACS) contains the HIV status of 283 homosexual men who were infected with HIV during the follow-up period between 1984 and 1991. Kaslow et al. (1987) presented the details for the related design, methods and medical implications of this study. The response variable is the CD4 cell percentage of a subject at a number of design time points after HIV infection. Three covariates were assessed in this study. The first one, "Smoking", takes the values of 1 or 0, according to whether a subject is a smoker or nonsmoker, respectively. The second covariate, "Age", is the age of a subject at the time of HIV infection. The third covariate, "PreCD4", is the last measured CD4 cell percentage level prior to HIV infection. All three covariates are time-independent and subject-specific. All subjects were scheduled to have clinical visits semi-annually for taking the measurements of CD4 cell percentage and other clinical status, but many subjects frequently missed their scheduled visits which resulted in unequal numbers of measurements and different measurement time points from different subjects in this longitudinal data set. We plotted the raw data from individual subjects and the simple pointwise mean of the data in Figure 1.4.

The aim of this study is to assess the effects of cigarette smoking, age at seroconversion and baseline CD4 cell percentage on the CD4 cell percentage depletion after HIV infection among the homosexual men population. From Figure 1.4, we can see that there was a trend of CD4 cell percentage depletion although the pointwise mean curve does not provide a good smooth estimate for this trend. Thus, a nonparametric modeling approach is required to characterize the CD4 cell depletion trend and to correlate this trend to the aforementioned covariates. In fact, Zeger and Diggle (1994), Wu and Chiang (2000), Fan and Zhang (2000), Rice and Wu (2001), Huang, Wu and Zhou (2002), among others have applied various nonparametric regression methods including time varying coefficient models to this data set. Similarly, we will use this data set to illustrate the proposed nonparametric regression models and smoothing methods in the succeeding chapters.

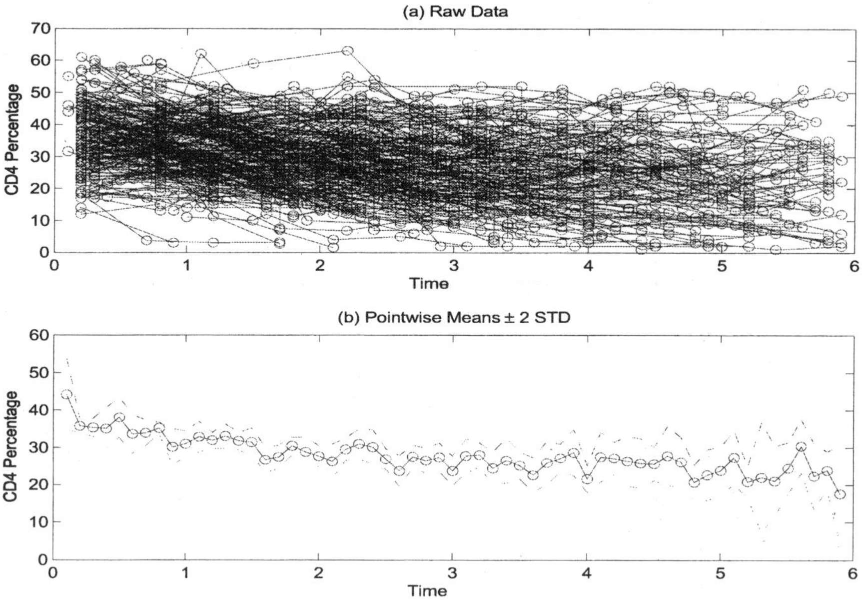

Fig. 1.4 The MACS data.

1.2 MIXED-EFFECTS MODELING: FROM PARAMETRIC TO NONPARAMETRIC

1.2.1 Parametric Mixed-Effects Models

For modeling longitudinal data, parametric mixed-effects models, such as linear and nonlinear mixed-effects models, are a natural tool. Linear or nonlinear mixed-effects models can be specified as hierarchical linear and nonlinear models from a Bayesian perspective.

Linear mixed-effects (LME) models are used when the relationship between a longitudinal response variable and its covariates can be expressed via a linear model. The LME model introduced by Harville (1976, 1977), and Laird and Ware (1982) can be generally written as

$$\begin{aligned} \mathbf{y}_i &= \mathbf{X}_i\boldsymbol{\beta} + \mathbf{Z}_i\mathbf{b}_i + \boldsymbol{\epsilon}_i, \\ \mathbf{b}_i \sim N(\mathbf{0}, \mathbf{D}),\ &\boldsymbol{\epsilon}_i \sim N(\mathbf{0}, \mathbf{R}_i),\quad i = 1, 2, \cdots, n, \end{aligned} \tag{1.1}$$

where $\mathbf{y}_i$ and $\boldsymbol{\epsilon}_i$ are, respectively, the vectors of responses and measurement errors for the i-th subject, $\boldsymbol{\beta}$ and $\mathbf{b}_i$ are, respectively, the vectors of fixed-effects (population parameters) and random-effects (individual parameters), and $\mathbf{X}_i$ and $\mathbf{Z}_i$ are the associated fixed-effects and random-effects design matrices. It is easy to notice that the mean and covariance matrix of $\mathbf{y}_i$ are given by

$$\mathrm{E}(\mathbf{y}_i) = \mathbf{X}_i\boldsymbol{\beta},\ \mathrm{Cov}(\mathbf{y}_i) = \mathbf{Z}_i\mathbf{D}\mathbf{Z}_i^T + \mathbf{R}_i,\quad i = 1, 2, \cdots, n.$$

Nonlinear mixed-effects (NLME) models are used when the relationship between a longitudinal response variable and its covariates can be expressed via a nonlinear model, which is known except for some parameters. A general hierarchical nonlinear model or NLME model may be written as (Davidian and Giltinan 1995, Vonesh and Chinchilli 1996):

$$\begin{array}{c} \mathbf{y}_i = \mathbf{f}(\mathbf{X}_i, \boldsymbol{\beta}_i) + \boldsymbol{\epsilon}_i, \ \boldsymbol{\beta}_i = \mathbf{d}(\mathbf{A}_i, \mathbf{B}_i, \boldsymbol{\beta}, \mathbf{b}_i), \\ \mathbf{b}_i \sim N(\mathbf{0}, \mathbf{D}), \ \boldsymbol{\epsilon}_i \sim N(\mathbf{0}, \mathbf{R}_i), \ \ i = 1, 2, \cdots, n, \end{array} \tag{1.2}$$

where $\mathbf{f}(\mathbf{X}_i, \boldsymbol{\beta}_i) = [f(\mathbf{x}_{i1}, \boldsymbol{\beta}_i), \cdots, f(\mathbf{x}_{in_i}, \boldsymbol{\beta}_i)]^T$ with $f(\cdot)$ being a known function, $\mathbf{X}_i = [\mathbf{x}_{i1}, \cdots, \mathbf{x}_{in_i}]^T$ a design matrix and $\boldsymbol{\beta}_i$ a subject-specific parameter for the i-th subject. In the above NLME model, the $\mathbf{d}(\cdot)$ is a known function of the design matrices $\mathbf{A}_i$ and $\mathbf{B}_i$, the fixed-effects vector $\boldsymbol{\beta}$ and the random-effects vector $\mathbf{b}_i$. As an example, a simple linear model for $\boldsymbol{\beta}_i$ can be written as $\boldsymbol{\beta}_i = \mathbf{A}_i\boldsymbol{\beta} + \mathbf{B}_i\mathbf{b}_i, \ i = 1, 2, \cdots, n$. The marginal mean and variance-covariance of $\mathbf{y}_i$ cannot be given for a general NLME model. They may be approximated using linearization techniques (Sheiner, Rosenberg and Melmon 1972, Sheiner and Beal 1982, and Lindstrom and Bates 1990, among others).

More detailed definitions of the LME and NLME models will be given in Chapter 2. In either a LME model or a NLME model, the between-subject and within-subject variations are separately quantified by the variance components $\mathbf{D}$ and $\mathbf{R}_i, \ i = 1, 2, \cdots, n$. In a longitudinal study, the data from different subjects are usually assumed to be independent, but the data from the same subject may be correlated. The correlations may be caused by the between-subject variation (heterogeneity across subjects) and/or the serially correlated measurement error. Ignoring the existing correlation of longitudinal data may lead to incorrect and inefficient inferences. Thus, a key requirement for longitudinal data analysis is to appropriately model and accurately estimate the variance components so that the underlying mean and individual functions can be efficiently modeled. This is the reason why longitudinal data analysis is more challenging in both theoretical development and practical implementation compared to cross-sectional data analysis.

The successful application of a LME or a NLME model to longitudinal data analysis strongly depends on the assumption of a proper linear or nonlinear model for the relationship between the response variable and the covariates. Sometimes this assumption may be invalid for a given longitudinal data set. In this case, the relationship between the response variable and the covariates has to be modeled nonparametrically. Therefore, we need to extend parametric mixed-effects models to nonparametric mixed-effects models.

1.2.2 Nonparametric Regression and Smoothing

A parametric regression model requires an assumption that the form of the underlying regression function is known except for the values of a finite number of parameters. The selection of a parametric model depends very much on the problem at hand. Sometimes the parametric model can be derived from mechanistic theories behind the scientific problem, whereas at other times the model is based on experience or is simply deduced from scatter plots of the data. A serious drawback of parametric modeling is that a parametric model may be too restrictive in some applications. If an inappropriate parametric model is used, it is possible to produce misleading conclusions from the regression analysis. In other situations, a parametric model may not be available to use. To overcome the difficulty caused by the restrictive assumption of a parametric form of the regression function, one may remove the restriction that the regression function belongs to a parametric family. This approach leads to so-called nonparametric regression.

There exist many nonparametric regression and smoothing methods. The most popular methods include kernel smoothing, local polynomial fitting, regression (polynomial) splines, smoothing splines, and penalized splines. Some other approaches such as locally weighted scatter plot smoothing (LOWESS), wavelet-based methods and other orthogonal series-based approaches are also frequently used in practice. The basic idea of these nonparametric approaches is to let the data determine the most suitable form of the functions. There are one or two so-called smoothing parameters in each of these methods for controlling the model complexity and the trade-off between the bias and variance of the estimator. For example, the bandwidth h in local kernel smoothing determines the smoothness of the regression function and the goodness-of-fit of the model to the data so that when $h = \infty$, the local nonparametric model becomes a global parametric model; whereas when $h = 0$, the resulting estimate essentially interpolates the data points. Thus, the boundary between parametric and nonparametric modeling may not be clear-cut if one takes the smoothing parameter into account. Nonparametric and parametric regression methods should not be regarded as competitors, instead they complement each other. In some situations, nonparametric techniques can be used to validate or suggest a parametric model. A combination of both nonparametric and parametric methods is more powerful than any single method in many practical applications.

There exists a vast literature on smoothing and nonparametric regression methods for cross-sectional data. Good surveys on these methods can be found in books by de Boor (1978), Eubank (1988), Härdle (1990), Wahba (1990), Green and Silverman (1994), Wand and Jones (1995), Fan and Gijbels (1996), and Ruppert, Wand and Carroll (2003), among others. However, very little effort has been made to develop nonparametric regression methods for longitudinal data analysis until recent years. Müller (1988) was the first to address longitudinal data analysis using nonparametric regression methods. However, in this earlier monograph, the basic approach is to estimate each individual curve separately, thus, the within-subject correlation of the longitudinal data was not considered in modeling. The methodologies in Müller (1988) are essentially similar to the nonparametric regression methods for cross-sectional data.

In recent years, there has been a boom in the development of nonparametric regression methods for longitudinal data analysis which include utilization of kernel-type smoothing methods (Hoover et al. 1998, Wu and Chiang 2000, Wu, Chiang and Hoover 1998, Fan and Zhang 2000, Lin and Carroll 2001a, b, Wu and Zhang 2002a, Welsh, Lin and Carroll 2002, Cai, Li and Wu 2003, Wang 2003, Wang, Carroll and Lin 2005), smoothing spline methods (Brumback and Rice 1998, Wang 1998a, b,

Zhang et al. 1998, Lin and Zhang 1999, Guo 2002a, b) and regression (polynomial) spline methods (Shi, Weiss and Taylor 1996, Rice and Wu 2001, Huang, Wu and Zhou 2002, Wu and Zhang 2002b, Liang, Wu and Carroll 2003). There is a vast amount of recent literature in this research area, and it is impossible for us to have an exhaustive list here. The importance of nonparametric modeling methods has been recognized in longitudinal data analysis and for practical applications, since nonparametric methods are flexible and robust against parametric assumptions. Such flexibility is useful for exploration and analysis of longitudinal data, when appropriate parametric models are unavailable. In this book, we do not intend to cover all nonparametric regression techniques. Instead we will focus on the most popular methods such as local polynomial smoothing, regression (polynomial) splines, smoothing splines and penalized splines (P-Splines) approaches. We incorporate these nonparametric smoothing procedures into mixed-effects models to propose nonparametric mixed-effects modeling techniques for longitudinal data analysis.

1.2.3 Nonparametric Mixed-Effects Models

A longitudinal data set such as the progesterone data and the ACTG 388 data presented in Section 1.1, can be expressed in a common form as

$$(t_{ij}, y_{ij}),\ j = 1, 2, \cdots, n_i;\ i = 1, 2, \cdots, n, \tag{1.3}$$

where t_{ij} denote the design time points (e.g., "day" in the progesterone data), y_{ij} the responses observed at t_{ij} (e.g. "log(Progesterone)" in the progesterone data), n_i the number of observations for the i-th subject, and n is the number of subjects.

For such a longitudinal data set, we do not assume a parametric model for the relationship between the response variable and the covariate time. Instead, we just assume that the individual and the population mean functions are smooth functions of time t, and let the data themselves determine the form of the underlying functions. Following Wu and Zhang (2002a), we introduce a nonparametric mixed-effects (NPME) model as

$$y_i(t) = \eta(t) + v_i(t) + \epsilon_i(t), \qquad i = 1, 2, \cdots, n, \tag{1.4}$$

where $\eta(t)$ models the population mean function of the longitudinal data set, called fixed-effect function, $v_i(t)$ models the departure of the i-th individual function from the population mean function $\eta(t)$, called the i-th random-effect function, and $\epsilon_i(t)$ the measurement errors that can not be explained by both the fixed-effect and the random-effect functions.

It is generally assumed that $v_i(t),\ i = 1, 2, \cdots, n$ are i.i.d realizations of an underlying smooth process (SP), $v(t)$, with mean function 0 and covariance function $\gamma(s, t)$, and $\epsilon_i(t)$ are i.i.d realizations of an uncorrelated white noise process, $\epsilon(t)$, with mean function 0 and variance function $\gamma_\epsilon(s, t) = \sigma^2(t)1_{\{s=t\}}$. That is, $v(t) \sim \mathrm{SP}(0, \gamma)$ and $\epsilon(t) \sim \mathrm{SP}(0, \gamma_\epsilon)$. Here $\gamma(s, t)$ quantifies the between-subject variation while the $\sigma^2(t)$ quantifies the within-subject variation. When discussing the likelihood-based inferences or Bayesian interpretation, for simplicity, we generally assume that the associated processes are Gaussian, i.e., $v(t) \sim \mathrm{GP}(0, \gamma)$, and $\epsilon \sim \mathrm{GP}(0, \gamma_\epsilon)$.

Under the NPME modeling framework, we need to accomplish the following tasks: (1) to estimate the fixed-effect (population mean) function $\eta(t)$; (2) to predict the random-effect functions $v_i(t)$ and individual functions $s_i(t) = \eta(t) + v_i(t),\ i = 1, 2, \cdots, n$; (3) to estimate the covariance function $\gamma(s,t)$; and (4) to estimate the noise variance function $\sigma^2(t)$.

The $\eta(t), \gamma(s,t)$ and $\sigma^2(t)$ characterize the population features of a longitudinal response while $v_i(t)$ and $s_i(t)$ capture the individual features. For simplicity, the population mean function $\eta(t)$ and the individual functions $s_i(t)$ are sometimes referred to as population and individual curves, respectively. Because in the NPME model (1.4), the target quantities $\eta(t), v_i(t), \gamma(s,t)$ and $\sigma^2(t)$ are all nonparametric, the combination of smoothing techniques and mixed-effects modeling approaches is necessary for estimating these unknown quantities. We will also extend this NPME modeling idea to semiparametric models, time varying coefficient models and models for analyzing discrete longitudinal data.

1.3 SCOPE OF THE BOOK

For longitudinal data analysis, a simple strategy is the so-called two-stage or derived variable approach (Diggle et al. 2002). The first step is to reduce the repeated measurements from individual subjects or units into one or two summary statistics, and the second step is to conduct the analysis for the summarized variables. This method may not be efficient if the repeatedly measured variables change significantly and informatively over time. In the book by Diggle et al. (2002), three alternative modeling strategies are discussed, i.e., the marginal modeling analysis, the random-effects modeling approach, and the transition modeling approach. For all three approaches, the dependence of the response on the explanatory variables and the autocorrelation among the responses are considered. In this book, the ideas from the three strategies will be used under the framework of nonparametric regression techniques although we may not explicitly use the same wording.

It is impossible to exhaustively survey all the nonparametric smoothing and regression methods for longitudinal data analysis in this monograph. Selection of covered materials is based on our experiences with practical data problems. We emphasize the introduction of basic ideas of methodologies and their applications to data analysis throughout the book. Since this book is an extension of nonparametric smoothing and regression methods for longitudinal data analysis, it is essential to combine the techniques from these two areas in an efficient way.

1.3.1 Building Blocks of the NPME Models

The building blocks of the NPME models include parametric mixed-effects models and nonparametric smoothing techniques. To better understand NPME models we begin with a review of LME and nonparametric smoothing techniques.

Two most popular parametric mixed-effects models are linear mixed-effects (LME) and nonlinear mixed-effects (NLME) models. LME models are the simplest mixed-effects models in which the responses are linear functions of the fixed-effects and random-effects. In Chapter 2, we shall briefly discuss how the models are specified, how the parameters are estimated and how the variance components are estimated. In particular, we briefly mention random-coefficients models as special cases of the usual LME models. In Chapter 2, we will also briefly review NLME models and related inference methods. In a NLME model, the responses are nonlinear functions of the fixed-effects and random-effects. It is a challenging task to estimate the parameters in NLME models. We will summarize the two-stage, first order approximation and conditional first order approximation methods in this chapter. Generalized linear and nonlinear mixed-effects models will also be briefly discussed.

In Chapter 3, we shall review some popular nonparametric regression techniques that include local polynomial smoothing, regression splines, smoothing splines, and penalized splines, among others. We will briefly discuss the basic ideas of these techniques, computational issues and smoothing parameter selections.

1.3.2 Fundamental Development of the NPME Models

Fundamental developments of the NPME modeling techniques will be presented in Chapters 4~7, and each chapter covers one popular nonparametric method. These are the core contents of this book and lay a good foundation for further extensions of the NPME models. Each of these chapters will also provide a review for the nonparametric population mean (NPM) model and naive smoothing methods before the mixed-effects modeling approach is introduced.

In Chapter 4, we will mainly investigate local polynomial mixed-effects models after a review of the NPM model and the local polynomial kernel-based generalized estimating equations (LPK-GEE) methods. The local polynomial smoothing approaches and LME modeling techniques are combined to estimate unknown functions and parameters for the NPME model (1.4). The key idea for this method is that for each fixed time point t, $\eta(t)$ and $v_i(t)$ in the NPME model (1.4) are approximated by a polynomial of some degree so that a local LME model is formed and solved. We will also study the bandwidth selection methods. Some theoretic results will be presented to provide theoretical justifications for the proposed methodologies. One of the advantages of this approach is that at each time point t, the associated local LME model can be solved by the existing statistical software such as the *lme* function in S-PLUS, or the procedure *PROC MIXED* in SAS.

In Chapter 5, we will introduce regression spline mixed-effects models. The main idea is to approximate $\eta(t)$ and $v_i(t)$ by regression splines so that the NPME model (1.4) can be transformed into a global parametric model, which can be solved using the existing statistical software such as S-PLUS or SAS. However, one needs to locate the knots for the regression splines and choose the number of knots using some model selection rules. The regression spline method is simple to implement and easy to understand. That is why it is the first NPME modeling technique studied in the literature (Shi, Weiss and Taylor 1996) and it is also attractive to practitioners.

In Chapter 6, we will focus on smoothing spline mixed-effects modeling techniques. The smoothing spline approach is one of the major nonparametric smoothing methods and has been well developed for cross-section i.i.d. data. For longitudinal data, several authors (Wang 1998a, Brumback and Rice 1998, Guo 2002a) have proposed some techniques using the LME model representation of a smoothing spline. Our idea is to incorporate the roughness of $\eta(t)$ and $v_i(t)$ of the NPME model (1.4) into NPME modeling in a natural way, and to develop new techniques for variance component estimation and smoothing parameter selection.

The penalized spline (P-spline) method recently became very popular in nonparametric modeling since it is computationally easier than the smoothing spline method, but still inherits all the advantages of the smoothing spline approach. We will combine the mixed-effects modeling ideas and the P-spline techniques for longitudinal data analysis in Chapter 7.

1.3.3 Further Extensions of the NPME Models

The fundamental NPME modeling methodologies introduced in Chapters 4$\sim$7 can be extended to semiparametric and varying-coefficient models. In a semiparametric mixed-effects model, part of the variations in the response variable can be explained by given parametric models of some covariates in the fixed effect component and/or the random-effect component, while the remaining is explained by a nonparametric function of time. In a time varying coefficient mixed-effects model, the coefficients of the fixed-effects and random-effects covariates are smooth functions of time. These two kinds of models are very important and useful in practical longitudinal data analysis. The challenging question is how to estimate the constant and time-varying parameters under a mixed-effects modeling framework. The fundamental NPME modeling methodologies can also be extended to discrete longitudinal data analysis. Chapters 8$\sim$10 will cover these extended models in details.

Chapter 8 will focus on semiparametric models for longitudinal data. In this chapter, we first provide a review of semiparametric population mean models before the semiparametric mixed-effects models are introduced. The local polynomial smoothing, regression spline, penalized spline, and smoothing spline methods will be covered. The methods that do not involve smoothing are also briefly discussed. The more sophisticated semiparametric nonlinear mixed-effects models will be presented.

In Chapter 9, we will introduce time varying coefficient models for longitudinal data. First the time varying coefficient nonparametric population mean (TVC-NPM) models are reviewed. The local polynomial smoothing, regression spline, penalized spline, and smoothing spline methods are introduced to fit the TVC-NPM models. The smoothing parameter selections are discussed and a backfitting algorithm is proposed. The two-step method of Fan and Zhang (2000) is also adapted to fit the TVC-NPM models. The TVC-NPM models with time-independent covariates are briefly discussed. The extension of the TVC-NPM models that include both parametric (linear or nonlinear) and nonparametric time varying coefficients is briefly explored. The time varying coefficient nonparametric mixed-effects (TVC-NPME) models, which is the focus of this chapter, are investigated in details. The aforementioned smoothing approaches are developed to fit the TVC-NPME models. The semiparametric

TVC-NPME models that include both constant and time varying coefficients are also introduced.

Chapter 10 will concentrate on an introduction to nonparametric regression methods for discrete longitudinal data. We first review the LPK-GEE methods for the generalized nonparametric and semiparametric population mean models proposed by Lin and Carroll (2000, 2001a,b), Wang (2003), Wang, Carroll and Lin (2005). We then introduce the generalized nonparametric mixed-effects models and generalized time varying coefficient nonparametric mixed-effects models as well as the local polynomial approach for fitting these models in details. The asymptotic properties of the estimators are also investigated. Finally the generalized semiparametric additive mixed-effects models initiated by Lin and Zhang (1999) are introduced in detail.

1.4 IMPLEMENTATION OF METHODOLOGIES

Most methodologies introduced in this book can be implemented using existing software such as S-PLUS and SAS, among others, although it may be more efficient to use Fortran, C or MATLAB codes. The latter usually requires intensive programming since Fortran, C or MATLAB subroutines for parametric mixed-effects modeling or nonparametric smoothing techniques are unavailable. We shall publish our MATLAB codes for most of the methodologies proposed in this book and the data analysis examples on our website: *http://www.urmc.rochester.edu/smd/biostat/people/faculty/WuSite/publications.htm* and keep updating the codes when it is necessary. We shall also make the data sets used in this book available through our website.

1.5 OPTIONS FOR READING THIS BOOK

Readers who are particularly interested in one or two of the nonparametric smoothing techniques for longitudinal data analysis may select the relevant chapters to read. For a lower level graduate course, Chapters 1~7 are recommended. If students already have some background in mixed-effects models and nonparametric smoothing techniques, Chapters 2 and 3 may be briefly reviewed or even skipped. Chapters 8~10 may be included in a higher level graduate course or can be used as individual research materials for those who may want to do research in this area.

1.6 BIBLIOGRAPHICAL NOTES

Nonparametric regression methods for longitudinal data analysis is still an active research area. In this book, we have not attempted to provide an exhaustive review of all the methodologies in the literature. Interested readers are strongly advised to read additional work by other authors whose methodologies have not been covered in this book. For nonparametric estimation of individual curves of longitudinal data without a mixed-effects framework, we refer readers to Müller (1988), and for nonparametric techniques for analyzing functional data (which may be regarded as longitudinal data with a very large number of measurements per subject), we refer readers to Ramsay and Silverman (1997, 2002). Important references for parametric mixed-effects modeling methods, nonparametric smoothing techniques and various nonparametric and semiparametric models are provided at the end of this book. Below, we briefly mention some important monographs on these subjects.

Various models and methods for dealing with longitudinal data analysis are given by Diggle, Liang and Zeger (1994), and Diggle, Heagerty, Liang and Zeger (2002). Monographs on linear and nonlinear mixed-effects models include Lindstrom and Bates (1990), Lindsey (1993), Davdian and Giltinan (1995), Vonesh and Chinchilli (1996), and Verbeke and Molenberghs (2000), among others. Longford (1993) surveys methods on random coefficient models for longitudinal data. Jones (1993) treats longitudinal data with serial correlation using a state-space approach. Pinheiro and Bates (2000) discuss the implementation of mixed-effects modeling in S and S-PLUS. Methods on variance components estimation are surveyed by Searle, Casella and McCulloch (1992), and by Cox and Solomon (2003). A recent monograph on theories and applications of mixed models is given by Demidenko (2004).

A good survey on kernel smoothing is provided by Wand and Jones (1995). A very readable introduction to local polynomial modeling and its applications is given by Fan and Gijbels (1996). A classical introduction to B-splines is given by de Boor (1978). Penalized splines and their applications in semiparametric models are investigated by Ruppert, Wand and Carroll (2003). Wahba (1990) and Green and Silverman (1994) are two monographs on smoothing spline approaches to nonparametric regression. Various nonparametric smoothing techniques and their applications can be found in books by Eubank (1988, 1999), Härdle (1990), Simonoff (1996), and Efromovich (1999), among others. Nonparametric lack-of-fit testing techniques are discussed by Hart (1997).

Generalized linear models are explored by McCullagh and Nelder (1989). Recent advances on theories and applications of semiparametric models for independent data are surveyed by Härdle, Liang and Gao (2000). Ruppert, Wand and Carroll (2003) survey methods for fitting semiparametric models using P-splines.

2

Parametric Mixed-Effects Models

2.1 INTRODUCTION

Parametric mixed-effects models or random-effects models are powerful tools for longitudinal data analysis. Linear and nonlinear mixed-effects models (including generalized linear and nonlinear mixed-effects models) have been widely used in many longitudinal studies. Good surveys on these approaches can be found in the books by Searle, Casella, and McCulloch (1992), Davidian and Giltinan (1995), Vonesh and Chinchilli (1996), Verbeke and Molenberghs (2000), Pinheiro and Bates (2000), Diggle et al. (2002), and Demidenko (2004), among others. In this chapter, we shall review various parametric mixed-effects models and emphasize the methods that we will use in later chapters. Since the focus of this book is to introduce the ideas of mixed-effects modeling in nonparametric smoothing and regression for longitudinal data analysis, it is important to understand the basic concepts and key properties of parametric mixed-effects models.

2.2 LINEAR MIXED-EFFECTS MODEL

2.2.1 Model Specification

Harville (1976, 1977) and Laird and Ware (1982) first proposed the following general linear mixed-effects (LME) model:

$$\begin{gathered} y_{ij} = \mathbf{x}_{ij}^T \beta + \mathbf{z}_{ij}^T \mathbf{b}_i + \epsilon_{ij}, \\ \mathbf{b}_i \sim N(\mathbf{0}, \mathbf{D}), \;\; \epsilon_i \sim N(\mathbf{0}, \mathbf{R}_i), \\ j = 1, 2, \cdots, n_i; \;\; i = 1, 2, \cdots, n, \end{gathered} \tag{2.1}$$

where $\epsilon_i = [\epsilon_{i1}, \cdots, \epsilon_{in_i}]^T$, y_{ij} and ϵ_{ij} denote the response and the measurement error of the j-th measurement of the i-th subject, the unknown parameters $\beta : p \times 1$ and $\mathbf{b}_i : q \times 1$ are usually called the fixed-effects vector and random-effects vectors, respectively (for simplicity, they are often referred to fixed-effects and random-effects parameters of the LME model), and $\mathbf{x}_{ij}$ and $\mathbf{z}_{ij}$ are the associated fixed-effects and random-effects covariate vectors. In the above expression, $\mathbf{D}$ and $\mathbf{R}_i,\ i = 1, 2, \cdots, n$ are known as the variance components of the LME model. In the above LME model, for simplicity, we assume that $\mathbf{b}_i$ and ϵ_i are independent with normal distributions, and the between-subject measurements are independent.

The LME model (2.1) is often written in the following form:

$$\begin{aligned} \mathbf{y}_i &= \mathbf{X}_i\beta + \mathbf{Z}_i\mathbf{b}_i + \epsilon_i, \\ \mathbf{b}_i &\sim N(\mathbf{0}, \mathbf{D}),\ \epsilon_i \sim N(\mathbf{0}, \mathbf{R}_i), \\ i &= 1, 2, \cdots, n, \end{aligned} \tag{2.2}$$

where $\mathbf{y}_i = [y_{i1}, \cdots, y_{in_i}]^T, \mathbf{X}_i = [\mathbf{x}_{i1}, \cdots, \mathbf{x}_{in_i}]^T$, and $\mathbf{Z}_i = [\mathbf{z}_{i1}, \cdots, \mathbf{z}_{in_i}]^T$.

The above LME model includes linear random coefficient models (Longford 1993) and models for repeated measurements as special cases. For example, a two-stage linear random-coefficient model for growth curves (Longford 1993) can be written as

$$\begin{aligned} \mathbf{y}_i &= \mathbf{Z}_i\beta_i + \epsilon_i,\ \beta_i = \mathbf{A}_i\beta + \mathbf{b}_i, \\ \mathbf{b}_i &\sim N(\mathbf{0}, \mathbf{D}),\ \epsilon_i \sim N(\mathbf{0}, \mathbf{R}_i), \\ i &= 1, 2, \cdots, n, \end{aligned} \tag{2.3}$$

where $\mathbf{y}_i, \mathbf{Z}_i, \mathbf{b}_i$ and ϵ_i are similarly defined as in (2.2), β_i is a $q \times 1$ vector of random coefficients of the i-th subject, and $\mathbf{A}_i$ is a $q \times p$ design matrix containing between-subject covariates. It is easy to see that the linear random-coefficient model (2.3) can be written into the form of the general LME model (2.2) once we set $\mathbf{X}_i = \mathbf{Z}_i\mathbf{A}_i,\ i = 1, 2, \cdots, n$.

In fact, we can write a general two-stage linear random coefficient model into the form of the general LME model (2.2). A general two-stage random coefficient model can be written as (Davidian and Giltinan 1995, Vonesh and Chinchilli 1996)

$$\begin{aligned} \mathbf{y}_i &= \mathbf{Z}_i\beta_i + \epsilon_i,\ \beta_i = \mathbf{A}_i\beta + \mathbf{B}_i\mathbf{b}_i, \\ \mathbf{b}_i &\sim N(\mathbf{0}, \mathbf{D}),\ \epsilon_i \sim N(\mathbf{0}, \mathbf{R}_i), \\ i &= 1, 2, \cdots, n, \end{aligned} \tag{2.4}$$

where $\mathbf{B}_i$ is a $q \times k$ design matrix with elements of 0's and 1's arranged to determine the components of β_i that are random, and $\mathbf{b}_i$ is the associated k-dimensional random-effects vector. This general two-stage random-coefficient model can be written into the form of the general LME model (2.2): $\mathbf{y}_i = \mathbf{X}_i\beta + \mathbf{Z}_i^*\mathbf{b}_i + \epsilon_i$ once we set $\mathbf{X}_i = \mathbf{Z}_i\mathbf{A}_i$ and $\mathbf{Z}_i^* = \mathbf{Z}_i\mathbf{B}_i,\ i = 1, 2, \cdots, n$. In fact, we can easily show that the general two-stage random coefficient model (2.4) is equivalent to the general LME model (2.2). In particular, when $\mathbf{B}_i = \mathbf{I}_q$, the general two-stage random coefficient model (2.4) reduces to the random coefficient model (2.3) for growth curves. Notice that the general two-stage random coefficient model (2.4) is also known as a two-stage mixed-effects model and the general LME model (2.2) is also called a hierarchical linear model.

In matrix notation, the general LME model (2.2) can be further written as

$$\begin{aligned} \mathbf{y} &= \mathbf{X}\beta + \mathbf{Z}\mathbf{b} + \epsilon, \\ \mathbf{b} &\sim N(\mathbf{0}, \tilde{\mathbf{D}}), \ \epsilon \sim N(\mathbf{0}, \mathbf{R}), \end{aligned} \tag{2.5}$$

where

$$\begin{array}{lllllll} \mathbf{y} &=& [\mathbf{y}_1^T, \cdots, \mathbf{y}_n^T]^T, & \mathbf{b} &=& [\mathbf{b}_1^T, \cdots, \mathbf{b}_n^T]^T, \\ \epsilon &=& [\epsilon_1^T, \cdots, \epsilon_n^T]^T, & \mathbf{X} &=& [\mathbf{X}_1^T, \cdots, \mathbf{X}_n^T]^T, \\ \mathbf{Z} &=& \text{diag}(\mathbf{Z}_1, \cdots, \mathbf{Z}_n), & \tilde{\mathbf{D}} &=& \text{diag}(\mathbf{D}, \cdots, \mathbf{D}), \\ \mathbf{R} &=& \text{diag}(\mathbf{R}_1, \cdots, \mathbf{R}_n). \end{array} \tag{2.6}$$

It is usually assumed that the repeated measurements from different subjects are independent and they are correlated only when they come from the same subject. Based on the general LME model (2.5), we have $\text{Cov}(\mathbf{y}) = \text{diag}(\text{Cov}(\mathbf{y}_1), \cdots, \text{Cov}(\mathbf{y}_n))$ where the covariance matrix of repeated measurement vector $\mathbf{y}_i$ for the i-th subject is $\text{Cov}(\mathbf{y}_i) = \mathbf{Z}_i\mathbf{D}\mathbf{Z}_i^T + \mathbf{R}_i$. We can see that the correlation among the repeated measurements can be induced either through the between-subject variation term $\mathbf{Z}_i\mathbf{D}\mathbf{Z}_i^T$ or through the within-subject covariance matrix $\mathbf{R}_i$. Thus, even if the intra-subject measurement errors ($\epsilon_i, \ i = 1, 2, \cdots, n$) are independent, the repeated measurements $\mathbf{y}_i$ may be still correlated due to the between-subject variation. In some problems, the correlation may come from both sources. However, for simplicity, we may assume that the correlation is induced solely via the between-subject variation or assume that $\mathbf{R}_i$ is diagonal in the development of methodologies.

2.2.2 Estimation of Fixed and Random-Effects

The inferences for β and $\mathbf{b}_i, \ i = 1, 2, \cdots, n$ for the general LME model (2.2) can be based on the likelihood method or generalized least squares method. For known $\mathbf{D}$ and $\mathbf{R}_i, \ i = 1, 2, \cdots, n$, the estimates of β and $\mathbf{b}_i, \ i = 1, 2, \cdots, n$ may be obtained by minimizing the following twice negative logarithm of the joint density function of $\mathbf{y}_i, \ i = 1, 2, \cdots, n$ and $\mathbf{b}_i, \ i = 1, 2, \cdots, n$ (up to a constant):

$$\begin{aligned} \text{GLL}(\beta, \mathbf{b}_i|\mathbf{y}) = \textstyle\sum_{i=1}^n \Big\{ & [\mathbf{y}_i - \mathbf{X}_i\beta - \mathbf{Z}_i\mathbf{b}_i]^T \mathbf{R}_i^{-1} [\mathbf{y}_i - \mathbf{X}_i\beta - \mathbf{Z}_i\mathbf{b}_i] \\ & + \mathbf{b}_i^T \mathbf{D}^{-1} \mathbf{b}_i + \log|\mathbf{D}| + \log|\mathbf{R}_i| \Big\}. \end{aligned} \tag{2.7}$$

Since $\mathbf{b}_i, \ i = 1, 2, \cdots, n$ are random-effects parameter vectors, the expression (2.7) is not a conventional log-likelihood. For convenience, from now on and throughout this book, we call (2.7) a generalized log-likelihood (GLL) of the mixed-effects parameters $(\beta, \mathbf{b}_i, \ i = 1, 2, \cdots, n)$. Note that the first term of the right-hand side of (2.7) is a weighted residual taking the within-subject variation into account, and the term $\mathbf{b}_i^T\mathbf{D}^{-1}\mathbf{b}_i$ is a penalty due to random-effects $\mathbf{b}_i$ taking the between-subject variation into account.

For given $\mathbf{D}$ and $\mathbf{R}_i, \ i = 1, 2, \cdots, n$, minimizing the GLL criterion (2.7) is equivalent to solving the so-called mixed model equations (Harville 1976, Robinson

1991):

$$\begin{bmatrix} \mathbf{X}^T\mathbf{R}^{-1}\mathbf{X} & \mathbf{X}^T\mathbf{R}^{-1}\mathbf{Z} \\ \mathbf{Z}^T\mathbf{R}^{-1}\mathbf{X} & \mathbf{Z}^T\mathbf{R}^{-1}\mathbf{Z} + \tilde{\mathbf{D}}^{-1} \end{bmatrix} \begin{bmatrix} \beta \\ \mathbf{b} \end{bmatrix} = \begin{bmatrix} \mathbf{X}^T\mathbf{R}^{-1}\mathbf{y} \\ \mathbf{Z}^T\mathbf{R}^{-1}\mathbf{y} \end{bmatrix},$$

where $\mathbf{y}, \mathbf{b}, \mathbf{X}, \mathbf{Z}, \tilde{\mathbf{D}}$ and $\mathbf{R}$ are defined in (2.6). Using matrix algebra, the mixed model equations yield

$$\hat{\beta} = (\mathbf{X}^T\mathbf{V}^{-1}\mathbf{X})^{-1}\mathbf{X}^T\mathbf{V}^{-1}\mathbf{y}, \tag{2.8}$$

$$\hat{\mathbf{b}}_i = \mathbf{D}\mathbf{Z}_i^T\mathbf{V}_i^{-1}(\mathbf{y}_i - \mathbf{X}_i\hat{\beta}), \quad i = 1, 2, \cdots, n, \tag{2.9}$$

where $\mathbf{V}_i = \mathbf{Z}_i\mathbf{D}\mathbf{Z}_i^T + \mathbf{R}_i,\ i = 1, 2, \cdots, n$ and $\mathbf{V} = \text{diag}(\mathbf{V}_1, \cdots, \mathbf{V}_n)$. The covariance matrices of $\hat{\beta}$ and $\hat{\mathbf{b}}_i$ are:

$$Cov(\hat{\beta}) = (\mathbf{X}^T\mathbf{V}^{-1}\mathbf{X})^{-1} = \left(\sum_{i=1}^{n} \mathbf{X}_i^T\mathbf{V}_i^{-1}\mathbf{X}_i\right)^{-1}, \tag{2.10}$$

$$\begin{aligned} Cov(\hat{\mathbf{b}}_i - \mathbf{b}_i) &= \mathbf{D} - \mathbf{D}\,(\mathbf{Z}_i^T\mathbf{V}_i^{-1}\mathbf{Z}_i)\mathbf{D} + \mathbf{D}(\mathbf{Z}_i^T\mathbf{V}_i^{-1}\mathbf{X}_i) \\ &\quad \times \left(\sum_{j=1}^{n} \mathbf{X}_j^T\mathbf{V}_j^{-1}\mathbf{X}_j\right)^{-1} (\mathbf{X}_i^T\mathbf{V}_i^{-1}\mathbf{Z}_i)\,\mathbf{D}, \\ &\quad i = 1, 2, \cdots, n. \end{aligned} \tag{2.11}$$

2.2.3 Bayesian Interpretation

It is well known that the general LME model (2.2) has a close connection with a Bayesian model in the sense that the solutions (2.8) and (2.9) are the posterior expectations of the parameters of a Bayesian model under non-informative priors.

Before we go further, we state the following two useful lemmas whose proofs can be found in some standard multivariate textbooks, e.g., Anderson (1984).

Lemma 2.1 *Let* $\mathbf{A}, \mathbf{B}$ *and* $\mathbf{X}$ *be* $p \times p, q \times q$ *and* $p \times q$ *matrices so that* $\mathbf{A}$ *and* $\mathbf{A} + \mathbf{X}\mathbf{B}\mathbf{X}^T$ *are invertible. Then*

$$(\mathbf{A} + \mathbf{X}\mathbf{B}\mathbf{X}^T)^{-1} = \mathbf{A}^{-1} - \mathbf{A}^{-1}\mathbf{X}\mathbf{B}(\mathbf{B} + \mathbf{B}\mathbf{X}^T\mathbf{A}^{-1}\mathbf{X}\mathbf{B})^{-1}\mathbf{B}\mathbf{X}^T\mathbf{A}^{-1}. \tag{2.12}$$

In particular, when $q = 1, \mathbf{B} = 1$ *and* $\mathbf{X} = \mathbf{x}$ *where* $\mathbf{x}$ *is a* $p \times 1$ *vector, we have*

$$(\mathbf{A} + \mathbf{x}\mathbf{x}^T)^{-1} = \mathbf{A}^{-1} - \mathbf{A}^{-1}\mathbf{x}\mathbf{x}^T\mathbf{A}^{-1}/(1 + \mathbf{x}^T\mathbf{A}^{-1}\mathbf{x}). \tag{2.13}$$

Lemma 2.2 *Let*

$$\begin{pmatrix} \mathbf{X}_1 \\ \mathbf{X}_2 \end{pmatrix} \sim N\left[\begin{pmatrix} \mu_1 \\ \mu_2 \end{pmatrix}, \begin{pmatrix} \Sigma_{11} & \Sigma_{12} \\ \Sigma_{21} & \Sigma_{22} \end{pmatrix}\right],$$

where Σ_{22} *is invertible. Then*

$$\mathbf{X}_1|\mathbf{X}_2 \sim N\left[\mu_1 + \Sigma_{12}\Sigma_{22}^{-1}(\mathbf{X}_2 - \mu_2), \Sigma_{11} - \Sigma_{12}\Sigma_{22}^{-1}\Sigma_{21}\right].$$

We now define the following Bayesian problem:

$$\mathbf{y}|\beta, \mathbf{b} \sim N(\mathbf{X}\beta + \mathbf{Z}\mathbf{b}, \mathbf{R}), \tag{2.14}$$

with prior distribution for β and $\mathbf{b}$:

$$\beta \sim N(\mathbf{0}, \mathbf{H}), \ \mathbf{b} \sim N(\mathbf{0}, \tilde{\mathbf{D}}), \tag{2.15}$$

where $\beta, \mathbf{b}$ and $\epsilon = \mathbf{y} - \mathbf{X}\beta - \mathbf{Z}\mathbf{b}$ are all independent of each other, and $\tilde{\mathbf{D}}$ is defined in (2.6).

Notice that specification of $\mathbf{H}$ is flexible. For example, we may let $\mathbf{H} = \lambda\mathbf{I}_p$. This indicates that the components of β are independent of each other. Moreover, when $\lambda \to \infty$, we have $\mathbf{H}^{-1} = \lambda^{-1}\mathbf{I}_p \to \mathbf{0}$. This indicates that the limit of the prior on β is non-informative.

Theorem 2.1 *The Best Linear Unbiased Predictors (2.8) and (2.9) that minimize the GLL criterion (2.7) are the same as the limit posterior expectations of the Bayesian problem defined in (2.14) and (2.15) as $\mathbf{H}^{-1} \to \mathbf{0}$. That is,*

$$\hat{\beta} = \lim_{\mathbf{H}^{-1}\to\mathbf{0}} E(\beta|\mathbf{y}), \quad \hat{\mathbf{b}} = \lim_{\mathbf{H}^{-1}\to\mathbf{0}} E(\mathbf{b}|\mathbf{y}). \tag{2.16}$$

Moreover, as $\mathbf{H}^{-1} \to \mathbf{0}$, we have the following posterior distributions:

$$\begin{aligned} \beta|\mathbf{y} &\to N\left(\hat{\beta}, (\mathbf{X}^T\mathbf{V}^{-1}\mathbf{X})^{-1}\right), \\ \mathbf{b}|\mathbf{y} &\to N\left(\hat{\mathbf{b}}, \tilde{\mathbf{D}} - \tilde{\mathbf{D}}\mathbf{Z}^T\mathbf{P}_\mathbf{V}\mathbf{Z}\tilde{\mathbf{D}}\right), \\ \epsilon|\mathbf{y} &\to N\left(\hat{\epsilon}, \mathbf{R} - \mathbf{R}\mathbf{P}_\mathbf{V}\mathbf{R}\right), \end{aligned} \tag{2.17}$$

where $\hat{\epsilon} = \mathbf{y} - \mathbf{X}\hat{\beta} - \mathbf{Z}\hat{\mathbf{b}}$ and

$$\mathbf{P}_\mathbf{V} = \mathbf{V}^{-1} - \mathbf{V}^{-1}\mathbf{X}(\mathbf{X}^T\mathbf{V}^{-1}\mathbf{X})^{-1}\mathbf{X}^T\mathbf{V}^{-1}. \tag{2.18}$$

Proof: see Appendix.

Notice that $\hat{\beta}$ and $\hat{\mathbf{b}}$ involve the unknown parameters $\tilde{\mathbf{D}}$ and $\mathbf{R}$. If we substitute point estimates of $\tilde{\mathbf{D}}$ and $\mathbf{R}$ (We shall discuss how to estimate them in the next subsections), the Bayesian estimates, $\hat{\beta}$ and $\hat{\mathbf{b}}$ are usually referred to as empirical Bayes estimates, although empirical Bayes estimation is conventionally applied only to the random-effects $\mathbf{b}_i$, $i = 1, 2, \cdots, n$.

Theorem 2.1 gives the limit posterior distributions of β, $\mathbf{b}$ and ϵ under the Bayesian framework (2.14) and (2.15) when $\mathbf{H}^{-1} \to \mathbf{0}$ or when the prior on β is non-informative. Sometimes, it is of interest to know the posterior distributions of $\mathbf{b}$ and ϵ when β is given, e.g., when $\beta = \hat{\beta}$. Actually, this knowledge is a basis for the maximum likelihood-based EM-algorithm that we shall review in the next subsection. The following theorem gives the related results.

Theorem 2.2 *Under the Bayesian framework (2.14) and (2.15), we have*

$$\begin{aligned} \mathbf{b}|\mathbf{y},\beta &\sim N\left[\tilde{\mathbf{D}}\mathbf{Z}^T\mathbf{V}^{-1}(\mathbf{y}-\mathbf{X}\beta), \tilde{\mathbf{D}}-\tilde{\mathbf{D}}\mathbf{Z}^T\mathbf{V}^{-1}\mathbf{Z}\tilde{\mathbf{D}}\right], \\ \epsilon|\mathbf{y},\beta &\sim N\left[\mathbf{R}\mathbf{V}^{-1}(\mathbf{y}-\mathbf{X}\beta), \mathbf{R}-\mathbf{R}\mathbf{V}^{-1}\mathbf{R}\right]. \end{aligned} \quad (2.19)$$

Proof: see Appendix.

It is worthwhile to notice that by Theorem 2.2, we have $\mathrm{E}(\mathbf{b}|\mathbf{y},\beta=\hat{\beta})=\hat{\mathbf{b}}$ and $\mathrm{E}(\epsilon|\mathbf{y},\beta=\hat{\beta})=\hat{\epsilon}$.

2.2.4 Estimation of Variance Components

If the covariance matrices, $\mathbf{D}$ and $\mathbf{R}_i$, are unknown, but their point estimates, say, $\hat{\mathbf{D}}$ and $\hat{\mathbf{R}}_i$, are available, then we can have $\hat{\mathbf{V}}_i = \mathbf{Z}_i\hat{\mathbf{D}}\mathbf{Z}_i^T + \hat{\mathbf{R}}_i$. The estimates of β and $\mathbf{b}_i$ thus can be obtained by substitution of $\hat{\mathbf{V}}_i$ and $\hat{\mathbf{D}}$ in (2.8) and (2.9). Their corresponding standard errors are given by (2.10) and (2.12) after replacing $\mathbf{V}_i$ and $\mathbf{D}$ by their estimates. However, these standard errors are underestimated since the estimation errors of $\hat{\mathbf{V}}_i$ and $\hat{\mathbf{D}}$ are not accounted for.

Under the normality assumption, the maximum likelihood (ML) method and the restricted maximum likelihood (REML) method are two popular techniques to estimate the unknown components of $\mathbf{D}$ and $\mathbf{R}_i$, although this may not be appropriate if the normality assumption is questionable.

Under the following normality assumptions,

$$\mathbf{y}|\beta,\mathbf{b} \sim N(\mathbf{X}\beta+\mathbf{Z}\mathbf{b},\mathbf{R}), \quad \mathbf{b}\sim N(\mathbf{0},\tilde{\mathbf{D}}),$$

the generalized likelihood function can be written as

$$\mathbf{L}(\beta,\mathbf{b},\tilde{\mathbf{D}},\mathbf{R}|\mathbf{y}) = (2\pi)^{-N/2}|\mathbf{R}|^{-1/2}\exp\left\{-\tfrac{1}{2}[\mathbf{y}-\mathbf{X}\beta-\mathbf{Z}\mathbf{b}]^T\mathbf{R}^{-1}\right.$$
$$\left.\times[\mathbf{y}-\mathbf{X}\beta-\mathbf{Z}\mathbf{b}]\right\}(2\pi)^{-qn/2}|\tilde{\mathbf{D}}|^{-1/2}\exp\left(-\tfrac{1}{2}\mathbf{b}^T\tilde{\mathbf{D}}^{-1}\mathbf{b}\right),$$

where qn is the dimension of $\mathbf{b}$ and $N=\sum_{i=1}^n n_i$. If the random-effects vector $\mathbf{b}$ is integrated out, we can obtain the following conventional likelihood function:

$$\begin{aligned} \mathbf{L}(\beta,\tilde{\mathbf{D}},\mathbf{R}|\mathbf{y}) &= \textstyle\int \mathbf{L}(\beta,\mathbf{b},\tilde{\mathbf{D}},\mathbf{R}|\mathbf{y})d\mathbf{b} \\ &= (2\pi)^{-N/2}|\mathbf{V}|^{-1/2}\exp\left\{-\tfrac{1}{2}(\mathbf{y}-\mathbf{X}\beta)^T\mathbf{V}^{-1}(\mathbf{y}-\mathbf{X}\beta)\right\}. \end{aligned}$$

The ML method for estimation of variance components is to maximize the following log-likelihood function:

$$\log\mathbf{L}(\beta,\tilde{\mathbf{D}},\mathbf{R}|\mathbf{y}) = -\frac{1}{2}N\log 2\pi - \frac{1}{2}\log|\mathbf{V}| - \frac{1}{2}(\mathbf{y}-\mathbf{X}\beta)^T\mathbf{V}^{-1}(\mathbf{y}-\mathbf{X}\beta), \quad (2.20)$$

with respect to the variance components for a given β. However, joint maximization with respect to the variance components $\tilde{\mathbf{D}}$, $\mathbf{R}$ and the fixed-effects parameter vector β also results in the estimate of β in (2.8).

The REML method is used to integrate out both $\mathbf{b}$ and β from $\mathbf{L}(\beta, \mathbf{b}, \tilde{\mathbf{D}}, \mathbf{R}|\mathbf{y})$ in order to adjust for loss of degrees of freedom due to estimating β from the ML method, i.e., to maximize

$$\mathbf{L}(\tilde{\mathbf{D}}, \mathbf{R}|\mathbf{y}) = \int\int \mathbf{L}(\beta, \mathbf{b}, \tilde{\mathbf{D}}, \mathbf{R}|\mathbf{y}) d\mathbf{b} d\beta.$$

It can be shown that

$$\mathbf{L}(\tilde{\mathbf{D}}, \mathbf{R}|\mathbf{y}) = \frac{(2\pi)^{p/2}|\mathbf{X}^T\mathbf{V}^{-1}\mathbf{X}|^{-1/2}}{(2\pi)^{N/2}|\mathbf{V}|^{1/2}} \exp\left(-\frac{1}{2}\mathbf{y}^T\mathbf{P}_V\mathbf{y}\right),$$

where $\mathbf{P}_V = \mathbf{V}^{-1} - \mathbf{V}^{-1}\mathbf{X}(\mathbf{X}^T\mathbf{V}^{-1}\mathbf{X})^{-1}\mathbf{X}^T\mathbf{V}^{-1}$ as defined in (2.18). Thus, we have

$$\mathbf{L}(\tilde{\mathbf{D}}, \mathbf{R}|\mathbf{y}) = (2\pi)^{p/2}|\mathbf{X}^T\mathbf{V}^{-1}\mathbf{X}|^{-1/2}\mathbf{L}(\hat{\beta}, \tilde{\mathbf{D}}, \mathbf{R}|\mathbf{y}).$$

The REML estimates of variance components can be obtained via maximizing

$$\log \mathbf{L}(\tilde{\mathbf{D}}, \mathbf{R}|\mathbf{y}) = \log \mathbf{L}(\hat{\beta}, \tilde{\mathbf{D}}, \mathbf{R}|\mathbf{y}) + \frac{1}{2}p \log 2\pi - \frac{1}{2}\log|\mathbf{X}^T\mathbf{V}^{-1}\mathbf{X}|. \quad (2.21)$$

More detailed derivations for these results can be found in Davidian and Giltinan (1995).

2.2.5 The EM-Algorithms

The implementation of the ML and REML methods is not trivial. To overcome this implementation difficulty, the EM-algorithm and Newton-Raphson methods have been proposed (Laird and Ware 1982, Dempster, Rubin and Tsutakawa 1981, Laird, Lange and Stram 1987, Jennrich and Schluchter 1986, Lindstrom and Bates 1990). The books by Searle et al. (1992), Davidian and Giltinan (1995), Vonesh and Chinchilli (1996) and Pinheiro and Bates (2000) also provide a good review of these implementation methods. The standard statistical software packages such as SAS and S-PLUS now offer convenient functions to implement these methods (e.g., the S-PLUS function *lme* and the SAS procedure *PROC MIXED*). We shall briefly review the EM-algorithm here.

Recall that we generally assume that $\mathbf{R}_i$ has the following simple form:

$$\mathbf{R}_i = \sigma^2\mathbf{I}_{n_i}, \quad i = 1, 2, \cdots, n. \quad (2.22)$$

When $\mathbf{b}_i$ and ϵ_i were known, under the normality assumption, the natural ML estimates of σ^2 and $\mathbf{D}$ would be

$$\hat{\sigma}^2 = N^{-1}\sum_{i=1}^{n}\epsilon_i^T\epsilon_i, \quad \hat{\mathbf{D}} = n^{-1}\sum_{i=1}^{n}\mathbf{b}_i\mathbf{b}_i^T. \quad (2.23)$$

This is the M-step of the EM-algorithm. Because ϵ_i and $\mathbf{b}_i$ are unknown, the above estimates are not computable. There are two ways for overcoming this difficulty, associated, respectively, with the ML or REML-based EM-algorithm.

Notice that the ML estimates of $\mathbf{D}$ and σ^2 are obtained via maximizing the log-likelihood function (2.20) with the fixed-effects parameter vector $\beta = \hat{\beta}$ given. Therefore, the key for the ML-based EM-algorithm is to replace the $\hat{\sigma}^2$ and $\hat{\mathbf{D}}$ in (2.23) with

$$\mathrm{E}(\hat{\sigma}^2|\mathbf{y}, \beta = \hat{\beta}) \quad \text{and} \quad \mathrm{E}(\hat{\mathbf{D}}|\mathbf{y}, \beta = \hat{\beta}), \tag{2.24}$$

respectively. The underlying rationale is that the variance components $\mathbf{D}$ and σ^2 are estimated based on the residuals after the estimated fixed-effects component $\mathbf{X}\hat{\beta}$ is removed from the raw data, and the estimation will not take the variation of $\mathbf{X}\hat{\beta}$ into account. This is the E-step of the ML-based EM-algorithm. Using Theorem 2.2, we can show the following theorem.

Theorem 2.3 *Assume the Bayesian model defined in (2.14) and (2.15) holds, and assume* $\mathbf{R}_i$, $i = 1, 2, \cdots, n$ *satisfy (2.22). Then we have*

$$\begin{aligned} E(\hat{\sigma}^2|\mathbf{y}, \beta = \hat{\beta}) &= N^{-1}\textstyle\sum_{i=1}^n \left\{\hat{\epsilon}_i^T\hat{\epsilon}_i + \sigma^2\left[n_i - \sigma^2 tr(\mathbf{V}_i^{-1})\right]\right\}, \\ E(\hat{\mathbf{D}}|\mathbf{y}, \beta = \hat{\beta}) &= n^{-1}\textstyle\sum_{i=1}^n \left\{\hat{\mathbf{b}}_i\hat{\mathbf{b}}_i^T + [\mathbf{D} - \mathbf{D}\mathbf{Z}_i^T\mathbf{V}_i^{-1}\mathbf{Z}_i\mathbf{D}]\right\}. \end{aligned} \tag{2.25}$$

Proof: see Appendix.

In the right-hand sides of the expressions (2.25), the variance components σ^2 and $\mathbf{D}$ are still unknown. However, when they are replaced by the current available values, the updated values for $\hat{\sigma}^2$ and $\hat{\mathbf{D}}$ can be obtained. In other words, provided some initial values for $\mathbf{D}$ and σ^2, we can update $\hat{\sigma}^2$ and $\hat{\mathbf{D}}$ using (2.25) until convergence. This is the main idea of the EM-algorithm. For simplicity, the initial values can be taken as $\hat{\mathbf{D}} = \mathbf{I}_q$ and $\hat{\sigma}^2 = 1$. The major cycle for the ML-based EM-algorithm is as follows:

(a) Given $\hat{\mathbf{D}}$ and $\hat{\sigma}^2$, compute $\hat{\beta}$ and $\hat{\mathbf{b}}_i$ using (2.8) and (2.9).

(b) Given $\hat{\beta}$ and $\hat{\mathbf{b}}_i$, update $\hat{\mathbf{D}}$ and $\hat{\sigma}^2$ using (2.25).

(c) Iterate between (a) and (b) until convergence.

Let $r = 0, 1, 2, \cdots$ index the sequence of the iterations, and $\hat{\beta}_{(r)}, \hat{\mathbf{b}}_{i(r)}$ the estimated values for β and $\mathbf{b}_i$ at iteration r. Other notations such as $\mathbf{V}_{i(r)}, \sigma^2_{(r)}$ are similarly defined. Then more formally, the ML-based EM-algorithm may be written as follows:

ML-Based EM-Algorithm

Step 0. Set $r = 0$. Let $\hat{\sigma}^2_{(r)} = 1$, and $\hat{\mathbf{D}}_{(r)} = \mathbf{I}_q$.

Step 1. Set $r = r + 1$. Update $\hat{\beta}_{(r)}$ and $\hat{\mathbf{b}}_{i(r)}$ using

$$\begin{aligned} \hat{\beta}_{(r)} &= \left(\sum_{i=1}^n \mathbf{X}_i^T\hat{\mathbf{V}}_{i(r-1)}^{-1}\mathbf{X}_i\right)^{-1}\left(\sum_{i=1}^n \mathbf{X}_i^T\hat{\mathbf{V}}_{i(r-1)}^{-1}\mathbf{y}_i\right), \\ \hat{\mathbf{b}}_{i(r)} &= \mathbf{D}_{(r-1)}\mathbf{Z}_i^T\hat{\mathbf{V}}_{i(r-1)}^{-1}\left(\mathbf{y}_i - \mathbf{X}_i\hat{\beta}_{(r)}\right), \; i = 1, 2, \cdots, n, \end{aligned}$$

where

$$\hat{\mathbf{V}}_{i(r-1)} = \mathbf{Z}_i \hat{\mathbf{D}}_{(r-1)} \mathbf{Z}_i^T + \hat{\sigma}^2_{(r-1)} \mathbf{I}_{n_i},\ i = 1, 2, \cdots, n.$$

Step 2. Update $\hat{\sigma}^2_{(r)}$ and $\hat{\mathbf{D}}_{(r)}$ using

$$\begin{aligned}
\hat{\sigma}^2_{(r)} &= N^{-1} \sum_{i=1}^{n} \left\{ \hat{\boldsymbol{\epsilon}}_{i(r)}^T \hat{\boldsymbol{\epsilon}}_{i(r)} + \hat{\sigma}^2_{(r-1)} [n_i - \hat{\sigma}^2_{(r-1)} \mathrm{tr}(\hat{\mathbf{V}}_{i(r-1)}^{-1})] \right\}, \\
\hat{\mathbf{D}}_{(r)} &= n^{-1} \sum_{i=1}^{n} \left\{ \hat{\mathbf{b}}_{i(r)} \hat{\mathbf{b}}_{i(r)}^T \right. \\
&\quad \left. + [\hat{\mathbf{D}}_{(r-1)} - \hat{\mathbf{D}}_{(r-1)} \mathbf{Z}_i^T \hat{\mathbf{V}}_{i(r-1)}^{-1} \mathbf{Z}_i \hat{\mathbf{D}}_{(r-1)}] \right\},
\end{aligned}$$

where $\hat{\boldsymbol{\epsilon}}_{i(r)} = \mathbf{y}_i - \mathbf{X}_i \hat{\boldsymbol{\beta}}_{(r)} - \mathbf{Z}_i \hat{\mathbf{b}}_{i(r)}$.

Step 3. Repeat Steps 1 and 2 until convergence.

The REML-based EM-algorithm can be similarly described. The main differences include:

(a) The REML-based EM-algorithm is developed to find the REML estimates of σ^2 and $\mathbf{D}$ that maximize (2.21).

(b) The key for the REML-based EM-algorithm is to replace $\hat{\sigma}^2$ and $\hat{\mathbf{D}}$ in (2.23) by $\mathrm{E}(\hat{\sigma}^2|\mathbf{y})$ and $\mathrm{E}(\hat{\mathbf{D}}|\mathbf{y})$ instead of their expectations conditional on $\mathbf{y}$ and $\boldsymbol{\beta}$ as given in (2.24). These conditional expectations can be easily obtained using Theorem 2.1 and we shall present them in Theorem 2.4 below for easy reference.

(c) The REML-based EM-algorithm can be simply obtained via replacing all the $\hat{\mathbf{V}}_{i(r-1)}^{-1}$ in the Step 2 of the ML-based EM-algorithm above with $\mathbf{P}_{\hat{\mathbf{V}}_{i(r-1)}}$, where

$$\mathbf{P}_{\hat{\mathbf{V}}_{i(r-1)}} = \hat{\mathbf{V}}_{i(r-1)}^{-1} - \hat{\mathbf{V}}_{i(r-1)}^{-1} \mathbf{X}_i \left(\sum_{j=1}^{n} \mathbf{X}_j^T \hat{\mathbf{V}}_{j(r-1)}^{-1} \mathbf{X}_j \right)^{-1} \mathbf{X}_i^T \hat{\mathbf{V}}_{i(r-1)}^{-1}.$$

Theorem 2.4 below is similar to Theorem 2.3 but it is based on Theorem 2.1.

Theorem 2.4 *Assume the Bayesian model defined in (2.14) and (2.15) holds, and assume* $\mathbf{R}_i,\ i = 1, 2, \cdots, n$ *satisfy (2.22). Then as* $\mathbf{H}^{-1} \to 0$,

$$\begin{aligned}
E(\hat{\sigma}^2|\mathbf{y}) &\to N^{-1} \textstyle\sum_{i=1}^{n} \left\{ \hat{\boldsymbol{\epsilon}}_i^T \hat{\boldsymbol{\epsilon}}_i + \sigma^2 \left[n_i - \sigma^2 tr(\mathbf{P}_{\mathbf{V}_i}) \right] \right\}, \\
E(\hat{\mathbf{D}}|\mathbf{y}) &\to n^{-1} \textstyle\sum_{i=1}^{n} \left\{ \hat{\mathbf{b}}_i \hat{\mathbf{b}}_i^T + [\mathbf{D} - \mathbf{D}\mathbf{Z}_i^T \mathbf{P}_{\mathbf{V}_i} \mathbf{Z}_i \mathbf{D}] \right\},
\end{aligned} \tag{2.26}$$

where $\mathbf{P}_{\mathbf{V}_i} = \mathbf{V}_i^{-1} - \mathbf{V}_i^{-1} \mathbf{X}_i \left(\sum_{i=1}^{n} \mathbf{X}_i^T \mathbf{V}_i^{-1} \mathbf{X}_i \right)^{-1} \mathbf{X}_i^T \mathbf{V}_i^{-1}$.

Proof: see Appendix.

2.3 NONLINEAR MIXED-EFFECTS MODEL

2.3.1 Model Specification

The nonlinear mixed-effects (NLME) model is a natural generalization from the LME model. Let $(\boldsymbol{\mu}, \boldsymbol{\Omega})$ denote a generic distribution with mean vector $\boldsymbol{\mu}$ and covariance matrix $\boldsymbol{\Omega}$. Then a NLME model can be written into a two-stage hierarchical form as follows:

Stage 1, intra-subject variation

$$\mathbf{y}_i = \mathbf{f}(\mathbf{X}_i, \boldsymbol{\beta}_i) + \boldsymbol{\epsilon}_i, \ \boldsymbol{\epsilon}_i|\boldsymbol{\beta}_i \sim (\mathbf{0}, \mathbf{R}_i(\boldsymbol{\beta}_i, \boldsymbol{\xi})), \ \ i = 1, 2, \cdots, n, \tag{2.27}$$

where $\mathbf{y}_i$ and $\boldsymbol{\epsilon}_i$ are $(n_i \times 1)$ vectors of the responses and the measurement errors for subject i, respectively; the vector function $\mathbf{f}(\mathbf{X}_i, \boldsymbol{\beta}_i) = [f(\mathbf{x}_{i1}, \boldsymbol{\beta}_i), \cdots, f(\mathbf{x}_{in_i}, \boldsymbol{\beta}_i)]^T$ where $f(\cdot)$ is a known function, $\boldsymbol{\beta}_i : p \times 1$ is an unknown parameter vector, and the design matrix $\mathbf{X}_i = [\mathbf{x}_{i1}, \cdots, \mathbf{x}_{in_i}]^T$; and the covariance matrix $\mathbf{R}_i(\cdot)$ is a known function with $\boldsymbol{\xi}$ being an unknown parameter vector.

Stage 2, inter-subject variation

$$\boldsymbol{\beta}_i = \mathbf{d}(\mathbf{a}_i, \boldsymbol{\beta}, \mathbf{b}_i), \ \mathbf{b}_i \sim (\mathbf{0}, \mathbf{D}), \ i = 1, 2, \cdots, n, \tag{2.28}$$

where $\mathbf{d}(\mathbf{a}_i, \boldsymbol{\beta}, \mathbf{b}_i)$ is a known p-dimensional function of the between-subject covariate vector $\mathbf{a}_i$, with the population parameter vector (known also as fixed-effects parameter vector) $\boldsymbol{\beta}$ and the random-effects vector $\mathbf{b}_i$. The function $\mathbf{d}(\cdot)$ may be linear or nonlinear. But in real applications, a simple linear model

$$\boldsymbol{\beta}_i = \mathbf{A}_i\boldsymbol{\beta} + \mathbf{b}_i, \ \mathbf{b}_i \sim (\mathbf{0}, \mathbf{D}), \ \ i = 1, 2, \cdots, n, \tag{2.29}$$

is often used. See Davidian and Giltinan (1995) and Vonesh and Chinchilli (1996) for details.

Inference methods for NLME models have been proposed based on two basic approaches. One is derived from the estimates from individual subjects and another is based on linearization approaches. We briefly introduce these methods in the following subsections.

2.3.2 Two-Stage Method

If sufficient data are available from individual subjects to allow construction of subject-specific parameter estimates, we may use such estimates as building blocks for further inference. The inference methods based on individual estimates are usually called "two-stage" methods since a two-stage scheme is used to obtain final inference results. The first stage is to obtain individual estimates and the second stage is to use the individual estimates as building blocks to conduct final inferences based on various procedures. We briefly introduce these methods in this subsection. More details can be found in the books by Davidian and Giltinan (1995) and Vonesh and Chinchilli (1996).

The first stage is to construct the initial individual estimates based on the first stage model (2.27) using standard nonlinear regression techniques. For example, we first assume $\mathbf{R}_i(\beta_i, \xi) = \sigma^2 \mathbf{S}_i(\gamma)$, where $\mathbf{S}_i(\cdot)$ is known but σ^2 and γ are unknown variance-covariance parameters. Then the generalized least squares (GLS) approach can be used to estimate the unknown parameters β_i and γ via minimizing the following GLS criterion:

$$[\mathbf{y}_i - \mathbf{f}(\mathbf{X}_i, \beta_i)]^T \mathbf{S}_i^{-1}(\gamma)[\mathbf{y}_i - \mathbf{f}(\mathbf{X}_i, \beta_i)], \quad i = 1, 2, \cdots, n.$$

The Newton-Raphson method or a quasi-Newton method may be used to estimate γ and β_i iteratively.

If we assume that the variance-covariance parameters σ^2 and γ are common for all subjects, we may pool all the data across subjects to get more efficient estimates for them. For example, the maximum pseudo-likelihood method can be used if the error terms ϵ_i, $i = 1, 2, \cdots, n$ are assumed to be normally distributed. That is, we minimize the following pooled log-likelihood function with respect to σ^2 and γ:

$$PL(\sigma^2, \gamma) = \sum_{i=1}^{n} \{ \log |\sigma^2 \mathbf{S}_i(\gamma)| + [\mathbf{y}_i - \mathbf{f}(\mathbf{X}_i, \hat{\beta}_i)]^T [\sigma^2 \mathbf{S}_i(\gamma)]^{-1} [\mathbf{y}_i - \mathbf{f}(\mathbf{X}_i, \hat{\beta}_i)] \},$$

to obtain the estimates of the variance-covariance parameters σ^2 and γ. Similarly, a restricted maximum likelihood (REML) approach can be employed via minimizing

$$REML(\sigma^2, \gamma) = PL(\sigma^2, \gamma) - \log \sigma^{2np} + \sum_{i=1}^{n} \{ \log |\mathbf{F}_i(\hat{\beta}_i)^T [\sigma^2 \mathbf{S}_i(\gamma)]^{-1} \mathbf{F}_i(\hat{\beta}_i)| \},$$

where $\mathbf{F}_i(\hat{\beta}_i)$ is the $(n_i \times p)$ matrix of derivatives of $\mathbf{f}(\mathbf{X}_i, \beta_i)$ evaluated at $\hat{\beta}_i$. The confidence intervals or regions and hypothesis testing of the unknown parameters can be constructed based on standard profile likelihood methods.

In the second stage of the estimation procedure, the population parameters are estimated based on the individual estimates from the first stage. Several methods have been proposed to obtain the population estimates of parameters which include the standard two-stage (STS) method, global two-stage (GTS) method, and Bayes method. We briefly introduce these methods as follows.

STS method. The standard two-stage (STS) method (Steimer et al. 1984) treats the estimates of β_i from the first stage, denoted by β_i^*, as if they were the true β_i. If the simple linear model (2.29) for β_i is assumed, approximately we have $\beta_i^* \approx \mathbf{A}_i\beta + \mathbf{b}_i$, where $\mathbf{b}_i \sim N(\mathbf{0}, \mathbf{D})$. The seemingly unrelated regression estimates (Zellner 1962, Davidian and Giltinan 1993) can be used to construct the population estimates as

$$\begin{aligned} \hat{\beta}_{STS} &= \left(\sum_{i=1}^{n} \mathbf{A}_i^T \mathbf{A}_i \right)^{-1} \left(\sum_{i=1}^{n} \mathbf{A}_i^T \beta_i^* \right), \\ \hat{\mathbf{D}}_{STS} &= (n-1)^{-1} \sum_{i=1}^{n} \left(\beta_i^* - \mathbf{A}_i \hat{\beta}_{STS} \right) \left(\beta_i^* - \mathbf{A}_i \hat{\beta}_{STS} \right)^T. \end{aligned}$$

If $\mathbf{A}_i = \mathbf{I}_{n_i}$, the $\hat{\beta}_{STS}$ reduces to the simple average of β_i^*. Although the STS method is simple and does not require the normality assumption, it ignores the uncertainty in the estimates of β_i which results in an upward bias in the estimate of $\mathbf{D}$ (Davidian and Giltinan 1995). In addition, there is no refinement for individual estimates β_i^*. These critical drawbacks make this method less attractive.

GTS method. The global two-stage (GST) method is more favorable since it is designed to avoid the problems in the STS method. The uncertainty of β_i^* can be modeled by assuming approximately $\beta_i^*|\beta_i \sim N(\beta_i, \mathbf{C}_i)$. Thus, a working model for β_i^* can be written as

$$\begin{aligned} \beta_i^* \approx \mathbf{A}_i\beta + \mathbf{b}_i + \mathbf{e}_i,\ \mathbf{b}_i \sim N(\mathbf{0}, \mathbf{D}),\ \mathbf{e}_i \sim N(\mathbf{0}, \mathbf{C}_i), \\ i = 1, 2, \cdots, n, \end{aligned} \tag{2.30}$$

where $\mathbf{C}_i$ are assumed to be fixed and known. In fact, this is a standard LME model. An iterative EM-algorithm (Steimer et al. 1984, Davidian and Giltinan 1993, Davidian and Giltinan 1995) can be used to estimate the population parameters as well as to refine the estimates of individual parameters. The EM-algorithm can be specified as follows. At $(k+1)$-th iteration,

1. *E-step*: Update the estimates of β_i, $i = 1, 2, \cdots, n$ as

$$\hat{\beta}_{i,(k+1)} = \left(\mathbf{C}_i^{-1} + \hat{\mathbf{D}}_{(k)}^{-1}\right)^{-1} \left(\mathbf{C}_i^{-1}\beta_i^* + \hat{\mathbf{D}}_{(k)}^{-1}\mathbf{A}_i\hat{\beta}_{(k)}\right),\quad i = 1, 2, \cdots, n.$$

2. *M-step*: Update the estimates of the population parameters as

$$\begin{aligned} \hat{\beta}_{(k+1)} &= \left(\sum_{i=1}^n \mathbf{A}_i^T\hat{\mathbf{D}}_{(k)}^{-1}\mathbf{A}_i\right)^{-1} \sum_{i=1}^n \mathbf{A}_i^T\hat{\mathbf{D}}_{(k)}^{-1}\hat{\beta}_{i,(k+1)}, \\ \hat{\mathbf{D}}_{(k+1)} &= n^{-1}\sum_{i=1}^n (\hat{\beta}_{i,(k+1)} - \mathbf{A}_i\hat{\beta}_{(k+1)})(\hat{\beta}_{i,(k+1)} - \mathbf{A}_i\hat{\beta}_{(k+1)})^T \\ &\quad + n^{-1}\sum_{i=1}^n (\mathbf{C}_i^{-1} + \hat{\mathbf{D}}_{(k)}^{-1})^{-1}. \end{aligned}$$

This EM-algorithm is iterated until the M-step converges (i.e., the difference between successive estimates of β and $\mathbf{D}$ is sufficiently small). We denote the final estimates by $\hat{\beta}_{GTS}, \hat{\mathbf{D}}_{GTS}$, and $\hat{\beta}_{i,GTS}$. Note that the starting values for the EM-algorithm can use the STS estimates $(\hat{\beta}_{STS}, \hat{\mathbf{D}}_{STS})$ introduced above.

It can be shown (Davidian and Giltinan 1995) that the final estimates $(\hat{\beta}_{GTS}, \hat{\mathbf{D}}_{GTS})$ produced from this EM-algorithm minimize twice the negative log-likelihood function under the normality assumption in model (2.30) with treating β_i^* as "data":

$$L_{GTS} = \sum_{i=1}^n \log|\mathbf{C}_i + \mathbf{D}| + \sum_{i=1}^n (\beta_i^* - \mathbf{A}_i\beta)^T(\mathbf{C}_i + \mathbf{D})^{-1}(\beta_i^* - \mathbf{A}_i\beta),$$

where $\mathbf{C}_i$ are assumed to be fixed and known. Although the GTS estimates are derived based on the maximum likelihood approach under the normality assumption for $\mathbf{b}_i$

and $\mathbf{e}_i$, it is still robust against the departure of the normality assumption as long as the first two moments of the asymptotic distribution of $\beta_i^*|\beta_i$ can be approximated with sufficient accuracy. This is because the same GTS estimators can be derived from the GLS approach. However, similar to a standard LME model, the GTS estimate of $\mathbf{D}$ may be biased downward due to loss of degrees of freedom for estimation of β. Standard inferences such as confidence intervals and hypothesis testing can be constructed based on asymptotic theory of maximum likelihood under normality assumptions.

Bayes Method Under similar assumptions as in the GTS method, Racine-Poon (1985) proposed a full Bayesian approach to estimate $\beta, \mathbf{D}$, and β_i. The key distribution assumptions include (1) $\beta_i^*|\beta_i \sim N(\beta_i, \mathbf{C}_i)$; (2) $\beta_i \sim N(\mathbf{A}_i\beta, \mathbf{D})$; and (3) β has a vague non-informative prior distribution and $\mathbf{D}^{-1}$ has a Wishart prior distribution. Under these assumptions, a full Bayesian method can be developed. Since the Bayesian approach is not the focus of this book, we will not discuss this further. The reader is referred to the books by Davidian and Giltinan (1995) and Vonesh and Chinchilli (1996) or the original paper by Racine-Poon (1985).

2.3.3 First-Order Linearization Method

Since the two-stage methods require sufficient data from each subject to obtain the individual estimates in the first step, this limitation restricts the popular use of these two-stage approaches. An alternative method is based on linearization of the NLME model. This approach allows us to efficiently deal with sparse individual data and easily utilize the inference techniques in the LME model. First we introduce the first-order linearization approach pioneered by Sheiner, Rosenberg and Melmon (1972), and then we discuss a conditional first-order linearization method suggested by Lindstrom and Bates (1990) which is a refinement of Sheiner et al. (1972)'s approach.

The linearization of the NLME model with respect to the random-effects was first proposed by Sheiner, Rosenberg and Melmon (1972), and was further studied by Sheiner and his colleagues (Sheiner, Rosenberg and Marathe 1977, Sheiner and Beal 1980, 1981, and Beal and Sheiner 1982).

Consider the general form of NLME models (2.27) and (2.28). For simplicity, we suppress the design matrices $\mathbf{X}_i$ via denoting $\mathbf{f}(\mathbf{X}_i, \beta_i)$ as $\mathbf{f}_i(\beta_i)$ and let $\epsilon_i = \mathbf{R}_i^{1/2}(\beta_i, \xi)\mathbf{e}_i$, where $\mathbf{e}_i \sim N(\mathbf{0}, \mathbf{I}_{n_i})$ and is independent of $\mathbf{b}_i$, and $\mathbf{R}_i^{1/2}(\beta_i, \xi)$ is the Cholesky decomposition of $\mathbf{R}_i(\beta_i, \xi)$. Substituting model (2.28) into model (2.27), we have

$$\begin{aligned}\mathbf{y}_i &= \mathbf{f}_i\{\mathbf{d}(\mathbf{a}_i, \beta, \mathbf{b}_i)\} + \mathbf{R}_i^{1/2}\{\mathbf{d}(\mathbf{a}_i, \beta, \mathbf{b}_i), \xi\}\mathbf{e}_i, \\ & \quad i = 1, 2, \cdots, n.\end{aligned} \tag{2.31}$$

Beal and Sheiner (1982) suggested approximating this model using a Taylor expansion in the random-effects vector $\mathbf{b}_i$ about the mean value $\mathrm{E}(\mathbf{b}_i) = \mathbf{0}$, which yields

$$\begin{aligned}\mathbf{y}_i &\approx \mathbf{f}_i\{\mathbf{d}(\mathbf{a}_i, \beta, \mathbf{0})\} + \mathbf{F}_i(\beta, \mathbf{0})\Delta_{bi}(\beta, \mathbf{0})\mathbf{b}_i + \mathbf{R}_i^{1/2}\{\mathbf{d}(\mathbf{a}_i, \beta, \mathbf{0}), \xi\}\mathbf{e}_i, \\ & \quad i = 1, 2, \cdots, n,\end{aligned} \tag{2.32}$$

where $\mathbf{F}_i(\boldsymbol{\beta}, \mathbf{0})$ is a $(n_i \times p)$ matrix and $\Delta_{bi}(\boldsymbol{\beta}, \mathbf{0})$ is a $(p \times q)$ matrix of the following derivative functions,

$$\mathbf{F}_i(\boldsymbol{\beta}, \mathbf{0}) = \frac{d}{d\boldsymbol{\beta}_i}\mathbf{f}_i(\boldsymbol{\beta}_i)|_{\boldsymbol{\beta}_i = \mathbf{d}(\mathbf{a}_i, \boldsymbol{\beta}, \mathbf{0})}, \quad \Delta_{bi}(\boldsymbol{\beta}, \mathbf{0}) = \frac{d}{d\mathbf{b}_i}\mathbf{d}(\mathbf{a}_i, \boldsymbol{\beta}, \mathbf{b}_i)|_{\mathbf{b}_i = \mathbf{0}}.$$

Note that the approximation to the mean function $\mathbf{f}_i\{\mathbf{d}(\mathbf{a}_i, \boldsymbol{\beta}, \mathbf{b}_i)\}$ is based on the first two terms of a Taylor expansion , whereas the approximation to the error term $\mathbf{R}_i^{1/2}\{\mathbf{d}(\mathbf{a}_i, \boldsymbol{\beta}, \mathbf{b}_i), \boldsymbol{\xi}\}\mathbf{e}_i$ is based on the first term, since the impact of misspecification of first-moment properties is more serious than that of second moments.

Define $\mathbf{Z}_i(\boldsymbol{\beta}, \mathbf{0}) = \mathbf{F}_i(\boldsymbol{\beta}, \mathbf{0})\Delta_{bi}(\boldsymbol{\beta}, \mathbf{0})$ and $\mathbf{e}_i^* = \mathbf{R}_i^{1/2}\{\mathbf{d}(\mathbf{a}_i, \boldsymbol{\beta}, \mathbf{0}), \boldsymbol{\xi}\}\mathbf{e}_i$, model (2.32) can be rewritten as

$$\mathbf{y}_i \approx \mathbf{f}_i\{\mathbf{d}(\mathbf{a}_i, \boldsymbol{\beta}, \mathbf{0})\} + \mathbf{Z}_i(\boldsymbol{\beta}, \mathbf{0})\mathbf{b}_i + \mathbf{e}_i^*. \tag{2.33}$$

From this model, we have

$$\begin{aligned} \mathrm{E}(\mathbf{y}_i) &\approx \mathbf{f}_i\{\mathbf{d}(\mathbf{a}_i, \boldsymbol{\beta}, \mathbf{0})\}, \\ \mathrm{Cov}(\mathbf{y}_i) &\approx \mathbf{R}_i\{\mathbf{d}(\mathbf{a}_i, \boldsymbol{\beta}, \mathbf{0}), \boldsymbol{\xi}\} + \mathbf{Z}_i(\boldsymbol{\beta}, \mathbf{0})\mathbf{D}\mathbf{Z}_i^T(\boldsymbol{\beta}, \mathbf{0}) \equiv \mathbf{V}_i(\boldsymbol{\beta}, \mathbf{0}, \boldsymbol{\omega}), \end{aligned}$$

where $\boldsymbol{\omega}$ denotes the vector of all parameters in the variance-covariance matrix. Compared to the standard LME model (2.2), model (2.33) is similar except that $\mathbf{Z}_i(\boldsymbol{\beta}, \mathbf{0})$ is not a given design matrix, but rather depends on the parameter $\boldsymbol{\beta}$; and the fixed-effect $\mathbf{f}_i\{\mathbf{d}(\mathbf{a}_i, \boldsymbol{\beta}, \mathbf{0})\}$ is a nonlinear function of $\boldsymbol{\beta}$. In addition, the error $\mathbf{e}_i^*$ may also depend on fixed-effects parameter $\boldsymbol{\beta}$ and covariance parameter $\boldsymbol{\xi}$.

Two estimation approaches have been proposed to estimate the unknown parameters based on the approximation model (2.33). One is the ML method and another is the GLS method. The ML method assumes the approximation model (2.33) is exact and that $\mathbf{b}_i$ and $\mathbf{e}_i^*$ are normally distributed (Beal and Sheiner 1982). The MLEs of $\boldsymbol{\beta}$ and $\boldsymbol{\omega}$ can be obtained by minimizing the objective function

$$\begin{aligned} L(\boldsymbol{\beta}, \boldsymbol{\omega}) &= \sum_{i=1}^{n}\Big([\mathbf{y}_i - \mathbf{f}_i\{\mathbf{d}(\mathbf{a}_i, \boldsymbol{\beta}, \mathbf{0})\}]^T\mathbf{V}_i^{-1}(\boldsymbol{\beta}, \mathbf{0}, \boldsymbol{\omega})[\mathbf{y}_i - \mathbf{f}_i\{\mathbf{d}(\mathbf{a}_i, \boldsymbol{\beta}, \mathbf{0})\}] \\ &\quad + \log|\mathbf{V}_i(\boldsymbol{\beta}, \mathbf{0}, \boldsymbol{\omega})|\Big). \end{aligned}$$

Numerical techniques such as the Newton-Raphson method can be used to implement this method. The starting value for $\boldsymbol{\beta}$ may be obtained by the ordinary least squares (OLS) method. The Wald-type or likelihood ratio inference approaches can be employed to conduct hypothesis testing and construct confidence intervals for the unknown parameters.

The GLS method, an alternative to the ML approach, is more robust to the normality assumption and misspecification of the individual covariance structure. Let $\mathbf{W}_i(\boldsymbol{\beta}, \mathbf{0}, \boldsymbol{\omega}) = \mathbf{V}_i^{-1}(\boldsymbol{\beta}, \mathbf{0}, \boldsymbol{\omega})$. A general GLS algorithm is given as follows:

1. Obtain a preliminary estimate $\hat{\boldsymbol{\beta}}^{(0)}$ for $\boldsymbol{\beta}$. The OLS estimator may be used.

2. Estimate the covariance parameters ω, denoted by $\hat{\omega}$, and obtain the estimated weight matrices $\mathbf{W}_i(\hat{\boldsymbol{\beta}}^{(0)}, \mathbf{0}, \hat{\omega}),\ \ i = 1, 2, \cdots, n$.

3. Using the estimates $\mathbf{W}_i(\hat{\boldsymbol{\beta}}^{(0)}, \mathbf{0}, \hat{\omega})$ from Step 2, re-estimate $\boldsymbol{\beta}$ via minimizing

$$\sum_{i=1}^{n} [\mathbf{y}_i - \mathbf{f}_i\{\mathbf{d}(\mathbf{a}_i, \boldsymbol{\beta}, \mathbf{0})\}]^T\, \mathbf{W}_i(\hat{\boldsymbol{\beta}}^{(0)}, \mathbf{0}, \hat{\omega})\, [\mathbf{y}_i - \mathbf{f}_i\{\mathbf{d}(\mathbf{a}_i, \boldsymbol{\beta}, \mathbf{0})\}]$$

with respect to $\boldsymbol{\beta}$.

4. Take the estimate of $\boldsymbol{\beta}$ from Step 3 as the new preliminary estimate $\hat{\boldsymbol{\beta}}^{(0)}$ and repeat Steps 2 and 3 until convergence.

Estimation of ω in Step 2 may use the pseudo-likelihood approach which is equivalent to minimizing

$$\begin{aligned} PL(\hat{\boldsymbol{\beta}}^{(0)}, \omega) &= \sum_{i=1}^{n} \Big([\mathbf{y}_i - \mathbf{f}_i\{\mathbf{d}(\mathbf{a}_i, \hat{\boldsymbol{\beta}}^{(0)}, \mathbf{0})\}]^T \mathbf{V}_i^{-1}(\hat{\boldsymbol{\beta}}^{(0)}, \mathbf{0}, \omega) \\ &\quad \times [\mathbf{y}_i - \mathbf{f}_i\{\mathbf{d}(\mathbf{a}_i, \hat{\boldsymbol{\beta}}^{(0)}, \mathbf{0})\}] + \log|\mathbf{V}_i(\hat{\boldsymbol{\beta}}^{(0)}, \mathbf{0}, \omega)| \Big). \end{aligned}$$

Alternatively a REML approach would minimize

$$\begin{aligned} REML(\hat{\boldsymbol{\beta}}^{(0)}, \omega) &= \sum_{i=1}^{n} \Big\{ \log|\mathbf{X}_i^T(\hat{\boldsymbol{\beta}}^{(0)}, \mathbf{0})\mathbf{V}_i^{-1}(\hat{\boldsymbol{\beta}}^{(0)}, \mathbf{0}, \omega)\mathbf{X}_i(\hat{\boldsymbol{\beta}}^{(0)}, \mathbf{0})| \Big\} \\ &\quad + PL(\hat{\boldsymbol{\beta}}^{(0)}, \omega), \end{aligned}$$

where $\mathbf{X}_i(\boldsymbol{\beta}, \mathbf{0}) = \mathbf{F}_i(\boldsymbol{\beta}, \mathbf{0})\Delta_{\beta i}(\boldsymbol{\beta}, \mathbf{0})$ and $\Delta_{\beta i}(\boldsymbol{\beta}, \mathbf{0})$ is the $(p \times q)$ matrix of derivatives of $\mathbf{d}(\mathbf{a}_i, \boldsymbol{\beta}, \mathbf{0})$ with respect to $\boldsymbol{\beta}$ evaluated at $\mathbf{b}_i = \mathbf{0}$.

Vonesh and Carter (1992) also proposed a variation of the GLS algorithm via replacing the estimates of ω in Step 2 in the above procedure with separate estimates of ξ and $\mathbf{D}$ in the two-component error structure model (2.33). Another variation of the GLS algorithm is to replace the estimates of ω in Step 2 with a generalized estimating equation (GEE) estimate. More details of these variations of the GLS algorithm can be found in Davidian and Giltinan (1995) or in the original literature.

2.3.4 Conditional First-Order Linearization Method

A refinement of the first-order linearization approach was proposed by Lindstrom and Bates (1990). In the first-order linearization method, the Taylor expansion in $\mathbf{b}_i$ about the mean value $\mathrm{E}(\mathbf{b}_i) = \mathbf{0}$ is used to approximate model (2.31). Lindstrom and Bates (1990) suggested improving the first-order linearization method by a Taylor expansion in $\mathbf{b}_i$ about an estimated $\mathbf{b}_i$. This method is referred to as the conditional first-order linearization.

For the model (2.31), similar to the standard first-order linearization approach, we can approximate the mean function and error term using the Taylor expansion in the random-effects ($\mathbf{b}_i$) about some estimated value ($\mathbf{b}_i^*$) which yields

$$\begin{aligned}\mathbf{y}_i &\approx \mathbf{f}_i[\mathbf{d}(\mathbf{a}_i, \boldsymbol{\beta}, \mathbf{b}_i^*)] + \mathbf{F}_i(\boldsymbol{\beta}, \mathbf{b}_i^*)\Delta_{bi}(\boldsymbol{\beta}, \mathbf{b}_i^*)(\mathbf{b}_i - \mathbf{b}_i^*) \\ &\quad + \mathbf{R}_i^{1/2}[\mathbf{d}(\mathbf{a}_i, \boldsymbol{\beta}, \mathbf{b}_i^*), \boldsymbol{\xi}]\mathbf{e}_i, \qquad (2.34)\end{aligned}$$

where $\mathbf{F}_i(\boldsymbol{\beta}, \mathbf{b}_i^*)$ is a $(n_i \times p)$ matrix and $\Delta_{bi}(\boldsymbol{\beta}, \mathbf{b}_i^*)$ is a $(p \times q)$ matrix of the following derivative functions,

$$\mathbf{F}_i(\boldsymbol{\beta}, \mathbf{b}_i^*) = \frac{d}{d\boldsymbol{\beta}_i}\mathbf{f}_i(\boldsymbol{\beta}_i)|_{\boldsymbol{\beta}_i = \mathbf{d}(\mathbf{a}_i, \boldsymbol{\beta}, \mathbf{b}_i^*)}, \quad \Delta_{bi}(\boldsymbol{\beta}, \mathbf{b}_i^*) = \frac{d}{d\mathbf{b}_i}\mathbf{d}(\mathbf{a}_i, \boldsymbol{\beta}, \mathbf{b}_i)|_{\mathbf{b}_i = \mathbf{b}_i^*}.$$

Note that the first two terms of the Taylor expansion about $\mathbf{b}_i = \mathbf{b}_i^*$ of $\mathbf{f}_i\{\mathbf{d}(\mathbf{a}_i, \boldsymbol{\beta}, \mathbf{b}_i)\}$ and the leading term of the Taylor expansion of $\mathbf{R}_i^{1/2}\{\mathbf{d}(\mathbf{a}_i, \boldsymbol{\beta}, \mathbf{b}_i), \boldsymbol{\xi}\}\mathbf{e}_i$ are retained in the approximation model. We expect that $\mathbf{b}_i^*$ is closer to $\mathbf{b}_i$ than its expectation $\mathbf{0}$. Thus, the approximation of model (2.34) is expected to be better than model (2.32).

Define $\mathbf{Z}_i(\boldsymbol{\beta}, \mathbf{b}_i^*) = \mathbf{F}_i(\boldsymbol{\beta}, \mathbf{b}_i^*)\Delta_{bi}(\boldsymbol{\beta}, \mathbf{b}_i^*)$ and $\mathbf{e}_i^* = \mathbf{R}_i^{1/2}\{\mathbf{d}(\mathbf{a}_i, \boldsymbol{\beta}, \mathbf{b}_i^*), \boldsymbol{\xi}\}\mathbf{e}_i$, model (2.34) can be rewritten as

$$\mathbf{y}_i \approx \mathbf{f}_i\{\mathbf{d}(\mathbf{a}_i, \boldsymbol{\beta}, \mathbf{b}_i^*)\} - \mathbf{Z}_i(\boldsymbol{\beta}, \mathbf{b}_i^*)\mathbf{b}_i^* + \mathbf{Z}_i(\boldsymbol{\beta}, \mathbf{b}_i^*)\mathbf{b}_i + \mathbf{e}_i^*. \qquad (2.35)$$

From this model, the conditional mean and covariance of $\mathbf{y}_i$ (given $\mathbf{b}_i^*$) is

$$\begin{aligned}\mathrm{E}(\mathbf{y}_i) &\approx \mathbf{f}_i\{\mathbf{d}(\mathbf{a}_i, \boldsymbol{\beta}, \mathbf{b}_i^*)\} - \mathbf{Z}_i(\boldsymbol{\beta}, \mathbf{b}_i^*)\mathbf{b}_i^*, \\ \mathrm{Cov}(\mathbf{y}_i) &\approx \mathbf{R}_i\{\mathbf{d}(\mathbf{a}_i, \boldsymbol{\beta}, \mathbf{b}_i^*), \boldsymbol{\xi}\} + \mathbf{Z}_i(\boldsymbol{\beta}, \mathbf{b}_i^*)\mathbf{D}\mathbf{Z}_i^T(\boldsymbol{\beta}, \mathbf{b}_i^*) \\ &\equiv \mathbf{V}_i(\boldsymbol{\beta}, \mathbf{b}_i^*, \boldsymbol{\omega}).\end{aligned}$$

The estimation approach for model (2.35) is similar to the LME model and the methods for the standard first-order linearization method (Lindstrom and Bates (1990). We refer readers to the books by Davidian and Giltinan (1995) and Vonesh and Chinchilli (1996) for detailed discussions.

2.4 GENERALIZED MIXED-EFFECTS MODEL

Many applications of longitudinal studies involve discrete rather than continuous outcomes (e.g., binary data, count data and ordinal data). In this case, the standard linear or nonlinear mixed-effects models for continuous data as we outlined above are not appropriate. In this section, we briefly review parametric generalized linear and nonlinear mixed-effects models for repeatedly measured discrete data or discrete longitudinal data.

2.4.1 Generalized Linear Mixed-Effects Model

Generalized linear mixed-effects (GLME) models are natural extensions of generalized linear models (GLM) introduced by Nelder and Wedderburn (1972). The GLMs

(Wedderburn 1974, McCullagh and Nelder 1989) broaden the standard Gaussian-based linear regression models to models for data with a more general exponential family of distributions that include many standard continuous and discrete distributions. This generalization unifies the modeling and analysis approaches for both discrete (e.g., binomial and Poisson) and continuous (e.g., normal and gamma) data. In order to model discrete repeated measurement data, mixed-effects were introduced into generalized linear models by a number of authors (e.g., Stiratelli, Laird and Ware 1984, Zeger, Liang and Albert 1988, Schall 1991, Breslow and Claton 1993). We give a brief overview of GLME models in this subsection.

Let $\mathbf{y}_i = [y_{i1}, \cdots, y_{in_i}]^T$ denote the repeated measurements of a response variable from the i-th of n subjects and assume that $\mathbf{y}_i$ are independent across subjects but the observations within subjects might be correlated. Let $\mathbf{b}_i$ denote the random-effects vector for the i-th subject. Then conditional to the $\mathbf{b}_i$, the response vector $\mathbf{y}_i$ follows a distribution from the exponential family (McCullagh and Nelder 1989) with density $f(\mathbf{y}_i|\mathbf{b}_i, \beta)$ where β is common for all the subjects and is called population parameter or fixed-effects parameter vector. Let $\boldsymbol{\mu}_i = \mathrm{E}(\mathbf{y}_i|\mathbf{b}_i)$, and $\mathrm{Cov}(\mathbf{y}_i|\mathbf{b}_i) = \phi\mathbf{V}_i(\boldsymbol{\mu}_i)$, where ϕ is an unknown variance parameter and $\mathbf{V}_i(\boldsymbol{\mu}_i) = \mathrm{diag}[V(\mu_{i1}), \cdots, V(\mu_{in_i})]$ with $V(\cdot)$ being a known variance function. Let $\boldsymbol{\eta}_i = g(\boldsymbol{\mu}_i)$ where $g(\boldsymbol{\mu}_i) = [g(\mu_{i1}), \cdots, g(\mu_{in_i})]^T$ with $g(\cdot)$ being a known link function. The GLME model can be written as

$$\eta_i = \mathbf{X}_i\beta + \mathbf{Z}_i\mathbf{b}_i, \ \mathbf{b}_i \sim (\mathbf{0}, \mathbf{D}), \ \ i = 1, 2, \cdots, n. \tag{2.36}$$

For estimation of the unknown parameters in GLME models, the two techniques introduced for GLMs (McCullagh and Nelder 1989), i.e., the conditional maximum likelihood (CML) method and the full maximum likelihood (FML) method, can be used. The basic idea of the CML method is to treat the random-effects $\mathbf{b}_i$ as nuisance parameters. Thus, only the fixed-effect parameter β are estimated using the conditional likelihood of the data given the sufficient statistics for $\mathbf{b}_i$. Since we prefer to estimate both fixed and random-effects, we will focus our discussion on the FML method as described below. We refer the readers to the book by Diggle et al. (2002) for the details of the CML method.

Let $\boldsymbol{\theta}$ denote all the unknown parameters in the GLME model (2.36). The marginal likelihood function of $\mathbf{y}$ for $\boldsymbol{\theta}$ is given by

$$L(\boldsymbol{\theta}; \mathbf{y}) = \prod_{i=1}^{n} \int \prod_{j=1}^{n_i} f(\mathbf{y}_i|\mathbf{b}_i, \beta) f(\mathbf{b}_i; \mathbf{D}) d\mathbf{b}_i. \tag{2.37}$$

For non-Gaussian models, the integration of the likelihood function does not have a closed form and numerical methods are required. We can use a score equation method to find the maximum likelihood estimate. First we assume that the "complete data" for an individual subject are $(\mathbf{y}_i, \mathbf{b}_i)$, then, under the canonical link function, the "complete data" score function for β can be simply written as

$$\mathbf{S}_\beta(\boldsymbol{\theta}|\mathbf{y}, \mathbf{b}) = \sum_{i=1}^{n} \sum_{j=1}^{n_i} \mathbf{x}_{ij} \left[y_{ij} - \mu_{ij}(\mathbf{b}_i)\right] = 0, \tag{2.38}$$

where $\mathbf{b} = [\mathbf{b}_1^T, \cdots, \mathbf{b}_n^T]^T$, $\mu_{ij}(\mathbf{b}_i) = \mathrm{E}(y_{ij}|\mathbf{b}_i) = g^{-1}(\eta_{ij})$, and $\eta_{ij} = \mathbf{x}_{ij}^T\beta + \mathbf{z}_{ij}^T\mathbf{b}_i$. The observed data score equations can be obtained by taking the expectation of (2.38) with respect to the conditional distribution of $\mathbf{b}_i$ given the data $\mathbf{y}$, i.e.,

$$\mathbf{S}_\beta(\boldsymbol{\theta}|\mathbf{y}) = \sum_{i=1}^{n}\sum_{j=1}^{n_i} \mathbf{x}_{ij}\left[y_{ij} - \mathrm{E}\left\{\mu_{ij}(\mathbf{b}_i)|\mathbf{y}_i\right\}\right] = 0. \tag{2.39}$$

Similarly, the score equations for $\mathbf{D}$ can be obtained as

$$\mathbf{S}_\mathbf{D}(\boldsymbol{\theta}|\mathbf{y}) = \frac{1}{2}\mathbf{D}^{-1}\left\{\sum_{i=1}^{n}\mathrm{E}(\mathbf{b}_i\mathbf{b}_i^T|\mathbf{y}_i)\right\}\mathbf{D}^{-1} - \frac{n}{2}\mathbf{D}^{-1} = 0. \tag{2.40}$$

Like the NLME models, there is no closed form solution to the score equations. To solve for the MLE of $\boldsymbol{\theta}$, the EM-algorithm or Newton-Raphson method needs to be used. If the dimension of $\mathbf{b}_i$ is not high (one or two), numerical integration techniques can be easily used (Crouch and Spiegelman 1990). However, for higher dimensional problems, Monte Carlo integration methods such as Gibbs sampling are required (Zeger and Karim 1991).

An alternative method is to avoid the integration by approximating the score equations via linearizations (Stirateli et al. 1984, Zeger and Karim 1991, Schall 1991, Breslow and Clayton 1993, Lindstrom and Bates 1990). To provide an outline for the algorithm, define $v_{ij} = \mathrm{Var}(y_{ij}|\mathbf{b}_i)$ and $\mathbf{R}_i = \mathrm{diag}[v_{ij}g'(\mu_{ij})^2]$, where $g'(\cdot)$ is the first derivative of $g(\cdot)$. Let $\mathbf{y}_i^*$ be the surrogate response defined to have elements

$$y_{ij}^* = g(\mu_{ij}) + (y_{ij} - \mu_{ij})g'(\mu_{ij}), j = 1, \cdots, n_i.$$

Define the $n_i \times n_i$ matrix $\mathbf{V}_i = \mathbf{R}_i + \mathbf{Z}_i\mathbf{D}\mathbf{Z}_i^T$. An iterative algorithm can be used to update β and $\mathbf{b}$, and then $\mathbf{D}$. For given $\mathbf{D}$, the β and $\mathbf{b}$ can be updated by iteratively solving

$$\hat{\beta} = \left(\sum_{i=1}^{n}\mathbf{X}_i^T\mathbf{V}_i^{-1}\mathbf{X}_i\right)^{-1}\sum_{i=1}^{n}\mathbf{X}_i^T\mathbf{V}_i^{-1}\mathbf{y}_i^*, \tag{2.41}$$

$$\hat{\mathbf{b}}_i = \mathbf{D}\mathbf{Z}_i\mathbf{V}_i^{-1}\left(\mathbf{y}_i^* - \mathbf{X}_i\hat{\beta}\right), \; i = 1, 2, \cdots, n. \tag{2.42}$$

These results are the solutions to the mixed model equations of the LME model (2.2) (Harville 1977). To update $\mathbf{D}$, the score equation (2.40) can be used,

$$\begin{aligned}\hat{\mathbf{D}} &= n^{-1}\sum_{i=1}^{n}\mathrm{E}(\mathbf{b}_i\mathbf{b}_i^T|\mathbf{y}_i)\\ &= n^{-1}\sum_{i=1}^{n}\mathrm{E}(\mathbf{b}_i|\mathbf{y}_i)\mathrm{E}(\mathbf{b}_i|\mathbf{y}_i)^T + n^{-1}\sum_{i=1}^{n}\mathrm{Cov}(\mathbf{b}_i|\mathbf{y}_i).\end{aligned}$$

A variety of slightly different algorithms for estimation of $\mathbf{D}$ have been proposed. See an example in Breslow and Clayton (1993).

The above method generally provides a good estimate for β, but the estimates of random-effects $\mathbf{b}_i$ and $\mathbf{D}$ are sensitive to the Gaussian approximation to the conditional distribution, in particular for sparse individual data. Breslow and Lin (1995) and Lin and Breslow (1996) have proposed a good bias correction method to improve the estimate of $\mathbf{D}$.

2.4.2 Examples of GLME Model

GLME models for binary data Mixed-effects models for longitudinal or clustered binary data were introduced to account for over-dispersion or so-called extra-binomial variation. Let $y_{i1}, \cdots, y_{in_i}$ denote the n_i binary responses from cluster or subject i where $i = 1, 2, \cdots, n$. The binomial-beta mixed-effects model can be specified as

1. Conditional on μ_i, the responses $y_{i1}, \cdots, y_{in_i}$ are independent and follow a Bernoulli distribution with a common parameter μ_i, i.e., $y_{ij} \sim \text{Binomial}(1, \mu_i)$, where $j = 1, 2, \cdots, n_i$, and Binomial(k, p) denotes a binomial distribution with parameters k and p.

2. The μ_i, $i = 1, 2, \cdots, n$ follow a beta distribution with mean μ and variance $\delta\mu(1-\mu)$, i.e., $\mu_i \sim \text{Beta}(\mu, \delta\mu(1-\mu))$, $i = 1, 2, \cdots, n$.

Unconditionally, $S_i = \sum_{j=1}^{n_i} y_{ij}$ has a binomial-beta distribution with

$$\text{E}(S_i) = n_i\mu, \;\; \text{Var}(S_i) = n_i\mu(1-\mu)[1 + (n_i - 1)\delta].$$

The over-dispersion parameter δ is the correlation for each pair of binary responses from the same cluster or subject. The extension of this model can include cluster-level or subject-specific covariates in a parametric model such as a logistic model $\text{logit}(\mu_i) = \mathbf{x}_i^T\beta$ where $\mathbf{x}_i$ is the covariate vector for the i-th subject or cluster. Other extensions are also available (Rosner 1984, Jones and Kenward 1987).

Another popular GLME model for clustered or longitudinal binary data is logistic models with Gaussian random-effects which can be specified as follows:

1. Conditional on μ_{ij}, the responses $y_{i1}, \cdots, y_{in_i}$ are independent and $y_{ij} \sim \text{Binomial}(1, \mu_{ij})$, $j = 1, 2, \cdots, n_i$.

2. The link function $g(\cdot)$ is a logit function, i.e., $\eta_{ij} = g(\mu_{ij}) = \log[\mu_{ij}/(1 - \mu_{ij})]$, $j = 1, 2, \cdots, n_i$; $i = 1, 2, \cdots, n$.

3. $\eta_{ij} = \mathbf{x}_{ij}^T\beta + \mathbf{z}_{ij}^T\mathbf{b}_i$, $\mathbf{b}_i \sim N(\mathbf{0}, \mathbf{D})$, $j = 1, 2, \cdots, n_i$; $i = 1, 2, \cdots, n$.

For this model, the likelihood function for β and $\mathbf{D}$ is

$$\begin{aligned} & L(\beta, \mathbf{D}|\mathbf{y}) \\ = \;& \prod_{i=1}^n \int \prod_{j=1}^{n_i} [\mu_{ij}(\beta, \mathbf{b}_i)]^{y_{ij}} [1 - \mu_{ij}(\beta, \mathbf{b}_i)]^{1-y_{ij}} f(\mathbf{b}_i; \mathbf{D}) d\mathbf{b}_i \\ = \;& \prod_{i=1}^n \Bigg[\int \exp\left\{ \beta^T \mathbf{u}_i + \mathbf{b}_i^T \mathbf{v}_i - \textstyle\sum_j \log[1 + \exp(\mathbf{x}_{ij}^T\beta + \mathbf{z}_{ij}^T\mathbf{b}_i)] \right\} \\ & \times (2\pi)^{-q/2} |\mathbf{D}|^{-1/2} \exp\left(-\tfrac{1}{2}\mathbf{b}_i^T \mathbf{D}^{-1} \mathbf{b}_i\right) d\mathbf{b}_i \Bigg], \end{aligned}$$

where q is the dimension of $\mathbf{b}_i$, $\mu_{ij}(\boldsymbol{\beta}, \mathbf{b}_i) = \mathrm{E}(y_{ij}|\boldsymbol{\beta}, \mathbf{b}_i)$, $\mathbf{u}_i = \sum_{j=1}^{n_i} \mathbf{x}_{ij} y_{ij}$ and $\mathbf{v}_i = \sum_{j=1}^{n_i} \mathbf{z}_{ij} y_{ij}$.

For GLME models with binary data, the numerical integration methods (Crouch and Spiegelman 1990, Zeger and Karim 1991) or the approximation method (Breslow and Clayton 1993) can be used. Both methods can be implemented in SAS to obtain the MLE for the GLME models.

GLME models for count data It is well known that the standard model for count data is the Poisson distribution. To account for the extra variation or over-dispersion in count data, a Poisson-Gamma mixed-effects model is usually used. Let $y_{i1}, \cdots, y_{in_i}$ denote the count responses for the i-th subject where $i = 1, 2, \cdots, n$, the Poisson-Gamma mixed-effects model can be specified as follows:

1. Conditional on μ_i, the responses $y_{ij}, j = 1, 2, \cdots, n_i$ follow a Poisson distribution with mean μ_i, i.e., $y_{ij} \sim \text{Poisson}(\mu_i), j = 1, 2, \cdots, n_i$.

2. The μ_i, $i = 1, 2, \cdots, n$ independently and identically follow a Gamma distribution with mean μ and variance $\phi\mu^2$, i.e., $\mu_i \sim \text{Gamma}(\mu, \phi\mu^2)$, $i = 1, 2, \cdots, n$.

Thus, the unconditional distribution of y_{ij} is negative binomial with

$$\mathrm{E}(y_{ij}) = \mu \text{ and } \mathrm{Var}(y_{ij}) = \mu + \phi\mu^2.$$

The extension of this model is to include subject-specific or cluster-level covariates $\mathbf{x}_i$ in a parametric model of μ_i, for example, the log-linear model $\log(\mu_i) = \mathbf{x}_i^T \boldsymbol{\beta}$. Note that this model limits the covariates as constants within a subject.

Morton (1987) extended the above model to incorporate the time varying covariates within subjects in the context of longitudinal data:

1. Conditional on an independent unobserved random variable δ_i, $\mathrm{E}(y_{ij}|\delta_i) = \exp(\mathbf{x}_{ij}^T \boldsymbol{\beta})\delta_i$, $j = 1, \cdots, n_i$.

2. $\mathrm{Var}(y_{ij}|\delta_i) = \phi \mathrm{E}(y_{ij}|\delta_i)$.

3. $\mathrm{E}(\delta_i) = 1$ and $\mathrm{Var}(\delta_i) = \sigma^2$.

These assumptions result in $\mathrm{E}(y_{ij}) = \exp(\mathbf{x}_{ij}^T \boldsymbol{\beta}) = \mu_{ij}$, and $\mathrm{Var}(y_{ij}) = \phi\mu_{ij} + \sigma^2\mu_{ij}^2$. An attractive feature of this model is that only the first two moments of δ_i need to be specified and the complete distribution of δ_i is not required. Thus, GEE-type techniques can be used for inference.

The more general form of mixed-effects models for count data can be written as

1. $\log \mathrm{E}(y_{ij}|\mathbf{b}_i) = \mathbf{x}_{ij}^T \boldsymbol{\beta} + \mathbf{z}_{ij}^T \mathbf{b}_i$.

2. Given random-effects $\mathbf{b}_i$, the responses $y_{i1}, \cdots, y_{in_i}$ are independent and $y_{ij} \sim \text{Poisson}[\mathrm{E}(y_{ij}|\mathbf{b}_i)], j = 1, 2, \cdots, n_i$.

3. The $\mathbf{b}_i$, $i = 1, 2, \cdots, n$ are independent realizations from a distribution with density function $f(\mathbf{b}_i; \mathbf{D})$.

The above model is more flexible to model within-subject variations, but requires the specification of a distribution for $\mathbf{b}_i$ in order to make inferences about the unknown parameters. For example, one can assume that $\mathbf{b}_i$ follow a Gaussian distribution. This gives a so-called Poisson-Gaussian mixed-effects model for longitudinal count data.

2.4.3 Generalized Nonlinear Mixed-Effects Model

GLME models can be extended to generalized nonlinear mixed-effects (GNLME) models. For simplicity, we shall use the same notations as in the previous subsection. Let $\mathbf{y}_i$ denote the repeated measurements of a response variable from the i-th of n subjects and assume that $\mathbf{y}_i$ follows a distribution from the exponential family. Let $\mathbf{b}_i$ denote the random-effects vector for the i-th subject. Let $\boldsymbol{\mu}_i = E(\mathbf{y}_i|\mathbf{b}_i)$, $\text{Cov}(\mathbf{y}_i|\mathbf{b}_i) = \phi\mathbf{V}_i(\boldsymbol{\mu}_i)$, where ϕ is an unkown variance parameter and $\mathbf{V}_i(\boldsymbol{\mu}_i) = \text{diag}[V(\mu_{i1}), \cdots, V(\mu_{in_i})]$ with $V(\cdot)$ being a known variance function. Let $\boldsymbol{\eta}_i = g(\boldsymbol{\mu}_i) = [g(\mu_{i1}), \cdots, g(\mu_{in_i})]^T$ where $g(\cdot)$ is a known link function. To specify the GNLME model, we only need to modify the systematic component of the GLME model as follows:

$$\begin{aligned} \boldsymbol{\eta}_i &= \mathbf{f}(\mathbf{X}_i, \boldsymbol{\beta}_i), \ \boldsymbol{\beta}_i = \mathbf{d}(\mathbf{a}_i, \boldsymbol{\beta}, \mathbf{b}_i), \\ \mathbf{b}_i &\sim (\mathbf{0}, \mathbf{D}), \quad i = 1, 2, \cdots, n, \end{aligned} \tag{2.43}$$

where $\mathbf{f}(\cdot)$ and $\mathbf{d}(\cdot)$ are vectors of known nonlinear functions. For example, $\mathbf{d}(\cdot)$ can be a simple linear function such that $\boldsymbol{\beta}_i = \mathbf{A}_i\boldsymbol{\beta} + \mathbf{B}_i\mathbf{b}_i, \ i = 1, 2, \cdots, n$ where $\mathbf{A}_i$ and $\mathbf{B}_i$ are design matrices for fixed-effects and random effects, respectively.

Notice that the standard NLME models and the GNLME models introduced in Vonesh and Chinchilli (1996, Chapter 8) can be written into the form of (2.43). The techniques for dealing with the NLME and GLME models can be applied to make inferences for the GNLME models. The fundamental details of these techniques have been summarized in the book by Vonesh and Chinchilli (1996, Chapter 8).

2.5 SUMMARY AND BIBLIOGRAPHICAL NOTES

In this Chapter, we reviewed linear/nonlinear mixed-effects models as well as generalized linear/nonlinear mixed-effects models. The basic inference methods were briefly discussed. Here we only covered the mixed-effects modeling materials that will be used in subsequent chapters. Many other topics, such as mixed-effects ANOVA models for repeated measurements, semi/nonparametric maximum likelihood approaches, a fully Bayesian approach for mixed-effects models, and missing data issues in mixed-effects models were not given a great deal of attention. We did not review other models or methods for longitudinal data analysis, for example, marginal models and transition models. However, these topics and techniques are very important for longitudinal data analysis although we do not use them in the development of nonparametric mixed-effects models in later chapters. We refer readers to several recent books by Davidian and Giltinan (1995), Vonesh and Chinchilli (1996), Verbeke and Molenberghs (2000), Pinheiro and Bates (2000), and Diggle et al. (2002) for the details of parametric modeling of longitudinal data.

We note that the inferences for mixed-effects models can be easily implemented using SAS or S-PLUS statistical packages. For example, the S-PLUS function *lme* or the SAS procedure *PROC MIXED* can be used to fit linear mixed-effects models. The SAS procedures *PROC GLIMMIX* and *PROC NLMIXED* and the S-PLUS function *nlme* (Pinheiro and Bates 2000) can be used to fit generalized linear or nonlinear mixed-effects models. Some other software packages such as *NONMEM* (Beal and Sheiner 1992) may also be used to fit mixed-effects models, but these packages are usually designed for some specific purposes such as pharmacokinetic/pharmacodynamic data analysis.

2.6 APPENDIX: PROOFS

Proof of Theorem 2.1 By (2.14) and (2.15), we have

$$\begin{pmatrix} \beta \\ \mathbf{b} \\ \epsilon \\ \mathbf{y} \end{pmatrix} \sim N\left[\mathbf{0}, \begin{pmatrix} \mathbf{H} & \mathbf{0} & \mathbf{0} & \mathbf{H}\mathbf{X}^T \\ \mathbf{0} & \tilde{\mathbf{D}} & \mathbf{0} & \tilde{\mathbf{D}}\mathbf{Z}^T \\ \mathbf{0} & \mathbf{0} & \mathbf{R} & \mathbf{R} \\ \mathbf{X}\mathbf{H} & \mathbf{Z}\tilde{\mathbf{D}} & \mathbf{R} & \mathbf{\Omega} \end{pmatrix}\right], \tag{2.44}$$

where $\mathbf{\Omega} = \mathbf{V} + \mathbf{X}\mathbf{H}\mathbf{X}^T$ with $\mathbf{V} = \mathbf{Z}\tilde{\mathbf{D}}\mathbf{Z}^T + \mathbf{R}$ as defined before. Then applying Lemma 2.2, we have

$$\beta|\mathbf{y} \sim N\left[\mathbf{H}\mathbf{X}^T\mathbf{\Omega}^{-1}\mathbf{y}, \mathbf{H} - \mathbf{H}\mathbf{X}^T\mathbf{\Omega}^{-1}\mathbf{X}\mathbf{H}\right].$$

Applying Lemma 2.1, we have

$$\begin{aligned} \mathbf{\Omega}^{-1} &= \left(\mathbf{V} + \mathbf{X}\mathbf{H}\mathbf{X}^T\right)^{-1} \\ &= \mathbf{V}^{-1} - \mathbf{V}^{-1}\mathbf{X}\mathbf{H}(\mathbf{H} + \mathbf{H}\mathbf{X}^T\mathbf{V}^{-1}\mathbf{X}\mathbf{H})^{-1}\mathbf{H}\mathbf{X}^T\mathbf{V}^{-1} \\ &= \mathbf{V}^{-1} - \mathbf{V}^{-1}\mathbf{X}(\mathbf{H}^{-1} + \mathbf{X}^T\mathbf{V}^{-1}\mathbf{X})^{-1}\mathbf{X}^T\mathbf{V}^{-1}. \end{aligned}$$

It follows that

$$\begin{aligned} \mathbf{H}\mathbf{X}^T\mathbf{\Omega}^{-1} &= \mathbf{H}\mathbf{X}^T\mathbf{V}^{-1} - \mathbf{H}\mathbf{X}^T\mathbf{V}^{-1}\mathbf{X}(\mathbf{H}^{-1} + \mathbf{X}^T\mathbf{V}^{-1}\mathbf{X})^{-1}\mathbf{X}^T\mathbf{V}^{-1} \\ &= \mathbf{H}\mathbf{H}^{-1}(\mathbf{H}^{-1} + \mathbf{X}^T\mathbf{V}^{-1}\mathbf{X})^{-1}\mathbf{X}^T\mathbf{V}^{-1} \\ &= (\mathbf{H}^{-1} + \mathbf{X}^T\mathbf{V}^{-1}\mathbf{X})^{-1}\mathbf{X}^T\mathbf{V}^{-1}. \end{aligned}$$

Moreover,

$$\begin{aligned} \mathbf{H} - \mathbf{H}\mathbf{X}^T\mathbf{\Omega}^{-1}\mathbf{X}\mathbf{H} &= \mathbf{H} - (\mathbf{H}^{-1} + \mathbf{X}^T\mathbf{V}^{-1}\mathbf{X})^{-1}\mathbf{X}^T\mathbf{V}^{-1}\mathbf{X}\mathbf{H} \\ &= (\mathbf{H}^{-1} + \mathbf{X}^T\mathbf{V}^{-1}\mathbf{X})^{-1}. \end{aligned}$$

It follows that as $\mathbf{H}^{-1} \to \mathbf{0}$, we have the first expression in (2.17) and the frist expression in (2.16) due to (2.8). Similarly, we can show the other expressions in (2.17) and (2.16). The theorem is then proved.

Proof of Theorem 2.2 Set $\tilde{\mathbf{y}} = \mathbf{y} - \mathbf{X}\beta$. Under the Bayesian framework (2.14) and (2.15), we have

$$\begin{pmatrix} \mathbf{b} \\ \epsilon \\ \tilde{\mathbf{y}} \end{pmatrix} \sim N \left[0, \begin{pmatrix} \tilde{\mathbf{D}} & \mathbf{0} & \tilde{\mathbf{D}}\mathbf{Z}^T \\ \mathbf{0} & \mathbf{R} & \mathbf{R} \\ \mathbf{Z}\tilde{\mathbf{D}} & \mathbf{R} & \mathbf{V} \end{pmatrix} \right].$$

Applying Lemma 2.2 directly, we have

$$\begin{aligned} \mathbf{b}|\tilde{\mathbf{y}} &\sim N\left[\tilde{\mathbf{D}}\mathbf{Z}^T\mathbf{V}^{-1}\tilde{\mathbf{y}}, \tilde{\mathbf{D}} - \tilde{\mathbf{D}}\mathbf{Z}^T\mathbf{V}^{-1}\mathbf{Z}\tilde{\mathbf{D}}\right], \\ \epsilon|\tilde{\mathbf{y}} &\sim N\left[\mathbf{R}\mathbf{V}^{-1}\tilde{\mathbf{y}}, \mathbf{R} - \mathbf{R}\mathbf{V}^{-1}\mathbf{R}\right]. \end{aligned}$$

Then (2.19) follows by replacing $\tilde{\mathbf{y}}$ by $\mathbf{y} - \mathbf{X}\beta$ in the above expressions. The theorem is proved.

Proof of Theorem 2.3 Because $\mathbf{R}$, $\mathbf{V}$, $\mathbf{Z}$ and $\tilde{\mathbf{D}}$ are all block diagonal, by Theorem 2.2, we have for $i = 1, 2, \cdots, n$ that

$$\begin{aligned} \epsilon_i|\mathbf{y}, \beta &\sim N\left[\mathbf{R}_i\mathbf{V}_i^{-1}(\mathbf{y}_i - \mathbf{X}_i\beta), \mathbf{R}_i - \mathbf{R}_i\mathbf{V}_i^{-1}\mathbf{R}_i\right], && (2.45) \\ \mathbf{b}_i|\mathbf{y}, \beta &\sim N\left[\mathbf{D}\mathbf{Z}_i^T\mathbf{V}_i^{-1}(\mathbf{y}_i - \mathbf{X}_i\beta), \mathbf{D} - \mathbf{D}\mathbf{Z}_i^T\mathbf{V}_i^{-1}\mathbf{Z}_i\mathbf{D}\right]. && (2.46) \end{aligned}$$

By Theorem 2.2 and (2.45),

$$\mathrm{E}(\epsilon_i|\mathbf{y}, \beta = \hat{\beta}) = \mathbf{R}_i\mathbf{V}_i^{-1}(\mathbf{y}_i - \mathbf{X}_i\hat{\beta}) = \hat{\epsilon}_i,$$

and

$$\mathrm{Cov}(\epsilon_i|\mathbf{y}, \beta = \hat{\beta}) = \mathbf{R}_i - \mathbf{R}_i\tilde{\mathbf{V}}_i^{-1}\mathbf{R}_i = \sigma^2(\mathbf{I}_{n_i} - \sigma^2\tilde{\mathbf{V}}_i^{-1}).$$

Therefore,

$$\begin{aligned} & \mathrm{E}(\epsilon_i^T\epsilon_i|\mathbf{y}, \beta = \hat{\beta}) = \mathrm{tr}\left[\mathrm{E}(\epsilon_i\epsilon_i^T|\mathbf{y}, \beta = \hat{\beta})\right] \\ = & \ \mathrm{tr}\left[\mathrm{E}(\epsilon_i|\mathbf{y}, \beta = \hat{\beta})\mathrm{E}(\epsilon_i|\mathbf{y}, \beta = \hat{\beta})^T + \mathrm{Cov}(\epsilon_i|\mathbf{y}, \beta = \hat{\beta})\right] \\ = & \ \mathrm{tr}\left[\hat{\epsilon}_i\hat{\epsilon}_i^T + \sigma^2(\mathbf{I}_{n_i} - \sigma^2\tilde{\mathbf{V}}_i^{-1})\right] \\ = & \ \hat{\epsilon}_i^T\hat{\epsilon}_i + \sigma^2[n_i - \sigma^2\mathrm{tr}(\tilde{\mathbf{V}}_i^{-1})]. \end{aligned}$$

The first claim in (2.25) then follows directly from the definition of $\hat{\sigma}^2$ in (2.23). Similarly, by Theorem 2.2 and (2.46), we have

$$\mathrm{E}(\mathbf{b}_i|\mathbf{y}, \beta = \hat{\beta}) = \mathbf{D}\mathbf{Z}_i^T\mathbf{V}_i^{-1}(\mathbf{y}_i - \mathbf{X}_i\hat{\beta}) = \hat{\mathbf{b}}_i,$$

and $\mathrm{Cov}(\mathbf{b}_i|\mathbf{y}, \beta = \hat{\beta}) = \mathbf{D} - \mathbf{D}\mathbf{Z}_i^T\mathbf{V}_i^{-1}\mathbf{Z}_i\mathbf{D}$. Therefore,

$$\begin{aligned} & \mathrm{E}(\mathbf{b}_i\mathbf{b}_i^T|\mathbf{y}, \beta = \hat{\beta}) \\ = & \ \mathrm{E}(\mathbf{b}_i|\mathbf{y}, \beta = \hat{\beta})\mathrm{E}(\mathbf{b}_i|\mathbf{y}, \beta = \hat{\beta})^T + \mathrm{Cov}(\mathbf{b}_i|\mathbf{y}, \beta = \hat{\beta}) \\ = & \ \hat{\mathbf{b}}_i\hat{\mathbf{b}}_i^T + [\mathbf{D} - \mathbf{D}\mathbf{Z}_i^T\mathbf{V}_i^{-1}\mathbf{Z}_i\mathbf{D}]. \end{aligned}$$

The second claim in (2.25) then follows. The theorem is proved.

Proof of Theorem 2.4 The proof is similar to that of Theorem 2.3 but it is now based on Theorem 2.1.

3

Nonparametric Regression Smoothers

3.1 INTRODUCTION

In Chapter 2, we reviewed parametric mixed-effects models for longitudinal data, including linear and nonlinear mixed-effects models, and generalized linear and nonlinear mixed-effects models. These parametric mixed-effects models have been extensively studied and applied to analyze longitudinal data in the literature (Lindsey 1993, Diggle, Liang and Zeger 1994, Davidian and Giltian 1995, Vonesh and Chinchilli 1996, Pinheiro and Bates 2000, Verbeke and Molenberghs 2000). One of the basic assumptions for these models is that the response variable (or via a known link function) is a known parametric function of both fixed-effects and random-effects. That is, for each individual, the underlying relationship between the response and the mixed-effects covariates is parametric. However, this assumption is not always satisfied in practical applications.

Take the progesterone data, introduced in Section 1.1.1 of Chapter 1, as an example. Figure 3.1 depicts the plot of the individual progesterone data points (circles) of a selected subject. Presented are examples of some lower-degree polynomial fits (solid curves) to the data. Panel (a) depicts a linear model fit, which does not adequately fit the data. This difficulty may be overcome by increasing the degree of the polynomials, say from linear to quadratic, cubic, or quartic shown as in Panels (b), (c), and (d), respectively. It is seen that the higher the degree of the polynomial, the more adequately the data is fitted. It is seen that both the cubic and quartic polynomial models generally fit the data well, but the fits are still poor prior to Day 0.

Similar results were obtained when we replaced the selected subject by some other chosen subjects. Therefore, a lower degree polynomial model may not fit the progesterone data well. These data are just one example of practical datasets that

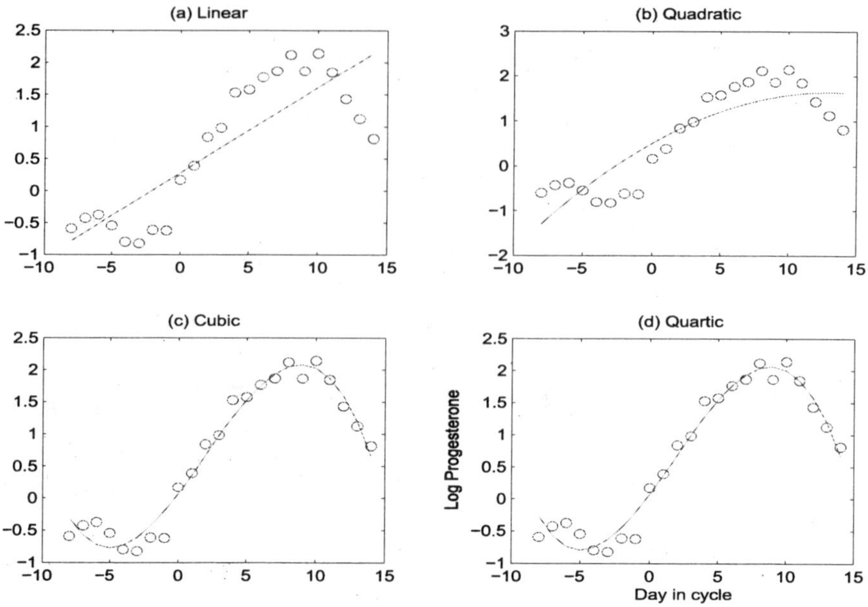

Fig. 3.1 Polynomial model fits of the progesterone data over time for a single subject.

can not be well fitted by lower degree polynomials. Härdle (1990), Fan and Gijbels (1996), Green and Silverman (1994), and Ramsay and Silverman (1997, 2002), among others, provide data examples where it is not possible to adequately fit the data using polynomials of any degree or any parametric models. In these cases, nonparametric modeling techniques are required.

The progesterone data for the selected subject, presented as circles in Figure 3.1, can be denoted as

$$(t_i, y_i), \quad i = 1, 2, \cdots, n, \tag{3.1}$$

where $t_i,\ i = 1, 2, \cdots, n$ are known as design time points, and $y_i,\ i = 1, 2, \cdots, n$ are the responses at the design time points. The design time points may be equally spaced in an interval of interest, or be regarded as a random sample from a continuous design density, namely, $\pi(t)$. For simplicity, let us denote the interval of interest, or the support of $\pi(t)$ as $\mathcal{T}$, which can be a finite interval, e.g., $[a, b]$ or the whole real line $(-\infty, \infty)$. The responses $y_i,\ i = 1, 2, \cdots, n$ are often observed with errors.

For a data set like the one above, a simple nonparametric regression model is usually written as

$$y_i = f(t_i) + \epsilon_i, \quad i = 1, .., n, \tag{3.2}$$

where $f(t)$ models the underlying regression function that we want to estimate, but can not be approximated using a parametric model adequately, and $\epsilon_i,\ i = 1, 2, \cdots, n$ denote the measurement errors that can not be explained by the regression function

$f(t)$. Mathematically, $f(t)$ is the conditional expectation of y_i, given $t_i = t$, i.e.,

$$f(t) = \mathrm{E}(y_i|t_i = t), \quad i = 1, 2, \cdots, n.$$

For longitudinal data, the data set (3.1) describes the data structure for an individual subject where $f(\cdot)$ is the individual function, and t_i, $i = 1, 2, \cdots, n$ are the design time points of the individual with n measurements.

There are many existing smoothers that can be used to estimate the $f(t)$ in (3.2). Different smoothers have different strengths in one aspect or another. For example, smoothing splines may be good for handling sparse data, while local polynomial smoothers may be computationally advantageous for handling dense designs. In this chapter, we shall review four of the most popular smoothers that include local polynomial smoothers (Wand and Jones 1995, Fan and Gijbels 1996), regression splines (Eubank 1988, 1999), smoothing splines (Wahba 1990, Green and Silverman 1994), and P-splines (Ruppert, Wand and Carroll 2003), respectively, in Sections 3.2, 3.3, 3.4, and 3.5. In later chapters, we shall develop nonparametric population mean and mixed-effects models for longitudinal data based on these smoothers. Section 3.6 briefly describes linear smoothers, which include the above four popular smoothers as special cases. Each linear smoother is indexed by a smoothing parameter or a smoothing parameter vector that controls the trade-off between the goodness-of-fit and model complexity (bias and variance). In Section 3.7, smoothing parameter selection criteria for linear smoothers are reviewed. We conclude this chapter with a summary and some bibliographical notes in Section 3.8.

3.2 LOCAL POLYNOMIAL KERNEL SMOOTHER

3.2.1 General Degree LPK Smoother

The main idea of local polynomial kernel (LPK) smoothing is to locally approximate the f in (3.2) by a polynomial of some degree. Its foundation is Taylor expansion, which states that any smooth function can be locally approximated by a polynomial of some degree.

Specifically, let t_0 be an arbitrary fixed time point where the function f in (3.2) will be estimated. Assume $f(t)$ has a $(p+1)$-st continuous derivative for some integer $p \geq 0$ at t_0. By Taylor expansion, $f(t)$ can be locally approximated by a polynomial of degree p. That is,

$$f(t) \approx f(t_0) + (t - t_0)f^{(1)}(t_0) + \cdots + (t - t_0)^p f^{(p)}(t_0)/p!,$$

in a neighborhood of t_0 that allows the above expansion where $f^{(r)}(t_0)$ denotes the r-th derivative of $f(t)$ at t_0.

Set $\beta_r = f^{(r)}(t_0)/r!$, $r = 0, \ldots, p$. Let $\hat{\beta}_r, r = 0, 1, 2, \cdots, p$ be the minimizers of the following weighted least squares (WLS) criterion:

$$\sum_{i=1}^{n} \{y_i - [\beta_0 + (t_i - t_0)\beta_1 + \cdots + (t_i - t_0)^p \beta_p]\}^2 K_h(t_i - t_0), \tag{3.3}$$

where $K_h(\cdot) = K(\cdot/h)/h$, which is obtained via re-scaling a kernel function $K(\cdot)$ with a constant $h > 0$, called the bandwidth or smoothing parameter. The bandwidth h is mainly used to specify the size of the local neighborhood, namely,

$$I_h(t_0) = [t_0 - h, t_0 + h], \tag{3.4}$$

where local fitting is conducted. The kernel function, $K(\cdot)$, determines how observations within $I_h(t_0)$ contribute to the fit at t_0. We will discuss kernel functions in Section 3.2.3. Denote the estimate of the r-th derivative $f^{(r)}(t_0)$ as $\hat{f}_h^{(r)}(t_0)$. Then

$$\hat{f}_h^{(r)}(t_0) = r!\hat{\beta}_r, r = 0, 1, \cdots, p.$$

In particular, the resulting p-th degree LPK estimator of $f(t_0)$ is $\hat{f}_h(t_0) = \hat{\beta}_0$.

An explicit expression for $\hat{f}_h^{(r)}(t_0)$ is useful and can be made via matrix notation. Let

$$\mathbf{X} = \begin{bmatrix} 1 & (t_1 - t_0) & \cdots & (t_1 - t_0)^p \\ \vdots & \vdots & \ddots & \vdots \\ 1 & (t_n - t_0) & \cdots & (t_n - t_0)^p \end{bmatrix},$$

and

$$\mathbf{W} = \text{diag}(K_h(t_1 - t_0), \cdots, K_h(t_n - t_0)),$$

be the design matrix and the weight matrix for the LPK fit around t_0. Then the WLS criterion (3.3) can be rewritten as

$$(\mathbf{y} - \mathbf{X}\boldsymbol{\beta})^T \mathbf{W} (\mathbf{y} - \mathbf{X}\boldsymbol{\beta}), \tag{3.5}$$

where $\mathbf{y} = (y_1, \cdots, y_n)^T$ and $\boldsymbol{\beta} = (\beta_0, \beta_1, \cdots, \beta_p)^T$. It follows that

$$\hat{f}_h^{(r)}(t_0) = r!\mathbf{e}_{r+1}^T \mathbf{S}_n^{-1} \mathbf{T}_n \mathbf{y}, \; r = 0, 1, \cdots, p,$$

where $\mathbf{e}_{r+1}$ denotes a $(p+1)$-dimensional unit vector whose $(r+1)$-st entry is 1 and the other entries are 0, and

$$\mathbf{S}_n = \mathbf{X}^T \mathbf{W} \mathbf{X}, \quad \mathbf{T}_n = \mathbf{X}^T \mathbf{W}.$$

When t_0 runs over the whole support $\mathcal{T}$ of the design time points, a whole range estimation of $f^{(r)}(t)$ is obtained. The derivative estimator $\hat{f}_h^{(r)}(t), t \in \mathcal{T}$ is usually called the LPK smoother of the underlying derivative function $f^{(r)}(t)$. The derivative smoother $\hat{f}_h^{(r)}(t_0)$ is usually calculated on a grid of t's in $\mathcal{T}$.

In this chapter, we only focus on the curve smoother

$$\hat{f}_h(t_0) = \mathbf{e}_1^T \mathbf{S}_n^{-1} \mathbf{T}_n \mathbf{y}, \tag{3.6}$$

unless we discuss derivative estimation. Set $\hat{y}_i = \hat{f}_h(t_i)$ to be the fitted value of $f(t_i)$. By (3.6), it is seen that

$$\hat{f}_h(t_i) = \mathbf{a}(t_i)^T \mathbf{y},$$

where $\mathbf{a}(t_i)$ is $\mathbf{e}_1^T \mathbf{S}_n^{-1} \mathbf{T}_n$ after replacing t_0 with t_i. Let $\hat{\mathbf{y}}_h = [\hat{y}_1, \cdots, \hat{y}_n]^T$ denote the fitted values at all the design time points. Then $\hat{\mathbf{y}}_h$ can be expressed as

$$\hat{\mathbf{y}}_h = \mathbf{A}_h \mathbf{y}, \tag{3.7}$$

where

$$\mathbf{A}_h = (\mathbf{a}(t_1), \cdots, \mathbf{a}(t_n))^T \tag{3.8}$$

is known as the smoother matrix of the LPK smoother. Since $\mathbf{A}_h$ does not depend on the response vector $\mathbf{y}$, the LPK smoother $\hat{f}_h$ is known as a linear smoother; see Section 3.6 for a definition of linear smoothers.

3.2.2 Local Constant and Linear Smoothers

Local constant and linear smoothers are the two simplest and most useful LPK smoothers. The local constant smoother is known as the Nadaraya-Watson estimator (Nadaraya 1964, Watson 1964). This smoother results from the LPK smoother $\hat{f}_h(t_0)$ (3.6) by simply taking $p = 0$:

$$\hat{f}_h(t_0) = \frac{\sum_{i=1}^n K_h(t_i - t_0) y_i}{\sum_{i=1}^n K_h(t_i - t_0)}. \tag{3.9}$$

Within a local neighborhood $I_h(t_0) = [t_0 - h, t_0 + h]$, it fits the data with a constant. That is, it is the minimizer $\hat{\beta}_0$ of the following WLS criterion:

$$\sum_{i=1}^n (y_i - \beta_0)^2 K_h(t_i - t_0).$$

The Nadaraya-Watson estimator is simple to understand and easy to compute. Let $I_A(t)$ denote the indicator function of some set A. When the kernel function K is the Uniform kernel

$$K(t) = I_{[-1,1]}(t)/2 = \begin{cases} \frac{1}{2}, & t \in [-1, 1], \\ 0, & \text{otherwise}, \end{cases} \tag{3.10}$$

as depicted in the left panel of Figure 3.2, the Nadaraya-Watson estimator (3.9) is exactly the local average of y_i's that are within the local neighborhood $I_h(t_0)$ (3.4):

$$\hat{f}_h(t_0) = \frac{\sum_{i=1}^n I_{[t_0-h, t_0+h]}(t_i) y_i}{\sum_{i=1}^n I_{[t_0-h, t_0+h]}(t_i)} = \left\{ \sum_{t_i \in I_h(t_0)} y_i \right\} / m_h(t_0),$$

where $m_h(t_0)$ denotes the number of the observations falling into the local neighborhood $I_h(t_0)$. However, when t_0 is at the boundary of $\mathcal{T}$, fewer design points are within the neighborhood $I_h(t_0)$ so that $\hat{f}_h(t_0)$ has a slower convergence rate than the case when t_0 is inside $\mathcal{T}$. For a detailed explanation of this boundary effect, the reader is referred to Fan and Gijbels (1996) and Cheng, Fan and Marron (1997).

The local linear smoother (Stone 1984, Fan 1992, 1993) is obtained via fitting a data set locally with a linear function. Let $(\hat{\beta}_0, \hat{\beta}_1)$ minimize the following WLS criterion:

$$\sum_{i=1}^{n} [y_i - \beta_0 - (t_i - t_0)\beta_1]^2 K_h(t_i - t_0).$$

Then the local linear smoother is $\hat{f}_h(t_0) = \hat{\beta}_0$. It can be easily obtained from the LPK smoother $\hat{f}_h(t_0)$ (3.6) by simply taking $p = 1$. It is known as a smoother with a free boundary effect (Cheng, Fan and Marron 1997). That is, it has the same convergence rate at any point in $\mathcal{T}$. It also exhibits many good properties that the other linear smoothers may lack. Good discussions on these properties can be found in Fan (1992, 1993), Hastie and Loader (1993), and Fan and Gijbels (1996, Chapter 2), among others. A local linear smoother can be simply expressed as

$$\hat{f}_h(t_0) = \frac{\sum_{i=1}^{n} [s_2(t_0) - s_1(t_0)(t_i - t_0)] K_h(t_i - t_0) y_i}{s_2(t_0)s_0(t_0) - s_1^2(t_0)}, \tag{3.11}$$

where

$$s_r(t_0) = \sum_{i=1}^{n} K_h(t_i - t_0)(t_i - t_0)^r, \; r = 0, 1, 2.$$

Usually, the choice of the LPK fitting degree, p, is not as important as the choice of the bandwidth, h. A local constant ($p = 0$) or a local linear ($p = 1$) smoother is often good enough for most application problems if the kernel function K and the bandwidth h are adequately determined. Fan and Gijbels (1996, Chapter 3) pointed out that for curve estimation (not valid for derivative estimation) an odd p is preferable. This is true since a LPK fit with $p = 2q + 1$, introduces an extra parameter compared to a LPK fit with $p = 2q$, but does not increase the variance of the associated LPK estimator. However, the associated bias may be significantly reduced especially in the boundary regions (Fan 1992, 1993, Hastie and Loader 1993, Fan and Gijbels 1996, Cheng, Fan and Marron 1997). Thus, the local linear smoother is strongly recommended for most problems in practice.

3.2.3 Kernel Function

The kernel function $K(\cdot)$ used in the LPK smoother (3.6) is usually a symmetric probability density function. While the bandwidth h specifies the size of the local neighborhood $I_h(t_0)$, the kernel $K(\cdot)$ specifies how the observations contribute to the LPK fit at t_0.

Figure 3.2 shows two widely-used kernel functions. The left panel shows the Uniform kernel (3.10) and the right panel shows the Gaussian kernel (standard normal probability density function)

$$K(t) = \exp(-t^2/2)/\sqrt{2\pi}. \tag{3.12}$$

When the Uniform kernel is used, all the t_i's within the local neighborhood $I_h(t_0)$ contribute equally (the weights are the same) in the LPK fit at t_0, while all the t_i's

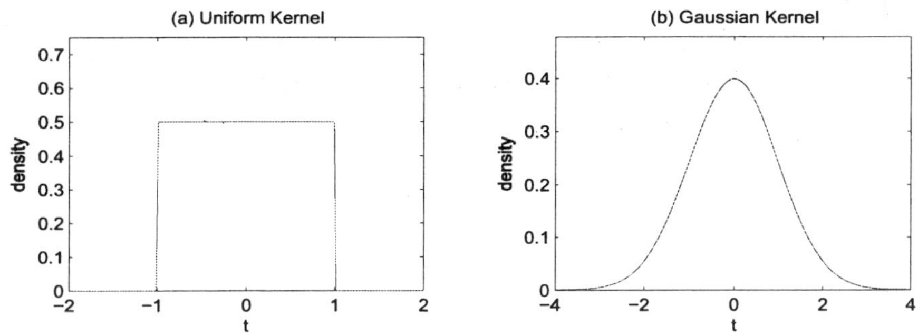

Fig. 3.2 Two widely-used kernel functions.

outside the neighborhood contribute nothing. When the Gaussian kernel is used, however, the contribution of the t_i's is determined by the distance of t_i from t_0, that is, the smaller the distance $(t - t_0)$, the larger the contribution. This is because the Gaussian kernel is bell-shaped and peaked at the origin. The Uniform kernel has a bounded support which allows the LPK fit to use the data only in the neighborhood $I_h(t_0)$. This makes a fast implementation of the LPK fit possible, which is advantageous especially for large data sets. The use of the Gaussian kernel often results in good visual effects of the LPK smoothers, but pays a price of requiring more computation effort.

The Uniform and Gaussian kernels are two special members of the following well-known symmetric Beta family (Marron and Nolan 1988):

$$K(t) = \frac{1}{\text{Beta}(1/2, 1+\gamma)}(1-t^2)_+^\gamma, \ \gamma = 0, 1, \cdots, \tag{3.13}$$

where $w_+^\gamma = [\max(0, w)]^\gamma$ and Beta(a, b) denotes a beta function with parameters a and b. Choices of $\gamma = 0, 1, 2$, and 3 lead to the Uniform, Epanechnikov, Biweight and Triweight kernel functions, respectively. The Gaussian kernel is the limit of the family (3.13) as $\gamma \to \infty$. The Epanechnikov kernel is known as the optimal kernel (Fan and Gijbels 1996) for LPK smoothing.

The choice of a kernel is usually not so crucial since it does not determine the convergence rate of the LPK smoother (3.6) to the underlying curve. However, it does determine the relative efficiency of the LPK smoother. For more discussion about kernel choice, see Gasser et al. (1985), Fan and Gijbels (1996), Zhang and Fan (2000) and references therein.

3.2.4 Bandwidth Selection

A smoother is considered to be good if it produces a small prediction error, usually measured by the Mean Squared Error (MSE) or the Mean Integrated Squared Error (MISE) of the smoother. For the LPK smoother $\hat{f}_h(t_0)$, its MSE and MISE are defined as

$$\begin{aligned}
\text{MSE}\left(\hat{f}_h(t_0)\right) &= \text{E}\left(\hat{f}_h(t_0) - f(t_0)\right)^2 \\
&= \text{Bias}^2\left(\hat{f}_h(t_0)\right) + \text{Var}\left(\hat{f}_h(t_0)\right), \\
\text{MISE}(\hat{f}_h) &= \int \text{MSE}\left(\hat{f}_h(t)\right) w(t)\, dt, \qquad (3.14)
\end{aligned}$$

where

$$\begin{aligned}
\text{Bias}\left(\hat{f}_h(t_0)\right) &= \text{E}\hat{f}_h(t_0) - f(t_0), \\
\text{Var}\left(\hat{f}_h(t_0)\right) &= \text{E}\left(\hat{f}_h(t_0) - \text{E}\hat{f}_h(t_0)\right)^2,
\end{aligned}$$

are known as the bias and variance of $\hat{f}_h(t_0)$, and $w(t)$ is a weight function, often used to specify a particular range of interest.

Under some regularity conditions including that t_0 is an interior point, we can show that as $n \to \infty$,

$$\text{Bias}\left(\hat{f}_h(t_0)\right) = \begin{cases} O_P(h^{p+1}), & \text{as } p \text{ odd}, \\ O_P(h^{p+2}), & \text{as } p \text{ even}, \end{cases} \qquad (3.15)$$

$$\text{Var}\left(\hat{f}_h(t_0)\right) = O_P\left((nh)^{-1}\right), \qquad (3.16)$$

where $X = O_P(Y)$ means X/Y is bounded in probability. See, for example, Fan and Gijbels (1996, Chapter 3) for more details. From this, we can see that the bandwidth h controls the trade-off between the squared bias and the variance of the LPK smoother $\hat{f}_h(t_0)$. When h is small, the squared bias is small but the variance is large. On the other hand, when h is large, the squared bias is large while the variance is small. A good choice of h will generally trade-off these two terms so that the associated MSE or MISE is minimized.

The role played by the bandwidth h can also be seen intuitively. As mentioned previously, the bandwidth h specifies the size of the local neighborhood $I_h(t_0) = [t_0 - h, t_0 + h]$. When h is small, $I_h(t_0)$ contains only a few observations so that $\hat{f}_h(t_0)$ can be well adjusted based on the WLS criterion (3.3) to closely approximate $f(t_0)$. This implies a small bias for $\hat{f}_h(t_0)$. However, since only a few observations are involved in the LPK fit, the variance of the estimator is large. With a similar reasoning, when h is large, $I_h(t_0)$ contains many observations so that $\hat{f}_h(t_0)$ has a large bias but a small variance.

It is then natural to select a global bandwidth h so that the MISE (MSE for a local bandwidth) of $\hat{f}_h(t_0)$ is minimized. Unfortunately, the MISE (3.14) is not computable since f is, after all, unknown and is the target to be estimated. This problem can be overcome by selecting h to minimize some estimator of the MISE. An estimator of the MISE may be obtained via estimating the unknown quantities in the asymptotic MISE expression using some higher degree LPK fit, resulting in the so-called plug-in bandwidth selectors (Fan and Gijbels 1992, Ruppert et al. 1995). The MISE can also be estimated by cross-validation or its modified versions: generalized cross-validation (Wahba 1985), Akaike information criterion (Akaike 1973) and Bayesian information criterion (Schwarz 1978), among others. Further details will be given in Section 3.7.

3.2.5 An Illustrative Example

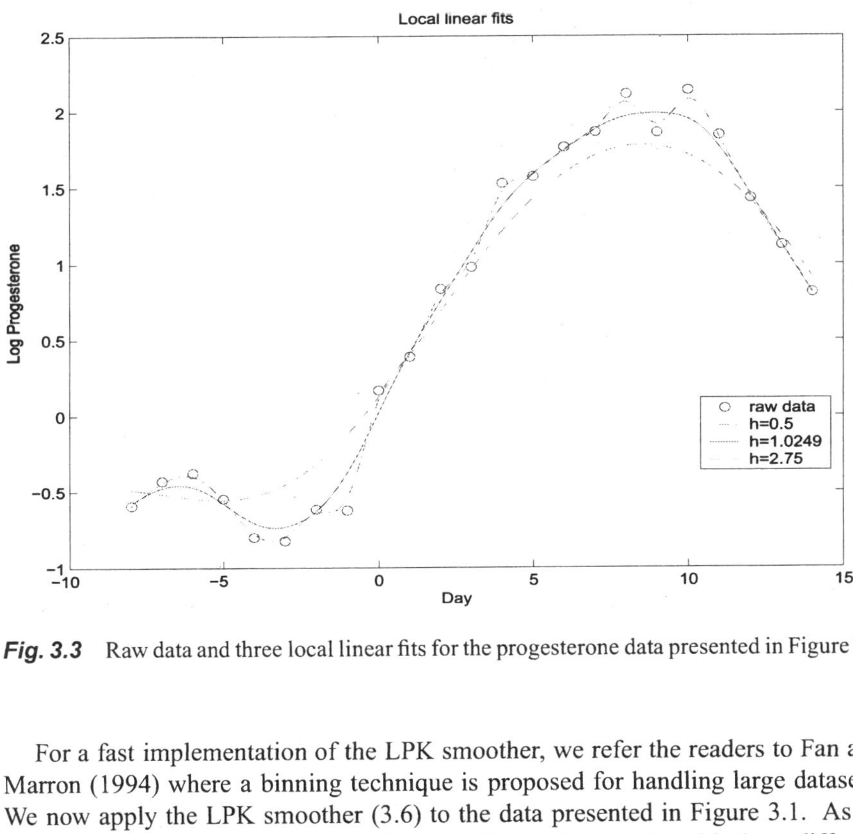

Fig. 3.3 Raw data and three local linear fits for the progesterone data presented in Figure 3.1.

For a fast implementation of the LPK smoother, we refer the readers to Fan and Marron (1994) where a binning technique is proposed for handling large datasets. We now apply the LPK smoother (3.6) to the data presented in Figure 3.1. As an illustrative example, we employed the local linear fit ($p = 1$) with three different bandwidths. In Figure 3.3, three local linear fits are presented. The dot-dashed curve almost interpolates the data since it uses a bandwidth $h = .5$ [$\log_{10}(.5) = -.3010$], which is too small. This is the case of undersmoothing. The dashed curve does not fit the data well since it uses a bandwidth $h = 2.75$ [$\log_{10}(2.75) = .4393$], which is too large. This is the case of oversmoothing. The solid curve produces a nice fit to the data since it uses a bandwidth $h = 1.0249$ [$\log_{10}(1.9249) = .0107$] selected by GCV, which is not too small or too large. The associated GCV curve against h in $\log_{10}$-scale is presented in Figure 3.4. The GCV curve allows us to see how the GCV value changes along the bandwidth h.

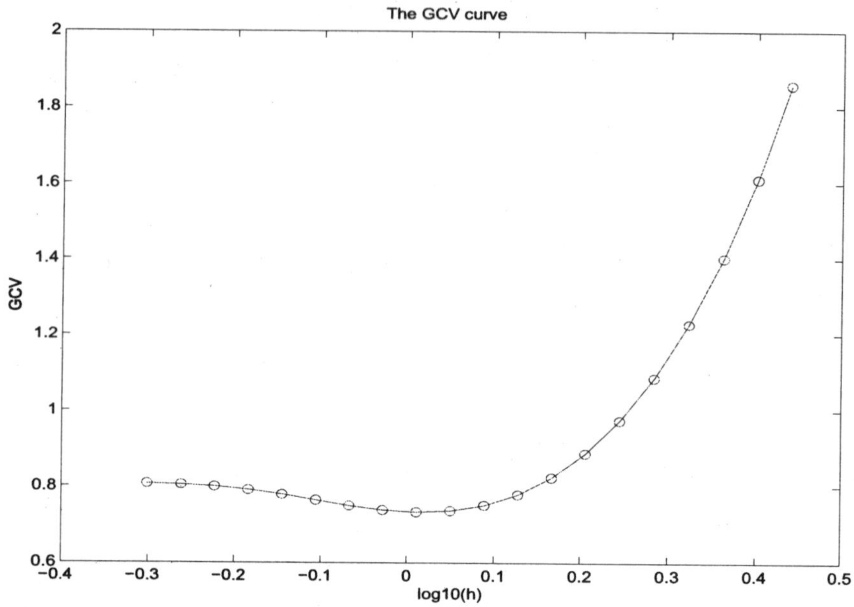

Fig. 3.4 The associated GCV curve for the local linear fits in the above figure.

3.3 REGRESSION SPLINES

Polynomials are not flexible in their ability to model data across a large range of values; see Figure 3.1. However, this is not the case when the range is small enough. In local polynomial kernel (LPK) smoothing introduced in Section 3.2, local neighborhoods were specified by a bandwidth h and a fixed time point t_0. In regression spline smoothing that we shall discuss in this section, local neighborhoods are specified by a group of locations, say,

$$\tau_0, \tau_1, \tau_2, \cdots, \tau_K, \tau_{K+1}, \tag{3.17}$$

in the range of interest, say, an interval $[a, b]$ where $a = \tau_0 < \tau_1 < \cdots < \tau_K < \tau_{K+1} = b$. These locations are known as knots, and $\tau_r, r = 1, 2, \cdots, K$ are called interior knots or simply knots. These knots divide the interval of interest, $[a, b]$, into K subintervals (local neighborhoods):

$$[\tau_r, \tau_{r+1}), r = 0, 1, \cdots, K,$$

so that within any two neighboring knots, a Taylor's expansion up to some degree is valid. In other words, a regression spline is a piecewise polynomial which is a polynomial of some degree within any two neighboring knots τ_r and τ_{r+1} for $r = 0, 1, \cdots, K$ and is joined together at knots properly but allows discontinuous derivatives at the knots.

3.3.1 Truncated Power Basis

A regression spline can be constructed using the following so-called k-th degree truncated power basis with K knots $\tau_1, \tau_2, \cdots, \tau_K$:

$$1, t, \cdots, t^k, (t-\tau_1)_+^k, \cdots, (t-\tau_K)_+^k, \tag{3.18}$$

where $w_+^k = [w_+]^k$ denotes power k of the positive part of w with $w_+ = \max(0, w)$. Note that the first $(k+1)$ basis functions of the truncated power basis (3.18) are polynomials of degree up to k, and the others are all the truncated power functions of degree k. Conventionally, the truncated power basis of degree "$k = 0, 1, 2$, and 3" is called "constant, linear, quadratic" and "cubic" truncated power basis, respectively.

Using the above truncated power basis (3.18), a regression spline can be expressed as

$$f(t) = \sum_{s=0}^{k} \beta_s t^s + \sum_{r=1}^{K} \beta_{k+r}(t-\tau_r)_+^k, \tag{3.19}$$

where $\beta_0, \beta_1, \cdots, \beta_{k+K}$ are the associated coefficients. For convenience, it may be called a regression spline of degree k with knots $\tau_1, \cdots, \tau_K$. The regression splines (3.19) associated with $k = 1, 2$ and 3 are usually called linear, quadratic, and cubic regression splines, respectively.

We can see that within any subinterval or local neighborhood $[\tau_r, \tau_{r+1})$, we have

$$f(t) = \sum_{s=0}^{k} \beta_s t^s + \sum_{l=1}^{r} \beta_{k+l}(t-\tau_l)^k,$$

which is a k-th degree polynomial. However, for $r = 1, 2, \cdots, K$,

$$\begin{aligned} f^{(k)}(\tau_r-) &= k!(\beta_k + \sum_{l=1}^{r-1} \beta_{k+l}), \\ f^{(k)}(\tau_r+) &= k!(\beta_k + \sum_{l=1}^{r} \beta_{k+l}). \end{aligned}$$

Therefore

$$f^{(k)}(\tau_r+) - f^{(k)}(\tau_r-) = k!\beta_{k+r}. \tag{3.20}$$

That is, $f^{(k)}(t)$ jumps at τ_r with amount $k!\beta_{k+r}$ for $r = 1, 2, \cdots, K$. In other words, a regression spline of degree k with knots $\tau_1, \cdots, \tau_K$ has continuous derivatives up to $k-1$ times, and has a discontinuous k-times derivative; the coefficient β_{k+r} of the r-th truncated power basis function measures how large the jump is (up to a constant multiplicity of $k!$).

Figure 3.5 (a) presents a cubic truncated power basis with knots .2, .4, .6, and .8. It includes the first four polynomials $1, t, t^2, t^3$, and the four truncated power functions with nonzero-values starting at knots .2, .4, .6 and .8, respectively. Figure 3.5 (b) displays three cubic regression splines as simple linear combinations (3.19) of the truncated power basis using coefficients generated randomly. It can be seen that the truncated power basis is flexible in describing functions from simple to complicated ones.

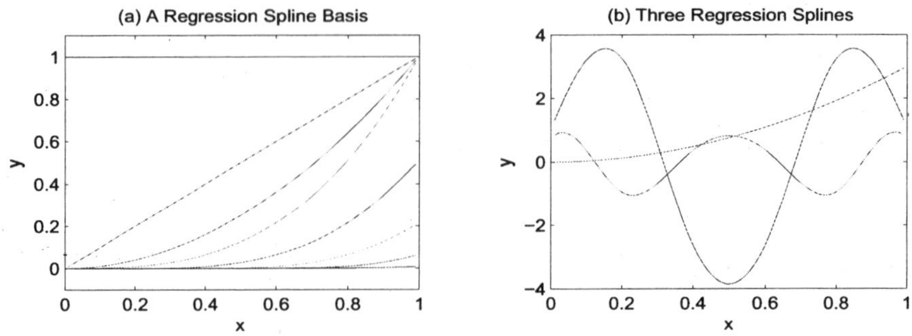

Fig. 3.5 Example of a cubic regression spline basis, and three cubic regression splines.

3.3.2 Regression Spline Smoother

For convenience, it is often useful to denote the truncated power basis (3.18) as

$$\mathbf{\Phi}_p(t) = [1, t, \cdots, t^k, (t-\tau_1)_+^k, \cdots, (t-\tau_K)_+^k]^T, \tag{3.21}$$

where $p = K + k + 1$ denotes the number of the basis functions involved. Similarly, denote the associated coefficients as

$$\boldsymbol{\beta} = [\beta_0, \cdots, \beta_k, \beta_{k+1}, \cdots, \beta_{k+K}]^T.$$

Then the regression spline (3.19) can be re-expressed as

$$f(t) = \mathbf{\Phi}_p(t)^T \boldsymbol{\beta}, \tag{3.22}$$

so that the model (3.2) can be re-written as

$$\mathbf{y} = \mathbf{X}\boldsymbol{\beta} + \boldsymbol{\epsilon}, \tag{3.23}$$

where,

$$\begin{aligned} \mathbf{y} &= (y_1, \cdots, y_n)^T, \\ \mathbf{X} &= (\mathbf{\Phi}_p(t_1), \cdots, \mathbf{\Phi}_p(t_n))^T, \\ \boldsymbol{\epsilon} &= (\epsilon_1, \cdots, \epsilon_n)^T. \end{aligned}$$

Since $\mathbf{\Phi}_p(t)$ is a basis, $\mathbf{X}$ is of full rank, and hence $\mathbf{X}^T\mathbf{X}$ is invertible when $n \geq p$. A natural estimator for $\boldsymbol{\beta}$, which solves the approximation linear model (3.23) by the ordinary least squares (OLS) method, is

$$\hat{\boldsymbol{\beta}} = (\mathbf{X}^T\mathbf{X})^{-1}\mathbf{X}^T\mathbf{y}. \tag{3.24}$$

It follows that the regression spline fit of the function $f(t)$ in (3.2) is

$$\hat{f}_p(t) = \mathbf{\Phi}_p(t)^T(\mathbf{X}^T\mathbf{X})^{-1}\mathbf{X}^T\mathbf{y}, \tag{3.25}$$

which is often called a regression spline smoother of f. In particular, the values of $\hat{f}_p(t)$ evaluated at the design time points $t_i, \ i = 1, 2, \cdots, n$ are collected in the following fitted response vector:

$$\hat{\mathbf{y}}_p = \mathbf{X}(\mathbf{X}^T\mathbf{X})^{-1}\mathbf{X}^T\mathbf{y} = \mathbf{A}_p\mathbf{y}, \tag{3.26}$$

where $\hat{\mathbf{y}}_p = (\hat{y}_1, \cdots, \hat{y}_n)^T$ with $\hat{y}_i = \hat{f}_p(t_i), \ i = 1, 2, \cdots, n$, and

$$\mathbf{A}_p = \mathbf{X}(\mathbf{X}^T\mathbf{X})^{-1}\mathbf{X}^T \tag{3.27}$$

is called the regression spline smoother matrix. It is easy to notice that $\mathbf{A}_p$ is a projection matrix satisfying $\mathbf{A}_p^T = \mathbf{A}_p, \mathbf{A}_p^2 = \mathbf{A}_p$ and $\text{tr}(\mathbf{A}_p) = p$. The trace of the smoother matrix $\mathbf{A}_p$ is often called the degrees of freedom of the regression spline smoother. It measures the complexity of the regression spline model used.

3.3.3 Selection of Number and Location of Knots

Good performance of the regression spline smoother (3.25) strongly depends on good knot locations, $\tau_1, \tau_2, \cdots, \tau_K$, and good choice of the number of knots, K. The degree of the regression spline basis (3.18), k, is usually less crucial, and it is often taken as $1, 2$ or 3 for computational convenience. Three widely-used methods for locating the knots are listed as follows:

Equally Spaced Method. This method takes K equally spaced points in the range of interest, say, $[a, b]$, as knots. That is, the K knots are defined as

$$\tau_r = a + (b - a)r/(K + 1), r = 1, 2, \cdots, K. \tag{3.28}$$

This method of knots placing is independent of the design time points. It is usually employed when the design time points are believed to be uniformly scattered in the range of interest.

Equally Spaced Sample Quantiles as Knots Method. This method uses equally spaced sample quantiles of the design time points $t_i, \ i = 1, 2, \cdots, n$ as knots. Let $t_{(1)}, \cdots, t_{(n)}$ be the order statistics of the design time points. Then the K knots are defined as

$$\tau_r = t_{(1+[rn/(K+1)])}, r = 1, 2, \cdots, K, \tag{3.29}$$

where $[a]$ denotes the integer part of a. This method of knots placing is design adaptive. It locates more knots where more design time points are scattered. When the design time points are uniformly scattered, it is approximately equivalent to the equally spaced method.

Model Selection Based Method. This method uses all the distinct design time points as knot candidates. Note that for the truncated power basis (3.18), deletion of a knot is equivalent to deletion of a truncated power basis function associated with the knot. This is equivalent to the deletion of a covariate in the approximation linear model (3.23). Therefore, knot selection can be done via model selection methods such as forward selection, backward elimination, or stepwise regression for introducing or deleting a knot from the knot candidates (Stone et al. 1997).

For the last method, the selection of the number of knots is done at the same time as knot introduction or deletion. But for the first two methods, after a knot placing method is specified, the number of knots, K, is generally chosen based on a smoothing parameter selection criterion such as those to be presented in Section 3.7 for linear smoothers. For more knot selection methods for regression splines, see, for example, Friedman and Silverman (1989), Friedman (1991), and Smith and Kohn (1996), among others.

3.3.4 General Basis-Based Smoother

A general basis-based smoother is obtained when one replaces the truncated power basis $\mathbf{\Phi}_p(t)$ in (3.22) with any other basis. Choice of $\mathbf{\Phi}_p(t)$ may include local bases such as B-spline bases, wavelet bases, or global bases such as Fourier Series, polynomial bases, among others. For example, when $\mathbf{\Phi}_p(t)$ is replaced by the polynomial basis $[1, t, \cdots, t^p]^T$, the model (3.22) will reduce to the usual polynomial model that leads to the polynomial fits in Figure 3.1. A comprehensive introduction to bases is given in Ramsay and Silverman (1997).

3.4 SMOOTHING SPLINES

For regression splines, when the knot locating method is specified, the remaining task is to choose the number of knots, K. In general, K is smaller than the sample size n. Since K must be an integer, the opportunity for adjusting K is limited, and the adjustment is usually rather crude. Alternatively, we can use all the distinct design time points as knots. This will generally result in undersmoothing since too many knots are in use. The resulting curve is usually quite rough, showing dramatic curve changes over a short interval. Smoothing splines overcome this difficulty by introducing a penalty for the roughness of the curve under consideration.

To be specific, without loss of generality, let us assume the range of interest of f in (3.2) is a finite interval, say, $[a, b]$ for some finite numbers a and b. The roughness of f is usually defined as the integral of its squared k-times derivative

$$\int_a^b \left\{ f^{(k)}(u) \right\}^2 du \tag{3.30}$$

for some $k \geq 1$. Then the smoothing spline smoother of the f in (3.2) is defined as the minimizer $\hat{f}_\lambda(t)$ of the following penalized least squares (PLS) criterion:

$$\sum_{i=1}^n [y_i - f(t_i)]^2 + \lambda \int_a^b \left\{ f^{(k)}(t) \right\}^2 dt, \tag{3.31}$$

over the k-th order Sobolev space $\mathcal{W}_2^k[a,b]$:

$$\left\{f : f^{(r)} \text{ absolute continuous for } 0 \leq r \leq k-1, \int_a^b \{f^{(k)}(t)\}^2\, dt < \infty\right\}, \quad (3.32)$$

where $\lambda > 0$ is a smoothing parameter controlling the size of the roughness penalty, and it is usually used to trade-off the goodness of fit, indicated by the first term in (3.31), and the roughness of the resulting curve. The $\hat{f}_\lambda(t)$ is known as a natural smoothing spline of degree $(2k-1)$. For example, when $k=2$, the associated $\hat{f}_\lambda(t)$ is a natural cubic smoothing spline (NCSS). For a detailed description of smoothing splines, see for example, Eubank (1988, 1999), Wahba (1990), Green and Silverman (1994) and Gu (2002), among others.

3.4.1 Cubic Smoothing Splines

To minimize the PLS criterion (3.31), we need to compute the integral that defines the roughness, and estimate parameters numbering up to the sample size n. This is a challenging aspect for computing a smoothing spline.

When $k=2$, however, the associated cubic smoothing spline is less computationally challenging. Actually, there is a way to compute the roughness term quickly as stated in Green and Silverman (1994). That is one of the reasons why cubic smoothing splines are popular in statistical applications.

Let $\tau_1, \cdots, \tau_K$ be all the distinct design time points and be sorted in an increasing order. They are all the knots of the cubic smoothing spline $\hat{f}_\lambda(t)$ that minimizes (3.31) when $k=2$. Set

$$h_r = \tau_{r+1} - \tau_r, r = 1, 2, \cdots, K-1.$$

Define $\mathbf{A} = (a_{rs})$ as a $K \times (K-2)$ matrix with all the entries being 0 except for $r = 1, 2, \cdots, K-2$,

$$a_{r,r} = h_r^{-1}, \quad a_{r+1,r} = -(h_r^{-1} + h_{r+1}^{-1}), \quad a_{r+2,r} = -h_{r+1}^{-1}.$$

Define $\mathbf{B} = (b_{rs})$ as a $(K-2) \times (K-2)$ matrix with all the entries being 0 except

$$b_{11} = (h_1 + h_2)/3, \quad b_{21} = h_2/6,$$

and for $r = 1, 2, \cdots, K-4$,

$$b_{r,r+1} = h_{(r+1)}/6, \quad b_{r+1,r+1} = (h_{(r+1)} + h_{(r+2)})/3, \quad b_{r+2,r+1} = h_{(r+2)}/6,$$

and

$$b_{K-3,K-2} = h_{(K-2)}/6, \quad b_{K-2,K-2} = (h_{(K-2)} + h_{(K-1)})/3.$$

Finally define a $K \times K$ matrix $\mathbf{G}$:

$$\mathbf{G} = \mathbf{A}\mathbf{B}^{-1}\mathbf{A}^T. \quad (3.33)$$

Let $\mathbf{f} = (f_1, \cdots, f_K)^T$ where $f_r = f(\tau_r), r = 1, 2, \cdots, K$. Then it is easy to show that the roughness, (3.30), of $f(t)$ for $k = 2$ can be expressed as

$$\int_a^b [f''(t)]^2 dt = \mathbf{f}^T \mathbf{G} \mathbf{f}. \tag{3.34}$$

Therefore, we often refer to $\mathbf{G}$ as a roughness matrix. It follows that the PLS criterion (3.31) can be written as

$$||\mathbf{y} - \mathbf{W}\mathbf{f}||^2 + \lambda \mathbf{f}^T \mathbf{G} \mathbf{f}, \tag{3.35}$$

where $\mathbf{W} = (w_{ir})$ is an $n \times K$ incidence matrix with $w_{ir} = 1$ if $t_i = \tau_r$ and 0 otherwise, and $||\mathbf{a}||^2 = \sum_{i=1}^n a_i^2$ denotes the usual L^2-norm of $\mathbf{a}$. Therefore an explicit expression for the cubic smoothing spline $\hat{\mathbf{f}}_\lambda$, evaluated at the knots $\tau_r, r = 1, 2, \cdots, K$, is as follows:

$$\hat{\mathbf{f}}_\lambda = (\mathbf{W}^T \mathbf{W} + \lambda \mathbf{G})^{-1} \mathbf{W}^T \mathbf{y}. \tag{3.36}$$

Moreover, the fitted response vector at the design time points is

$$\hat{\mathbf{y}}_\lambda = \mathbf{A}_\lambda \mathbf{y}, \tag{3.37}$$

where

$$\mathbf{A}_\lambda = \mathbf{W}(\mathbf{W}^T \mathbf{W} + \lambda \mathbf{G})^{-1} \mathbf{W}^T \tag{3.38}$$

is known as the cubic smoothing spline smoother matrix. The expression (3.36) indicates that the cubic smoothing spline smoother is a linear smoother as defined later in Section 3.6.

When all the design time points are distinct,

$$\hat{\mathbf{y}}_\lambda = \hat{\mathbf{f}}_\lambda = (\mathbf{I}_n + \lambda \mathbf{G})^{-1} \mathbf{y},$$

since $\mathbf{W} = \mathbf{I}_n$, an identity matrix of size n. For more details about cubic smoothing splines, see Green and Silverman (1994), among others.

3.4.2 General Degree Smoothing Splines

When $k \neq 2$, computation of the roughness term in (3.31) is quite challenging. However, it becomes easier if a good basis is provided for the underlying function $f(t)$ in (3.2). Let $\mathbf{\Phi}_p(t) = (\phi_1(t), \cdots, \phi_p(t))^T$ denote a basis so that $\phi_r^{(k)}(t), r = 1, 2, \cdots, p$ are squared integrable. Then we can express the $f(t)$ in (3.2) as $f(t) = \mathbf{\Phi}_p(t)^T \boldsymbol{\alpha}$ where $\boldsymbol{\alpha}$ is a p-dimensional coefficient vector, and the roughness

$$\int_a^b [f^{(k)}(t)]^2 dt = \boldsymbol{\alpha}^T \mathbf{G} \boldsymbol{\alpha},$$

where obviously

$$\mathbf{G} = \int_a^b \mathbf{\Phi}_p^{(k)}(t) \mathbf{\Phi}_p^{(k)}(t)^T dt. \tag{3.39}$$

It follows that the PLS criterion (3.31) can be expressed as

$$||\mathbf{y} - \mathbf{W}\boldsymbol{\alpha}||^2 + \lambda\boldsymbol{\alpha}^T\mathbf{G}\boldsymbol{\alpha}, \tag{3.40}$$

where $\mathbf{W} = (\boldsymbol{\Phi}_p(t_1), \cdots, \boldsymbol{\Phi}_p(t_n))^T$. Therefore, $\hat{\mathbf{y}}_\lambda$ can be expressed as

$$\hat{\mathbf{y}}_\lambda = \mathbf{A}_\lambda\mathbf{y}, \tag{3.41}$$

where

$$\mathbf{A}_\lambda = \mathbf{W}(\mathbf{W}^T\mathbf{W} + \lambda\mathbf{G})^{-1}\mathbf{W}^T \tag{3.42}$$

is the associated smoother matrix.

Note that $\boldsymbol{\Phi}_p(t)$ can be a truncated power basis (3.18) of degree $(2k-1)$ with knots at all the distinct design time points among $t_i, i = 1, 2, \cdots, n$, a B-spline basis (de Boor 1978) or a reproducing kernel Hilbert space basis (Wahba 1990) or any other basis.

3.4.3 Connection between a Smoothing Spline and a LME Model

There is a close connection between a smoothing spline and a LME (linear mixed-effects) model as pointed out by Speed (1991), and used by Brumback and Rice (1998), and Wang (1998a, b), among others in analysis of longitudinal data and functional data. The key for establishing such a connection is a singular value decomposition (SVD) of the roughness matrix $\mathbf{G}$.

For a natural cubic smoothing spline (NCSS), the $\mathbf{G}$ is defined in (3.33) while for a general $(2k-1)$-st degree smoothing spline, the $\mathbf{G}$ is defined in (3.39) based on a given basis $\boldsymbol{\Phi}_p(t)$. In both cases, the $\mathbf{G}$ is symmetric, and nonnegative definite. Hence it has a SVD. In what follows, we just show how this connection is established for a NCSS based on the roughness matrix $\mathbf{G}$ (3.33). The case for a general $(2k-1)$-st degree smoothing spline based on the roughness matrix $\mathbf{G}$ (3.39) is essentially the same, and will be omitted for space saving purposes.

Denote the SVD of $\mathbf{G}$ (3.33) as $\mathbf{G} = \mathbf{U}\boldsymbol{\Omega}\mathbf{U}^T$ where $\mathbf{U}$ is an orthonormal matrix, containing all the eigenvectors of $\mathbf{G}$, and $\boldsymbol{\Omega}$ is a diagonal matrix. Since the $\mathbf{G}$ is the roughness matrix of a NCSS, it has 2 zero eigenvalues and $K = p - 2$ nonzero eigenvalue where K is the number of knots for the NCSS. Therefore, we have $\boldsymbol{\Omega} = \text{diag}(\mathbf{0}, \boldsymbol{\Omega}_2)$ where $\mathbf{0}$ is a zero matrix of size 2×2 and $\boldsymbol{\Omega}_2$ is a diagonal matrix with all the nonzero eigenvalues of $\mathbf{G}$ as its diagonal entries.

Divide $\mathbf{U}$ as $[\mathbf{U}_1, \mathbf{U}_2]$ where $\mathbf{U}_1$ consists of the first 2 columns of $\mathbf{U}$, and $\mathbf{U}_2$ the remaining columns. Let

$$\begin{array}{rclrcl} \mathbf{X} & = & \mathbf{W}\mathbf{U}_1, & \mathbf{Z} & = & \mathbf{W}\mathbf{U}_2, \\ \boldsymbol{\beta} & = & \mathbf{U}_1^T\mathbf{f}, & \mathbf{b} & = & \mathbf{U}_2^T\mathbf{f}. \end{array}$$

Then the PLS criterion (3.35) becomes

$$||\mathbf{y} - \mathbf{X}\boldsymbol{\beta} - \mathbf{Z}\mathbf{b}||^2 + \lambda\mathbf{b}^T\boldsymbol{\Omega}_2\mathbf{b}. \tag{3.43}$$

It follows that under some regularity conditions that include normality assumption for the measurement errors, the NCSS (evaluated at the design time points) that minimizes the PLS criterion (3.35) can be expressed as

$$\hat{\mathbf{y}}_\lambda = \mathbf{X}\hat{\beta} + \mathbf{Z}\hat{\mathbf{b}}, \tag{3.44}$$

where $\hat{\beta}$ and $\hat{\mathbf{b}}$ are the Best Linear Unbiased Predictors (BLUPs) of the following LME model:

$$\mathbf{y} = \mathbf{X}\beta + \mathbf{Z}\mathbf{b} + \epsilon, \tag{3.45}$$

where β is the vector of fixed-effects, $\mathbf{b}$ is the vector of random-effects following $N(0, \sigma^2/\lambda\mathbf{\Omega}_2^{-1})$, and $\epsilon \sim N(0, \sigma^2\mathbf{I}_n)$.

Note that $\mathbf{\Omega}_2$ is completely known since $\mathbf{G}$ is known. A further simplification of the LME model (3.45) can be easily made. Set $\tilde{\mathbf{Z}} = \mathbf{Z}\mathbf{\Omega}_2^{-1/2}$ and $\tilde{\mathbf{b}} = \mathbf{\Omega}_2^{1/2}\mathbf{b}$. Then (3.43) becomes

$$\|\mathbf{y} - \mathbf{X}\beta - \tilde{\mathbf{Z}}\tilde{\mathbf{b}}\|^2 + \lambda\tilde{\mathbf{b}}^T\tilde{\mathbf{b}}, \tag{3.46}$$

so that the NCSS (evaluated at the design time points) that minimizes (3.46) can then be expressed as $\hat{\mathbf{y}}_\lambda = \mathbf{X}\hat{\beta} + \tilde{\mathbf{Z}}\hat{\tilde{\mathbf{b}}}$ with the BLUPs of the following LME model:

$$\mathbf{y} = \mathbf{X}\beta + \tilde{\mathbf{Z}}\tilde{\mathbf{b}} + \epsilon, \tag{3.47}$$

where $\tilde{\mathbf{b}} \sim N(0, \sigma^2/\lambda\mathbf{I}_K)$.

The idea of regarding a smoothing spline as the BLUP of a LME model is interesting. Moreover it transfers the problem of selecting a smoothing parameter λ to the problem of choosing the variance components σ^2/λ and σ^2 of the LME model (3.45). It is well-known that the variance components of a LME model can be easily found via the EM-algorithm with maximum likelihood method or restricted maximum likelihood method (see Chapter 2). However, it is worthwhile to mention that the random-effects here are somewhat artificial and are different from those subject-specific random-effects that we have discussed in Chapter 2. We shall discuss this point in greater detail in later chapters.

3.4.4 Connection between a Smoothing Spline and a State-Space Model

It is well known that the minimizer, namely $\hat{f}_\lambda(t)$ as an estimator of the underlying function $f(t)$, of the PLS criterion (3.31) is a smoothing spline of degree $(2k-1)$ where λ is the associated smoothing parameter. Wahba (1978) showed that $\hat{f}_\lambda(t)$ can be obtained through a signal extraction approach by introducing the following stochastic process $P(t)$ for $f(t)$:

$$\begin{aligned} P(t) &= \textstyle\sum_{r=0}^{k-1} \alpha_r \frac{t^r}{r!} + \lambda^{-1/2}\sigma \int_a^t \frac{(t-h)^{k-1}}{(k-1)!} dW(h), \\ y(t_i) &= P(t_i) + \epsilon_i, \quad i = 1, 2, \cdots, n, \end{aligned} \tag{3.48}$$

where $\alpha = [\alpha_0, \cdots, \alpha_{k-1}]^T \sim N(0, \lambda_0^{-1}\mathbf{I}_k)$, $W(h)$ is a standard Wiener process, $\epsilon = [\epsilon_1, \cdots, \epsilon_n]^T \sim N(0, \sigma^2\mathbf{I}_n)$. Under some regularity conditions, Wahba showed that the $\hat{f}_\lambda(t)$ is the Bayesian estimate of $P(t)$ as $\lambda_0 \to 0$. That is,

$$\hat{f}_\lambda(t) = \lim_{\lambda_0 \to 0} E_{\lambda_0}\{P(t)|y(t_i) = y_i, i = 1, \cdots, n\}.$$

Based on this relationship, Wahba (1978) and Craven and Wahba (1979) developed the cross-validation (CV) criterion for estimating the smoothing parameter λ. However, computing the CV criterion requires $O(n^3)$ operations, which is quite computationally intensive for large sample sizes. To overcome this difficulty, Wecker and Ansley (1983) found that the smoothing spline $\hat{f}_\lambda(t)$, as the minimizer of the PLS criterion (3.31), can be obtained via using the Kalman filter with $O(n)$ operations, which achieves significant computational advantage over Wahba's approach.

To use the Kalman filter, Wecker and Ansley (1983) re-wrote (3.48) in a state-space form. Assume that the design time points are distinct and ordered so that $t_1 < t_2 < \cdots < t_n$. Let $\mathbf{z}(t) = [f(t), f'(t), \cdots, f^{(k-1)}(t)]^T$. Then (3.48) can be written as the following state-space form:

$$\begin{aligned} \mathbf{z}(t_i) &= \mathbf{U}(t_i - t_{i-1})\mathbf{z}(t_{i-1}) + \lambda^{-1/2}\sigma\mathbf{u}(t_i - t_{i-1}), \\ y(t_i) &= \mathbf{e}_1^T\mathbf{z}(t_i) + \epsilon_i, \quad i = 1, 2, \cdots, n, \end{aligned} \tag{3.49}$$

where $\mathbf{e}_1$ denotes a k-dimensional unit vector whose first entry is 1 and 0 otherwise,

$$\mathbf{U}(\Delta) = \begin{bmatrix} 1 & \Delta & \frac{\Delta^2}{2!} & \cdots & \frac{\Delta^{k-1}}{(k-1)!} \\ 0 & 1 & \Delta & \cdots & \frac{\Delta^{k-2}}{(k-2)!} \\ \vdots & \vdots & \vdots & \vdots & \vdots \\ 0 & 0 & 0 & \cdots & 1 \end{bmatrix},$$

and $\mathbf{u}(\Delta)$ is a k-dimensional vector, following $N(\mathbf{0}, \mathbf{\Sigma}(\Delta))$, where $\mathbf{\Sigma}(\Delta)$ is a size $k \times k$ matrix whose (r, s)-th entry is

$$\frac{\Delta^{(2k-1)-(r+s)}}{[(2k-1)-(r+s)](k-r)!(k-s)!}, \quad r, s = 1, 2, \cdots, k.$$

In the state-space form (3.49), the first equation is known as the "state transition equation" and the second the "observation equation". This state-space equation system can be solved using the Kalman filter with $O(n)$ operation. This approach was also used by Guo et al. (1999) and Guo (2002a) who expressed the cubic smoothing spline in a state-space form and successfully applied the Kalman filter for modeling hormone time series and functional data.

3.4.5 Choice of Smoothing Parameters

The smoothing parameter λ plays an important role in the smoothing spline smoother (3.41). It trades-off the bias with the variance of the smoothing spline smoother $\hat{\mathbf{f}}_\lambda$ (3.41). Since $\hat{\mathbf{f}}_\lambda$ is a linear smoother, good choice of λ can be obtained by applying the smoothing parameter selectors such as CV, GCV, AIC or BIC as will be discussed in Section 3.7.

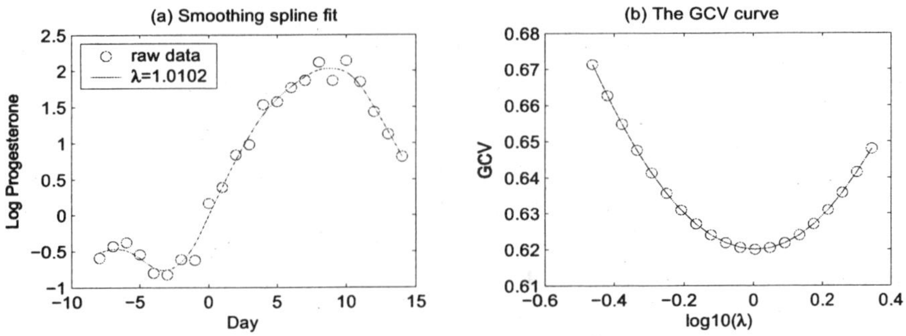

Fig. 3.6 (a) A cubic smoothing spline fit to the progesterone data from a subject. (b) The associated GCV curve.

Figure 3.6 (a) presents a NCSS fit (solid curve) of the progesterone data (dots) from a subject. The fit was obtained using the cubic smoothing spline smoother (3.37) with a smoothing parameter $\lambda = 1.0102$, selected by GCV. Figure 3.6 (b) presents the GCV curve against λ in a $\log_{10}$-scale.

3.5 PENALIZED SPLINES

Good performance of regression splines strongly depends on the location of knots and the number of knots. Smoothing splines overcome these drawbacks via taking all of the distinct time points as knots, and using a roughness penalty to control the smoothness of the resulting smoothers. When the number of distinct time points is large, the number of the parameters to be estimated is large. This is one of the reasons why smoothing splines are expensive to compute. Another reason is that smoothing splines need to compute the integral that defines the roughness of the underlying function. That is to compute the roughness matrix (3.33) for cubic smoothing splines or the roughness matrix (3.39) based on a pre-specified basis for a general degree smoothing spline. A penalized spline (P-spline) is designated to overcome these drawbacks by using the truncated power basis (3.18) (or a B-spline basis, see Eilers and Marx 1996 but we shall not discuss this kind of P-splines throughout this book; see Section 3.5.4 for more details.) of some degree with a pre-specified number of knots that is much smaller than the total number of distinct design time points. At the same time a P-spline avoids direct calculation of the roughness matrix (3.33) or (3.39) by specifying a simple roughness matrix, say, a diagonal matrix indicating that some coefficients of the basis functions are penalized and some are not (Ruppert, Wand and Carroll 2003).

3.5.1 Penalized Spline Smoother

Consider the PLS criterion (3.31). When $k \neq 2$, it is difficult to compute the integral that defines the roughness of the underlying function f. In Section 3.4.2, we suggested the use of a basis $\mathbf{\Phi}_p(t)$ to compute the integral and then the roughness matrix (3.39). Instead of using a general basis $\mathbf{\Phi}_p(t)$ and computing the roughness matrix (3.39) directly, a P-spline is obtained via using a special basis, e.g., the truncated power basis (3.18) and a special matrix in the place of the associated roughness matrix in (3.40).

Specifically, let $\mathbf{\Phi}_p(t)$ be the truncated power basis (3.21) of degree k with K knots $\tau_1, \tau_2, \cdots, \tau_K$. Then $f(t)$ can be first written as a regression spline $\mathbf{\Phi}_p(t)^T\boldsymbol{\alpha}$ where $\boldsymbol{\alpha} = [\alpha_0, \alpha_1, \cdots, \alpha_{k+K}]^T$ is the associated coefficient vector. Let $\mathbf{G}$ be a $p \times p$ diagonal matrix with its first $k+1$ diagonal entries being 0 and other diagonal entries 1. That is,

$$\mathbf{G} = \begin{pmatrix} \mathbf{0} & \mathbf{0} \\ \mathbf{0} & \mathbf{I}_K \end{pmatrix}. \tag{3.50}$$

Then a P-spline smoother of the function $f(t)$ in (3.2) is defined as $\hat{f}_\lambda(t) = \mathbf{\Phi}_p(t)^T\hat{\boldsymbol{\alpha}}$ where $\hat{\boldsymbol{\alpha}}$ is the minimizer of the following PLS criterion:

$$\sum_{i=1}^{n} \left(y_i - \mathbf{\Phi}_p(t_i)^T\boldsymbol{\alpha}\right)^2 + \lambda\boldsymbol{\alpha}^T\mathbf{G}\boldsymbol{\alpha}. \tag{3.51}$$

Note that

$$\boldsymbol{\alpha}^T\mathbf{G}\boldsymbol{\alpha} = \sum_{r=1}^{K} \alpha_{k+r}^2,$$

and α_{k+r} is the coefficient of the r-th truncated power basis function $(t-\tau_r)_+^k$ in the basis $\mathbf{\Phi}_p(t)$. The coefficient α_{k+r} measures the jump of the k-times derivative of $f(t)$ at the knot τ_r. It follows that the penalty in (3.51) is imposed just for the derivative jumps of the resulting P-spline at the knots. Therefore, the larger the coefficients $\alpha_{k+r}, r = 1, \cdots, K$, the rougher the resulting P-spline. Thus, the term $\boldsymbol{\alpha}^T\mathbf{G}\boldsymbol{\alpha}$ is a measure of the roughness of the resulting P-spline. For convenience, we call $\mathbf{G}$ the P-spline roughness matrix.

Since there is essentially no difference between (3.51) and (3.40), it is expected that a P-spline has essentially the same formula as that for a smoothing spline (3.41). Actually, the P-spline smoother $\hat{f}_\lambda(t)$ can be expressed is

$$\hat{f}_\lambda(t) = \mathbf{\Phi}_p(t)^T(\mathbf{W}^T\mathbf{W} + \lambda\mathbf{G})^{-1}\mathbf{W}^T\mathbf{y}, \tag{3.52}$$

where $\mathbf{W} = (\mathbf{\Phi}_p(t_1), \mathbf{\Phi}_p(t_2), \cdots, \mathbf{\Phi}_p(t_n))^T$.

Based on (3.52), the fitted response vector at the design time points is

$$\hat{\mathbf{y}}_\lambda = \mathbf{A}_\lambda\mathbf{y}, \tag{3.53}$$

where the smoother matrix is

$$\mathbf{A}_\lambda = \mathbf{W}(\mathbf{W}^T\mathbf{W} + \lambda\mathbf{G})^{-1}\mathbf{W}^T. \tag{3.54}$$

Therefore, a P-spline is also a linear smoother as defined later in Section 3.6.

3.5.2 Connection between a Penalized Spline and a LME Model

In Section 3.4.3, we established a connection between a smoothing spline and a LME model. A similar connection between a P-spline and a LME model can also be established. Consider the PLS criterion (3.51) with the roughness matrix $\mathbf{G}$ defined in (3.50). Let $\mathbf{X}$ and $\mathbf{Z}$, respectively be the design matrices consisting of the first $(k+1)$ columns and the remaining (last K) columns of the design matrix $\mathbf{W}$ induced by the basis $\mathbf{\Phi}_p(t)$, and let $\boldsymbol{\beta}$ and $\mathbf{b}$, respectively be the coefficient vectors consisting of the first $(k+1)$ entries and the remaining entries of the coefficient vector $\boldsymbol{\alpha}$. Then the PLS criterion (3.51) can be rewritten as

$$||\mathbf{y} - \mathbf{X}\boldsymbol{\beta} - \mathbf{Z}\mathbf{b}||^2 + \lambda \mathbf{b}^T\mathbf{b}.$$

Assume the measurement errors in the nonparametric regression model (3.2) follow a normal distribution with mean 0 and same variance σ^2, i.e., $\epsilon_1, \ldots, \epsilon_n$ i.i.d $N(0, \sigma^2)$. Then the fitted response vector (3.53) can be expressed as

$$\hat{\mathbf{y}}_\lambda = \mathbf{X}\hat{\boldsymbol{\beta}} + \mathbf{Z}\hat{\mathbf{b}},$$

where $\hat{\boldsymbol{\beta}}$ and $\hat{\mathbf{b}}$ are the BLUPs of the following LME model:

$$\begin{aligned} \mathbf{y} &= \mathbf{X}\boldsymbol{\beta} + \mathbf{Z}\mathbf{b} + \boldsymbol{\epsilon}, \\ \mathbf{b} &\sim N(\mathbf{0}, \sigma^2/\lambda \mathbf{I}_K), \ \boldsymbol{\epsilon} \sim N(\mathbf{0}, \sigma^2 \mathbf{I}_n), \end{aligned}$$

where $\boldsymbol{\beta}$ is the vector of fixed-effects and $\mathbf{b}$ the vector of random-effects. Notice that the LME model defined above is much simpler than those LME models defined in Section 3.4.3 for a smoothing spline. This is due to the special structure of the roughness matrix $\mathbf{G}$ as given in (3.50).

3.5.3 Choice of the Knots and Smoothing Parameter Selection

Choice of the knots for P-splines is not so crucial as long as the number of knots K is large enough so that when $\lambda = 0$, the P-spline smoother (3.52) is undersmoothing. One may use the equally spaced method or the equally spaced sample quantiles as knots method to locate the knots as described in Section 3.3. The number of knots K may be pre-specified subjectively, for example, taken as a third or a quarter of the total number of distinct design time points. Since a P-spline is a linear smoother (3.53), λ can be chosen using one of the methods presented later in Section 3.7 for linear smoothers.

3.5.4 Extension

So far, we just consider P-splines using the truncated power basis (3.18) studied by Ruppert and Carroll (2000), and Ruppert, Wand and Carroll (2003). We now consider some possible extensions.

First of all, the basis $\mathbf{\Phi}_p(t)$ is not necessarily restricted to be the truncated power basis (3.21). It can be other bases too. In fact, in the pioneering work on P-splines by Eilers and Marx (1996), B-spline bases (de Boor 1978) were employed for $\mathbf{\Phi}_p(t)$ with equally spaced knots, tracing the original idea about P-splines back to O'Sullivan (1986, 1988). A drawback for Eilers and Marx's P-splines is that the B-spline basis functions have no closed form formulas, and the equal knot placing rule may not work well for unequally spaced design time points.

Secondly, the matrix $\mathbf{G}$ can be other nonnegative definite matrices. For example, we may take $\mathbf{G} = \text{diag}(a_1, \cdots, a_p)$ where $a_1, \cdots, a_p$ are pre-specified, so that the penalty may be imposed unequally for the coefficients. Moreover, the penalty term $\boldsymbol{\alpha}^T\mathbf{G}\alpha$ in (3.51) can be replaced by a general form $P(\boldsymbol{\alpha})$ (Wang 1998a,b). For example, we can use $P(\boldsymbol{\alpha}) = \sum_r |\alpha_r|$ or $P(\boldsymbol{\alpha}) = \sum_{r\in S} |\alpha_r|$ where S consists of some of the subscripts of $\boldsymbol{\alpha}$ that are of interest. However, the resulting smoothers may not be linear smoothers, and hence may be computationally intensive.

3.6 LINEAR SMOOTHER

Linear smoothers express the fitted response vector $\hat{\mathbf{y}}$ at the design time points as linear combinations of the response vector $\mathbf{y}$ (Buja et al. 1989). Specifically, a smoother $\hat{f}_\rho(t)$ is called a linear smoother if it makes the fitted response vector $\hat{\mathbf{y}}$ to be connected with $\mathbf{y}$ via the following simple formula:

$$\hat{\mathbf{y}} = \mathbf{A}_\rho \mathbf{y}, \tag{3.55}$$

where $\hat{\mathbf{y}} = [\hat{y}_1, \cdots, \hat{y}_n]^T$ with $\hat{y}_i = \hat{f}_\rho(t_i)$, $\mathbf{A}_\rho$ is the so-called $n \times n$ smoother matrix, determined by the smoother with a smoothing parameter ρ. The smoother matrix $\mathbf{A}_\rho$ usually depends on the design time points but is required to be independent of the response vector $\mathbf{y}$. It follows that

$$\begin{aligned} \text{E}(\hat{\mathbf{y}}|t_1, \cdots, t_n) &= \mathbf{A}_\rho \text{E}(\mathbf{y}|t_1, \cdots, t_n), \\ \text{Cov}(\hat{\mathbf{y}}|t_1, \cdots, t_n) &= \mathbf{A}_\rho \text{Cov}(\mathbf{y}|t_1, \cdots, t_n)\mathbf{A}_\rho^T. \end{aligned}$$

These two formulas can be used to derive the biases and variances of the fitted response vector, and hence allow construction of pointwise standard deviation bands for the underlying function $f(t)$ based on $\hat{\mathbf{y}}$. Moreover, in next section, we shall see that we can develop unified smoothing parameter selection methods for linear smoothers.

From the previous discussions, it is clear that the local polynomial, regression spline, smoothing spline and P-spline smoothers are all linear smoothers as shown in (3.7), (3.26), (3.37) and (3.53), respectively. Other linear smoothers include the Gasser-Müller smoother (Gasser and Müller 1979). See Buja et al. (1989) for more discussions on linear smoothers.

3.7 METHODS FOR SMOOTHING PARAMETER SELECTION

Smoothing parameter selection can be handled in a unified manner for all the linear smoothers $\hat{f}_\rho(t)$ defined in (3.55). A good selector usually tries to select a good smoothing parameter ρ to trade-off the goodness of fit of the smoother and its model complexity.

3.7.1 Goodness of Fit

A measure for goodness of fit of a linear smoother (3.55) with a smoothing parameter ρ is its Sum of Squared Errors (SSE):

$$\text{SSE}_\rho = \sum_{i=1}^{n}(y_i - \hat{y}_i)^2 = \mathbf{y}^T(\mathbf{I}_n - \mathbf{A}_\rho)^T(\mathbf{I}_n - \mathbf{A}_\rho)\mathbf{y}, \tag{3.56}$$

where $\hat{y}_i = \hat{f}_\rho(t_i)$ denotes the fitted value for the response y_i using the smoother, and $y_i - \hat{y}_i$ is the residual at t_i that can not be explained by the smooth. It is clear that SSE_ρ indicates how close the model fits the data. When $\rho = 0$, i.e., no smoothing is actually involved, the fitted model often interpolates the data points so that $\text{SSE}_\rho = 0$.

Another measure for goodness of fit of a smoother is its log-likelihood . Assume that the measurement errors of the model (3.2) follow a normal distribution with variance σ^2, i.e.,

$$\epsilon_1, \epsilon_2, \ldots, \epsilon_n \text{ i.i.d} \sim N(0, \sigma^2). \tag{3.57}$$

Let $\mathbf{t} = [t_1, \cdots, t_n]'$. Then the associated likelihood function is

$$\begin{aligned} L(\sigma^2, f|\mathbf{y}, \mathbf{t}) &= p(\mathbf{y}|\sigma^2, f, \mathbf{t}) \\ &= (2\pi\sigma^2)^{-n/2} \exp\left(-\tfrac{1}{2\sigma^2}\textstyle\sum_{i=1}^n [y_i - f(t_i)]^2\right). \end{aligned} \tag{3.58}$$

Let $\hat{f}_\rho(t)$ be a nonparametric fit of $f(t)$ using some linear smoother with a smoothing parameter ρ. Then the associated maximum likelihood estimate of σ^2 is

$$\hat{\sigma}^2 = n^{-1}\sum_{i=1}^{n}[y_i - \hat{f}_\rho(t_i)]^2 = \text{SSE}_\rho/n.$$

Taking a logarithm of (3.58) after replacing the parameters σ^2 and f by their estimates defined above, we have the following so-called log-likelihood associated with the linear smoother $\hat{f}_\rho(t)$:

$$\text{Loglik} = -\frac{n}{2}\log\left(\frac{2\pi e}{n}\text{SSE}_\rho\right), \tag{3.59}$$

where $e = \exp(1)$. Under normality assumption (3.57), the Loglik (3.59) is equivalent to the SSE_ρ (3.56) in the sense that whenever the SSE_ρ decreases, the Loglik increases. In particular, when $\rho \to 0$, we generally have the $\text{SSE}_\rho \to 0$ while the Loglik $\to \infty$.

3.7.2 Model Complexity

Following Buja et al. (1989), the trace of the smoother matrix $\mathbf{A}_\rho$ is called the degrees of freedom (df) of the linear smoother (3.55):

$$\text{df} = \text{tr}(\mathbf{A}_\rho) = \sum_{i=1}^{n} a_{ii}, \tag{3.60}$$

where a_{ii} are the diagonal entries of $\mathbf{A}_\rho$. The df is often used as a measure for the model complexity of the linear smoother (3.55) and it indicates approximately how many effective parameters are involved in modeling. For example, for the regression spline smoother where the smoothing parameter ρ is the number of basis functions, p, the associated smoother matrix is defined in (3.27). It is easy to show that

$$\text{df} = \text{tr}(\mathbf{A}_p) = \text{tr}[\mathbf{X}(\mathbf{X}^T\mathbf{X})^{-1}\mathbf{X}^T] = p,$$

indicating that there are exactly p parameters (coefficients or basis functions) involved for regression spline smoothing. When $p = 1$, the associated fitting model is simple since only one basis function is involved; when $p = 20$, the associated fitting model is complicated since there are 20 basis functions in use.

It is worthwhile to mention that for other linear smoothers such as the LPK smoother (3.7), the smoothing spline smoother (3.37), and the P-spline smoother (3.53), the associated df is usually not an integer. Moreover, as the smoothing parameter ρ becomes smaller, the df is usually increasing, indicating a more complicated model. In contrast, the SSE_ρ is decreasing in this case (or Loglik is increasing), indicating a closer fit to the data.

When $\mathbf{A}_\rho$ is a symmetric, nonnegative definite matrix, the above statement can be partially explained as follows. Denote the singular value decomposition of $\mathbf{A}_\rho$ as $\mathbf{U}\boldsymbol{\Lambda}\mathbf{U}^T$ where $\mathbf{U}$ is orthonormal, and $\boldsymbol{\Lambda} = \text{diag}(\lambda_1, \cdots, \lambda_n)$ where $\lambda_i \geq 0,\ i = 1, 2, \cdots, n$. It follows that

$$\begin{aligned}
\text{SSE}_\rho &= \mathbf{y}^T(\mathbf{I}_n - \mathbf{U}\boldsymbol{\Lambda}\mathbf{U}^T)^2\mathbf{y} \\
&= \mathbf{z}^T(\mathbf{I}_n - \boldsymbol{\Lambda})^2\mathbf{z} \\
&= \textstyle\sum_{i=1}^n (1 - \lambda_i)^2 z_i^2, \\
\text{df} &= \text{tr}(\mathbf{U}\boldsymbol{\Lambda}\mathbf{U}^T) = \text{tr}(\boldsymbol{\Lambda}) \\
&= \textstyle\sum_{i=1}^n \lambda_i,
\end{aligned}$$

where $\mathbf{z} = \mathbf{U}^T\mathbf{y} = (z_1, \cdots, z_n)^T$. Thus, when the df increases, some λ_i will increase so that SSE_ρ will decrease provided $\mathbf{z}$ is unchanged. Therefore, a good ρ should be chosen to balance the goodness of fit and the model complexity of the associated linear smoother.

3.7.3 Cross-Validation

Cross-validation is a widely-used smoothing parameter selector due to its simplicity of construction. Although its construction is not directly based on the relationship between the goodness of fit and the model complexity of a smoother, its development follows a similar idea.

Consider the linear smoother $\hat{f}_\rho(t)$ in (3.55) for the nonparametric regression model (3.2). A good smoothing parameter ρ should be chosen so that the $\hat{f}_\rho(t)$ at least gives good fitted values at the design time points $t_1, \cdots, t_n$. That is, it should produce a small Average Squared Error (ASE) at all the design time points:

$$\text{ASE}(\hat{f}_\rho) = n^{-1}\sum_{i=1}^{n} \text{MSE}\left(\hat{f}_\rho(t_i)\right) = n^{-1}\sum_{i=1}^{n} \text{E}\left(\hat{f}_\rho(t_i) - f(t_i)\right)^2. \tag{3.61}$$

The above ASE is not computable due to the fact that f is the unknown function. It has to be estimated. A good estimator for the above ASE is the following cross-validation (CV) score:

$$\text{CV}(\rho) = n^{-1}\sum_{i=1}^{n}\left(y_i - \hat{f}_\rho^{(-i)}(t_i)\right)^2, \tag{3.62}$$

where $\hat{f}_\rho^{(-i)}(t_i)$ is the fitted value at time point t_i, using all the data except (t_i, y_i), with the smoothing parameter ρ for $i = 1, 2, \cdots, n$. The so-called CV method selects the ρ which minimizes the above CV score (3.62).

Direct computation of the above CV score (3.62) is quite intensive since for each i, a local fit $\hat{f}_\rho^{(-i)}(t_i)$ is required. Thus, evaluation of the CV score (3.62) for a value of ρ needs n local fits. However, for the linear smoother $\hat{f}_\rho(t)$ (3.55), it is easy to show that the CV score (3.62) can be simplified as

$$\text{CV}(\rho) = n^{-1}\sum_{i=1}^{n}\left\{\frac{y_i - \hat{f}_\rho(t_i)}{1 - a_{ii}}\right\}^2 = n^{-1}\sum_{i=1}^{n}\left\{\frac{y_i - \hat{y}_i}{1 - a_{ii}}\right\}^2, \tag{3.63}$$

where a_{ii} is the i-th diagonal element of $\mathbf{A}_\rho$ as defined in the subsection above. In this way, for each possible value of ρ, evaluation of the CV score (3.62) needs only one local fit.

Note that the CV score (3.63) is not directly related to the goodness of fit (say, measured by SSE_ρ) and the model complexity (say, measured by df). However, it is not difficult to see that it is related to the components that construct the SSE_ρ and the df, i.e., the residuals $y_i - \hat{y}_i$, $i = 1, 2, \cdots, n$ and the diagonal entries a_{ii}, $i = 1, 2, \cdots, n$ of $\mathbf{A}_\rho$. That is, the CV score (3.63) balances the goodness of fit and the model complexity of the smoother indirectly.

3.7.4 Generalized Cross-Validation

Generalized cross-validation (GCV) was first proposed and studied by Wahba (1977) and Craven and Wahba (1979) as an approximation of the cross-validation. The GCV score can be obtained from the simplified CV score (3.63) via substituting each of $a_{ii}, i = 1, 2, \cdots, n$ by their average

$$n^{-1}\sum_{i=1}^{n} a_{ii} = n^{-1}\text{tr}(\mathbf{A}_\rho) = \text{df}/n.$$

That is,

$$\text{GCV}(\rho) = \frac{n^{-1}\sum_{i=1}^{n}[y_i - \hat{y}_i]^2}{\{1 - \text{tr}(\mathbf{A}_\rho)/n)\}^2} = n^{-1}\text{SSE}_\rho/(1 - \text{df}/n)^2. \tag{3.64}$$

It is clear that the numerator represents the goodness of fit, and the denominator indicates the model complexity of the linear smoother. The GCV method selects ρ so that (3.64) is minimized, i.e., the goodness of fit and the model complexity are balanced.

Figure 3.4 shows an example of using the GCV selector for selecting a good bandwidth h for a local linear smoother. Another example for selecting a smoothing parameter λ for a cubic smoothing spline is displayed in Panel (b) of Figure 3.6.

3.7.5 Generalized Maximum Likelihood

Wahba (1985, 1990) proposed a generalized maximum likelihood (GML) method for selecting a smoothing parameter for her smoothing spline smoother. This method is generally applicable for selecting a smoothing parameter for a linear smoother. The GML method is the same as the REML method based on the connection between a smoothing spline and a LME model.

Consider the linear smoother $\hat{f}_\rho(t)$ defined in (3.55) for the nonparametric regression model (3.2) so that $\hat{\mathbf{y}} = \mathbf{A}_\rho \mathbf{y}$ where $\mathbf{A}_\rho$ is the smoother matrix and ρ is the associated smoothing parameter. Then Wahba's GML criterion can be written as

$$\text{GML}(\rho) = \frac{\mathbf{y}^T(\mathbf{I}_n - \mathbf{A}_\rho)\mathbf{y}}{|\mathbf{I}_n - \mathbf{A}_\rho|_+^{1/(n-m)}}, \tag{3.65}$$

where $|\mathbf{I}_n - \mathbf{A}_\rho|_+$ equals to the product of the $(n - m)$ nonzero eigenvalues of $(\mathbf{I}_n - \mathbf{A}_\rho)$. The GML smoothing parameter is obtained by minimizing the GML criterion (3.65) with respect to ρ.

Results in Wahba (1990, Section 4.9) suggest that both the GML and GCV criteria are comparable in performance. Kohn, Ansley and Tharm (1991) found that in many situations, GML is competitive with, and often superior to, GCV.

3.7.6 Akaike Information Criterion

Akaike Information Criterion (AIC) (Akaike 1973) is a model selection criterion widely used in linear models, time series and other areas. Its basic idea is to correct the log-likelihood of a fitted model (i.e., the goodness of fit) by the number of the parameters involved in parametric modeling (i.e., the model complexity). For the linear smoother $\hat{f}_\rho$ (3.55), it can be defined as

$$\text{AIC}(\rho) = -2\text{Loglik} + 2\text{df}, \tag{3.66}$$

where the Loglik and df are defined in (3.59) and (3.60), respectively. Alternatively, we can regard the second term of the right hand side of (3.66) as the penalty of the first term. Note that, when ρ is smaller, the first term of the AIC is also smaller (indicating better fit to the data), but the second term is larger (a more complex model involved). A good choice of ρ will trade-off the first term with the second term on the right-hand side of (3.66), i.e., minimize the AIC score (3.66).

Note that under the normality assumption (3.57), the AIC (3.66) is approximately equivalent to the GCV (3.64). Actually, taking the logarithm of the GCV (3.64), we have

$$\begin{aligned}
\log[\mathrm{GCV}(\rho)] &= \log(\mathrm{SSE}_\rho) - 2\log(1 - \mathrm{df}/n) \\
&\approx (-2\mathrm{Loglik} + 2\mathrm{df})/n - \log(2\pi e/n) \\
&= \mathrm{AIC}(\rho)/n - \log(2\pi e/n),
\end{aligned}$$

using a simple Taylor expansion and the definition (3.59) of the Loglik.

3.7.7 Bayesian Information Criterion

It can be observed that AIC may lead to a too small smoothing parameter ρ for a large data set. This is a result of the penalty term in (3.66) not being large enough. A larger penalty term is given in the following Bayesian Information Criterion (BIC)(Schwarz 1978):

$$\mathrm{BIC}(\rho) = -2\mathrm{Loglik} + \log(n)\mathrm{df}, \tag{3.67}$$

obtained via replacing the factor 2 in the AIC score (3.66) with $\log(n)$.

Since $\log(n) > 2$ when $n > 9$, the BIC rule (3.67) gives heavier penalty to the model complexity when $n > 9$. Therefore, the AIC rule is more conservative than the BIC rule in the sense that it favors a larger or more complex model.

3.8 SUMMARY AND BIBLIOGRAPHICAL NOTES

In this chapter, we reviewed four of the most popular linear smoothers, including the local polynomial, regression spline, smoothing spline, and P-spline smoothers. We mainly focused on how they are constructed, and how the smoothing parameters are selected. These four linear smoothers are known to perform well for uncorrelated data, and we shall extend their use to smoothing longitudinal data in later chapters.

Good surveys on local polynomial smoothing include the books by Eubank (1988, 1999), Härdle (1990), Wand and Jones (1995), Fan and Gijbels (1996) and Simonoff (1996), among others. Good surveys for smoothing splines include the books by Eubank (1988, 1999), Wahba (1990) and Gu (2002), among others. Eubank (1988, 1999) also provided a nice account on regression splines. A recent survey on P-splines was given by Ruppert, Wand and Carroll (2003).

There are some other linear and nonlinear smoothers well developed in the statistical literature, and also widely used. For example, the Gasser-Müller smoother (Gasser and Müller 1979) was popular in the early statistical literature of nonparametric smoothing. For comparison of this smoother to the local constant and linear smoothers, see Fan (1992, 1993), and Fan and Gijbels (1996) among others.

Linear smoothers are simple to construct and easy to implement. They also have some disadvantages. For example, linear smoothers are usually not robust against unusual observations/outliers. This is a result of being weighted least-squares (WLS) based smoothers. Nonlinear smoothers were then introduced to overcome this drawback, paying a price for more computation efforts when unusual observations are present in the data set. An example of nonlinear smoothers is provided by Cleveland (1979). He used a locally WLS estimate to fit a nonparametric regression function in a re-weighting manner. The resulting estimate is nonlinear in the sense that it can not be written in the form of (3.55), and it is robust against unusual observations or outliers. Wavelet-based thresholded smoothers (Donoho and Johnstone 1994, Fan 1996) are also examples of nonlinear smoothers.

4

Local Polynomial Methods

4.1 INTRODUCTION

Local polynomial smoothing techniques have been well developed for iid or cross-sectional data (Wand and Jones 1995, Fan and Gijbels 1996). In order to apply these techniques to longitudinal data analysis, substantial efforts have been made to incorporate the features of longitudinal data in kernel smoothing methods (Hoover et al. 1998, Wu, Chiang and Hoover 1998, Fan and Zhang 2000, Lin and Carroll 2000, Wu and Chiang 2000, Wu and Zhang 2002a, Welsh, Lin and Carroll 2002, Wang 2003, Park and Wu 2005). In longitudinal studies, the data collected from the same subject over time are likely to be correlated although the data from different subjects can be assumed to be independent. The within-subject and between-subject variations are different and need to be modeled appropriately.

Hoover et al. (1998), Wu, Chiang and Hoover (1998) and Wu and Chiang (2000) first proposed the kernel estimation method for time-varying coefficient models with longitudinal data. However, the features of longitudinal data are not directly incorporated in their methods, although the "leave-one-subject-out" cross-validation criterion is proposed for smoothing parameter selection in which the data in subject-based clusters are recognized. To nonparametrically model correlated data, such as longitudinal data, Diggle and Hutchison (1989), Altman (1990), Hart (1991), Rice and Silverman (1991) and others have proposed modifications to smoothing parameter selection criteria such as cross-validation (CV) or generalized cross-validation (GCV) or using "leave-one-subject-out" CV or GCV to indirectly account for the correlations within the data. Lin and Carroll (2000) proposed a local polynomial kernel generalized estimating equation (LPK-GEE) method for clustered or longitudinal data. They showed that the best strategy is to ignore the correlation structure of longitudinal data (pretend as if the data within a cluster or subject were independent) in the LPK-GEE estimator. However, their conclusions are based on the asymptotic results under the condition that the number of subjects or clusters goes to infinity and the number of measurements from each subject is finite. Lin and Carroll's working-independence estimator may not be the best for finite sample cases. In fact, some new results have shown that it is necessary to incorporate longitudinal data correlations in the estimator in order to achieve better efficiency in finite sample situations (Wu and Zhang 2002a, Welsh, Lin and Carroll 2002, Wang 2003). Fan and Zhang (2000) suggested a two-step approach (local averaging or regression first, then smoothing) to indirectly account for the data correlation. A local polynomial mixed-effects modeling approach, which more appropriately models the within-subject correlations, was proposed by Wu and Zhang (2002a). This method will be one of the focuses of this chapter.

We extend the linear and nonlinear mixed-effects models (Chapter 2) to a more general nonparametric model setting in this chapter. The rest of this chapter is organized as follows. First we review the methods for estimation of the population mean function for longitudinal data in Section 4.2. A simple local polynomial method, a LPK-GEE method and a two-step method are briefly surveyed. Section 4.3 introduces a nonparametric mixed-effects (NPME) model and Section 4.4 presents the local polynomial mixed-effects modeling approach. We discuss different bandwidth selection strategies in Section 4.5 and introduce a backfitting algorithm in Section 4.6. The asymptotic properties of the estimators are studied in Section 4.7 while the finite sample properties are investigated via Monte Carlo simulations in Section 4.8. To illustrate the methodologies, an application to the progesterone data is presented in Section 4.9. A brief summary and some bibliographical notes are given in Section 4.10. We outline the technical proofs of the theoretical results in Section 4.11. Most of the materials in Sections 4.3~4.9 come from two articles by Wu and Zhang (2002a) and Park and Wu (2005).

4.2 NONPARAMETRIC POPULATION MEAN MODEL

A longitudinal data set, e.g., the progesterone data introduced in Section 1.1.1 of Chapter 1, is usually collected by repeated measurements from a number of subjects over a period of time. The design time points may be different for different subjects and so are the number of measurements. Let n denote the number of subjects, and let (t_{ij}, y_{ij}) be the j-th design time point for the i-th subject and the associated response where $j = 1, 2, \cdots, n_i$ with n_i denoting the number of measurements for the i-th subject. Such a longitudinal data set can be symbolically expressed as

$$(t_{ij}, y_{ij}),\ j = 1, 2, \cdots, n_i;\ i = 1, 2, \cdots, n. \tag{4.1}$$

If no parametric model is available for modeling the population mean function of the above longitudinal data, it is natural to model it nonparametrically. That is, we just assume that the population mean function is smooth. Such a nonparametric population mean (NPM) model can be written as

$$y_{ij} = \eta(t_{ij}) + e_i(t_{ij}),\ j = 1, 2, \cdots, n_i;\ i = 1, 2, \cdots, n, \tag{4.2}$$

where $\eta(t)$ is the smooth population mean function, and $e_i(t_{ij})$ are the departures of the longitudinal measurements from the population mean function. This model is comparable with the standard nonparametric regression model (3.2) of Chapter 3, but differs in that the errors in the NPM model (4.2) are generally not independent.

Since no parametric form is available for modeling $\eta(t)$, nonparametric smoothing techniques are needed to be used. In fact, several nonparametric techniques have been proposed for time-varying coefficients models which include the NPM model (4.2) as a special case. In this section, we review three such techniques: a naive local polynomial kernel (LPK) method (Hoover et al. 1998); a LPK-GEE method (Lin and Carroll 2000), and a two-step method (Fan and Zhang 2000).

4.2.1 Naive Local Polynomial Kernel Method

The LPK method for time-varying coefficient models for longitudinal data was first proposed and studied by Hoover et al. (1998). As was the case with LPK smoothing of independent data reviewed in Section 3.2 of Chapter 3, the main idea of this LPK method is to fit a polynomial of some degree to $\eta(t)$ locally.

Let t be an arbitrary fixed time point. Assume that $\eta(t)$ has up to $(p+1)$-st continuous derivatives for some integer $p \geq 0$ at t. Then by Taylor expansion, $\eta(t_{ij})$ can be locally approximated by a polynomial of degree p. That is,

$$\eta(t_{ij}) \approx \mathbf{x}_{ij}^T \boldsymbol{\beta}, \; j = 1, 2, \cdots, n_i; \; i = 1, 2, \cdots, n, \tag{4.3}$$

where $\mathbf{x}_{ij} = [1, t_{ij} - t, \cdots, (t_{ij} - t)^p]^T$ and $\boldsymbol{\beta} = [\beta_0, \beta_1, \cdots, \beta_p]^T$ with $\beta_r = \eta^{(r)}(t)/r!$, $r = 0, \cdots, p$. Let $\hat{\boldsymbol{\beta}} = [\hat{\beta}_0, \hat{\beta}_1, \cdots, \hat{\beta}_p]^T$ be the estimator of $\boldsymbol{\beta}$ obtained by minimizing the following weighted least squares (WLS) criterion:

$$\sum_{i=1}^{n} \sum_{j=1}^{n_i} \left(y_{ij} - \mathbf{x}_{ij}^T \boldsymbol{\beta}\right)^2 K_h(t_{ij} - t), \tag{4.4}$$

where $K_h(\cdot) = K(\cdot/h)/h$ with K a kernel function and h a bandwidth. As with smoothing independent data described in Section 3.2, the bandwidth h is used to specify the size of the local neighborhood $[t-h, t+h]$ and the kernel K is used to specify the effect of the data points according to the distance between t_{ij} and t. Usually, the closer the distance is, the larger the effect is.

To give an explicit expression for $\hat{\boldsymbol{\beta}}$ in matrix notation, let

$$\mathbf{X}_i = [\mathbf{x}_{i1}, \mathbf{x}_{i2}, \cdots, \mathbf{x}_{in_i}]^T, \quad \mathbf{K}_{ih} = \text{diag}\left(K_h(t_{i1} - t), \cdots, K_h(t_{in_i} - t)\right),$$

be the design matrix and the weight matrix for the i-th subject, respectively. In addition, denote $\mathbf{X} = [\mathbf{X}_1^T, \cdots, \mathbf{X}_n^T]^T$ and $\mathbf{K}_h = \text{diag}(\mathbf{K}_{1h}, \cdots, \mathbf{K}_{nh})$. Then the WLS criterion (4.4) can be rewritten as

$$(\mathbf{y} - \mathbf{X}\boldsymbol{\beta})^T \mathbf{K}_h (\mathbf{y} - \mathbf{X}\boldsymbol{\beta}), \tag{4.5}$$

where $\mathbf{y} = [\mathbf{y}_1^T, \cdots, \mathbf{y}_n^T]^T$ with $\mathbf{y}_i = [y_{i1}, \cdots, y_{in_i}]^T$ being the i-th subject response vector. It follows from minimizing (4.5) with respect to $\boldsymbol{\beta}$ that

$$\hat{\boldsymbol{\beta}} = (\mathbf{X}^T \mathbf{K}_h \mathbf{X})^{-1} \mathbf{X}^T \mathbf{K}_h \mathbf{y}. \tag{4.6}$$

Let $\mathbf{e}_r$ be a $(p+1)$-dimensional unit vector whose r-th entry is 1 and others are 0. Then it is easy to see from the definitions of $\beta_r, r = 0, 1, \cdots, p$ that the estimators of the derivatives $\eta^{(r)}(t), r = 0, 1, \cdots, p$ are

$$\hat{\eta}^{(r)}(t) = r!\mathbf{e}_{r+1}^T\hat{\boldsymbol{\beta}}, r = 0, 1, \cdots, p. \tag{4.7}$$

In particular, the LPK estimator for the population mean function is $\hat{\eta}(t) = \mathbf{e}_1^T\hat{\boldsymbol{\beta}}$.

As with smoothing iid data described in Section 3.2, p can be taken as 0 and 1 for simplicity. For example, when $p = 0$, we have $\mathbf{X} = \mathbf{1}_N$, an N-dimension vector of 1's and the resulting LPK estimator $\hat{\eta}(t)$ is usually referred to as the so-called local constant kernel estimator of $\eta(t)$ where $N = \sum_{i=1}^n n_i$ is the number of the total measurements for all the subjects. From (4.6), the local constant kernel estimator of $\eta(t)$ has the following simple expression:

$$\hat{\eta}(t) = \frac{\sum_{i=1}^n \sum_{j=1}^{n_i} K_h(t_{ij} - t)y_{ij}}{\sum_{i=1}^n \sum_{j=1}^{n_i} K_h(t_{ij} - t)}. \tag{4.8}$$

When $n_i \equiv 1$, i.e., there is only one measurement per subject, the estimator (4.8) reduces to the iid data estimator in (3.9). The estimator (4.8) is called a local constant kernel estimator since it equals the minimizer, $\hat{\beta}_0$, of the following WLS criterion:

$$\sum_{i=1}^n \sum_{j=1}^{n_i} (y_{ij} - \beta_0)^2 K_h(t_{ij} - t). \tag{4.9}$$

In other words, $\hat{\beta}_0$ is the best constant that approximates y_{ij} within the local neighborhood $[t - h, t + h]$ with regards to minimizing (4.9).

When $p = 1$, the associated LPK estimator $\hat{\eta}(t)$ is usually referred to as the local linear kernel estimator of $\eta(t)$. From (4.6), the local linear kernel estimator may be expressed as

$$\hat{\eta}(t) = \frac{\sum_{i=1}^n \sum_{j=1}^{n_i} \left[s_2(t) - s_1(t)(t_{ij} - t)\right] K_h(t_{ij} - t)y_{ij}}{s_2(t)s_0(t) - s_1^2(t)}, \tag{4.10}$$

where

$$s_r(t) = \sum_{i=1}^n \sum_{j=1}^{n_i} K_h(t_{ij} - t)(t_{ij} - t)^r, \; r = 0, 1, 2.$$

Similarly, the estimator (4.10) is called a local linear kernel estimator since it is obtained by approximating y_{ij} within a local neighborhood using a linear function $\beta_0 + \beta_1(t_{ij} - t)$, i.e., minimizing the following WLS criterion:

$$\sum_{i=1}^n \sum_{j=1}^{n_i} [y_{ij} - \beta_0 - \beta_1(t_{ij} - t)]^2 K_h(t_{ij} - t). \tag{4.11}$$

Based on the results of Hoover et al. (1998), it is easy to show that when $p = 0$ or 1, under some regularity conditions, we have

$$\text{Bias}[\hat{\eta}(t)] = O_P(h^2), \quad \text{Var}[\hat{\eta}(t)] = O_P\left[1/(Nh) + \sum_{i=1}^n n_i^2/N^2\right], \tag{4.12}$$

where the first order term $O_P[1/(Nh)]$ in the expression of Var$[\hat{\eta}(t)]$ is related to the within-subject variation only, while the second order term $O_P\left[\sum_{i=1}^n n_i^2/N^2\right]$ is associated with the between-subject variation. It follows that the asymptotic properties of $\hat{\eta}(t)$ are different when n_i are bounded, compared to when n_i are not bounded. In fact, when all the n_i are bounded, the Var$[\hat{\eta}(t)]$ in (4.12) is dominated by its first order term so that MSE$[\hat{\eta}(t)] = O_P\left(N^{-4/5}\right)$; when all the n_i tend to infinity, the Var$[\hat{\eta}(t)]$ is dominated by the second order term $O_P\left(\sum_{i=1}^n n_i^2/N^2\right)$ so that MSE$(\hat{\eta}(t)) = O_P\left(\sum_{i=1}^n n_i^2/N^2\right)$. In particular, assume $n_i = m, i = 1, 2, \cdots, n$, then as $m \to \infty$, we have MSE$[\hat{\eta}(t)] = O_P(n^{-1})$. In this case, $\hat{\eta}(t)$ is $\sqrt{n}$-consistent.

From (4.12), the theoretical optimal bandwidth that minimizes MSE$[\hat{\eta}(t)] =$ Bias$^2[\hat{\eta}(t)]$ + Var$[\hat{\eta}(t)]$ is of order $O_P\left(N^{-1/5}\right)$ when n_i are bounded. Rice and Silverman (1991) proposed a "leave-one-subject-out" cross-validation method for selecting a proper bandwidth for longitudinal data. This bandwidth selection strategy was employed by Hoover et al. (1998).

4.2.2 Local Polynomial Kernel GEE Method

The LPK-GEE method was proposed and studied by Lin and Carroll (2000). For the NPM model (4.2), based on the notation such as $\mathbf{X}, \mathbf{K}_h, \mathbf{y}$ and β defined in the above subsection, the associated LPK-GEE is

$$\mathbf{X}^T\mathbf{K}_h^{1/2}\mathbf{\Omega}^{-1}\mathbf{K}_h^{1/2}(\mathbf{y} - \mathbf{X}\beta) = 0, \tag{4.13}$$

where $\mathbf{\Omega} = \mathbf{S}^{1/2}\mathbf{C}\mathbf{S}^{1/2}$ with $\mathbf{S} = \text{diag}(\text{Cov}(\mathbf{y}))$ and $\mathbf{C}$ being a user-specified working correlation matrix. When $\mathbf{\Omega} = \mathbf{I}_N$, the LPK-GEE (4.13) can be obtained via differentiating the WLS criterion (4.5) with respect to β and setting it equal to 0. Solving the above LPK-GEE with respect to β leads to the so-called LPK-GEE estimator

$$\hat{\beta} = \left(\mathbf{X}^T\mathbf{K}_h^{1/2}\mathbf{\Omega}^{-1}\mathbf{K}_h^{1/2}\mathbf{X}\right)^{-1}\mathbf{X}^T\mathbf{K}_h^{1/2}\mathbf{\Omega}^{-1}\mathbf{K}_h^{1/2}\mathbf{y}. \tag{4.14}$$

The estimators for $\eta(t)$ and its derivatives can be obtained easily using (4.7).

The working correlation matrix $\mathbf{C}$ in the LPK-GEE formulation (4.13) is used to partially account for the underlying correlation structure of $\mathbf{y}$. In particular, when we take $\mathbf{C} = \text{Cor}(\mathbf{y})$, we have $\mathbf{S}^{1/2}\mathbf{C}\mathbf{S}^{1/2} = \text{Cov}(\mathbf{y})$ so that the true correlation structure is taken into account although this is almost impossible in real applications.

The counter-intuitive result of Lin and Carroll (2000) is that the most efficient LPK-GEE estimator is obtained by ignoring the within-subject correlation instead of correctly specifying the within-subject correlation, i.e., assuming $\mathbf{C} = \mathbf{I}_N$. They argued that, asymptotically, there is no need to take the correlation into account because when the bandwidth is shrunk to 0 as the sample size $n \to \infty$, the chance for more than two observations from the same subject is small and hence the data involved in the local estimate are from different subjects which are assumed to be independent. This implies that the true covariance matrix for the data contributing to the local estimate is asymptotically diagonal. Thus, the "working independence" LPK-GEE estimator is asymptotically optimal (Lin and Carroll 2000). This is in contrast with the usual parametric GEE (Liang and Zeger 1986) in which the best strategy is to use the true correlation of the data. As mentioned in Hoover et al. (1998), we should interpret the asymptotic results with caution since in real data applications, the proper bandwidth selected by a bandwidth selector is usually not so small and the asymptotic results may not apply. In other words, properly taking the correlation into account may be needed for finite sample data analysis.

One may note that the LPK-GEE method uses kernel weight to control biases. In order to reduce the biases, all the data located far from the estimation point are down weighted although these data may contain useful information due to the correlation with the data near the estimation point from the same subject. Thus, the estimation efficiency may be lost since it is difficult to control the biases and to reduce the variance simultaneously. To tackle this problem, Wang (2003) proposed a two-step procedure. The basic idea is as follows: To efficiently use all the correlated data within a subject, once a data point from one subject or cluster is near the estimation point (say, within $\pm h$) and significantly contributes to the local estimation, all data points from this subject or cluster will be used. To avoid bias, the contributions of all these data points except the data point near the local estimation point are through their residuals. Define $\mathbf{G}_{ij}$ to be an $n_i \times (p+1)$ matrix with the j-th row to be $\mathbf{x}_{ij}^T = [1, t_{ij} - t, \cdots, (t_{ij} - t)^p]$ and 0 otherwise. The two-step procedure for the NPM model (4.2) can be described as follows (Wang 2003):

Step 1. Obtain an initial consistent estimator of $\eta(t)$, say $\tilde{\eta}(t)$. For example, the working independence estimator can be taken as $\tilde{\eta}(t)$.

Step 2. Obtain the final estimate of $\eta(t)$, say $\hat{\eta}(t) = \hat{\beta}_0(t)$, by solving the kernel weighted estimating equation

$$\sum_{i=1}^{n}\sum_{j=1}^{n_i} K_h(t_{ij} - t)\mathbf{G}_{ij}^T \mathbf{\Omega}_i^{-1}[\mathbf{y}_i - \boldsymbol{\eta}_{ij}^*(\boldsymbol{\beta})] = 0, \tag{4.15}$$

where the l-th element of $\boldsymbol{\eta}_{ij}^*(\boldsymbol{\beta})$ is $\mathbf{x}_{ij}^T\boldsymbol{\beta}$ when $l = j$ with t_{ij} being within $\pm h$ of the time point t; and the l-th element of $\boldsymbol{\eta}_{ij}^*(\boldsymbol{\beta})$ is $\tilde{\eta}(t_{il})$ when $l \neq j$.

The structure of $\boldsymbol{\eta}_{ij}^*(\boldsymbol{\beta})$ is so designed for that, for a y_{il} whose measurement time t_{il} is not within h of t, the residual $y_{il} - \tilde{\eta}(t_{il})$, rather than y_{il}, contributes to the local estimate $\hat{\eta}(t) = \hat{\beta}_0(t)$. This will guarantee the proposed estimator to be asymptotically unbiased at worst.

For the NPM model (4.2), we can express the two-step estimator as

$$\hat{\eta}(t) = \frac{\sum_{i=1}^{n}\sum_{j=1}^{n_i} K_h(t_{ij} - t)[w_i^{jj} y_{ij} + \sum_{l\neq j} w_i^{jl}\{y_{il} - \tilde{\eta}(t_{il})\}]}{\sum_{i=1}^{n}\sum_{j=1}^{n_i} K_h(t_{ij} - t) w_i^{jj}}, \tag{4.16}$$

where w_i^{jl} denotes the (j,l)-th entry of $\mathbf{\Omega}_i^{-1}$ with $\mathbf{\Omega}_i$ being the working covariance matrix for the i-th subject. Comparing (4.16) to the working independence estimator $\tilde{\eta}(t)$, namely

$$\tilde{\eta}(t) = \frac{\sum_{i=1}^{n}\sum_{j=1}^{n_i} K_h(t_{ij} - t) w_i^{jj} y_{ij}}{\sum_{i=1}^{n}\sum_{j=1}^{n_i} K_h(t_{ij} - t) w_i^{jj}}, \tag{4.17}$$

we see that the correlated data but not within h of t are incorporated in the two-step estimator by adding their weighted residuals obtained from the first step, and the weight is their correlation (covariance) to the j-th data point which is within h of t. The advantage of the two-step estimator is a reduction of variance without enlarging biases at least asymtotically. The above two-step method may be further improved by iterating the two steps. However, theoretical investigations show, to the first order, that the two-step estimator achieves similar asymptotic properties to the fully iterated estimator. Wang (2003) shown that the two-step estimator uniformly outperforms the working independence estimator (Lin and Carroll 2000) in terms of the asymptotic variance if the true covariance is correctly specified.

Wang's two-step method provides a smart way to incorporate within-subject correlations of longitudinal data in order to efficiently use the available data to improve the working independence estimator. However, the use of within $\pm h$ of t_0 for determining whether the data or their residuals should be used to estimate $\eta(t_0)$ is quite arbitrary. We do not know how this would affect the bandwidth selection. In order to implement Wang's method, the working covariance needs to be separately estimated. In Section 4.4, we will introduce the mixed-effect modeling approach to incorporate the within-subject correlations in a more natural way.

Chen and Jin (2005) recently proposed to simply use the local generalized least squares (GLS) method to accout for longitudinal data correlations. Their method is nothing new and can be considered as a special case of the local polynomial mixed-effects model described in Section 4.4. In addition, their method also requires to determine or estimate the covariance matrix separately, and an accurate estimate of the covariance matrix is usually difficult to obtain.

4.2.3 Fan-Zhang's Two-Step Method

Fan and Zhang (2000) proposed another two-step method to estimate time-varying coefficients in a time-varying coefficient model. For the NPM model (4.2), the two-step method essentially consists of an "averaging" step and a "smoothing" step.

Specifically, let $\tau_1, \tau_2, \cdots, \tau_M$ be the distinct design time points for all the subjects. For each τ_r, define the set $\mathcal{D}_r = \{(i,j)|t_{ij} = \tau_r\}$ and let m_r denote the number of elements in $\mathcal{D}_r$. It is clear that $m_r \geq 1$ and typically, most m_r are larger than 1. Then, we can define the average of the responses as

$$\bar{y}_r = \sum_{(i,j)\in\mathcal{D}_r} y_{ij}/m_r, \quad r = 1, 2, \cdots, M.$$

This is the "averaging " step. The "smoothing" step is done via smoothing

$$(\tau_r, \bar{y}_r), r = 1, 2, \cdots, M, \tag{4.18}$$

using a smoothing method such as LPK smoothing discussed in Section 3.2 of Chapter 3. Fan and Zhang (2000) suggested that the heteroscedasticity of the working data (4.18) should also be taken into account properly. When all the subjects have the same design timc points, i.e., $t_{ij} = \tau_j, j = 1, 2, \cdots, M, i = 1, 2, \cdots, n$, the two-step method used here essentially reduces to the method proposed by Hart and Wehrly (1986) for repeated measurements.

4.3 NONPARAMETRIC MIXED-EFFECTS MODEL

In the above section, we reviewed three popular nonparametric techniques for fitting the NPM model (4.2) to longitudinal data. A critical problem of the above techniques is that the features of longitudinal data are not incorporated in the estimators directly and estimates of the individual functions are not considered. In many longitudinal studies, estimation and inference of individual functions are as important as the population mean function. In this section, we extend the NPM model (4.2) to a model that incorporates the population mean function and the individual functions of the longitudinal data simultaneously. The new model can be expressed as

$$\begin{array}{l} y_{ij} = \eta(t_{ij}) + v_i(t_{ij}) + \epsilon_i(t_{ij}), \\ j = 1, 2, \cdots, n_i; \ i = 1, 2, \cdots, n, \end{array} \tag{4.19}$$

where as in the NPM model (4.2), $\eta(t)$ models the smooth population mean function of the longitudinal data, also called a fixed-effect function; $v_i(t)$ models the departure of the i-th individual function from the population mean function $\eta(t)$, called the i-th individual (subject-specified) effects or random-effect function; and $\epsilon_i(t)$ the measurement error function that can not be explained by either the fixed-effect or the random-effect functions. It is easy to see that the error term, $e_i(t)$, of the model (4.2), now becomes two terms, $v_i(t)$ and $\epsilon_i(t)$, of the new model (4.19). The model (4.19) is called a nonparametric mixed-effects (NPME) model since both the fixed-effect and random-effect functions are nonparametric.

For convenience, we often assume that the unobservable random-effect functions $v_i(t), \ i = 1, 2, \cdots, n$ are i.i.d copies of an underlying smooth process (SP) $v(t)$ with a mean function 0 and a covariance function $\gamma(s, t)$, and that the unobservable measurement error processes $\epsilon_i(t)$ are i.i.d copies of an uncorrelated white noise process $\epsilon(t)$ with a mean function 0 and a covariance function $\gamma_\epsilon(s, t) = \sigma^2(t)1_{\{s=t\}}$. That is, $v(t) \sim \text{SP}(0, \gamma)$ and $\epsilon(t) \sim \text{SP}(0, \gamma_\epsilon)$. In this book, when we deal with likelihood-based or Bayesian inferences, we usually assume that the associated processes are Gaussian, i.e.,

$$v(t) \sim \text{GP}(0, \gamma), \quad \epsilon \sim \text{GP}(0, \gamma_\epsilon). \tag{4.20}$$

Notice that $\eta(t), \gamma(s, t)$, and $\sigma^2(t)$ characterize the overall features of a longitudinal population so that they are "population features", while the random-effect functions $v_i(t), i = 1, 2, \cdots, n$ and the individual functions $s_i(t), \ i = 1, 2, \cdots, n$ are subject-specific so that they are "individual features". The main aim of NPME modeling is to estimate the population effect and to predict the individual effects for a longitudinal study. For simplicity, the population mean function $\eta(t)$ and the individual functions $s_i(t)$ are also referred to as population and individual curves. Because the target quantities $\eta(t)$, $\gamma(s, t)$ and $\sigma^2(t)$ are all nonparametric, NPME modeling requires a combination of a smoothing technique and a mixed-effects modeling approach.

4.4 LOCAL POLYNOMIAL MIXED-EFFECTS MODELING

In the rest of this chapter, we apply local polynomial kernel (LPK) smoothing techniques to the NPME model (4.19) to analyze longitudinal data. The local likelihood principles (Tibshirani and Hastie 1987) will be used to guide the development of the methodologies.

4.4.1 Local Polynomial Approximation

The target quantities $\eta(t), \gamma(s,t)$ and $\sigma^2(t)$ can be estimated via locally approximating the NPME model (4.19) by a polynomial-based LME model. This can be achieved via the Taylor expansions of $\eta(t)$ and $v_i(t)$ around a neighborhood of interest.

Assume $\eta(t)$ and $v_i(t)$ in the NPME model (4.19) to be smooth, say, they have up to $(p+1)$-times continuous derivatives at each time point within some interval of interest, namely $\mathcal{T}$, where p is a nonnegative integer. By Taylor expansion, for any fixed $t \in \mathcal{T}$, $\eta(t)$ and $v_i(t)$ at t_{ij} can be approximated by a p-th degree polynomial within a neighborhood of t:

$$\begin{aligned} \eta(t_{ij}) &\approx \eta(t) + \eta'(t)(t_{ij} - t) + \cdots + \frac{\eta^{(p)}(t)}{p!}(t_{ij} - t)^p = \mathbf{x}_{ij}^T\beta, \\ v_i(t_{ij}) &\approx v_i(t) + v_i'(t)(t_{ij} - t) + \cdots + \frac{v_i^{(p)}(t)}{p!}(t_{ij} - t)^p = \mathbf{x}_{ij}^T\mathbf{b}_i, \end{aligned} \tag{4.21}$$

where $\mathbf{x}_{ij} = [1, t_{ij} - t, \cdots, (t_{ij} - t)^p]^T, \;\; j = 1, 2, \cdots, n_i; \; i = 1, 2, \cdots, n$, and

$$\begin{aligned} \beta &= \left[\eta(t), \eta'(t), \cdots, \frac{\eta^{(p)}(t)}{p!}\right]^T, \\ \mathbf{b}_i &= \left[v_i(t), v_i'(t), \cdots, \frac{v_i^{(p)}(t)}{p!}\right]^T, \;\; i = 1, 2, \cdots, n. \end{aligned} \tag{4.22}$$

It follows that, within a neighborhood of t, the NPME model (4.19) can be reasonably approximated by a LME model:

$$\begin{gathered} y_{ij} = \mathbf{x}_{ij}^T\beta + \mathbf{x}_{ij}^T\mathbf{b}_i + \epsilon_{ij}, \\ j = 1, 2, \cdots, n_i; \; i = 1, 2, \cdots, n, \end{gathered} \tag{4.23}$$

where ϵ_{ij} denote the measurement and model approximation errors, and $\mathbf{b}_i$ denote the random-effects. Under the Gaussian assumption (4.20),

$$\epsilon = [\epsilon_{i1}, \cdots, \epsilon_{in_i}] \sim N(0, \mathbf{R}_i), \;\; \mathbf{b}_i \sim N(0, \mathbf{D}).$$

Based on the NPME model (4.19), the variance components $\mathbf{R}_i = \mathrm{E}(\epsilon_i\epsilon_i^T) = \mathrm{diag}[\sigma^2(t_{i1}), \cdots, \sigma^2(t_{in_i})]$ and $\mathbf{D} = \mathbf{D}(t) = \mathrm{E}(\mathbf{b}_i\mathbf{b}_i^T)$. Notice that since the fixed-effects vector β and the covariance matrix $\mathbf{D}$ are the functions of the local location t, for convenience, we call them the localized fixed-effects vector and the localized covariance matrix, respectively, or generally the localized parameters.

4.4.2 Local Likelihood Approach

Tibshirani and Hastie (1987) first proposed the local likelihood method. Staniswalis (1989) and Fan, Farmen, and Gijbels (1998) further studied the properties of local kernel-weighted likelihood estimators. In this subsection, we apply the local likelihood method to longitudinal data in which within-subject correlations commonly exist (Park and Wu 2005).

Suppose that $\mathbf{y}_i = [y_{i1}, \cdots, y_{in_i}]^T$ is a $(n_i \times 1)$ vector of observations obtained from the i-th subject at time points $t_{i1}, \cdots, t_{in_i}$ and has a probability density function $p_i(\cdot)$ for $i = 1, 2, \cdots, n$. Then the contribution from the i-th subject to the overall log-likelihood is $l_i(\boldsymbol{\theta}_i; \mathbf{y}_i) = \log p_i(\mathbf{y}_i; \boldsymbol{\theta}_i)$, where $\boldsymbol{\theta}_i$ are unknown parameter vectors to be estimated. The log-likelihood of the observations from all the n subjects is then given by

$$\sum_{i=1}^{n} l_i(\boldsymbol{\theta}_i; \mathbf{y}_i).$$

When $\boldsymbol{\theta}_i$ are localized parameters, e.g., the localized fixed-effects vector β and the localized covariance matrix $\mathbf{D}$ described in the previous subsection, it is more natural to define the local log-likelihood. A way for doing so is to use the kernel-weighted log-likelihood as discussed by Staniswalis (1989) and Fan et al. (1998), among others.

Let $K_h(\cdot) = K(\cdot/h)/h$ where K is a kernel function and h is a bandwidth. Let $\mathbf{K}_{ih} = \text{diag}[K_h(t_{i1} - t), \cdots, K_h(t_{in_i} - t)]$ be the diagonal matrix of kernel weights in the neighborhood of t for the i-th subject where $i = 1, 2, \cdots, n$. Then the kernel-weighted log-likelihood is defined by

$$l^*(\boldsymbol{\theta}; \mathbf{y}) = \sum_{i=1}^{n} l_i(\boldsymbol{\theta}_i; \mathbf{y}_i, \mathbf{K}_{ih}), \tag{4.24}$$

which is a function of $\mathbf{K}_{ih}$, $i = 1, 2, \cdots, n$ where $\boldsymbol{\theta} = [\boldsymbol{\theta}_1^T, \cdots, \boldsymbol{\theta}_n^T]^T$ and $\mathbf{y} = [\mathbf{y}_1^T, \cdots, \mathbf{y}_n^T]^T$.

As an example, if $n_i \equiv 1$ and $l_i(\boldsymbol{\theta}_i; \mathbf{y}_i, \mathbf{K}_{ih}) = l_i(\boldsymbol{\theta}_i; \mathbf{y}_i)\mathbf{K}_{ih}$, then the kernel-weighted log-likelihood can be written as

$$l^*(\boldsymbol{\theta}; \mathbf{y}) = \sum_{i=1}^{n} \left\{ \log p_i(y_i, \boldsymbol{\theta}_i) K_h(t_i - t) \right\},$$

which is the standard local log-likelihood function for independent data as discussed by Staniswalis (1989) and Fan et al. (1998). In the case of no within-subject correlation, the local weighted log-likelihood (4.24) can be written as

$$l^*(\boldsymbol{\theta}; \mathbf{y}) = \sum_{i=1}^{n} \sum_{j=1}^{n_i} l_i(\boldsymbol{\theta}_i; y_{ij}) K_h(t_{ij} - t)..$$

This coincides with the cases considered by Hoover et al. (1998) and Lin and Carroll (2000).

In general, the form of the local log-likelihood (4.24) is problem-specific. The application of the kernel weight in different ways may result in different estimators. The following subsections illustrate applications of the kernel-weighted log-likelihood (4.24) under different scenarios for NPME models.

4.4.3 Local Marginal Likelihood Estimation

In this subsection, we introduce a local marginal likelihood method for estimating the population mean function $\eta(t)$ (Park and Wu 2005). For the approximation LME model (4.23), let $\mathbf{X}_i = [\mathbf{x}_{i1}, \ldots, \mathbf{x}_{in_i}]^T$ and assume that the Gaussian assumption (4.20) is satisfied. Then, the local marginal distribution of $\mathbf{y}_i$ under the approximation LME model (4.23) is normal with a mean of $\mathbf{X}_i\beta$ and variance of $\mathbf{V}_i = \mathbf{X}_i\mathbf{D}\mathbf{X}_i^T + \mathbf{R}_i$. It thus yields the log-likelihood function for β:

$$L(\beta; \mathbf{y}) = -\frac{1}{2}\sum_{i=1}^{n}\left\{[\mathbf{y}_i - \mathbf{X}_i\beta]^T\mathbf{V}_i^{-1}[\mathbf{y}_i - \mathbf{X}_i\beta] + c_{i1}\right\}, \tag{4.25}$$

where $c_{i1} = \log(|\mathbf{V}_i|) + n_i\log(2\pi)$. Based on the above expression and applying (4.24), we can write the local marginal log-likelihood function for estimating β as

$$l^*(\beta; \mathbf{y}) = -\frac{1}{2}\sum_{i=1}^{n}\left\{[\mathbf{y}_i - \mathbf{X}_i\beta]^T\mathbf{\Psi}_{ih}[\mathbf{y}_i - \mathbf{X}_i\beta] + (\mathbf{1}_{n_i}^T\mathbf{K}_{ih}\mathbf{1}_{n_i})c_{i1}\right\}, \tag{4.26}$$

where $\mathbf{\Psi}_{ih} = \mathbf{K}_{ih}^{1/2}\mathbf{V}_i^{-1}\mathbf{K}_{ih}^{1/2}$ with the kernel weight matrix $\mathbf{K}_{ih}$ weighting the residual vector $(\mathbf{y}_i - \mathbf{X}_i\beta)$ symmetrically.

For given variance matrices $\mathbf{V}_i$, $i = 1, 2, \cdots, n$, the differentiation of (4.26) with respect to β yields the estimating equation for β:

$$\mathbf{X}^T\mathbf{\Psi}_h\mathbf{X}\beta = \mathbf{X}^T\mathbf{\Psi}_h\mathbf{y}, \tag{4.27}$$

where $\mathbf{X} = [\mathbf{X}_1^T, \ldots, \mathbf{X}_n^T]^T$, $\mathbf{y} = [\mathbf{y}_1^T, \ldots, \mathbf{y}_n^T]^T$, and $\mathbf{\Psi}_h = \mathrm{diag}[\mathbf{\Psi}_{1h}, \ldots, \mathbf{\Psi}_{nh}]$. Thus, a closed-form estimator for β is

$$\hat{\beta}_M = \left(\mathbf{X}^T\mathbf{\Psi}_h\mathbf{X}\right)^{-1}\left(\mathbf{X}^T\mathbf{\Psi}_h\mathbf{y}\right). \tag{4.28}$$

When $\mathbf{V}_i$, $i = 1, 2, \cdots, n$ are known, the estimator (4.28) can be obtained by fitting the following model:

$$\mathbf{V}^{-1/2}\mathbf{K}_h^{1/2}\mathbf{y} = \mathbf{V}^{-1/2}\mathbf{K}_h^{1/2}\mathbf{X}\beta + \mathbf{e}, \tag{4.29}$$

using the S-PLUS function *lm* or the SAS procedure *PROC GLM*, where $\mathbf{V} = \mathrm{diag}(\mathbf{V}_1, \cdots, \mathbf{V}_n)$, $\mathbf{K}_h = \mathrm{diag}(\mathbf{K}_{1h}, \cdots, \mathbf{K}_{nh})$, and $\mathbf{e}$ has mean 0 and variance $\sigma_e^2\mathbf{I}_N$. The model (4.29) is a standard linear regression model with the response variable $\mathbf{V}^{-1/2}\mathbf{K}_h^{1/2}\mathbf{y}$ and the covariate $\mathbf{V}^{-1/2}\mathbf{K}_h^{1/2}\mathbf{X}$.

The local marginal likelihood estimator of $\eta(t)$ can be found as

$$\hat{\eta}_M(t) = \mathbf{e}_1^T\hat{\beta}_M, \tag{4.30}$$

where $\mathbf{e}_1$ is a $(p+1)$-dimensional vector with the first element being 1, and 0 elsewhere.

The covariance matrices $\mathbf{V}_i,\ i = 1, 2, \cdots, n$ have been assumed to be known in order to obtain the closed-form estimator (4.28). In practice, we commonly encounter real examples where the covariance matrices are unknown and need to be estimated. The estimation of covariance matrices as well as random-effect curves will be introduced in later sections. When $\mathbf{V}_i,\ i = 1, 2, \cdots, n$ are known diagonal matrices, the estimator $\hat{\beta}$ reduces to the LPK-GEE estimator proposed by Lin and Carroll (2000).

4.4.4 Local Joint Likelihood Estimation

In this section, an alternative estimation approach is proposed to estimate the parameters in the localized LME model (4.23) with longitudinal data (Park and Wu 2005). Under the Gaussian assumption (4.20), we have $\mathbf{y}_i|\mathbf{b}_i \sim N(\mathbf{X}_i\beta + \mathbf{X}_i\mathbf{b}_i, \mathbf{R}_i)$ and $\mathbf{b}_i \sim N(\mathbf{0}, \mathbf{D})$. Thus, the logarithm of the joint density function of $(\mathbf{y}_i, \mathbf{b}_i,\ i = 1, 2, \cdots, n)$ is

$$l(\beta, \mathbf{b}; \mathbf{y}) = -\frac{1}{2}\sum_{i=1}^{n}\left\{\mathbf{r}_i^T\mathbf{R}_i^{-1}\mathbf{r}_i + \mathbf{b}_i^T\mathbf{D}^{-1}\mathbf{b}_i + c_{i3}\right\}, \tag{4.31}$$

where $\mathbf{r}_i = \mathbf{y}_i - \mathbf{X}_i\beta - \mathbf{X}_i\mathbf{b}_i$, $\mathbf{b} = [\mathbf{b}_1^T, \cdots, \mathbf{b}_n^T]^T$ and $c_{i3} = \log(|\mathbf{R}_i|) + \log(|\mathbf{D}|) + 2n_i\log(2\pi)$. Since $\mathbf{b}_i,\ i = 1, 2, \cdots, n$ are random-effects parameter vectors, the $l(\beta, \mathbf{b}; \mathbf{y})$ is not a usual log-likelihood. For convenience, from now on and throughout this book, we call $l(\beta, \mathbf{b}; \mathbf{y})$ a generalized log-likelihood (GLL) of $(\beta, \mathbf{b})$.

Then the localized generalized log-likelihood (LGLL) at the neighborhood of time t can be considered in two different ways:

$$l_1^*(\beta, \mathbf{b}; \mathbf{y}) = -\frac{1}{2}\sum_{i=1}^{n}\left\{\mathbf{r}_i^T\Theta_{ih}\mathbf{r}_i + \mathbf{b}_i^T\mathbf{D}^{-1}\mathbf{b}_i + c_{i3}\right\}, \tag{4.32}$$

and

$$l_2^*(\beta, \mathbf{b}; \mathbf{y}) = -\frac{1}{2}\sum_{i=1}^{n}\left\{\mathbf{r}_i^T\Theta_{ih}\mathbf{r}_i + \mathbf{1}_{n_i}^T\mathbf{K}_{ih}\mathbf{1}_{n_i}[\mathbf{b}_i^T\mathbf{D}^{-1}\mathbf{b}_i + c_{i3}]\right\}, \tag{4.33}$$

where $\Theta_{ih} = \mathbf{K}_{ih}^{1/2}\mathbf{R}_i^{-1}\mathbf{K}_{ih}^{1/2}$, and $\mathbf{1}_{n_i}$ is a $(n_i \times 1)$ vector with all elements being 1's. In (4.32), the kernel weights are symmetrically applied only to the residual terms $\mathbf{r}_i = \mathbf{y}_i - \mathbf{X}_i(\beta + \mathbf{b}_i),\ i = 1, 2, \cdots, n$ of the GLL function, while, in (4.33), the kernel weights are applied to the entire GLL function of (4.31) in which the random-effect terms $\mathbf{b}_i^T\mathbf{D}^{-1}\mathbf{b}_i$ are also multiplied by the kernel weights. These two different kernel weighting methods result in two different estimators.

Minimizing the LGLL criterion (4.32) results in the exact local polynomial mixed-effects (LPME) estimators proposed by Wu and Zhang (2002a), and the associated modeling is termed as LPME modeling. For given $\mathbf{D}$, $\mathbf{R}$ and $\mathbf{K}_h$, solving the minimization problem (4.32) is equivalent to solving the so-called mixed model equation

(Davidian and Giltinan 1995, Zhang et al. 1998):

$$\begin{bmatrix} \mathbf{X}^T\boldsymbol{\Theta}_h\mathbf{X} & \mathbf{X}^T\boldsymbol{\Theta}_h\mathbf{Z} \\ \mathbf{Z}^T\boldsymbol{\Theta}_h\mathbf{X} & \mathbf{Z}^T\boldsymbol{\Theta}_h\mathbf{Z}+\tilde{\mathbf{D}}^{-1} \end{bmatrix} \begin{bmatrix} \boldsymbol{\beta} \\ \mathbf{b} \end{bmatrix} = \begin{bmatrix} \mathbf{X}^T\boldsymbol{\Theta}_h\mathbf{y} \\ \mathbf{Z}^T\boldsymbol{\Theta}_h\mathbf{y} \end{bmatrix},$$

where $\mathbf{y}$ and $\mathbf{X}$ are defined as in the previous subsection, and $\mathbf{Z} = \text{diag}[\mathbf{X}_1, \cdots, \mathbf{X}_n]$, $\tilde{\mathbf{D}} = \text{diag}[\mathbf{D}, \cdots, \mathbf{D}]$, and $\boldsymbol{\Theta}_h = \text{diag}[\Theta_{1h}, \cdots, \Theta_{nh}]$.

Then the resulting LPME estimators for $\boldsymbol{\beta}, \mathbf{b}_i$, $i = 1, 2, \cdots, n$ are

$$\left\{ \begin{array}{rcl} \hat{\boldsymbol{\beta}} & = & \left\{\sum_{i=1}^n \mathbf{X}_i^T\boldsymbol{\Omega}_{ih}\mathbf{X}_i\right\}^{-1}\left\{\sum_{i=1}^n \mathbf{X}_i^T\boldsymbol{\Omega}_{ih}\mathbf{y}_i\right\}, \\ \hat{\mathbf{b}}_i & = & \left\{\mathbf{X}_i^T\boldsymbol{\Theta}_{ih}\mathbf{X}_i+\mathbf{D}^{-1}\right\}^{-1}\mathbf{X}_i^T\boldsymbol{\Theta}_{ih}(\mathbf{y}_i-\mathbf{X}_i\hat{\boldsymbol{\beta}}), \\ & & i=1,2,\cdots,n, \end{array} \right. \tag{4.34}$$

where

$$\left\{ \begin{array}{rcl} \boldsymbol{\Omega}_{ih} & = & \mathbf{K}_{ih}^{1/2}\mathbf{V}_i^{-1}\mathbf{K}_{ih}^{1/2}, \\ \mathbf{V}_i & = & \mathbf{K}_{ih}^{1/2}\mathbf{X}_i\mathbf{D}\mathbf{X}_i^T\mathbf{K}_{ih}^{1/2}+\mathbf{R}_i. \end{array} \right. \tag{4.35}$$

In matrix notation, the above estimators can be written in a more compact form:

$$\left\{ \begin{array}{rcl} \hat{\boldsymbol{\beta}} & = & \left(\mathbf{X}^T\boldsymbol{\Omega}_h\mathbf{X}\right)^{-1}\mathbf{X}^T\boldsymbol{\Omega}_h\mathbf{y}, \\ \hat{\mathbf{b}} & = & \left(\mathbf{Z}^T\boldsymbol{\Theta}_h\mathbf{Z}+\tilde{\mathbf{D}}^{-1}\right)^{-1}\mathbf{Z}^T\boldsymbol{\Theta}_h(\mathbf{y}-\mathbf{X}\hat{\boldsymbol{\beta}}), \end{array} \right. \tag{4.36}$$

where $\boldsymbol{\Omega}_h = \text{diag}(\boldsymbol{\Omega}_{1h}, \cdots, \boldsymbol{\Omega}_{nh})$. In the subsequent sections, we will focus on these estimators.

Similarly we can derive the LPME estimators based on the LGLL criterion (4.33). In fact, for given $\mathbf{D}$, $\mathbf{R}$, and $\mathbf{K}_h$, the LPME estimators obtained by maximizing (4.32) and (4.33) can be written in a unified form, which is the solution to the following mixed-model normal equations:

$$\left\{ \begin{array}{l} \sum_{i=1}^n \mathbf{X}_i^T\boldsymbol{\Theta}_{ih}\mathbf{X}_i(\boldsymbol{\beta}+\mathbf{b}_i) = \sum_{i=1}^n \mathbf{X}_i^T\boldsymbol{\Theta}_{ih}\mathbf{y}_i, \\ \mathbf{X}_i^T\boldsymbol{\Theta}_{ih}\mathbf{X}_i\boldsymbol{\beta} + (\mathbf{X}_i^T\boldsymbol{\Theta}_{ih}\mathbf{X}_i+\omega_{ih}\mathbf{D}^{-1})\mathbf{b}_i = \mathbf{X}_i^T\boldsymbol{\Theta}_{ih}\mathbf{y}_i, \\ i=1,2,\cdots,n, \end{array} \right. \tag{4.37}$$

where $\omega_{ih} = 1$ and $\omega_{ih} = \mathbf{1}_{n_i}^T\mathbf{K}_{ih}\mathbf{1}_{n_i}$ corresponding to the estimators derived from the LGLL criteria (4.32) and (4.33), respectively. By solving the above normal equations (4.37), the LPME estimators for $\boldsymbol{\beta}$ and $\mathbf{b}_i$, under the assumptions of known $\mathbf{D}$ and $\mathbf{R}, \mathbf{K}_h$, can be written as the following closed forms:

$$\hat{\boldsymbol{\beta}}_J = \left[\sum_{i=1}^n \mathbf{X}_i^T\mathbf{K}_{ih}^{1/2}\boldsymbol{\Sigma}_{ih}^{-1}\mathbf{K}_{ih}^{1/2}\mathbf{X}_i\right]^{-1}\left[\sum_{i=1}^n \mathbf{X}_i^T\mathbf{K}_{ih}^{1/2}\boldsymbol{\Sigma}_{ih}^{-1}\mathbf{K}_{ih}^{1/2}\mathbf{y}_i\right], \tag{4.38}$$

and

$$\hat{\mathbf{b}}_i = \left[\mathbf{X}_i^T\boldsymbol{\Theta}_{ih}\mathbf{X}_i+\omega_{ih}\mathbf{D}^{-1}\right]^{-1}\left[\mathbf{X}_i^T\boldsymbol{\Theta}_{ih}(\mathbf{y}_i-\mathbf{X}_i\hat{\boldsymbol{\beta}})\right], \tag{4.39}$$

where $\mathbf{\Sigma}_{ih} = \omega_{ih}^{-1}\mathbf{K}_{ih}^{\frac{1}{2}}\mathbf{X}_i\mathbf{D}\mathbf{X}_i^T\mathbf{K}_{ih}^{\frac{1}{2}} + \mathbf{R}_i$. Thus, the estimators of $\eta(t)$ and $v_i(t)$ can be found as

$$\hat{\eta}_{_J}(t) = \mathbf{e}_1^T\hat{\boldsymbol{\beta}}_J \text{ and } \hat{v}_i(t) = \mathbf{e}_1^T\hat{\mathbf{b}}_i \quad i = 1, 2, \cdots, n. \tag{4.40}$$

One may notice that the difference between the local marginal likelihood estimator (4.28) and the estimator (4.38) for the population parameter $\boldsymbol{\beta}$ is due to different weight functions. In the estimates of random-effects parameters (4.39), the population parameter $\boldsymbol{\beta}$ can be replaced by any consistent estimators, such as (4.28) or (4.38). In fact, $\mathbf{b}_i$ is an empirical Bayes estimator or a best linear unbiased predictor (BLUP), see Davidian and Giltinan (1995) and Vonesh and Chinchilli (1996) for details. The estimates of the random-effects allow us to capture the individual response curves, $s_i(t) = \eta(t) + v_i(t)$, which is a great advantage of NPME modeling. One can also easily see that, from (4.37) with $\omega_{ih} = 1$ and $\omega_{ih} = \mathbf{1}_{n_i}^T\mathbf{K}_{ih}\mathbf{1}_{n_i}$, the application of different kernel weights may result in different local likelihood estimators. These estimators may have different properties and efficiencies.

In the subsequent discussions, we focus our attention on the LPME estimators (4.34). However, the developed methodologies may similarly apply to the general estimators (4.38) and (4.39). One of the advantages of LPME modeling is that it can be easily implemented using existing software for LME models. In fact, for each given t, the LPME estimators (4.34) can be obtained via operationally fitting the following standard LME model:

$$\begin{gathered} \tilde{y}_{ij} = \tilde{\mathbf{x}}_{ij}^T\boldsymbol{\beta} + \tilde{\mathbf{x}}_{ij}^T\mathbf{b}_i + \epsilon_{ij}, \\ \mathbf{b}_i \sim N(\mathbf{0}, \mathbf{D}),\ \boldsymbol{\epsilon}_i = [\epsilon_{i1}, \cdots, \epsilon_{in_i}]^T \sim N(\mathbf{0}, \mathbf{R}_i), \\ j = 1, 2, \cdots, n_i;\ i = 1, 2, \cdots, n, \end{gathered} \tag{4.41}$$

where $\tilde{y}_{ij} = K_h^{1/2}(t_{ij} - t)y_{ij}$ and $\tilde{\mathbf{x}}_{ij} = K_h^{1/2}(t_{ij} - t)\mathbf{x}_{ij}$. The former is treated as the response variable, while the latter is treated as the fixed-effects and random-effects covariates. They are actually the localized response variable, fixed-effects and random-effects covariates at the given time point t. The LPME estimators (4.34) and their standard deviations can then be obtained via fitting (4.41) using the S-PLUS function *lme* or the SAS procedure *PROC MIXED*.

4.4.5 Component Estimation

From (4.22) and (4.34), one easily obtains the LPME estimators of $\eta(t), v_i(t)$ and their q-th derivatives:

$$\left\{ \begin{array}{rcl} \hat{\eta}^{(q)}(t) & = & q!\mathbf{e}_{q+1}^T\hat{\boldsymbol{\beta}}, \\ \hat{v}_i^{(q)}(t) & = & q!\mathbf{e}_{q+1}^T\hat{\mathbf{b}}_i, \quad i = 1, 2, \cdots, n, \end{array} \right. \tag{4.42}$$

for $q = 0, 1, \cdots, p$. In particular, $\hat{\eta}(t) = \hat{\eta}^{(0)}(t)$ and $\hat{v}_i(t) = \hat{v}_i^{(0)}(t)$ are the LPME estimators of $\eta(t)$ and $v_i(t)$. Using some simple algebra, we have developed the following proposition.

Proposition 4.1 *The LPME estimators of $\eta(t)$ and $v_i(t)$ from (4.34) can be written as*

$$\begin{cases} \hat{\eta}(t) = \mathbf{e}_1^T \left\{ \sum_{i=1}^n (\mathbf{I} + \mathbf{G}_i \mathbf{D})^{-1} \mathbf{G}_i \right\}^{-1} \sum_{i=1}^n (\mathbf{I} + \mathbf{G}_i \mathbf{D})^{-1} \mathbf{c}_i, \\ \hat{v}_i(t) = \mathbf{e}_1^T \mathbf{D} (\mathbf{I} + \mathbf{G}_i \mathbf{D})^{-1} \mathbf{d}_i, \quad i = 1, 2, \cdots, n, \end{cases} \tag{4.43}$$

where

$$\mathbf{G}_i = \begin{pmatrix} s_{i0} & s_{i1} & \cdots & s_{ip} \\ s_{i1} & s_{i2} & \cdots & s_{i(p+1)} \\ \vdots & \vdots & \vdots & \vdots \\ s_{ip} & s_{i(p+1)} & \cdots & s_{i(2p)} \end{pmatrix},$$

$$\mathbf{c}_i = \begin{pmatrix} c_{i0} \\ c_{i1} \\ \vdots \\ c_{ip} \end{pmatrix}, \quad \mathbf{d}_i = \begin{pmatrix} d_{i0} \\ d_{i1} \\ \vdots \\ d_{ip} \end{pmatrix},$$

$$\begin{aligned} s_{ir} &= \sum_{j=1}^{n_i} K_h(t_{ij} - t)(t_{ij} - t)^r \sigma^{-2}(t_{ij}), \quad r = 0, \cdots, 2p, \\ c_{ir} &= \sum_{j=1}^{n_i} K_h(t_{ij} - t)(t_{ij} - t)^r \sigma^{-2}(t_{ij}) y_{ij}, \quad r = 0, \cdots, p, \\ d_{ir} &= \sum_{j=1}^{n_i} K_h(t_{ij} - t)(t_{ij} - t)^r \sigma^{-2}(t_{ij}) [y_{ij} - \mathbf{x}_{ij}^T \hat{\boldsymbol{\beta}}], \quad r = 0, \cdots, p. \end{aligned}$$

The estimate of $\sigma^2(t)$ can be obtained directly from fitting the model (4.41), and we may estimate $\gamma(s,t)$ by the method of moments, e.g.,

$$\hat{\gamma}(s,t) = (n-1)^{-1} \sum_{i=1}^n \hat{v}_i(s) \hat{v}_i(t). \tag{4.44}$$

Based on $\hat{\sigma}^2(t)$ and $\hat{\gamma}(s,t)$, further inferences can be made. For example, one can perform principal component analysis (PCA) on the longitudinal data based on the singular value decomposition of $\hat{\gamma}(s,t)$. Moreover, $\hat{\sigma}^2(t)$ and $\hat{\gamma}(s,t)$ may be used to conduct hypothesis testing about $\eta(t)$. Further research in this direction is warranted.

4.4.6 A Special Case: Local Constant Mixed-Effects Model

Local constant mixed-effects (LCME) modeling is a special case of the general LPME modeling of Section 4.4.1 when the associated local polynomial degree $p = 0$. We single out this special case for further study because it provides more insight on the LPME modeling approach.

In this special case, for a given t, in the Taylor expansions (4.21), we have $\mathbf{x}_{ij} = 1$, $\boldsymbol{\beta} = \eta(t)$ and $\mathbf{b}_i = v_i(t)$. That is, $\eta(t_{ij})$ and $v_i(t_{ij})$ are locally approximated by two constants $\eta(t)$ and $v_i(t)$, respectively so that within a neighborhood of t, the NPME model (4.19) is no longer approximated by (4.41) but by the following LME model:

$$y_{ij} = \eta(t) + v_i(t) + \epsilon_{ij}, \quad j = 1, 2, \cdots, n_i;\ i = 1, 2, \cdots, n, \tag{4.45}$$

where $v_i(t) \sim N(\mathbf{0}, \mathbf{D})$ with $\mathbf{D} = \delta^2(t)$ being a scalar function of t. It follows that the LGLL criterion (4.32) now becomes

$$\begin{aligned}\sum_{i=1}^n \Big\{ [\mathbf{y}_i - \mathbf{1}_{n_i}\eta(t) - \mathbf{1}_{n_i}v_i(t)]^T \mathbf{K}_{ih}^{1/2}\mathbf{R}_i^{-1}\mathbf{K}_{ih}^{1/2}[\mathbf{y}_i - \mathbf{1}_{n_i}\eta(t) - \mathbf{1}_{n_i}v_i(t)] \\ + v_i^2(t)/\delta^2(t) + \log|\delta^2(t)| + \log|\mathbf{R}_i| \Big\}.\end{aligned} \tag{4.46}$$

By Proposition 4.1 and noticing that now

$$\begin{aligned}\mathbf{G}_i &= s_{i0} = \sum_{j=1}^{n_i} K_h(t_{ij} - t)\sigma^{-2}(t_{ij}), \\ \mathbf{c}_i &= c_{i0} = \sum_{j=1}^{n_i} K_h(t_{ij} - t)\sigma^{-2}(t_{ij})y_{ij}, \\ \mathbf{d}_i &= d_{i0} = \sum_{j=1}^{n_i} K_h(t_{ij} - t)\sigma^{-2}(t_{ij})[y_{ij} - \hat{\eta}(t)],\end{aligned}$$

we have the LCME estimators:

$$\begin{cases} \hat{\eta}(t) = \dfrac{\sum_{i=1}^n \sum_{j=1}^{n_i} w_{ij}(t)y_{ij}}{\sum_{i=1}^n \sum_{j=1}^{n_i} w_{ij}(t)}, \\ \hat{v}_i(t) = \dfrac{\sum_{j=1}^{n_i} w_{ij}(t)[y_{ij} - \hat{\eta}(t)]}{\sum_{j=1}^{n_i} w_{ij}(t)}, \quad i = 1, 2, \cdots, n, \end{cases} \tag{4.47}$$

where the weights

$$w_{ij}(t) = \frac{\delta^2(t)K_h(t_{ij} - t)\sigma^{-2}(t_{ij})}{1 + \delta^2(t)\sum_{l=1}^{n_i} K_h(t_{il} - t)\sigma^{-2}(t_{il})}. \tag{4.48}$$

We can see that the LCME estimators are linear estimators with weights constructed locally by taking both between-subject variation $[\delta^2(t)]$ and the within-subject variation $[\sigma^2(t_{ij})]$ into account. When $\delta^2(t)$ is large and $\sigma^2(\cdot)$ is a constant, the weights $w_{ij}(t) \approx K_h(t_{ij} - t)$ so that $\hat{\eta}(t)$ reduces to the usual kernel estimator (4.8) for $\eta(t)$ as discussed briefly in Section 4.2.1 and in greater detail in Hoover et al. (1998). It is also seen that $\hat{\eta}(t)$ involves the whole data set while $\hat{v}_i(t)$ involves mainly the data from the i-th subject, and it borrows information from other subjects mainly through $\hat{\eta}(t)$.

Set

$$A_i(t) = \sum_{j=1}^{n_i} w_{ij}(t), \quad B_i(t) = \sum_{j=1}^{n_i} w_{ij}(t)y_{ij}. \tag{4.49}$$

Then in a compact manner, we have

$$\begin{cases} \hat{\eta}(t) &= \sum_{i=1}^n B_i(t)/\sum_{i=1}^n A_i(t), \\ \hat{v}_i(t) &= B_i(t)/A_i(t) - \hat{\eta}(t), \quad i = 1, 2, \cdots, n. \end{cases} \tag{4.50}$$

These expressions allow fast implementation of the LCME estimation. They also can be used to estimate the individual functions $\hat{s}_i(t) = \hat{\eta}(t) + \hat{v}_i(t) = B_i(t)/A_i(t)$ quickly.

4.5 CHOOSING GOOD BANDWIDTHS

For simplicity of discussion, in the previous section, the kernel K and the bandwidth h are assumed to be given and fixed. In practice, h needs to be carefully chosen. When h is too small, the resulting LPME estimators $\hat{\eta}(t)$ and $\hat{v}_i(t)$ are usually very noisy, and when h is too large, $\hat{\eta}(t)$ and $\hat{v}_i(t)$ may oversmooth the data so that some important information in the data is not sufficiently captured. In this section, we shall discuss how to choose good bandwidths for LPME estimators.

First of all, by (4.34), it is easy to see that the whole data set is involved in the population estimators $\hat{\eta}(t)$ while only the data from subject i are mainly involved in the individual curve estimators for the i-th subject, i.e., $\hat{s}_i(t) = \hat{\eta}(t) + \hat{v}_i(t)$. Therefore, different bandwidths for estimating $\eta(t)$ and $s_i(t)$ should be used to account for the different amounts of data involved. Following Rice and Silverman (1991), the "leave-one-subject-out" cross-validation (SCV) criterion may be used to select a proper bandwidth for estimating $\eta(t)$. For a longitudinal data set, it is known that, conditional to a particular subject, say subject i, the measurements from subject i are uncorrelated or independent; moreover, the measurements' conditional mean function is exactly the individual curve $s_i(t)$. In this case, the usual "leave-one-point-out" cross-validation (PCV) criterion, which is traditionally proposed for uncorrelated or independent data, seems to be appropriate for selecting good bandwidths for estimating $s_i(t)$. For simplicity, a common bandwidth for estimating $s_i(t)$ for all subjects will be used because the $s_i(t)$ are assumed to be from the same underlying process and hence can be assumed to have similar smoothness in general.

4.5.1 Leave-One-Subject-Out Cross-Validation

The SCV score is defined as

$$\text{CV}_s(h) = \sum_{i=1}^{n} \sum_{j=1}^{n_i} \left\{ y_{ij} - \hat{\eta}^{(-i)}(t_{ij}) \right\}^2 /(nn_i), \tag{4.51}$$

where $\hat{\eta}^{(-i)}(t)$ stands for the estimator of $\eta(t)$ based on the data with the measurements from subject i entirely excluded, and the weights $1/(nn_i)$ take the number of measurements from individual subjects into account. The optimal SCV bandwidth h_s^* is defined as the minimizer of $\text{CV}_s(h)$. Rice and Silverman (1991) pointed out that the SCV is more appropriate for estimating the population (mean) curve than the PCV. Hart and Wehrly (1993) showed that the SCV bandwidth is consistent.

It is computationally intensive to compute the SCV criterion (4.51) since we need to repeatedly compute the LPME model fit n times in order to obtain $\hat{\eta}^{(-i)}(t_{ij})$; each fit takes about the same amount of computational effort as for computing $\hat{\eta}(t)$ using the whole data set. To overcome this problem, an approximation for $\hat{\eta}^{(-i)}(t)$ may be used. For a given bandwidth h, all data may be used to estimate $\mathbf{V}_i$ or $\mathbf{\Omega}_{ih}$ (4.35), namely $\hat{V}_i$ or $\hat{\mathbf{\Omega}}_{ih}$, then $\hat{\eta}^{(-i)}(t)$ is approximately obtained from the closed-form solution (4.42) for the estimate of $\eta(t)$ by deleting the term that involves the i-th subject. That is,

$$\hat{\eta}^{(-i)}(t) \approx \mathbf{e}_1^T \left\{ \sum_{k \neq i} \mathbf{X}_k^T \hat{\mathbf{\Omega}}_{kh} \mathbf{X}_k \right\}^{-1} \left\{ \sum_{k \neq i} \mathbf{X}_k^T \hat{\mathbf{\Omega}}_{kh} \mathbf{y}_k \right\}.$$

In particular, for the LCME estimator (when $p = 0$), by (4.50), this can be written as

$$\hat{\eta}^{(-i)}(t) \approx \frac{\sum_{k \neq i} B_k(t)}{\sum_{k \neq i} A_k(t)},$$

where the formulas for $A_k(t)$ and $B_k(t)$ are given in (4.49), with $\delta^2(t)$ and $\sigma^2(t)$ being estimated from all of the data. Thus, the approximation only requires to fit the LPME model once to compute the SCV score (4.51) for all subjects, and hence the computational effort is much less.

4.5.2 Leave-One-Point-Out Cross-Validation

The PCV criterion is defined as follows. Assume $\tau_1, \tau_2, \cdots, \tau_M$ to be all the distinct design time points for the whole data set. For a given τ_r, we assume subjects $i_1, i_2, \cdots, i_{m_r}$ to have measurements at τ_r:

$$y_{i_l}(\tau_r) = s_{i_l}(\tau_r) + \epsilon_{i_l}(\tau_r), l = 1, 2, \cdots, m_r.$$

Let $\hat{s}_{i_l}^{(-r)}(\tau_r)$ be the estimators of $s_{i_l}(\tau_r)$ when all the data at the design time point τ_r are excluded. Then the PCV-score is defined as

$$\mathrm{CV}_p(h) = \sum_{r=1}^{M} \sum_{l=1}^{m_r} \left[y_{i_l}(\tau_l) - \hat{s}_{i_l}^{(-r)}(\tau_r) \right]^2 / (Mm_r), \tag{4.52}$$

where the weights $1/(Mm_r)$ take the number of measurements at τ_r into account. The optimal PCV bandwidth h_p^* is defined as the minimizer of $\mathrm{CV}_p(h)$.

4.5.3 Bandwidth Selection Strategies

As pointed out by Diggle and Hutchinson (1989), Zeger and Diggle (1994), Rice and Silverman (1991), Hart and Wehrly (1993) and others, use of the usual PCV bandwidth tracks the individual subject's curve closely while use of the SCV bandwidth tracks the population (mean) curve better. To more efficiently estimate both population and individual curves, combining the PCV and SCV bandwidth selection strategies for estimating $\eta(t)$ and $s_i(t)$ is necessary. The following methods are based on various combinations of the PCV and SCV bandwidths for estimating $\eta(t), v_i(t)$ and $s_i(t)$.

Method 1 (the SCV method): Use the SCV bandwidth h_s^* for estimating $\eta(t), v_i(t)$ and $s_i(t)$ simultaneously, resulting in the estimators $\hat{\eta}_{\text{subj}}(t), \hat{v}_{\text{subj},i}(t)$ and $\hat{s}_{\text{subj},i}(t)$.

Method 2 (the PCV method): Use the PCV bandwidth h_p^* for estimating $\eta(t), v_i(t)$ and $s_i(t)$ simultaneously, resulting in the estimators $\hat{\eta}_{\text{pt}}(t), \hat{v}_{\text{pt},i}(t)$ and $\hat{s}_{\text{pt},i}(t)$.

Method 3 (the hybrid bandwidth (HB) method): The resulting estimators are defined as $\hat{\eta}_{\text{subj+pt}}(t) = \hat{\eta}_{\text{subj}}(t)$, $\hat{v}_{\text{subj+pt}}(t) = \hat{v}_{\text{pt},i}(t)$, and

$$\hat{s}_{\text{subj+pt},i}(t) = \hat{\eta}_{\text{subj}}(t) + \hat{v}_{\text{pt},i}(t).$$

Method 4 (the bias-corrected hybrid bandwidth (BCHB) method):

Step 1 Construct the residuals:

$$r_{ij} = y_{ij} - \hat{\eta}_{\text{subj}}(t_{ij}), \quad j = 1, 2, \cdots, n_i;\ i = 1, 2, \cdots, n. \tag{4.53}$$

Step 2 Fit the following NPME model:

$$r_{ij} = \Delta(t_{ij}) + v_i(t_{ij}) + \epsilon_{ij}, \tag{4.54}$$

using the associated PCV bandwidth, where $\Delta(t_{ij})$ denotes the mean bias of the population curve estimate. The resulting estimators are denoted by $\hat{\Delta}_{\text{subj/pt}}(t)$ and $\hat{v}_{\text{subj/pt},i}(t)$, respectively.

Step 3 Define $\hat{\eta}_{\text{subj/pt}}(t) = \hat{\eta}_{\text{subj}}(t) + \hat{\Delta}_{\text{subj/pt}}(t)$, and $\hat{s}_{\text{subj/pt},i}(t) = \hat{\eta}_{\text{subj/pt}}(t) + \hat{v}_{\text{subj/pt},i}(t)$.

The rationale of the above BCHB method can be stated as follows. Notice that $\hat{\eta}_{\text{subj}}(t)$ and $\hat{v}_{\text{subj},i}(t)$ are affected mutually through the estimate of V_i where $\hat{v}_{\text{subj},i}(t)$ are inappropriately obtained using the SCV bandwidth. The resulting $\hat{\eta}_{\text{subj}}(t)$ may be biased. The bias may be corrected via fitting the NPME model (4.54) for $\Delta(t)$ and $v_i(t)$ using the more appropriate PCV bandwidth where $\Delta(t)$ denotes the underlying bias of $\hat{\eta}_{\text{subj}}(t)$. Since $\Delta(t)$ is very close to 0, a big bandwidth such as the PCV bandwidth does not affect its estimation too much, and hence both $\Delta(t)$ and $v_i(t)$ can be well estimated using the PCV bandwidth. The bias-corrected estimator of $\eta(t)$ is then obtained as $\hat{\eta}_{\text{subj}}(t) + \hat{\Delta}_{\text{subj/pt}}(t)$.

Notice that the backfitting algorithm (Buja, Hastie and Tishirani 1989) may be employed to estimate $\eta(t)$ and $v_i(t)$ using the SCV and PCV bandwidths iteratively. This has been proposed by Park and Wu (2005) which will be briefly reviewed in the next section.

4.6 LPME BACKFITTING ALGORITHM

As mentioned earlier, one of the advantages of LPME modeling is that it can be easily implemented using the existing statistical software such as S-PLUS or SAS. That is, for each given t, to obtain the LPME estimators (4.34), we only need to operationally fit the standard LME model (4.41) using the S-PLUS function *lme* or the SAS procedure *PROC MIXED*. However, each of the LPME modeling fits essentially uses the same bandwidth for both the fixed-effect function and random-effect functions. As argued in Section 4.5, this is inefficient since different bandwidths should be used for the fixed-effect function and the random-effect functions. Although the BCHB method described in Section 4.5.3 can be used to tackle this problem, it can be better resolved using the backfitting algorithm proposed by Park and Wu (2005).

The LPME estimators for the localized fixed-effect and random-effects are given in (4.34) for known variance and covariance parameters. To estimate the variance components of $\mathbf{R}_i,\ i = 1, 2, \cdots, n$ and $\mathbf{D}$ in the standard LME model (4.41), Park and Wu (2005) assumed $\mathbf{R}_i = \sigma^2(t)\mathbf{I}_{n_i},\ i = 1, 2, \cdots, n$, and employed some ideas from the REML-based EM-algorithm (see Section 2.2.5). For easy presentation, set $\sigma^2 = \sigma^2(t)$ and $\mathbf{D} = \mathbf{D}(t)$. Assume that the current values for the variance components are σ^2 and $\mathbf{D}$. Then, as in Park and Wu (2005), these variance components are updated by

$$\begin{aligned} \hat{\sigma}^2 &= N^{-1}\textstyle\sum_{i=1}^n \left\{\hat{\boldsymbol{\epsilon}}_i^T\hat{\boldsymbol{\epsilon}}_i + \sigma^2[n_i - \sigma^2\mathrm{tr}(\mathbf{P}_{\mathbf{V}_i})]\right\}, \\ \hat{\mathbf{D}} &= n^{-1}\textstyle\sum_{i=1}^n \left\{\hat{\mathbf{b}}_i\hat{\mathbf{b}}_i^T + (\mathbf{D} - \mathbf{D}\mathbf{X}_i^T\mathbf{K}_{ih}^{1/2}\mathbf{P}_{\mathbf{V}_i}\mathbf{K}_{ih}^{1/2}\mathbf{X}_i\mathbf{D})\right\}, \end{aligned} \tag{4.55}$$

where $N = \sum_{i=1}^n n_i$, and

$$\mathbf{P}_{\mathbf{V}_i} = \mathbf{V}_i^{-1} - \mathbf{V}_i^{-1}\mathbf{K}_{ih}^{1/2}\mathbf{X}_i\Big(\sum_{i=1}^n \mathbf{X}_i^T\boldsymbol{\Omega}_{ih}\mathbf{X}_i\Big)^{-1}\mathbf{X}_i\mathbf{K}_{ih}^{1/2}\mathbf{V}_i^{-1},$$

with $\mathbf{V}_i$ defined in (4.35).

The above updating formulas are essentially the applications of those in (2.26) to the standard LME model (4.41) with the kernel-weighted response vectors $\tilde{\mathbf{y}}_i = \mathbf{K}_{ih}^{1/2}\mathbf{y}_i$, and the kernel-weighted covariates $\tilde{\mathbf{X}}_i = \mathbf{K}_{ih}^{1/2}\mathbf{X}_i$.

Park and Wu (2005)'s backfitting procedure is an iterative algorithm which can be written as follows.

REML-Based Backfitting Algorithm Let r index the sequence of the iteration.

Step 0. Set $r = 0$. Specify the starting values, say, let $\hat{\sigma}^2_{(0)} = 1$, and $\hat{\mathbf{D}}_{(0)} = \mathbf{I}_{p+1}$ where $p+1$ is the number of parameters in $\mathbf{b}_1$. Compute $\mathbf{V}_{i(0)} = \mathbf{K}_{ih}^{1/2}\mathbf{X}_i\hat{\mathbf{D}}_{(0)}\mathbf{X}_i^T\mathbf{K}_{ih}^{1/2} + \hat{\sigma}^2_{(0)}\mathbf{I}_{n_i},\ i = 1, 2, \cdots, n$ with a given bandwidth h. For example, the h may be taken as the SCV bandwidth h_s^* or the PCV bandwidth h_p^*.

Step 1. Set $r = r + 1$. Compute/update $\hat{\beta}_{(r)}$ using

$$\hat{\beta}_{(r)} = \left(\sum_{i=1}^{n} \mathbf{X}_i^T \mathbf{\Omega}_{ih(r-1)} \mathbf{X}_i\right)^{-1} \left(\sum_{i=1}^{n} \mathbf{X}_i^T \mathbf{\Omega}_{ih(r-1)} \mathbf{y}_i\right),$$

where $\mathbf{\Omega}_{ih(r-1)} = \mathbf{K}_{ih}^{1/2} \mathbf{V}_{i(r-1)}^{-1} \mathbf{K}_{ih}^{1/2}$ but with a new bandwidth, h_s^*, determined by the SCV rule. Then compute/update $\hat{\mathbf{b}}_{i(r)}$ using

$$\hat{\mathbf{b}}_{i(r)} = \hat{\mathbf{D}}_{(r-1)} \mathbf{X}_i^T \mathbf{\Omega}_{ih(r-1)} \left(\mathbf{y}_i - \mathbf{X}_i \hat{\beta}_{(r)}\right), \quad i = 1, 2, \cdots, n,$$

where $\mathbf{\Omega}_{ih(r-1)}$ is defined as above, but now h is replaced by the PCV bandwidth h_p^*.

Step 2. Update $\hat{\sigma}_{(r)}^2$ and $\hat{\mathbf{D}}_{(r)}$ using (4.55), i.e.,

$$\begin{aligned} \hat{\sigma}_{(r)}^2 &= N^{-1} \sum_{i=1}^{n} \left\{ \hat{\epsilon}_{i(r)}^T \hat{\epsilon}_{i(r)} + \hat{\sigma}_{(r-1)}^2 [n_i - \hat{\sigma}_{(r-1)}^2 \text{tr}(\mathbf{P}_{\mathbf{V}_{i(r-1)}})] \right\}, \\ \hat{\mathbf{D}}_{(r)} &= n^{-1} \sum_{i=1}^{n} \left\{ \hat{\mathbf{b}}_{i(r)} \hat{\mathbf{b}}_{i(r)}^T \right. \\ &\quad \left. + [\hat{\mathbf{D}}_{(r-1)} - \hat{\mathbf{D}}_{(r-1)} \mathbf{X}_i^T \mathbf{K}_{ih}^{1/2} \mathbf{P}_{\mathbf{V}_{i(r-1)}} \mathbf{K}_{ih}^{1/2} \mathbf{X}_i \hat{\mathbf{D}}_{(r-1)}] \right\}, \end{aligned}$$

where $\hat{\epsilon}_{i(r)} = \mathbf{K}_{ih}^{1/2}(\mathbf{y}_i - \mathbf{X}_i \hat{\beta}_{(r)} - \mathbf{X}_i \hat{\mathbf{b}}_{i(r)})$, and $\mathbf{P}_{\mathbf{V}_{i(r-1)}}$ is computed as $\mathbf{V}_{i(r-1)}^{-1} - \mathbf{V}_{i(r-1)}^{-1} \mathbf{K}_{ih}^{1/2} \mathbf{X}_i \left(\sum_{i=1}^{n} \mathbf{X}_i^T \mathbf{\Omega}_{ih(r-1)} \mathbf{X}_i \right)^{-1} \mathbf{X}_i \mathbf{K}_{ih}^{1/2} \mathbf{V}_{i(r-1)}^{-1}$. Then compute/update

$$\mathbf{V}_{i(r)} = \mathbf{K}_{ih}^{1/2} \mathbf{X}_i \hat{\mathbf{D}}_{(r)} \mathbf{X}_i^T \mathbf{K}_{ih}^{1/2} + \hat{\sigma}_{(r)}^2 \mathbf{I}_{n_i}, \; i = 1, 2, \cdots, n.$$

Within this step, the bandwidth h is the same as the one for computing $\hat{\mathbf{b}}_{i(r)}$, i.e., the PCV bandwidth h_p^*.

Step 3. When the difference of parameter estimates between the two iterations is sufficiently small, stop; otherwise, go back to Step 1.

Similarly, the ML-based backfitting algorithm can be easily obtained via replacing $\mathbf{P}_{\mathbf{V}_{i(r-1)}}$ by $\mathbf{V}_{i(r-1)}^{-1}$ in Step 2 of the above algorithm. We ignore the details to avoid redundancy.

It is seen that the above ML or REML-based backfitting algorithm not only allows us to estimate the fixed-effect function and the random-effect functions separately using different bandwidths at each iteration but also provides the flexibility to incorporate the ML or REML-based EM algorithm for estimating the variance components σ^2 and $\mathbf{D}$. Park and Wu (2005) carried out simulation studies and showed that the proposed REML-based backfitting algorithm performs similarly to the BCHB method proposed in Wu and Zhang (2002a) for estimating the fixed-effect function but outperforms the latter for estimating the individual functions (see Section 4.8 for more details). Compared with Wu and Zhang (2002a)'s procedure, this backfitting algorithm does pay a price of more intensive computation due to the selection of bandwidths at each iteration step; moreover, the backfitting procedure can not directly utilize the existing statistical packages for LME fitting as Wu and Zhang (2002a)'s procedure does.

4.7 ASYMPTOTICAL PROPERTIES OF THE LPME ESTIMATORS

In this section, we investigate the asymptotic properties of $\hat{\eta}(t)$, i.e., the LPME estimator of the fixed-effect curve $\eta(t)$ in (4.43). We mainly consider two cases: (i) when both n_i and n tend to infinity; and (ii) when n_i is bounded while n tends to infinity. The second case is more important since in many longitudinal studies, the number of subjects is much larger than the number of measurements per subject.

We first deal with Case (i). The main results are summarized in Theorems 4.1 and 4.2 below. The required Condition A and the associated proofs are given in Sections 4.11.1 and 4.11.2 of the Appendix, respectively. For convenience, throughout the rest of the chapter, $\tau^2(t)$ denotes the overall variance of $y(t)$ at time t, i.e., $\tau^2(t) = \gamma(t,t) + \sigma^2(t)$.

Before we present the results, we would like to review some basic concepts. For a given time point t, the bias and variance of $\hat{\eta}(t)$ are defined as follows:

$$\text{Bias}\{\hat{\eta}(t)\} = \text{E}\hat{\eta}(t) - \eta(t), \quad \text{Var}\{\hat{\eta}(t)\} = \text{E}\{\hat{\eta}(t) - \text{E}\hat{\eta}(t)\}^2.$$

The bias measures the overall departure of the estimator $\hat{\eta}(t)$ from its target $\eta(t)$, while the variance measures the variation of the estimator $\hat{\eta}(t)$. The Mean Squared Error (MSE) of the estimator $\hat{\eta}(t)$ is defined as

$$\text{MSE}\{\hat{\eta}(t)\} = \text{E}\{\hat{\eta}(t) - \eta(t)\}^2,$$

which measures the accuracy of $\hat{\eta}(t)$, taking into account both the bias and variance of the estimator. Actually, it is easy to show that

$$\text{MSE}\{\hat{\eta}(t)\} = \text{Bias}^2\{\hat{\eta}(t)\} + \text{Var}\{\hat{\eta}(t)\}.$$

To measure the accuracy of $\hat{\eta}(t)$ globally, the Mean Integrated Squared Errors (MISE) of the estimator $\hat{\eta}(t)$ is usually used. It is defined as

$$\text{MISE}_w\{\hat{\eta}(t)\} = \text{E}\int [\hat{\eta}(t) - \eta(t)]^2 w(t)dt,$$

where $w(\cdot)$ is a nonnegative weight function, often used to specify a range of interest or to limit the boundary effect. It is easy to see that

$$\begin{aligned}\text{MISE}_w\{\hat{\eta}(t)\} &= \int \text{MSE}\{\hat{\eta}(t)\}w(t)dt \\ &= \int \text{Bias}^2\{\hat{\eta}(t)\}w(t)dt + \int \text{Var}\{\hat{\eta}(t)\}w(t)dt.\end{aligned}$$

For simplicity, we shall use the following functionals of a kernel K:

$$B(K) = \int K(u)u^2du, \quad V(K) = \int K^2(u)du. \tag{4.56}$$

Moreover, let

$$\tilde{m} = \left[\sum_{i=1}^{n} n_i^{-1}/n\right]^{-1}. \tag{4.57}$$

Theorem 4.1 *(Wu and Zhang 2002a) Under Condition A given in Section 4.11.1 in the Appendix, when p = 0, we have:*
(a) the asymptotic bias and variance of $\hat{\eta}(t)$ are, respectively,

$$\begin{aligned} Bias\{\hat{\eta}(t)\} &= \frac{h^2}{2}[\eta''(t) + 2\eta'(t)f'(t)/f(t)]B(K)[1 + o_p(1)], & (4.58)\\ Var\{\hat{\eta}(t)\} &= \frac{\gamma(t,t)}{n} + \frac{\tau^2(t)}{n\tilde{m}hf(t)}V(K) + o_p\left[n^{-1} + (n\tilde{m}h)^{-1}\right]. & (4.59)\end{aligned}$$

(b) the asymptotic MSE of $\hat{\eta}(t)$ and the asymptotic local optimal bandwidth at t are, respectively,

$$\begin{aligned} MSE\{\hat{\eta}(t)\} &= \frac{\gamma(t,t)}{n} + \frac{\tau^2(t)}{n\tilde{m}hf(t)}V(K) \\ &\quad + \frac{h^4}{4}[\eta''(t) + 2\eta'(t)f'(t)/f(t)]^2 B^2(K) \\ &\quad + o_p\left[n^{-1} + (n\tilde{m}h)^{-1} + h^4\right], \end{aligned} \tag{4.60}$$

and

$$h_{opt}(t) = \left\{\frac{V(K)B^{-2}(K)\tau^2(t)/f(t)}{n\tilde{m}[\eta''(t) + 2\eta'(t)f'(t)/f(t)]^2}\right\}^{1/5} [1 + o_p(1)]. \tag{4.61}$$

(c) the asymptotic MISE of $\hat{\eta}(t)$ and the asymptotic global optimal bandwidth are, respectively,

$$\begin{aligned} MISE_w\{\hat{\eta}(t)\} &= n^{-1}\int \gamma(t,t)w(t)dt + \frac{V(K)}{n\tilde{m}h}\int w(t)\tau^2(t)/f(t)dt \\ &\quad + \frac{h^4}{4}B^2(K)\int [\eta''(t) + 2\eta'(t)f'(t)/f(t)]^2 w(t)dt \\ &\quad + o_p[n^{-1} + (n\tilde{m}h)^{-1} + h^4], \end{aligned}$$

and

$$h_{opt} = \left\{\frac{V(K)B^{-2}(K)\int w(t)\tau^2(t)/f(t)dt}{n\tilde{m}\int [\eta''(t) + 2\eta'(t)f'(t)/f(t)]^2 w(t)dt}\right\}^{1/5} [1 + o_p(1)]. \tag{4.62}$$

When p = 1, the above results still hold after the term $\eta''(t) + 2\eta'(t)f'(t)/f(t)$ in the above expressions is replaced with $\eta''(t)$.

Proof: see Appendix.

Remark 4.1 *As expected, when the within-subject correlations are* 0 *so that* $\gamma(t,t) \equiv 0$, *the asymptotic variance and the asymptotic MSE of* $\hat{\eta}(t)$ *reduce to those with independent data. This can be seen from (4.59) and (4.60).*

Remark 4.2 *Note that under Condition A,* h^4 *and* $(n\tilde{m}h)^{-1}$ *in (4.60) are of higher orders than* n^{-1}. *It follows that when* $\gamma(t,t) \neq 0$, $\hat{\eta}(t)$ *is* $n^{1/2}$*-consistent and asymptotically unbiased. This implies that in this case, the effect of smoothing is second ordered, and the within-subject correlation slows down the convergence of* $\hat{\eta}(t)$ *(the parametric rate* $n^{1/2}$ *is slower than the usual nonparametric rate* $(n\tilde{m}h)^{1/2}$ *when* $\tilde{m}h \to \infty$*).*

Remark 4.3 *The optimal bandwidth* $h_{opt} = O_p\left\{(n\tilde{m})^{-1/5}\right\}$ *for* $\hat{\eta}(t)$ *is much smaller than the usual optimal bandwidth* $h_{opt,i} = O_p\left\{n_i^{-1/5}\right\}$ *for smoothing the* i*-th individual curve* $s_i(t)$. *Thus, different bandwidths are needed for estimating* $\eta(t)$ *and* $s_i(t)$.

Under some additional conditions, we can show that $\hat{\eta}(t)$ is asymptotically Gaussian. This result is useful for statistical inference about the mean function $\eta(t)$. However, we must keep in mind that here we require both n and n_i to tend to infinity.

Theorem 4.2 *(Wu and Zhang 2002a) Assume that the within-subject correlations are not 0 so that* $\gamma(t,t) \neq 0$. *Consider both the* $p = 0$ *or* $p = 1$ *cases. Assume all the conditions for Theorem 4.1 are satisfied. Furthermore, either assume* $y_{ij} - \eta(t_{ij}) = v_i(t_{ij}) + \epsilon_{ij}$ *are uniformly bounded, i.e.,*

$$|y_{ij} - \eta(t_{ij})| \leq C < \infty, \; j = 1, 2, \cdots, n_i; \; i = 1, 2, \cdots, n, \tag{4.63}$$

for some constant C, *or assume the number of measurements is the same for all subjects, i.e.,*

$$n_i = m, \; i = 1, 2, \cdots, n. \tag{4.64}$$

Then we have,
(a) for any given t *in the interior of the support of* f, $\hat{\eta}(t)$ *is asymptotically normally distributed with mean* $\eta(t)$ *and variance* $\gamma(t,t)/n$, *i.e., as* $n \to \infty$,

$$n^{1/2}\{\hat{\eta}(t) - \eta(t)\} \sim AN[0, \gamma(t,t)]. \tag{4.65}$$

(b) within the interior of the support of f, $\hat{\eta}(t)$ *is asymptotically a Gaussian process with mean function* $\eta(t)$ *and covariance function* $\gamma(s,t)/n$. *That is, for any given* M *distinct time points* $s_1, \cdots, s_M$ *in the interior of the support of* f, *let* $\mathbf{s} = (s_1, \cdots, s_M)^T$, $\eta(\mathbf{s}) = [\eta(s_1), \cdots, \eta(s_M)]^T$ *and* $\hat{\eta}(\mathbf{s}) = [\hat{\eta}(s_1), \cdots, \hat{\eta}(s_M)]^T$, *then as* $n \to \infty$,

$$n^{1/2}\{\hat{\eta}(\mathbf{s}) - \eta(\mathbf{s})\} \sim AN[0, \mathbf{\Gamma_s}], \tag{4.66}$$

where

$$\mathbf{\Gamma_s} = \begin{bmatrix} \gamma(s_1, s_1) & \cdots & \gamma(s_1, s_M) \\ \gamma(s_2, s_1) & \cdots & \gamma(s_2, s_M) \\ \gamma(s_M, s_1) & \cdots & \gamma(s_M, s_M) \end{bmatrix}. \tag{4.67}$$

Proof: see Appendix.

We now deal with Case (ii). The main results are presented in Theorems 4.3 and 4.4 below. The required Condition B and the associated proofs are given in Sections 4.11.1 and 4.11.2 in the Appendix. First, let U be the uniform kernel:

$$U(u) = \frac{1_{\{|u| \le 1\}}}{2} = \begin{cases} \frac{1}{2}, & \text{when } |u| \le 1, \\ 0, & \text{else.} \end{cases} \tag{4.68}$$

Then $B(U) = \int_{-1}^{1} U(u)u^2 du = 1/3$ and $V(U) = \int_{-1}^{1} U^2(u)du = 1/2$. Moreover, let $N = \sum_{i=1}^{n} n_i$.

Theorem 4.3 *(Wu and Zhang 2002a) Under Condition B in Section 4.11.1 in the Appendix and n_i bounded, when $p = 0$, we have:*
(a) the asymptotic bias and variance of $\hat{\eta}(t)$ are, respectively,

$$\begin{aligned} \mathit{Bias}\{\hat{\eta}(t)\} &= \frac{h^2}{2}\left[\eta''(t) + 2\eta'(t)f'(t)/f(t)\right] B(U)[1 + o_p(1)], \qquad (4.69) \\ \mathit{Var}\{\hat{\eta}(t)\} &= \left\{N^{-2}\sum_{i=1}^{n} n_i^2\right\}\gamma(t,t) + \frac{\tau^2(t)}{Nhf(t)}V(U) \\ &\quad + o_p\left[N^{-2}\sum_{i=1}^{n} n_i^2 + (Nh)^{-1}\right]. \qquad (4.70) \end{aligned}$$

(b) the asymptotic MSE of $\hat{\eta}(t)$ and the asymptotic local optimal bandwidth at t are, respectively,

$$\begin{aligned} \mathit{MSE}\{\hat{\eta}(t)\} &= \left\{N^{-2}\sum_{i=1}^{n} n_i^2\right\}\gamma(t,t) + \frac{\tau^2(t)}{Nhf(t)}V(U) \\ &\quad + \frac{h^4}{4}\left[\eta''(t) + 2\eta'(t)f'(t)/f(t)\right]^2 B^2(U) \\ &\quad + o_p\left[N^{-2}\sum_{i=1}^{n} n_i^2 + (Nh)^{-1} + h^4\right], \qquad (4.71) \end{aligned}$$

and

$$h_{opt}(t) = \left\{\frac{V(U)B^{-2}(K)\tau^2(t)/f(t)}{N[\eta''(t) + 2\eta'(t)f'(t)/f(t)]^2}\right\}^{1/5} [1 + o_p(1)]. \tag{4.72}$$

(c) the asymptotic MISE of $\hat{\eta}(t)$ and the asymptotic global optimal bandwidth are, respectively,

$$\begin{aligned} \mathit{MISE}_w\{\hat{\eta}(t)\} &= \left\{N^{-2}\sum_{i=1}^{n} n_i^2\right\}\int \gamma(t,t)w(t)dt + \frac{V(U)}{Nh}\int w(t)\tau^2(t)/f(t)dt \\ &\quad + \frac{h^4}{4}B^2(U)\int \left[\eta''(t) + 2\eta'(t)f'(t)/f(t)\right]^2 w(t)dt \\ &\quad + o_p\left[N^{-2}\sum_{i=1}^{n} n_i^2 + (Nh)^{-1} + h^4\right], \qquad (4.73) \end{aligned}$$

$$h_{opt} = \left\{ \frac{V(U)B^{-2}(U) \int w(t)\tau^2(t)/f(t)dt}{N \int [\eta''(t) + 2\eta'(t)f'(t)/f(t)]^2 w(t)dt} \right\}^{1/5} [1 + o_p(1)]. \tag{4.74}$$

When p = 1, the above results still hold after the term $\eta''(t) + 2\eta'(t)f'(t)/f(t)$ in the above expressions is replaced by $\eta''(t)$.

Proof: see Appendix.

Remark 4.4 *Since under Condition B, $N^{-2}\sum_{i=1}^n n_i^2$ is of higher order than $(Nh)^{-1}$, it follows that $\hat{\eta}(t)$ converges at the same nonparametric rate $O_p[(Nh)^{1/2}]$ as that of the usual kernel estimator (Hoover et al. 1998) and that of the LPK-GEE estimator (Lin and Carroll 2000).*

Remark 4.5 *It is interesting to note from Theorem 4.3 that the asymptotic bias, variance and MSE of $\hat{\eta}(t)$ are independent of the kernel K used in $\hat{\eta}(t)$, and are only related to the uniform kernel U.*

Theorem 4.4 *(Wu and Zhang 2002a) Assume $E|\epsilon(t)|^{2+\omega} < \infty$ for some $\omega > 0$ and notice that $V(U) = 1/2$. Under Condition B in Section 4.11.1 in the Appendix, for both local constant (p = 0) or local linear (p = 1) LPME estimator, we have:*
(a) for any given t in the interior of the support of f, $\hat{\eta}(t)$ is asymptotically normally distributed with mean, $\eta(t) + Bias\{\hat{\eta}(t)\}$, and variance, $\tau^2(t)/[2Nhf(t)]$, where $Bias\{\hat{\eta}(t)\}$ is given in Theorem 4.3. That is, as $n \to \infty$,

$$(Nh)^{1/2}\left[\hat{\eta}(t) - \eta(t) - Bias\{\hat{\eta}(t)\}\right] \sim AN[0, \tau^2(t)/(2f(t))]. \tag{4.75}$$

(b) within the interior of the support of f, $\hat{\eta}(t)$ is asymptotically an uncorrelated Gaussian process whose mean and variance functions are $\eta(t) + Bias\{\hat{\eta}(t)\}$ and $\tau^2(t)/[2Nhf(t)]$, respectively. That is, for any given M distinct time points $s_1, \cdots, s_M$ in the interior of the support of f, let $\mathbf{s} = (s_1, \cdots, s_M)^T$, $\eta(\mathbf{s}) = [\eta(s_1), \cdots, \eta(s_M)]^T$ and $\hat{\eta}(\mathbf{s}) = [\hat{\eta}(s_1), \cdots, \hat{\eta}(s_M)]^T$, then as $n \to \infty$,

$$(Nh)^{1/2}\left[\hat{\eta}(\mathbf{s}) - \eta(\mathbf{s}) - Bias\{\hat{\eta}(\mathbf{s})\}\right] \sim AN[0, \mathbf{\Phi_s}], \tag{4.76}$$

where

$$\mathbf{\Phi_s} = 2^{-1} diag\left[\tau^2(s_1)/f(s_1), \tau^2(s_2)/f(s_2), \cdots, \tau^2(s_M)/f(s_M)\right]. \tag{4.77}$$

Proof: see Appendix.

Similar results for a general LPME estimator $\hat{\eta}(t)$ (for $p > 1$) may also hold, but stronger conditions are needed and proofs are tedious. Park and Wu (2005) also provide similar asymptotic results for the local marginal likelihood estimator (4.30).

4.8 FINITE SAMPLE PROPERTIES OF THE LPME ESTIMATORS

In this section, three simulation studies are presented. The first study aims to compare the estimators with different combinations of the bandwidth selection methods introduced in Section 4.5.3. The second is to evaluate the performance of the LPME estimator of $\eta(t)$ versus estimators derived from three other methods: the naive kernel method (Hoover et al. 1998), the two-step method (Fan and Zhang 2000), and the LPK-GEE method (Lin and Carroll 2000). These three methods were briefly reviewed in Section 4.2. The third simulation study is to compare the two LPME modeling implementation methods, one based on the BCHB bandwidth selection method given in Section 4.5, and the other based on the backfitting algorithm of Park and Wu (2005) described in Section 4.6. For simplicity, in this section, we focus on the LCME estimators, a special case of the general LPME estimators.

The simulation model is designed as

$$\begin{aligned} y_i(t) &= a_{i0} + a_{i1}\cos(2\pi t) + a_{i2}\sin(2\pi t) + \epsilon_i(t), \\ \mathbf{a}_i &= [a_{i0}, a_{i1}, a_{i2}]^T \sim N[(a_0, a_1, a_2)^T, \mathrm{diag}(\sigma_0^2, \sigma_1^2, \sigma_2^2)], \\ \epsilon_i(t) &\sim N[0, \sigma_\epsilon^2(1+t)], \quad i = 1, 2, \cdots, n, \end{aligned} \tag{4.78}$$

where n is the number of subjects, and $\mathbf{a}_i$ and $\epsilon_i(t)$ are independent. For simplicity, the design time points are first scheduled as $t_j = j/(m+1), j = 1, 2, \cdots, m$ where m is a given integer. To obtain an imbalanced design, which is more realistic, some responses on a subject are randomly removed at a rate r_{miss}. There are about $m(1 - r_{\text{miss}})$ measurements on average per subject, and $nm(1 - r_{\text{miss}})$ measurements in total for all simulated subjects.

The simulation mean function, covariance function, and noise variance function of the simulation model (4.78) are, respectively,

$$\eta(t) = a_0 + a_1\cos(2\pi t) + a_2\sin(2\pi t), \tag{4.79}$$

$$\gamma(s,t) = \sigma_0^2 + \sigma_1^2\cos(2\pi s)\cos(2\pi t) + \sigma_2^2\sin(2\pi s)\sin(2\pi t), \tag{4.80}$$

and

$$\gamma_\epsilon(t,t) = \sigma_\epsilon^2(1+t). \tag{4.81}$$

From the above design, it is easy to notice that when $\sigma_1 \neq \sigma_2$, the covariance function γ is not stationary; and when $\sigma_\epsilon \neq 0$, the noise variance function γ_ϵ is not homogeneous. Therefore, the simulation model (4.78) represents a variety of practical situations. Using (4.80) and (4.81), it is easy to compute the correlation coefficients between $y_i(s)$ and $y_i(t)$. In particular, when $\sigma_1^2 = \sigma_2^2 = \sigma^2$,

$$\mathrm{Cor}\{y_i(s), y_i(t)\} = \frac{\sigma_0^2 + \sigma^2\cos\{2\pi(s-t)\}}{\{[\sigma_0^2 + \sigma^2 + \sigma_\epsilon^2(1+s)][\sigma_0^2 + \sigma^2 + \sigma_\epsilon^2(1+t)]\}^{1/2}}, \quad s \neq t.$$

Therefore,

$$\frac{\sigma_0^2 - \sigma^2}{\sigma_0^2 + \sigma^2 + 2\sigma_\epsilon^2} \leq \mathrm{Cor}\{y_i(s), y_i(t)\} \leq \frac{\sigma_0^2 + \sigma^2}{\sigma_0^2 + \sigma^2 + \sigma_\epsilon^2}, \quad s \neq t.$$

For simplicity, in the following simulation studies, we set $\sigma_1^2 = \sigma_2^2 = \sigma_\epsilon^2 = 1$ and $\sigma_0^2 = 4, 9$ and 16. The associated lower and upper limits of the correlation coefficients between $y_i(s)$ and $y_i(t)$ $(s \neq t)$ are then $(0.43, 0.83)$, $(0.67, 0.91)$, and $(0.79, 0.94)$, respectively, reflecting three different correlation levels.

Let (t_{ij}, y_{ij}), $j = 1, 2, \cdots, n_i$; $i = 1, 2, \cdots, n$ denote the simulated sample. Based on this simulated sample, different estimators may be constructed for the mean function $\eta(t)$ and individual curves $s_i(t) = \eta(t) + v_i(t)$. In order to assess the performance of the population and individual curve estimators, the following so-called Average Squared Error (ASE) will be used:

$$\text{ASE}_\eta = \sum_{i=1}^{n}\sum_{j=1}^{n_i}[\eta(t_{ij}) - \hat{\eta}(t_{ij})]^2/(nn_i), \tag{4.82}$$

$$\text{ASE}_s = \sum_{i=1}^{n}\sum_{j=1}^{n_i}[s_i(t_{ij}) - \hat{s}_i(t_{ij})]^2/(nn_i). \tag{4.83}$$

To evaluate the relative performance of a method (e.g., a new method) against a standard method in N simulation runs, the following summary statistics will be used:

$$\text{ASER}(\%) = \text{AVG}\left\{\frac{\text{ASE}_{std} - \text{ASE}_{new}}{\text{ASE}_{std}}\right\} \times 100\%, \tag{4.84}$$

$$\text{ASEB}(\%) = \text{AVG}\left\{\text{ASE}_{new} \le \text{ASE}_{std}\right\} \times 100\%, \tag{4.85}$$

where AVG represents "average", while ASE_{new} and ASE_{std} denote the ASE of the new method and the standard method, respectively. The ASER is the average of ASE percentage reduction by the new method compared to the standard method, and the ASEB is the percentage that the new method is better (the ASE is smaller) than the standard method in N simulation runs.

4.8.1 Comparison of the LPME Estimators in Section 4.5.3

Recall that $\hat{\eta}_{\text{subj}}(t)$, $\hat{\eta}_{\text{pt}}(t)$ and $\hat{\eta}_{\text{subj/pt}}(t)$ are the estimators of $\eta(t)$ obtained using the SCV(or HB), PCV, and BCHB methods, respectively, and $\hat{s}_{\text{subj},i}(t)$, $\hat{s}_{\text{pt},i}(t)$, $\hat{s}_{\text{subj+pt},i}(t)$ and $\hat{s}_{\text{subj/pt},i}(t)$ are the estimators of $s_i(t)$ using the SCV, PCV, HB, and BCHB methods, respectively; see method details in Section 4.5.3. To compare these estimators, $N = 150$ simulation samples were generated, with parameters $n = 50, m = 30, r_{\text{miss}} = 20\%$ and $(\sigma_0^2, \sigma^2, \sigma_\epsilon^2) = (9, 1, 1)$ according to the simulation model (4.78).

Figure 4.1 (a) presents the results for comparing the three estimators, $\hat{\eta}_{\text{subj/pt}}(t)$, $\hat{\eta}_{\text{subj}}(t)$ and $\hat{\eta}_{\text{pt}}(t)$ in boxplots of the ASEs. It seems that the first two perform similarly, and better than the third one. Actually, the average ASEs of $\hat{\eta}_{\text{subj/pt}}(t)$, $\hat{\eta}_{\text{subj}}(t)$, and $\hat{\eta}_{\text{pt}}(t)$ are 0.230, 0.234, and 0.286, and the median ASEs are 0.129, 0.130, and 0.181, respectively. Moreover, the first two methods reduce the ASEs of the third one by 26% on average, and are better (i.e. smaller ASEs) than the third one by 89% in $N = 150$ simulation runs. Thus, it seems inefficient to estimate $\eta(t)$ using the PCV bandwidth.

Figure 4.1 (b) presents the results for comparing the four estimators of individual curves, $\hat{s}_{\text{subj/pt},i}(t)$, $\hat{s}_{\text{subj+pt},i}(t)$, $\hat{s}_{\text{pt},i}(t)$ and $\hat{s}_{\text{subj},i}(t)$. It seems that these estimators performed quite differently: the best one is $\hat{s}_{\text{subj/pt},i}(t)$ while the poorest is

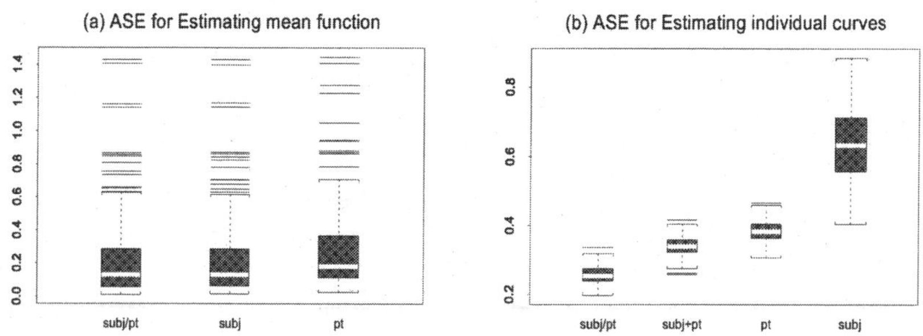

Fig. 4.1 Boxplots of the ASEs for estimating $\eta(t)$ and $s_i(t)$. Reprinted with permission from *The Journal of the American Statistical Association.*

$\hat{s}_{\text{subj},i}(t)$. Actually, the average ASEs of $\hat{s}_{\text{subj/pt},i}(t)$, $\hat{s}_{\text{subj+pt},i}(t)$, $\hat{s}_{\text{pt},i}(t)$ and $\hat{s}_{\text{subj},i}(t)$, are $0.255, 0.338, 0.382$, and 0.633, respectively. Moreover, the first three reduce the ASEs of the last one by 59%, 45% and 38% on average, respectively, and have smaller ASEs than the last one in all 150 simulation runs.

From this simulation study, it is easy to conclude that, (1) different bandwidth selection methods should be used for estimating the population curve and individual curves; (2) when estimating the population curve, the SCV bandwidth selection method is good enough (comparable to the BCHB method), but the PCV bandwidth selection method yields bandwidths which are too large to be used; and (3) when estimating the individual curves, the BCHB method outperforms all the other three methods. The BCHB method is the one that should be used in practical data analysis.

4.8.2 Comparison of Different Smoothing Methods

The simulation model (4.78) was also used to compare the LCME estimator with three non-LPME estimators for the population curve $\eta(t)$: the naive local constant kernel estimator proposed by Hoover et al. (1998), the two-step estimator by Fan and Zhang (2000), and the local constant kernel GEE estimator by Lin and Carroll (2000), which are denoted by HRWY, FZ and LC, respectively.

A brief review of these three non-LPME modeling methods is provided in Section 4.2. The SCV bandwidth selection method was used in all comparisons (note that the LCME estimator with the SCV bandwidth is comparable to that with the BCHB bandwidth as shown in the last subsection). To reduce the computational burden, true variances which are known based on the simulated model (4.78), were used for the LC estimator, and which in general will favor the LC estimator.

In the simulation study, there were 27 scenarios of different combinations of sample size ($n = 30, 50, 70$), measurements per subject ($m = 20$), missing rates ($r_{\text{miss}} = 20\%, 50\%$ and 80%), and smallest correlation levels ($\rho_{\min} = 0.43, 0.67, 0.79$). Note that these correlation levels correspond to $(\sigma_0^2, \sigma^2, \sigma_\epsilon^2) = (4,1,1), (9,1,1), (16,1,1)$, respectively in the simulation model (4.78). The expected number of measurements for a subject is then $m(1 - r_{\text{miss}}) = 16, 10$ and 4, respectively associated with the three missing rates. Note that larger missing rates usually imply that the resulting

Table 4.1 The ASER(%) and ASEB(%) ± 2 standard errors (%) for the FZ, LC, and LCME estimators of $\eta(t)$ over the HRWY estimator, $N = 250$ (In the table, $r = r_{\text{miss}}$ and $\rho = \rho_{\min}$). Reprinted with permission from *The Journal of the American Statistical Association.*

$(n,\ m,\ r)$	ρ	FZ/HRWY ASER% (ASEB%)	LC/HRWY ASER% (ASEB%)	LCME/HRWY ASER% (ASEB%)
(30, 20, 20%)	.43	−.28± .50 (56± 6)	−.04± .10 (55± 6)	13.09± 2.08 (76± 5)
	.67	−.65± .77 (37± 6)	−.29± .11 (31± 6)	10.75± 2.69 (76± 5)
	.79	−1.02± .42 (44± 6)	−.02± .04 (51± 6)	20.89± 3.29 (80± 5)
(50, 20, 20%)	.43	−.21± .39 (38± 6)	−.07± .10 (43± 6)	10.15± 1.70 (80± 5)
	.67	.70± .20 (50± 6)	−.07± .08 (52± 6)	9.52± 3.00 (70± 6)
	.79	.14± .37 (50± 6)	−.06± .04 (37± 6)	7.15± 2.68 (62± 6)
(70, 20, 20%)	.43	.21± .39 (46± 6)	−.30± .13 (34± 6)	−1.03± 1.98 (54± 6)
	.67	.15± .33 (46± 6)	−.14± .04 (39± 6)	6.55± 2.58 (63± 6)
	.79	.23± .47 (49± 6)	.07± .04 (44± 6)	8.07± 2.79 (66± 6)
(30, 20, 50%)	.43	−3.10± 1.12 (30± 6)	−.15± .18 (44± 6)	16.17± 2.56 (74± 5)
	.67	−1.08± 1.15 (49± 6)	−.15± .09 (38± 6)	9.92± 4.75 (56± 6)
	.79	−1.32± .80 (38± 6)	−.11± .06 (69± 6)	6.39± 5.28 (62± 6)
(50, 20, 50%)	.43	3.79± .67 (68± 6)	−.13± .10 (44± 6)	4.83± 2.68 (63± 6)
	.67	.50± .88 (55± 6)	−.23± .08 (39± 6)	6.83± 3.90 (64± 6)
	.79	− .28± 1.05 (61± 6)	.03± .04 (56± 6)	18.50± 4.85 (81± 5)
(70, 20, 50%)	.43	.05± .95 (61± 6)	.02± .17 (45± 6)	10.76± 2.54 (79± 5)
	.67	.11± .81 (63± 6)	−.03± .09 (37± 6)	13.22± 2.79 (79± 5)
	.79	1.37± .73 (55± 6)	.04± .03 (60± 6)	18.58± 3.92 (80± 5)
(30, 20, 80%)	.43	−10.81± 4.02 (37± 6)	.21± .10 (51± 6)	−19.45± 5.00 (37± 6)
	.67	−21.92± 6.14 (43± 6)	−.03± .12 (37± 6)	8.81± 5.14 (69± 6)
	.79	−9.96± 3.76 (45± 6)	.08± .04 (69± 6)	5.58± 4.22 (56± 6)
(50, 20, 80%)	.43	−3.62± 1.97 (44± 6)	.15± .20 (50± 6)	.59± 4.60 (55± 6)
	.67	−7.56± 2.13 (39± 6)	−.15± .07 (44± 6)	2.18± 4.66 (69± 6)
	.79	−14.53± 3.41 (45± 6)	.04± .05 (70± 6)	12.00± 4.02 (82± 5)
(70, 20, 80%)	.43	−3.53± 1.79 (54± 6)	−.01± .19 (39± 6)	.46± 4.39 (56± 6)
	.67	−3.14± 1.92 (51± 6)	.10± .07 (45± 6)	−.07± 4.49 (56± 6)
	.79	−8.07± 2.49 (34± 6)	.12± .06 (65± 6)	10.95± 3.43 (65± 6)

observed data are less correlated although the underlying model is the same.

For each scenario, $N = 250$ simulation runs were conducted. To evaluate the relative performance of the estimators under consideration, the HRWY estimator was used as the comparison basis (the standard estimator) and the ASER (4.84) and ASEB (4.85) with ±2 standard errors were computed for all other estimators. The results for the 27 scenarios are reported in Table 4.1. Note that according to the definitions of ASER and ASEB, when the ASER is negative, it implies that the new method is worse than the standard method (its ASE is larger than that of the standard method);

when the ASEB is larger than 50%, it implies that the new method performs better than the standard method.

Table 4.1 shows some interesting results. First of all, it appears that the LCME estimator generally outperforms the HRWY estimator. The largest ASER is about $20.89\% \pm 3.29\%$ while the largest ASEB is about $82\% \pm 5\%$. The HRWY estimator significantly outperforms the LCME estimator in only one case out of the 27 scenarios, i.e., the sample size is very small ($n = 30$), the missing rate is very high ($r_{\text{miss}} = 80\%$), and the data are less correlated ($\rho_{\min} = .43$). In this case, the correlation is too small to be accurately estimated using the sparse data and hence the resulting LCME estimator may not be preferable compared to the simple HRWY estimator. However, as shown by the other scenarios, when the within-subject correlation can be accurately estimated using the data, the LCME estimator which accounts for the within-subject correlation is more efficient, and hence is definitely preferred. Secondly, although the LC estimator uses the true variances (which in general reduces the variation of the LC estimator of Lin and Carroll (2000)) in its construction, the LC estimator performed no better than the HRWY estimator in most simulated cases. Finally, it seems that the FZ estimator performed slightly worse than the HRWY estimator, especially when the missing rate is high ($r_{\text{miss}} = 80\%$) or the sample size is small ($n = 30$).

In summary, the LCME estimator of $\eta(t)$ is generally more efficient than the aforementioned non-LPME modeling estimators based on our finite-sample simulation results. One possible reason is that it takes the correlation of longitudinal data into account properly whereas the non-LPME modeling methods do not.

4.8.3 Comparisons of BCHB-Based versus Backfitting-Based LPME Estimators

Park and Wu (2005) used the same simulation model as the one in (4.78) for comparing the two different LPME modeling implementation methods. The first method is based on the BCHB bandwidth selection method in Section 4.5 (denoted by LCME), and the other is based on the backfitting algorithm (denoted by LCBF) as briefly reviewed in Section 4.6. As in the last subsection, both methods are local constant based.

Park and Wu (2005) considered 8 scenarios of different combinations of sample size ($n = 30, 70$), measurements per subject ($m = 6, 24$), and smallest correlation levels ($\rho_{\min} = 0.15, 0.67$). They fixed the missing rate as $r_{\text{miss}} = 15\%$ so that the expected number of measurements on a subject is about $m(1 - r_{\text{miss}}) = 5$ and 20, respectively associated with the two different sizes of m. Notice that the two correlation levels can be obtained via setting $(\sigma_0^2, \sigma^2, \sigma_\epsilon^2) = (1.69, 1, 1), (9, 1, 1)$, respectively in the simulation model (4.78). For each scenario, $N = 500$ simulation runs were conducted.

To evaluate the relative performance of the LCBF estimator over the LCME estimator, the associated ASER (4.84) and ASEB (4.85) with ± 2 standard errors were computed. For the estimators of the mean function $\eta(t)$, the results for the 8 scenarios are given in Table 4.2; for the estimators of the individual functions $s_i(t),\ i = 1, 2, \cdots, n$, the results are presented in Table 4.3. Notice that for discussion convenience, Tables 4.2 and 4.3 were not directly copied from Park and Wu (2005) but instead

Table 4.2 The ASER(%) and ASEB(%) $\pm$ 2 standard errors (%) for the LCBF estimator of $\eta(t)$ over the associated LCME estimator, $N = 500$.

$(n,\ m,\ r_{\text{miss}})$	$\rho_{\min}$	ASER% (ASEB%)
(30,6,15%)	.15	−2.9± 2.4 (52.2± 4.4)
	.67	−4.2± 4.4 (54.2± 4.4)
(30,24,15%)	.15	−5.0± 1.8 (45.2± 4.4)
	.67	−4.5± 2.2 (45.6± 4.4)
(70,6,15%)	.15	0.2± 2.8 (58.6± 4.4)
	.67	−13.6± 5.6 (47.8± 4.4)
(70,24,15%)	.15	−5.2± 1.8 (40.8± 4.4)
	.67	−8.5± 2.4 (39.4± 4.4)

they were computed using the simulation results from Park and Wu (2005) and were formated following the structure of Table 4.1.

From Table 4.2, we can see that for estimating the mean function $\eta(t)$, the LCME estimator slightly outperforms the LCBF estimator since most of the associated ASER(%) are negative, and most of the associated ASEB(%) are less than 50%. However, Park and Wu (2005) pointed out that in both cases, the differences in their ASEs are not statistically significant by the Wilcoxon signed rank test. Therefore, we may say that the LCME and LCBF estimators for the mean function $\eta(t)$ perform similarly.

Table 4.3 The ASER(%) and ASEB(%) $\pm$ 2 standard errors (%) for the LCBF estimators of $s_i(t)$ over the associated LCME estimators, $N = 500$.

$(n,\ m,\ r_{\text{miss}})$	$\rho_{\min}$	ASER% (ASEB%)
(30,6,15%)	.15	64.6± 15 (86.6± 3.0)
	.67	33.8± 5.2 (91.8± 2.4)
(30,24,15%)	.15	20.1± 3.2 (84.0± 3.2)
	.67	5.6± 1.2 (80.6± 3.6)
(70,6,15%)	.15	60.9± 7.4 (84.0± 3.2)
	.67	36.0± 6.0 (85.0± 3.2)
(70,24,15%)	.15	17.3± 2.6 (83.6± 3.4)
	.67	8.7± 2.2 (73.0± 4.0)

For estimating the individual functions $s_i(t),\ i = 1, 2, \cdots, n$, the story is different. From Table 4.3, we can see that in each scenario, the LCBF estimator outperforms the LCME estimator since all the ASER(%) are significantly positive and all the ASEB(%) are significantly more than 50%. As pointed out by Park and Wu (2005), the LCBF estimator uses different bandwidths for the mean function and the individual functions in each iteration, which may help to improve the LCBF performance.

4.9 APPLICATION TO THE PROGESTERONE DATA

The progesterone data set introduced in Chapter 1 has been carefully studied by Brumback and Rice (1998) as an interesting illustration of their smoothing spline-based functional ANOVA models. The need for intensive computation represents a large challenge for their method. Fan and Zhang (2000) re-analyzed the data using a two-step method. In this section, we apply the LCME method to this data set as an illustration of the methodologies introduced earlier in this chapter.

The progesterone data consist of two groups of urinary metabolite progesterone curves (see Figures 1.1 and 1.2). One is known as the nonconceptive group with 69 women's menstrual cycles; the other as the conceptive group with 22 women's menstrual cycles. Approximately 8.3% of the data were missing. Both groups of curves are highly correlated with correlation coefficients larger than .70 and .50, respectively. In this example of high correlation and low missing rate, as shown in the simulation

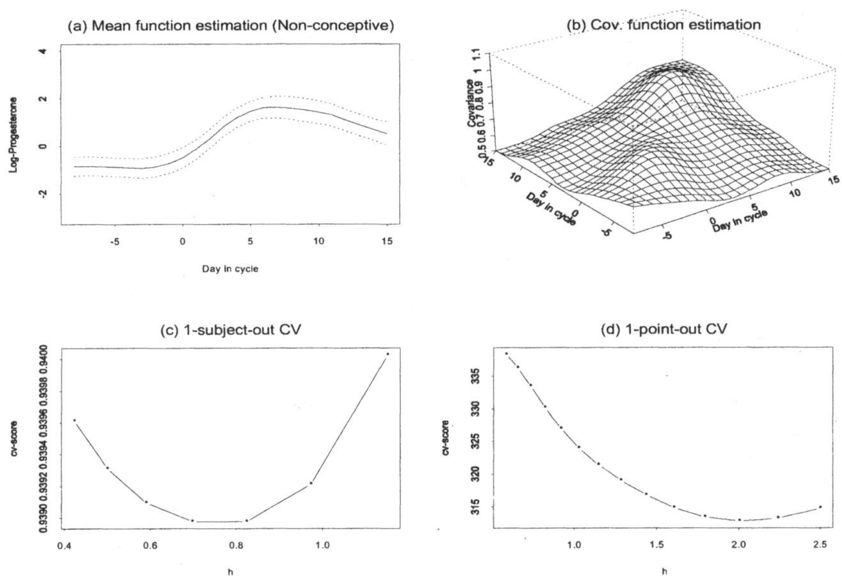

Fig. 4.2 Analysis of the nonconceptive progesterone data: (a) mean function estimation with ± 2 standard deviation pointwise bands; (b) covariance function estimation; (c) SCV for estimating the mean function; and (d) PCV for estimating the individual functions. Reprinted with permission from *The Journal of the American Statistical Association.*

studies presented in Section 4.8, the LCME method is more appropriate and more efficient to estimate the population curves and individual curves than other methods. Because the conceptive and non-conceptive groups seem to exhibit differences, they should be analyzed separately. To save space, we only report the results for the nonconceptive group data or equivalently the nonconceptive progesterone data.

The details for fitting the LCME model (4.45) to the nonconceptive progesterone data are as follows. For a fixed t, the S-PLUS function *lme* was used to fit the model (4.45) locally. The normal density kernel function was used in the LCME estimators. The BCHB procedure introduced in Section 4.5.3 was strictly followed. First the SCV bandwidth $h_s^* = .70$ was obtained (see Figure 4.2 (c)). Then this bandwidth was used to obtain the population curve estimate $\hat{\eta}_{\text{subj}}(t)$. After that, the residuals (4.53) were computed. Based on these residuals, the PCV bandwidth $h_p^* = 2$ was determined using the BCHB model (4.54) (see Figure 4.2(d)). Then this bandwidth was used to fit the model (4.54). Finally, the corresponding BCHB estimators for the population curve $\hat{\eta}_{\text{subj/pt}}(t)$ and the individual curves $\hat{s}_{\text{subj/pt},i}(t)$ were obtained.

The population curve estimate $\hat{\eta}_{\text{subj/pt}}(t)$ with ± 2 standard deviation pointwise bands (the S-PLUS LME function also gives the standard deviations of the estimators) is displayed in Figure 4.2 (a). The estimated covariance function of random-effect curves $v_i(t)$, computed using (4.44), is displayed in Figure 4.2 (b). From Figure 4.2 (b), it seems that the correlation of the progesterone data is large and cannot be ignored.

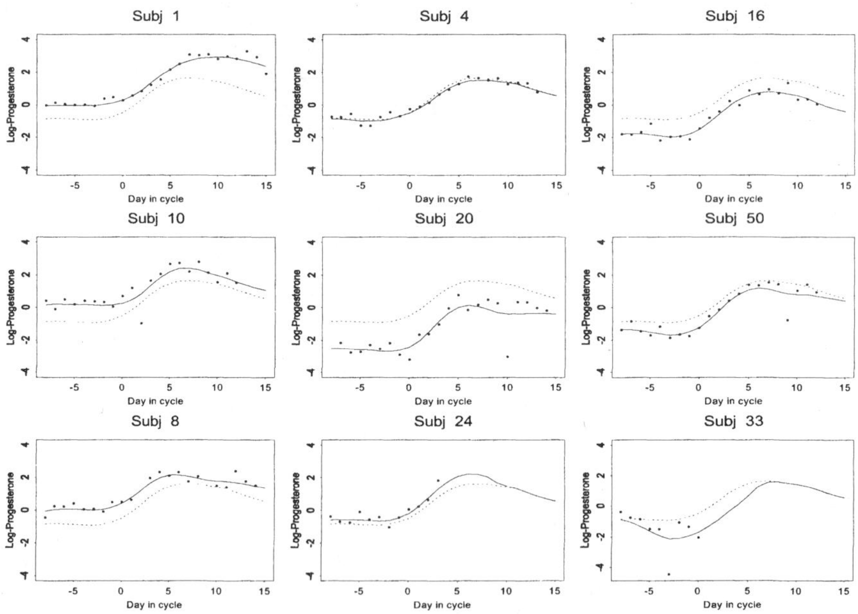

Fig. 4.3 Selected individual curve estimates (solid lines) from the nonconceptive progesterone data. The population curve estimate (dotted lines) is also plotted for comparison. Note that outliers appear in Row 2 and missing data in Row 3. Reprinted with permission from *The Journal of the American Statistical Association.*

In general, the individual curves were well fitted, even for some individual curves that have outliers or missing data (see Figure 4.3). It seems that the LCME individual curve estimators are pretty robust against outliers, and they handle the missing data pretty well. This is because (1) the LCME individual curve estimators borrow information across subjects via LCME model fitting so that one or two outliers from one individual may not be too influential; and (2) the missing values for an individual can be estimated or predicted based on the data from this particular individual and the information across subjects, i.e., the whole data set, as indicated in the individual curve estimators in (4.34).

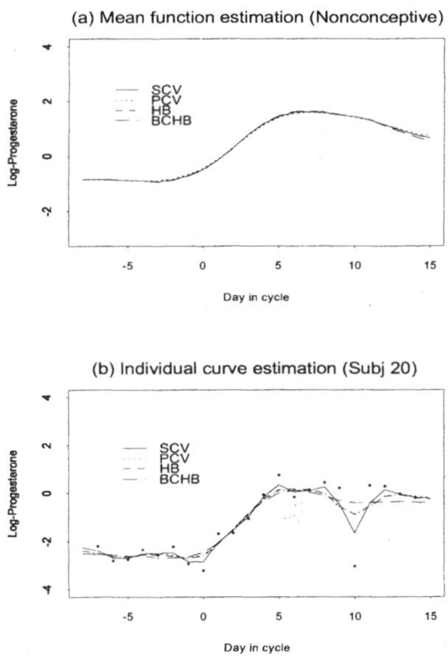

Fig. 4.4 Comparisons of the different bandwidth selection methods for the nonconceptive progesterone data: (a) population curve estimation; and (b) individual curve estimation (Subj 20).

Comparisons of the SCV, PCV, HB and BCHB methods in Section 4.5.3 for estimating the population and individual curves are given in Figures 4.4 (a) and (b), respectively. Visually the population curve estimates are nearly identical for the different methods, however for the individual curve estimates, the PCV and HB methods outperform the SCV method, while the BCHB method performs the best in the sense that the BCHB individual curve estimate is much more smooth and less affected by outliers than those obtained by the others.

4.10 SUMMARY AND BIBLIOGRAPHICAL NOTES

In this chapter, we first reviewed the naive LPK method (Hoover et al. 1998), the LPK-GEE method (Lin and Carroll 2000) and the two-step method (Fan and Zhang 2000) in which the data correlations are ignored. We then reviewed the LPME modeling approach (Wu and Zhang 2002a) for fitting the NPME model (4.19) in which the data correlations are accounted for. The key idea for LPME modeling is to combine LME modeling techniques and LPK smoothing to take the local correlation structure of longitudinal data into account. Simulation studies show that the LPME modeling method produces more efficient estimators for the population curve than the naive LPK method, the LPK-GEE method and the two-step method. The backfitting algorithm in Park and Wu (2005) for NPME modeling was also reviewed. Its key idea is to use different bandwidths in each step for estimating the fixed-effect and the random-effect functions.

Investigation of local polynomial methods for longitudinal data analysis dates back over two decades. Hart and Wehrly (1986) proposed a kernel smoothing method for repeated measurements. Hoover et al. (1998), Fan and Zhang (2000), Wu and Chiang (2000), Wu, Chiang and Hoover (1998), among others studied LPK smoothing for longitudinal data with time-varying coefficient models. The LPK smoothing technique using the generalized least squares for time series data was considered in Altman (1990) and Hart (1991). The generalized estimating equation (GEE) method using local polynomial smoothing was considered in Lin and Carroll (2000). A survey of methods on generalized local polynomial smoothing, smoothing splines, and orthogonal series was given by Opsomer, Wang and Yang (2001). The LPME modeling idea was pioneered by Wu and Zhang (2002a), which is the focus of Section 4.4.1. Since then, more work has been done in this direction, including Cai, Li and Wu (2003), Liang and Wu (2004, 2005), Liang, Wu and Carroll (2003), and Park and Wu (2005), among others. Park and Wu (2005) proposed a backfitting algorithm for NPME modeling. They showed, via a small-scale simulation, that their backfitting algorithm performs similarly to the LPME modeling method of Wu and Zhang (2002a) in population mean estimation, but ourperforms the latter in individual curve estimation. Liang and Wu (2005), on the other hand, considered the problem of a time-varying coefficient mixed-effects model with covariates containing measurement errors. Liang, Wu and Carroll (2003) applied this technique to an AIDS clinical trial. Cai, Li and Wu (2003) investigated a generalized nonparametric mixed-effects model. More details about these models and methods can be found in Chapters 8~10.

4.11 APPENDIX: PROOFS

4.11.1 Conditions

We first list the following conditions.
Condition A:

1. The design time points t_{ij}, $j = 1, 2, \cdots, n_i$; $i = 1, 2, \cdots, n$ are iid with density $f(\cdot)$. Within each subject, there are no ties among the design time points. That is, $t_{ij} \neq t_{ik}$ if $j \neq k$ for $i = 1, 2, \cdots, n$.

2. The given time point t is in the interior of the support of f where $f(t) \neq 0$ and $f'(t)$ exists.

3. The fixed-effects curve $\eta(t)$ has twice-continuous derivatives at t, i.e., $\eta''(t)$ exists and is continuous.

4. The covariance function $\gamma(s, t)$ has twice-continuous derivatives for both s and t.

5. The variance function of the measurement errors, $\sigma^2(t)$, is continuous at t.

6. The kernel K is a bounded symmetrical probability density function with bounded support, $[-1, 1]$, say, so that $\int_{-1}^{1} K(u)du = 1$, $\int_{-1}^{1} K(u)udu = 0$, $B(K) < \infty$ and $V(K) < \infty$.

7. This condition is stated separately for the $p = 0$ and $p = 1$ cases. When $p = 0$, as $n \to \infty$,

$$n_i = O(n^{\Delta}), \quad 1/4 < \Delta < 3/2, \quad h = O[(n\tilde{m})^{-1/5}], \tag{4.86}$$

so that $n_i \to \infty, h \to 0, n_i h \to \infty$ and $n_i h^3 \to 0$, where $\tilde{m}$ is defined in (4.57).

When $p = 1$, as $n \to \infty$,

$$n_i = O(n^{\Delta}), \quad 2/3 < \Delta < 3/2, \quad h = O[(n\tilde{m})^{-1/5}], \tag{4.87}$$

so that $n_i \to \infty, h \to 0, n_i h^2 \to \infty$ and $n_i h^3 \to 0$.

8. When $p = 0$, $\mathbf{D}$ is a scale function, denoted as $\delta^2(t)$ with $\delta^2(t) > 0$. When $p = 1$, $\mathbf{D}$ is assumed to be a diagonal matrix, say $\mathbf{D} = \text{diag}\{\delta_1^2(t), \delta_2^2(t)\}$ with $\delta_1^2(t) > 0, \delta_2^2(t) > 0$.

Conditions A1–6 are regular but not the weakest ones. Condition A7 allows $n_i \to \infty$ with a moderate rate compared with n. The minimum rate $1/4$ for $p = 0$ guarantees $n_i h \to \infty$ and the validity of expression (4.90). The minimum rate $2/3$ for $p = 1$, guarantees $n_i h^2 \to \infty$ and the validity of expression (4.94). Note that $1/4 < 2/3$. This is reasonable since there are two local parameters to be estimated when $p = 1$ while there is only one when $p = 0$. Condition A8 for the case $p = 1$

is given for technical convenience. In practice, for fast implementation, one often assumes that $\mathbf{D} = \mathbf{D}(t)$ is a diagonal matrix.

Condition B below is assumed for Theorems 4.3 and 4.4.

Condition B:

1. Conditions A1–A6 and Condition A8 are satisfied.

2. $K(0) > 0$.

3. n_i are bounded. That is, for some $0 < C_0 < \infty$,

$$n_i \leq C_0, i = 1, 2, \cdots, n. \tag{4.88}$$

4. As $n \to \infty$, $h = O(N^{-1/5})$ where $N = \sum_{i=1}^{n} n_i$.

4.11.2 Proofs

We now give the proofs:

Proof of Theorem 4.1 First, consider the MELC estimator $\hat{\eta}(t)$ (i.e., $p = 0$). We only need to prove (a) since (b) and (c) follow immediately from (a). Note that for any random variable R with finite first two moments, we have

$$R = \mathrm{E}(R) + O_p[\mathrm{Var}^{1/2}(R)]. \tag{4.89}$$

Applying (4.89) and Condition A7 leads to

$$\sum_{j=1}^{n_i} K_h(t_{ij} - t)\sigma^{-2}(t_{ij}) = \frac{n_i f(t)}{\sigma^2(t)}[1 + O_p((n_i h)^{-1/2})]. \tag{4.90}$$

From (4.50), we have

$$\hat{\eta}(t) = \frac{\sum_{i=1}^{n} B_i(t)}{\sum_{i=1}^{n} A_i(t)} = \eta(t) + \sum_{i=1}^{n} c_i(t) Z_i(t),$$

where $c_i(t) = A_i(t)/\sum_{k=1}^{n} A_k(t)$ with $A_i(t)$ and $B_i(t)$ defined in (4.49), and

$$Z_i(t) = \frac{\sum_{j=1}^{n_i} K_h(t_{ij} - t)\sigma^{-2}(t_{ij})[y_{ij} - \eta(t)]}{\sum_{j=1}^{n_i} K_h(t_{ij} - t)\sigma^{-2}(t_{ij})}. \tag{4.91}$$

The $Z_i(t)$'s are independent of each other because measurements from different subjects are independent.

Applying (4.90) and noticing that $\delta^2(t) > 0$, we have

$$a_i(t) \quad = \quad \delta^2(t) \sum_{j=1}^{n_i} K_h(t_{ij} - t)\sigma^{-2}(t_{ij})$$

$$
\begin{aligned}
&= \frac{n_i\delta^2(t)f(t)}{\sigma^2(t)}[1+O_p((n_ih)^{-1/2})],\\
A_i(t) &= a_i(t)/[1+a_i(t)] = 1+O_p(n_i^{-1}),\\
\sum_{k=1}^{n_i} A_k(t) &= n[1+O_p(\tilde{m}^{-1})].
\end{aligned}
$$

It follows that

$$c_i(t) = n^{-1}[1+O_p(\tilde{m}^{-1})],$$

using $O(n_i^{-1}) = O(\tilde{m}^{-1}) = O[n^{-\Delta}]$ where $1/4 < \Delta < 3/2$ and $\tilde{m}$ is defined in (4.57). We thus have

$$\hat{\eta}(t) = \eta(t) + n^{-1}\sum_{i=1}^{n} Z_i(t)[1+O_p(\tilde{m}^{-1})], \tag{4.92}$$

Let $\mathcal{D} = \{t_{ij},\ j=1,2,\cdots,n_i;\ i=1,2,\cdots,n\}$ denote the collection of the design time points. Then applying (4.90) and Condition A7, and using Taylor expansion ignoring the higher-order terms, we have

$$
\begin{aligned}
\mathrm{E}\{Z_i(t)|\mathcal{D}\} &= \frac{\sum_{j=1}^{n_i} K_h(t_{ij}-t)\sigma^{-2}(t_{ij})[\eta(t_{ij})-\eta(t)]}{\sum_{j=1}^{n_i} K_h(t_{ij}-t)\sigma^{-2}(t_{ij})}\\
&= \frac{n_i\int K(u)f(t+hu)\sigma^{-2}(t+hu)[\eta(t+hu)-\eta(t)]du[1+O_p((n_ih)^{-1/2}]}{n_i\int K(u)f(t+hu)\sigma^{-2}(t+hu)du[1+O_p((n_ih)^{-1/2}]}\\
&= \frac{h^2}{2}[\eta''(t)+2\eta'(t)f'(t)/f(t)]B(K)[1+O_p((n_ih)^{-1/2}],
\end{aligned} \tag{4.93}
$$

and

$$
\begin{aligned}
\mathrm{Var}\{Z_i(t)|\mathcal{D}\} &= \frac{\sum_{j=1}^{n_i} K_h^2(t_{ij}-t)\sigma^{-4}(t_{ij})[\gamma(t_{ij},t_{ij})+\sigma^2(t_{ij})]}{\{\sum_{j=1}^{n_i} K_h(t_{ij}-t)\sigma^{-2}(t_{ij})\}^2}\\
&\quad + \frac{\sum_{j=1}^{n_i}\sum_{l\neq j}^{n_i} K_h(t_{ij}-t)K_h(t_{il}-t)\sigma^{-2}(t_{ij})\sigma^{-2}(t_{il})\gamma(t_{ij},t_{il})}{\{\sum_{j=1}^{n_i} K_h(t_{ij}-t)\sigma^{-2}(t_{ij})\}^2}\\
&= \frac{\tau^2(t)}{n_ihf(t)}V(K) + \{\gamma(t,t)+h^2\gamma_{20}(t,t)B(K)\} + o_p[(n_ih)^{-1}+h^2]\\
&= \gamma(t,t) + \frac{\tau^2(t)}{n_ihf(t)}V(K) + o_p[(n_ih)^{-1}].
\end{aligned}
$$

Here γ_{20} represents the second derivative of the first argument of $\gamma(\cdot,\cdot)$, and we have dropped the h^2 term because it is a higher-order term than $(n_ih)^{-1}$ due to Condition A7. Therefore, we have

$$
\begin{aligned}
\mathrm{Bias}\{\hat{\eta}(t)|\mathcal{D}\} &= \mathrm{E}\{\hat{\eta}(t)|\mathcal{D}\} - \eta(t)\\
&= n^{-1}\sum_{i=1}^{n}\mathrm{E}\{Z_i(t)|\mathcal{D}\}[1+O_p(\tilde{m}^{-1})]
\end{aligned}
$$

$$\begin{aligned} &= n^{-1}\sum_{i=1}^{n}\frac{h^2}{2}[\eta''(t)+2\eta'(t)f'(t)/f(t)]B(K) \\ &\quad\times\left[1+O_p((n_ih)^{-1/2}\right]\left[1+O_p(\tilde{m}^{-1})\right] \\ &= \frac{h^2}{2}[\eta''(t)+2\eta'(t)f'(t)/f(t)]B(K)[1+o_p(1)], \end{aligned}$$

and

$$\begin{aligned} \text{Var}\{\hat{\eta}(t)|\mathcal{D}\} &= n^{-2}\sum_{i=1}^{n}\text{Var}\{Z_i(t)|\mathcal{D}\}\left[1+O_p(\tilde{m}^{-2})\right] \\ &= \gamma(t,t)/n+\frac{\tau^2(t)}{n\tilde{m}hf(t)}V(K)+o_p\left[n^{-1}+(n\tilde{m}h)^{-1}\right], \end{aligned}$$

because $(n\tilde{m}^2)^{-1}$ is a higher-order term than $(n\tilde{m}h)^{-1}$. Since the conditional bias and variance do not depend on $\mathcal{D}$, they are asymptotically unconditional. This completes the proof of Theorem 4.1 (a). Simple algebra shows that (b) and (c) follow immediately when $p = 0$.

When $p = 1$, (4.89) and Condition A7, for $r = 0, 1, 2$, lead to

$$s_{ir} = \sum_{j=1}^{n_i} K_h(t_{ij}-t)(t_{ij}-t)^r\sigma^{-2}(t_{ij}) = \frac{n_ih^rs_rf(t)}{\sigma^2(t)}\left[1+O_p((n_ih)^{-1/2})\right], \tag{4.94}$$

where $s_r = \int K(u)u^r du$ with $s_0 = 1, s_1 = 0, s_2 = B(K)$. It follows that

$$\mathbf{G}_i = \begin{pmatrix} s_{i0} & s_{i1} \\ s_{i1} & s_{i2} \end{pmatrix} = \frac{f(t)}{\sigma^2(t)}\begin{pmatrix} n_i & 0 \\ 0 & n_ih^2B(K) \end{pmatrix}\left[1+O_p((n_ih)^{-1/2})\right].$$

Because $\mathbf{D} = \text{diag}\{\delta_1^2(t), \delta_2^2(t)\}$ with $\delta_1^2(t) > 0, \delta_2^2(t) > 0$ because of Condition A8, we have

$$\begin{aligned} \mathbf{G}_i^{-1}\mathbf{D}^{-1} &= \begin{pmatrix} [n_i\delta_1^2(t)]^{-1} & 0 \\ 0 & [n_ih^2\delta_2^2(t)B(K)]^{-1} \end{pmatrix} \\ &\quad\times\frac{\sigma^2(t)}{f(t)}\left[1+O_p((n_ih)^{-1/2})\right], \\ n^{-1}\sum_{k=1}^{n}\mathbf{G}_k^{-1}\mathbf{D}^{-1} &= \begin{pmatrix} [\tilde{m}\delta_1^2(t)]^{-1} & 0 \\ 0 & [\tilde{m}h^2\delta_2^2(t)B(K)]^{-1} \end{pmatrix} \\ &\quad\times\frac{\sigma^2(t)}{f(t)}\left[1+O_p((\tilde{m}h)^{-1/2})\right], \end{aligned} \tag{4.95}$$

using $O(n_i^{-1}) = O(\tilde{m}^{-1}) = O[n^{-\Delta}]$ where $2/3 < \Delta < 3/2$. By (4.43), the local linear MELP estimator $\hat{\eta}(t)$ can be expressed as:

$$\hat{\eta}(t) = \eta(t)+\mathbf{e}_1^T\left\{\sum_{k=1}^{n}(\mathbf{I}+\mathbf{G}_k\mathbf{D})^{-1}\mathbf{G}_k\right\}^{-1}$$

$$
\begin{aligned}
& \times \left\{ \sum_{i=1}^{n} (\mathbf{I} + \mathbf{G}_i \mathbf{D})^{-1} \mathbf{X}_i^T \boldsymbol{\Theta}_{ih} [\mathbf{y}_i - \mathbf{X}_i \beta] \right\} \\
= \;& \eta(t) + \sum_{i=1}^{n} \mathbf{c}_i(t)^T \mathbf{G}_i^{-1} \mathbf{X}_i^T \boldsymbol{\Theta}_{ih} [\mathbf{y}_i - \mathbf{X}_i \beta] \\
= \;& \eta(t) + \sum_{i=1}^{n} \sum_{j=1}^{n_i} \mathbf{c}_i(t)^T \mathbf{G}_i^{-1} \begin{pmatrix} 1 \\ t_{ij} - t \end{pmatrix} \frac{K_h(t_{ij} - t)}{\sigma^2(t_{ij})} \\
& \times [y_{ij} - \eta(t) - \eta'(t)(t_{ij} - t)],
\end{aligned}
$$

where $\boldsymbol{\beta} = [\eta(t), \eta'(t)]$, and

$$
\mathbf{c}_i(t)^T = \mathbf{e}_1^T \left\{ \sum_{k=1}^{n} (\mathbf{I} + \mathbf{G}_k \mathbf{D})^{-1} \mathbf{G}_k \right\}^{-1} (\mathbf{I} + \mathbf{G}_i \mathbf{D})^{-1} \mathbf{G}_i .
$$

Because

$$
\begin{aligned}
(\mathbf{I} + \mathbf{G}_i \mathbf{D})^{-1} \mathbf{G}_i &= \left[(\mathbf{G}_i \mathbf{D})^{-1} - (\mathbf{G}_i \mathbf{D})^{-2} + \cdots \right] \mathbf{G}_i \\
&= \mathbf{D}^{-1} - \mathbf{D}^{-1} \mathbf{G}_i^{-1} \mathbf{D}^{-1} + \cdots
\end{aligned}
$$

and

$$
\begin{aligned}
& \left\{ \sum_{k=1}^{n} (\mathbf{I} + \mathbf{G}_k \mathbf{D})^{-1} \mathbf{G}_k \right\}^{-1} \\
= \;& \left\{ \sum_{k=1}^{n} (\mathbf{D}^{-1} - \mathbf{D}^{-1} \mathbf{G}_k^{-1} \mathbf{D}^{-1} + \cdots) \right\}^{-1} \\
= \;& n^{-1} \mathbf{D} \left\{ \mathbf{I} - \mathbf{D}^{-1} (n^{-1} \sum_{k=1}^{n} \mathbf{G}_k^{-1}) + \cdots \right\}^{-1} \\
= \;& n^{-1} \mathbf{D} + n^{-2} \sum_{k=1}^{n} \mathbf{G}_k^{-1} + \cdots ,
\end{aligned}
$$

by (4.95) we have,

$$
\begin{aligned}
\mathbf{c}_i(t)^T &= \mathbf{e}_1^T \left\{ \sum_{k=1}^{n} (\mathbf{I} + \mathbf{G}_k \mathbf{D})^{-1} \mathbf{G}_k \right\}^{-1} (\mathbf{I} + \mathbf{G}_i \mathbf{D})^{-1} \mathbf{G}_i \\
&= \mathbf{e}_1^T \left(n^{-1} \mathbf{D} + n^{-2} \sum_{k=1}^{n} \mathbf{G}_k^{-1} + \cdots \right) \\
&\quad \times (\mathbf{D}^{-1} - \mathbf{D}^{-1} \mathbf{G}_i^{-1} \mathbf{D}^{-1} + \cdots) \\
&= n^{-1} \mathbf{e}_1^T \left(\mathbf{I} + n^{-1} \sum_{k=1}^{n} \mathbf{G}_k^{-1} \mathbf{D}^{-1} - \mathbf{G}_i^{-1} \mathbf{D}^{-1} + \cdots \right) \\
&= n^{-1} \left[1 + O_p(\tilde{m}^{-1}) \right] \mathbf{e}_1^T .
\end{aligned}
$$

It follows that

$$\begin{aligned}
\hat{\eta}(t) &= \eta(t) + n^{-1} \sum_{i=1}^{n} \sum_{j=1}^{n_i} \mathbf{e}_1^T \mathbf{G}_i^{-1} \begin{pmatrix} 1 \\ t_{ij} - t \end{pmatrix} \\
&\quad \times \frac{K_h(t_{ij} - t)}{\sigma^2(t_{ij})} [y_{ij} - \eta(t) - \eta'(t)(t_{ij} - t)] \left[1 + O_p(\tilde{m}^{-1})\right] \\
&= \eta(t) + n^{-1} \sum_{i=1}^{n} Z_i(t) \left[1 + O_p(\tilde{m}^{-1})\right],
\end{aligned}$$

where

$$Z_i(t) = \sum_{j=1}^{n_i} \mathbf{e}_1^T G_i^{-1} \begin{pmatrix} 1 \\ t_{ij} - t \end{pmatrix} \frac{K_h(t_{ij} - t)}{\sigma^2(t_{ij})} [y_{ij} - \eta(t) - \eta'(t)(t_{ij} - t)],$$

which are independent of each other. Similar to the proof when $p = 0$, applying equation (4.94) and Condition A7, and using Taylor expansion with ignoring the higher order terms, we can show that

$$\begin{aligned}
\mathrm{E}\{Z_i(t)|\mathcal{D}\} &= \sum_{j=1}^{n_i} \mathbf{e}_1^T \mathbf{G}_i^{-1} \begin{pmatrix} 1 \\ t_{ij} - t \end{pmatrix} \\
&\quad \times \frac{K_h(t_{ij} - t)}{\sigma^2(t_{ij})} [\eta(t_{ij}) - \eta(t) - \eta'(t)(t_{ij} - t)] \\
&= \sum_{j=1}^{n_i} \frac{K_h(t_{ij} - t)\sigma^{-2}(t_{ij})}{n_i f(t)\sigma^{-2}(t)} \\
&\quad \times \eta''[t + \xi_{ij}(t_{ij} - t)] \frac{(t_{ij} - t)^2}{2} [1 + O_p((n_i h)^{-1/2})] \\
&= \frac{h^2}{2} \eta''(t) B(K) [1 + O_p((n_i h)^{-1/2})]
\end{aligned}$$

and

$$\begin{aligned}
\mathrm{Var}\{Z_i(t)|\mathcal{D}\} &= \left\{ \sum_{j=1}^{n_i} \mathbf{e}_1^T \mathbf{G}_i^{-1} \begin{pmatrix} 1 \\ t_{ij} - t \end{pmatrix} \frac{K_h(t_{ij} - t)}{\sigma^2(t_{ij})} \right\} \\
&\quad \times \gamma(t_{ij}, t_{il}) \left\{ \sum_{l=1}^{n_i} \mathbf{e}_1^T \mathbf{G}_i^{-1} \begin{pmatrix} 1 \\ t_{il} - t \end{pmatrix} \frac{K_h(t_{il} - t)}{\sigma^2(t_{il})} \right\} \\
&\quad + \sum_{j=1}^{n_i} \left\{ \mathbf{e}_1^T \mathbf{G}_i^{-1} \begin{pmatrix} 1 \\ t_{ij} - t \end{pmatrix} \frac{K_h(t_{ij} - t)}{\sigma^2(t_{ij})} \right\}^2 \sigma^2(t_{ij}) \\
&= \gamma(t, t) + \frac{\tau^2(t)}{n_i h f(t)} V(K) + o_p[(n_i h)^{-1}],
\end{aligned}$$

where ξ_{ij} are real numbers in $[0, 1]$. The rest of the proof progresses as those when $p = 0$. Thus Theorem 4.1 is proved.

Proof of Theorem 4.2 First, consider $p = 0$. Under Condition A, using (4.90) and similar arguments as for Theorem 4.1, when $p = 0$, we have

$$\hat{\eta}(t) - \mathrm{E}\{\hat{\eta}(t)|\mathcal{D}\} = n^{-1}\sum_{i=1}^{n} W_i(t)[1 + O_p(\tilde{m}^{-1})],$$

where

$$W_i(t) = \frac{\sum_{j=1}^{n_i} K_h(t_{ij} - t)\sigma^{-2}(t_{ij})[y_{ij} - \eta(t_{ij})]}{\sum_{j=1}^{n_i} K_h(t_{ij} - t)\sigma^{-2}(t_{ij})}. \tag{4.96}$$

Note that $\mathrm{E}\{W_i(t)|\mathcal{D}\} = 0$,

$$\mathrm{Var}\{W_i(t)|\mathcal{D}\} = \mathrm{Var}\{Z_i(t)|\mathcal{D}\} = \gamma(t, t) + \frac{\tau^2(t)}{n_i h f(t)} V(K)[1 + o_p(1)],$$

where $Z_i(t)$ is defined in (4.91). When Condition (4.64) holds, $W_i(t)$ are iid because $\mathbf{t}_i = (t_{i1}, \cdots, t_{im})^T$, $i = 1, 2, \cdots, n$ and $\mathbf{y}_i = (y_{i1}, \cdots, y_{im})^T$ are iid; when Condition (4.63) holds, $W_i(t)$ are uniformly bounded, i.e.,

$$|W_i(t)| \le \frac{\sum_{j=1}^{n_i} K_h(t_{ij} - t)\sigma^{-2}(t_{ij})C}{\sum_{j=1}^{n_i} K_h(t_{ij} - t)\sigma^{-2}(t_{ij})} = C.$$

In either case, by the celebrated Liapounov's central limit theorem, we have

$$n^{1/2}[\hat{\eta}(t) - \mathrm{E}\{\hat{\eta}(t)|\mathcal{D}\}] \sim AN[0, \gamma(t, t)].$$

Using (4.92) and (4.93), we have $\mathrm{E}\{\hat{\eta}(t)|\mathcal{D}\} = \eta(t) + O_p(h^2)$, and thus,

$$n^{1/2}[\eta(t) - \mathrm{E}\{\hat{\eta}(t)|\mathcal{D}\}] = O_p[n^{1/2}h^2] = O_p[n^{-2(\Delta - 3/2)/5}] = o_p(1),$$

due to Condition A7. It follows that

$$n^{1/2}[\hat{\eta}(t) - \eta(t)] \sim AN[0, \gamma(t, t)]$$

as desired. Thus Theorem 4.2(a) for $p = 0$ is proved.

When $p = 0$, to prove (b), let $\boldsymbol{\alpha} = (\alpha_1, \cdots, \alpha_m)^T$ be any m-dimensional real vector. It is then sufficient to prove that as $n \to \infty$,

$$n^{1/2}\boldsymbol{\alpha}^T\{\hat{\eta}(\mathbf{s}) - \eta(\mathbf{s})\} \sim AN[0, \boldsymbol{\alpha}^T\Gamma_{\mathbf{s}}\boldsymbol{\alpha}]. \tag{4.97}$$

Note that

$$\begin{aligned}
\boldsymbol{\alpha}^T[\hat{\eta}(\mathbf{s}) - \mathrm{E}\{\hat{\eta}(\mathbf{s})|\mathcal{D}\}] &= n^{-1}\sum_{i=1}^{n}\sum_{j=1}^{m} \alpha_j W_i(s_j)[1 + O_p(\tilde{m}^{-1})] \\
&= n^{-1}\sum_{i=1}^{n} H_i(\mathbf{s})[1 + O_p(\tilde{m}^{-1})],
\end{aligned}$$

where $H_i(\mathbf{s}) = \sum_{j=1}^m \alpha_j W_i(s_j)$, and

$$\begin{aligned} \mathrm{E}\{H_i(\mathbf{s})\} &= \sum_{j=1}^m \alpha_j \mathrm{E}\{W_i(s_j)\} = 0, \\ n^{1/2}[\eta(\mathbf{s}) - \mathrm{E}\{\hat{\eta}(\mathbf{s})|\mathcal{D}\}] &= \sum_{j=1}^m \alpha_j n^{1/2}[\eta(s_j) - \mathrm{E}\{\hat{\eta}(s_j)|\mathcal{D}\}] = o_p(1). \end{aligned}$$

When Condition (4.64) holds, $H_i(\mathbf{s})$ are iid; when Condition (4.63) holds,

$$|H_i(\mathbf{s})| \leq \sum_{j=1}^m |\alpha_j||W_i(s_j)| \leq C \sum_{j=1}^m |\alpha_j|.$$

Therefore (4.97) follows if

$$\begin{aligned} \mathrm{Var}\{H_i(\mathbf{s})|\mathcal{D}\} &= \sum_{j=1}^m \sum_{l=1}^m \alpha_j \alpha_l \mathrm{Cov}\{W_i(s_j), W_i(s_l)\} \\ &= \alpha^T \Gamma_{\mathbf{s}} \alpha [1 + o_p(1)]. \end{aligned} \tag{4.98}$$

Simple calculation shows that

$$\begin{aligned} &\mathrm{Cov}\{W_i(s_j), W_i(s_l)|\mathcal{D}\} = \gamma(s_j, s_l) \\ &+ \frac{\gamma(s_l, s_l) + \sigma^2(s_l)}{n_i h f(s_l)} \int K(u) K\left(u + \frac{s_j - s_l}{h}\right) du + o_p[1 + (n_i h)^{-1}]. \end{aligned}$$

Equation (4.98) follows and the proof for $p = 0$ is completed.

When $p = 1$, the proof is along the same lines.

Proof of Theorem 4.3 For $p = 0$, from (4.50) and under Condition B3, when $\delta^2(t) > 0$, we have

$$\hat{\eta}(t) = \frac{\sum_{i=1}^n B_i(t)}{\sum_{i=1}^n A_i(t)} = \frac{\sum_{i=1}^n \sum_{j=1}^{n_i} w_{ij}(t) y_{ij}}{\sum_{i=1}^n \sum_{j=1}^{n_i} w_{ij}(t)},$$

where

$$w_{ij}(t) = (2h)^{-1} \frac{\delta^2(t) K_h(t_{ij} - t) \sigma^{-2}(t_{ij})}{1 + \delta^2(t) \sum_{l=1}^{n_i} K_h(t_{il} - t) \sigma^{-2}(t_{il})}.$$

Notice that the weights $w_{ij}(t)$ here are slightly different from those in (4.48). If we can show that under Condition B,

$$w_{ij}(t) = (2h)^{-1} 1_{\{|\frac{t_{ij}-t}{h}| \leq 1\}} [1 + o(1)] = U_h(t_{ij} - t)[1 + o(1)], \tag{4.99}$$

where $U_h(u) = U(u/h)/h$, and U is the uniform kernel as defined in (4.68), then

$$\hat{\eta}(t) = \frac{\sum_{i=1}^n \sum_{j=1}^{n_i} U_h(t_{ij} - t) y_{ij}}{\sum_{i=1}^n \sum_{j=1}^{n_i} U_h(t_{ij} - t)} [1 + o(1)]. \tag{4.100}$$

It follows that $\hat{\eta}(t)$ is asymptotically equivalent to the usual kernel estimator of $\eta(t)$ with the uniform kernel U. Theorem 4.3 for $p = 0$ then follows immediately using standard arguments as those in Hoover et al (1998).

Actually, under Condition B, n_i are bounded while $n \to \infty, h \to 0$. Note that for any fixed time point t at the interior of the support of the density $f(t)$, there are only two cases about the relationship between $t_{ij}, j = 1, \cdots, n_i$ and t:

(1) $t_{ij} \neq t$ for all $j = 1, \cdots, n_i$;

(2) $t_{ij_0} = t$ for only one $1 \leq j_0 \leq n_i$ since there are no ties among the design time points of subject i.

For the first case, when $h < \min_{1 \leq j \leq n_i} |t_{ij} - t|$, $K_h(t_{ij} - t) = 0, j = 1, \cdots, n_i$ since K has a bounded support $[-1, 1]$. It follows that

$$(2h) w_{ij}(t) = \frac{\delta^2(t) K_h(t_{ij} - t)\sigma^{-2}(t_{ij})}{1 + \delta^2(t) \sum_{l=1}^{n_i} K_h(t_{il} - t)\sigma^{-2}(t_{il})} \to \frac{0}{1+0} = 0,$$

as $h \to 0$. Thus, in this case, $w_{ij}(t)$ can be expressed as in (4.99).

For the second case, similarly, when $h < \min_{1 \leq j \leq n_i, j \neq j_0} |t_{ij} - t|$, we have

$$\begin{aligned} K\left(\frac{t_{ij_0} - t}{h}\right) &= K(0) > 0, \\ K\left(\frac{t_{ij} - t}{h}\right) &= 0, j = 1, \cdots, j_0 - 1, j_0 + 1, \cdots, n_i, \\ \sum_{j=1}^{n_i} K\left(\frac{t_{ij} - t}{h}\right) &= K(0). \end{aligned}$$

It follows that, as $h \to 0$, we have

$$\begin{aligned} (2h) w_{ij_0}(t) &= \frac{\delta^2(t) h^{-1} K(0) \sigma^{-2}(t)}{1 + \delta^2(t) h^{-1} K(0) \sigma^{-2}(t)} \to 1, \\ (2h) w_{ij}(t) &= \frac{0}{1 + \delta^2(t) h^{-1} K(0) \sigma^{-2}(t)} \to 0, \\ j &= 1, \cdots, j_0 - 1, j_0 + 1, \cdots, n_i. \end{aligned}$$

Thus, in this case (and hence in both cases), $w_{ij}(t)$ can be expressed as in (4.99). Theorem 4.3 for $p = 0$ is then proved.

When $p = 1$, similar arguments lead to

$$\begin{aligned} \hat{\eta}(t) &= \mathbf{e}_1^T \left\{ \sum_{i=1}^{n} \sum_{j=1}^{n_i} U_h(t_{ij} - t) \begin{pmatrix} 1 \\ t_{ij} - t \end{pmatrix} (1 \quad t_{ij} - t) \right\}^{-1} \\ & \left\{ \sum_{i=1}^{n} \sum_{j=1}^{n_i} U_h(t_{ij} - t) \begin{pmatrix} 1 \\ t_{ij} - t \end{pmatrix} y_{ij} \right\} [1 + o(1)]. \end{aligned}$$

Theorem 4.3 for $p = 1$ then follows immediately using standard arguments as those in Hoover et al (1998). Theorem 4.3 is proved.

Proof of Theorem 4.4 When $p = 0$, by (4.100), $\hat{\eta}(t)$ is asymptotically equivalent to the usual kernel estimator of $\eta(t)$ with the uniform kernel U, Theorem 4.4 for $p = 0$ follows immediately by standard arguments for asymptotic normality of the local average kernel estimator for longitudinal data as in Wu, Chiang and Hoover (1998). For $p = 1$, similar arguments apply. Theorem 4.4 is then proved.

5

Regression Spline Methods

5.1 INTRODUCTION

In the last chapter, we described local polynomial methods for longitudinal data analysis. In this chapter, the focus is on regression spline methods. Regression splines (Wold 1974, Smith 1979, Eubank 1988, 1999) are piecewise polynomials that are specified by a group of breakpoints (knots) and some continuity conditions. A brief review was given in Section 3.3 of Chapter 3. Regression splines can be defined as linear combinations of a regression spline basis. Truncated power basis (Eubank 1988, 1999) and B-spline basis (de Boor 1978) are two of the most popular regression spline bases.

Regression splines are the first technique that is applied to smooth longitudinal data since they are natural extensions of polynomials. In this chapter, we provide an overview of regression spline methods for longitudinal data. The existing methods include naive regression splines (Huggins and Loesch 1998, Huang, Wu and Zhou 2002), generalized regression splines (Wang and Taylor 1995, Zhang 1997), and mixed-effects regression splines (Shi, Taylor and Wang 1996, Rice and Wu 2001). The first two methods aim to estimate the population mean function of a longitudinal data set, while the third method can be used to estimate the population mean function and individual functions simultaneously.

5.2 NAIVE REGRESSION SPLINES

Huggins and Loesch (1998) fit the population mean of growth curves using B-splines based on the ordinary least squares (OLS) method. Huang, Wu and Zhou (2002) extended this approach to fit a varying coefficient model. In both cases, the regression spline smoother (3.25) in Chapter 3 is applied to longitudinal data in a naive manner. That is, the within-subject correlation is not taken into account. We refer to this technique as a naive regression spline (NRS) method.

The NRS method aims to fit the following nonparametric population mean (NPM) model:

$$y_{ij} = \eta(t_{ij}) + e_{ij},\ j = 1, 2, \cdots, n_i;\ i = 1, 2, \cdots, n, \tag{5.1}$$

to a longitudinal data set:

$$(t_{ij}, y_{ij}),\ j = 1, 2, \cdots, n_i;\ i = 1, 2, \cdots, n, \tag{5.2}$$

where

$$t_{ij},\ j = 1, 2, \cdots, n_i;\ i = 1, 2, \cdots, n, \tag{5.3}$$

are the design time points. The NPM model (5.1) emphasizes the population mean function $\eta(t)$ and leaves everything else as errors e_{ij}.

5.2.1 The NRS Smoother

The key for the NRS method is to approximate the population mean function $\eta(t)$ of (5.1) by a linear combination of a regression spline basis $\mathbf{\Phi}_p(t) = [\phi_1(t), \cdots, \phi_p(t)]^T$. That is, we can approximately express $\eta(t)$ as a regression spline:

$$\eta(t) \approx \mathbf{\Phi}_p(t)^T \boldsymbol{\beta}, \tag{5.4}$$

where $\boldsymbol{\beta} = [\beta_1, \cdots, \beta_p]^T$ denotes the associated coefficient vector. When the basis $\mathbf{\Phi}_p(t)$ and the coefficient vector $\boldsymbol{\beta}$ are well chosen, the regression spline approximation (5.4) to the underlying population mean function $\eta(t)$ can be very accurate. As demonstrated in Section 3.3 of Chapter 3, we may estimate $\boldsymbol{\beta}$ by the OLS method. For this purpose, we replace $\eta(t)$ of (5.1) by the right-hand side of (5.4) and obtain a standard linear regression model:

$$y_{ij} = \mathbf{x}_{ij}^T \boldsymbol{\beta} + e_{ij},\quad j = 1, 2, \cdots, n_i;\ i = 1, 2, \cdots, n, \tag{5.5}$$

where $\mathbf{x}_{ij} = \mathbf{\Phi}_p(t_{ij})$.

For the i-th subject, denote the response vector, the design matrix, and the error vector, respectively, as

$$\mathbf{y}_i = [y_{i1}, \cdots, y_{in_i}]^T,\ \mathbf{X}_i = [\mathbf{x}_{i1}, \cdots, \mathbf{x}_{in_i}]^T,\ \mathbf{e}_i = [e_{i1}, \cdots, e_{in_i}]^T. \tag{5.6}$$

Then the model (5.5) can be written as

$$\mathbf{y}_i = \mathbf{X}_i \boldsymbol{\beta} + \mathbf{e}_i,\ i = 1, 2, \cdots, n. \tag{5.7}$$

For the whole data set, denote the response vector, the design matrix, and the error vector, respectively, as

$$\mathbf{y} = [\mathbf{y}_1^T, \cdots, \mathbf{y}_n^T]^T,\ \mathbf{X} = [\mathbf{X}_1^T, \cdots, \mathbf{X}_n^T]^T,\ \mathbf{e} = [\mathbf{e}_1^T, \cdots, \mathbf{e}_n^T]^T. \tag{5.8}$$

Then the model (5.7) can be further written as

$$\mathbf{y} = \mathbf{X}\beta + \mathbf{e}.$$

Then the OLS criterion is written as

$$(\mathbf{y} - \mathbf{X}\beta)^T(\mathbf{y} - \mathbf{X}\beta). \tag{5.9}$$

Minimizing the above with respect to β leads to the OLS estimator of β:

$$\hat{\beta}_{nrs} = (\mathbf{X}^T\mathbf{X})^{-1}\mathbf{X}^T\mathbf{y}. \tag{5.10}$$

From this, we obtain the NRS estimator of $\eta(t)$ at any time point t as,

$$\hat{\eta}_{nrs}(t) = \mathbf{\Phi}_p(t)^T\hat{\beta}_{nrs} = \mathbf{\Phi}_p(t)^T(\mathbf{X}^T\mathbf{X})^{-1}\mathbf{X}^T\mathbf{y}. \tag{5.11}$$

In particular, we have

$$\begin{aligned} \hat{y}_{nrs,ij} &= \mathbf{x}_{ij}^T(\mathbf{X}^T\mathbf{X})^{-1}\mathbf{X}^T\mathbf{y}, \\ \hat{\mathbf{y}}_{nrs,i} &= \mathbf{X}_i(\mathbf{X}^T\mathbf{X})^{-1}\mathbf{X}^T\mathbf{y}, \\ \hat{\mathbf{y}}_{nrs} &= \mathbf{X}(\mathbf{X}^T\mathbf{X})^{-1}\mathbf{X}^T\mathbf{y} = \mathbf{A}_{nrs}\mathbf{y}, \end{aligned} \tag{5.12}$$

where the so-called smoother matrix of the NRS estimator $\hat{\eta}(t)$ is

$$\mathbf{A}_{nrs} = \mathbf{X}(\mathbf{X}^T\mathbf{X})^{-1}\mathbf{X}^T. \tag{5.13}$$

Although the smoother matrix $\mathbf{A}_{nrs}$ (5.13) is based on the longitudinal data (5.2), it is still a projection matrix. That is, it has the following properties:

$$\mathbf{A}_{nrs}^T = \mathbf{A}_{nrs},\ \mathbf{A}_{nrs}^2 = \mathbf{A}_{nrs},\ \text{tr}(\mathbf{A}_{nrs}) = p.$$

Moreover, the expressions in (5.12) imply that the NRS estimator (5.11) is a linear smoother of $\eta(t)$ (Buja et al. 1989). From this, it is seen that the fitted response at any design time point t_{ij} is a linear combination of the response vector $\mathbf{y}$:

$$\hat{y}_{nrs,ij} = \sum_{k=1}^{n}\sum_{l=1}^{n_k} a_{nrs}(ij,kl)y_{kl},\ \ j = 1, 2, \cdots, n_i;\ i = 1, 2, \cdots, n, \tag{5.14}$$

where $a_{nrs}(ij, kl)$ is the (ij, kl)-th entry of the smoother matrix $\mathbf{A}_{nrs}$. In particular, $a_{nrs}(ij, ij)$ is the ij-th diagonal entry of $\mathbf{A}_{nrs}$.

5.2.2 Variability Band Construction

The pointwise standard deviation (SD) band of $\eta(t)$ allows us to assess how accurate the estimator $\hat{\eta}_{nrs}(t)$ is at different locations within the range of interest. To construct this pointwise SD band, we need to estimate the variance of $\hat{\eta}_{nrs}(t)$. For this purpose, we need to assume some structure about the covariance of the errors e_{ij}. The simplest working independence assumption is that

$$\text{the errors } e_{ij} \text{ are i.i.d.} \tag{5.15}$$

The above working independence assumption may not be valid for a "real-life" longitudinal data set but is made for convenience. The assumption is often appropriate when longitudinal data are sparse. Under this assumption, we have $\text{Cov}(\mathbf{y}) = \text{Cov}(\mathbf{e}) = \sigma^2 \mathbf{I}_N$ where σ^2 is the working variance. Thus

$$\begin{aligned} \text{Cov}(\hat{\beta}_{nrs}) &= \sigma^2 (\mathbf{X}^T \mathbf{X})^{-1}, \\ \text{Cov}[\hat{\eta}_{nrs}(s), \hat{\eta}_{nrs}(t)] &= \sigma^2 \mathbf{\Phi}_p(s)^T (\mathbf{X}^T \mathbf{X})^{-1} \mathbf{\Phi}_p(t). \end{aligned}$$

In particular, $\text{Var}(\hat{\eta}_{nrs}(t)) = \sigma^2 \mathbf{\Phi}_p(t)^T (\mathbf{X}^T \mathbf{X})^{-1} \mathbf{\Phi}_p(t)$. The estimated variance, denoted as $\widehat{\text{Var}}(\hat{\eta}_{nrs}(t))$, is then obtained via replacing the working variance σ^2 by its estimator, say,

$$\hat{\sigma}^2 = \|\mathbf{y} - \hat{\mathbf{y}}_{nrs}\|^2/(N-p) = \sum_{i=1}^{n} \sum_{j=1}^{n_i} (y_{ij} - \hat{y}_{nrs,ij})^2/(N-p), \tag{5.16}$$

where here and throughout this chapter, $N = \sum_{i=1}^{n} n_i$. Then the pointwise SD bands of $\hat{\eta}_{nrs}(t)$ can be constructed easily. For example, the 95% pointwise SD band of $\eta(t)$ can be constructed as

$$\hat{\eta}_{nrs}(t) \pm 1.96 \sqrt{\widehat{\text{Var}}[\hat{\eta}_{nrs}(t)]}. \tag{5.17}$$

5.2.3 Choice of the Bases

The NRS smoother (5.11) is constructed based on a given regression spline basis $\mathbf{\Phi}_p(t)$. In principle, $\mathbf{\Phi}_p(t)$ can be constructed using any basis such as the truncated power basis (Wold 1974, Smith 1979, Eubank 1999), B-spline basis (de Boor 1978), reproducing kernel Hilbert space basis (Wahba 1990), or wavelet basis, among others. The simplest one is the truncated power basis.

The truncated power basis, which has been defined in (3.21) of Chapter 3, can be written as

$$\mathbf{\Phi}_p(t) = [1, t, \cdots, t^k, (t-\tau_1)_+^k, \cdots, (t-\tau_K)_+^k]^T, \tag{5.18}$$

where $w_+^k = [w_+]^k$, with $w_+ = \max(0, w)$, is used to define the truncated power functions, and

$$\tau_1, \tau_2, \cdots, \tau_K, \tag{5.19}$$

are K given knots (in an increasing order) scattered in the range of interest. The number of the basis functions, $p = K + k + 1$, with k being the degree of the truncated power basis (5.18).

When a truncated power basis or a B-spline basis is used in the NRS smoother (5.11), we need to locate the knots (5.19) and select the number of the basis functions, p, for good performance of the NRS smoother. For the truncated power basis (5.18), $p = K + k + 1$, and thus when k is fixed, selecting p is equivalent to selecting K.

5.2.4 Knot Locating Methods

In Section 3.3.3, three methods for locating the knots for a truncated power basis or a B-spline basis were reviewed. They can also be applied for locating the knots (5.19) for a truncated power basis or a B-spline basis, $\mathbf{\Phi}_p(t)$, with the longitudinal data (5.2). The third method is called the "model selection based method" which can be applied directly to the current longitudinal data setup. For the first method, i.e., the "equally spaced method", we only need to specify the interval of interest $[a, b]$ for the knot construction formula (3.28). For the longitudinal data (5.2), we can use the minimum and maximum values of the pooled design time points (5.3) as a and b, respectively. That is,

$$a = \min_{i=1}^{n} \min_{j=1}^{n_i} t_{ij}, \quad b = \max_{i=1}^{n} \max_{j=1}^{n_i} t_{ij}.$$

For the second method, the "equally spaced sample quantiles method," the key is to specify the order statistics $t_{(l)}, l = 1, 2, \cdots, N$ of the pooled design time points (5.3). Then we can apply the knot construction formula (3.29) to these order statistics to construct the knots (5.19) via replacing n by N. For the longitudinal data (5.2), some knots may be tied. This will cause the design matrix $\mathbf{X}$ defined in (5.8) to degenerate. This problem can be easily solved via removing those extra tied knots.

5.2.5 Selection of the Number of Basis Functions

For the third knot locating method, the choice of the number of basis functions, p, is done at the same time as the knot introduction or deletion. But for the other two methods, p is generally chosen using a smoothing parameter selector. The "leave-one-point-out" cross-validation (PCV) and the "leave-one-subject-out" cross-validation (SCV) for the naive LPK smoothers of Chapter 4 can be adopted here.

Based on the NRS smoother (5.11), the PCV score may be written as

$$\mathrm{PCV}_{nrs}(p) = \sum_{i=1}^{n} (\mathbf{y}_i - \hat{\mathbf{y}}_{nrs,i}^{(-pt)})^T \mathbf{W}_i (\mathbf{y}_i - \hat{\mathbf{y}}_{nrs,i}^{(-pt)}), \tag{5.20}$$

where $\mathbf{W} = \mathrm{diag}(\mathbf{W}_1, \cdots, \mathbf{W}_n)$ is a user-specified weight matrix,

$$\hat{\mathbf{y}}_{nrs,i}^{(-pt)} = [\hat{\mathbf{y}}_{nrs,i1}^{(-i1)}, \cdots, \hat{\mathbf{y}}_{nrs,in_i}^{(-in_i)}]^T,$$

with $\hat{y}_{nrs,ij}^{(-ij)}$ denoting the NRS fit (5.11) computed using all the data (5.3) except the data point (t_{ij}, y_{ij}).

The weight matrix $\mathbf{W}$ is often specified by one of the following three methods:

1. Take $\mathbf{W}_i = N^{-1}\mathbf{I}_{n_i}$ so that each of the measurements is treated equally.

2. Take $\mathbf{W}_i = (nn_i)^{-1}\mathbf{I}_{n_i}$ so that each of the measurements within a subject is treated equally.

3. Take $\mathbf{W}_i = \mathbf{V}_i^{-1}$ where $\mathbf{V}_i = \text{Cov}(\mathbf{y}_i)$ so that the within-subject correlation is taken into account.

For the NRS smoother (5.11), we often use Method 1 for simplicity.

As we might expect, directly computing the PCV score (5.20) is intensive since we need to compute the NRS fit (5.11) N times. Using (5.12), the PCV score (5.20) can be simplified as

$$\text{PCV}_{nrs}(p) = \sum_{i=1}^{n} (\mathbf{y}_i - \hat{\mathbf{y}}_{nrs,i})^T \mathbf{S}_i^{-1} \mathbf{W}_i \mathbf{S}_i^{-1} (\mathbf{y}_i - \hat{\mathbf{y}}_{nrs,i}), \tag{5.21}$$

where $\mathbf{S}_i = \mathbf{I}_{n_i} - \text{diag}(a_{nrs}(i1, i1), \cdots, a_{nrs}(in_i, in_i))$ with $a_{nrs}(ij, ij)$ being the coefficient of y_{ij} in the expression (5.14) and being the (ij)-th diagonal entry of $\mathbf{A}_{nrs}$ as defined in (5.13).

To further reduce the computational effort of the PCV score (5.21), following Wahba (1983), we may replace the diagonal entries $a_{nrs}(ij, ij)$ in the PCV score (5.20) by their overall average

$$\sum_{i=1}^{n} \sum_{j=1}^{n_i} a_{nrs}(ij, ij)/N = \text{tr}(\mathbf{A}_{nrs})/N = p/N,$$

so that the $\mathbf{S}_i$ can be replaced by $(1 - p/N)\mathbf{I}_{n_i}$. Then we have the following so-called GCV score:

$$\text{GCV}_{nrs}(p) = \sum_{i=1}^{n} (\mathbf{y}_i - \hat{\mathbf{y}}_{nrs,i})^T \mathbf{W}_i (\mathbf{y}_i - \hat{\mathbf{y}}_{nrs,i}) / (1 - p/N)^2. \tag{5.22}$$

Notice that the numerator in the GCV score (5.22) is the weighted SSE (sum of squared errors), representing the goodness of fit, and the denominator is associated with the model complexity of the NRS smoother (5.11).

Following Rice and Silverman (1991), we may define the SCV score as follows:

$$\text{SCV}_{nrs}(p) = \sum_{i=1}^{n} (\mathbf{y}_i - \hat{\mathbf{y}}_{nrs,i}^{(-i)})^T \mathbf{W}_i (\mathbf{y}_i - \hat{\mathbf{y}}_{nrs,i}^{(-i)}) \tag{5.23}$$

where $\hat{\mathbf{y}}_{nrs,i}^{(-i)}$ is computed using (5.12) with all the data except the data from the i-th subject. That is, in the definition of the SCV score above, the predictions at all the design time points for a subject are computed using the whole data set with the data for that subject completely excluded.

In practice, we also need to specify a proper range for p. We should restrict p so that the design matrix $\mathbf{X}$, as defined in (5.8), is of full rank, i.e., not degenerated. A general rule for p is that p should be less than the number of the distinct design time points, M.

When the truncated power basis (5.18) is used and when k is fixed, specification of p is equivalent to specification of the number of knots, K. In this case, we limit $p = K + k + 1 \leq M$ so that

$$K_{\max} = M - k - 1. \tag{5.24}$$

Then, we select K over $\{0, 1, \cdots, K_{\max}\}$ so that the PCV, GCV or SCV score is minimized. In most applications, it is sufficient to focus on some small k's, say, $k = 1, 2, 3$ or 4. For example, we often focus on $k = 2$ or 3, i.e., the quadratic or cubic regression splines.

5.2.6 Example and Model Checking

As an example, we applied the NRS fit (5.11) to the ACTG 388 data, introduced in Chapter 1. For simplicity, we adopted the truncated power basis (5.18) as $\mathbf{\Phi}_p(t)$ with $k = 2$, and adopted the "equally spaced sample quantiles as knots" method to specify the knots. The number of knots, K, or equivalently, the number of basis functions, $p = K + k + 1$, was selected by the GCV rule (5.22). Figure 5.1 displays the results. Figure 5.1 (a) shows the NRS fit (solid curve), together with the raw CD4 counts (dots), and the 95% pointwise SD band (dashed curves). It is seen that the estimated population mean function was smooth. It increased sharply during the first several weeks, and continued to increase at a slower rate until about week 110, and then dropped at the end of the study. This shows that with anti-viral treatment, the overall CD4 counts increased dramatically during the first few weeks, but the drug effect faded over time and completely disappeared after about week 110 so that the CD4 counts began to drop.

The NRS fit was computed using (5.11) with $p = 7$. The degree of the truncated power basis $\mathbf{\Phi}_p(t)$ is $k = 2$ so that $K = 4$ knots were actually used. Figure 5.1 (b) shows how this best p was selected using the GCV rule (5.22), where the GCV score was plotted against the number of basis functions, p (instead of the number of knots, K). It is seen that $p = 7$ corresponds to the smallest GCV score.

From Figure 5.1 (a), it is seen that for each fixed time point t, the NRS fit is not at the center of the responses from all the subjects. Is this reasonable? This question may be answered by looking at Figure 5.1 (c) where the NRS fit, together with its 95% pointwise SD band, was superimposed with the raw pointwise confidence intervals (CIs) at each of the distinct design time points. The raw pointwise CIs were computed in a way described briefly as follows. At a given time point t, the mean and SD are computed using all the responses at t; when there is only one response at t, the global SD is applied; see Chapter 1 for some details about constructing these raw pointwise CIs. It is seen that

(1). When there are enough measurements at t, the raw pointwise CI and the associated NRS SD interval are comparable.

(2). When there are few measurements at t, the raw pointwise CI is much wider than the associated NRS SD interval. This is due to the fact that the raw pointwise CI just used the data at t while the NRS SD interval used the data nearby.

(3). The raw mean curve obtained via connecting the centers of the raw pointwise CIs is very rough.

This indicates that the NRS method is effective in smoothing the population mean function.

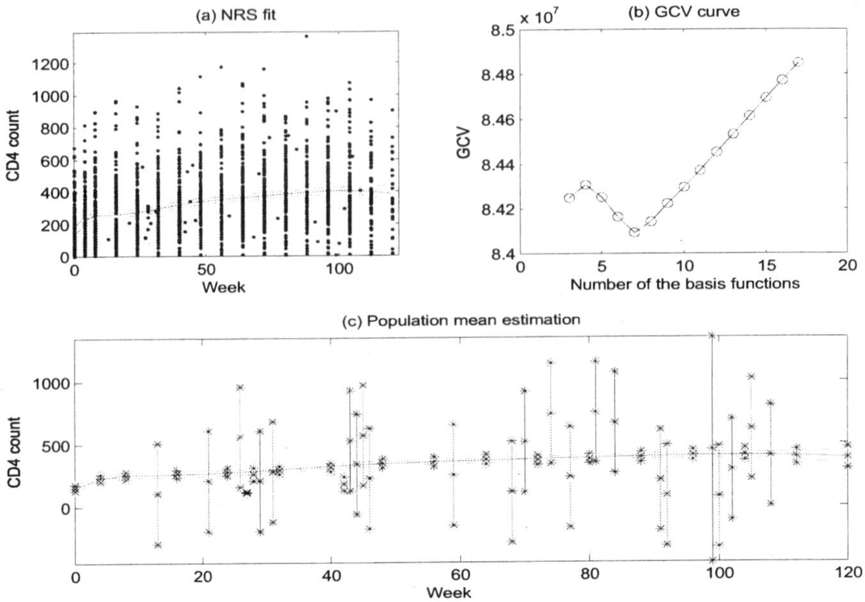

Fig. 5.1 NRS fit to the ACTG 388 data. (a) raw CD4 counts (dots), NRS fit (solid curve), and 95% pointwise SD band (dashed curves); (b) GCV curve; and (c) NRS fit (solid curve), 95% pointwise SD band (dashed curves) and raw pointwise CIs (vertical intervals with stars at the ends and the center).

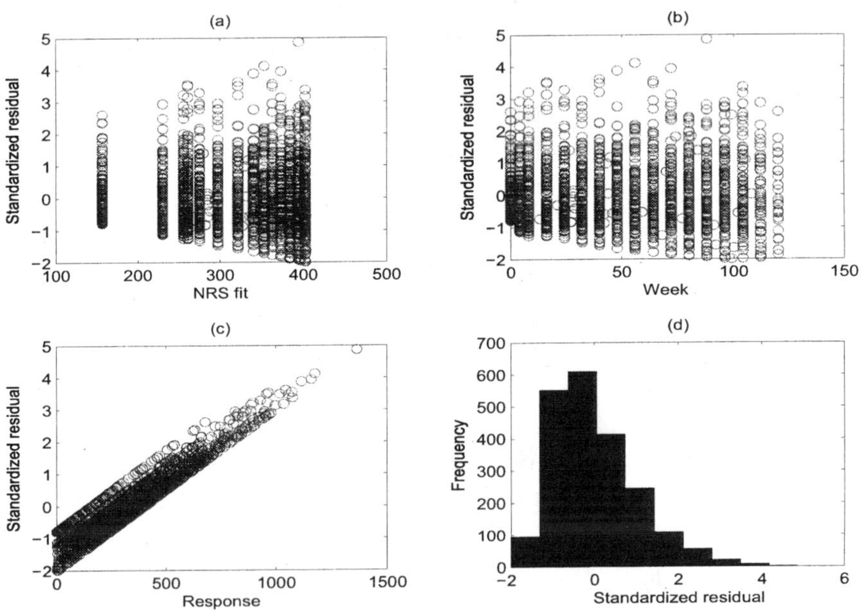

Fig. 5.2 Residual analysis of the quadratic NRS fit to the ACTG 388 data.

We conducted a residual analysis to assess the adequacy of the NRS fit to the population mean function of the ACTG 388 data. Figure 5.2 displays the results. Figure 5.2 (a), (b) and (c) show the standardized residual against the NRS fit, time and response, respectively. Figure 5.2 (d) displays the histogram of the standardized residuals. From Figures 5.2 (a) and (b), it is seen that the population mean function had been adequately fitted although both the residual plots indicate that the noise is not very homogeneous and some transformation of the data is needed. The linear trend in Figure 5.2 (c) indicates that there is some information left in the standardized residuals, which was not accounted for by the NPM model (5.1). In a later section (Section 5.4.6), we will show that this linear trend disappears after we account for the between-subject variation (within-subject correlation) in the data. The histogram in Figure 5.2 (d) shows that the standardized residuals are not symmetric, and are seriously skewed.

A similar analysis of the ACTG 388 data was also conducted by using the SCV rule (5.23). The number of basis functions, p, selected by the SCV rule is 8 instead of 7, but the estimated mean function (not shown) is about the same as the one using the GCV rule. The residual plots are also quite similar to those presented in Figure 5.2, resulting in similar conclusions.

5.2.7 Comparing GCV against SCV

In the ACTG 388 data example, the truncated power basis (5.18) was used. The results produced by SCV were about the same as those by GCV except that the number of basis functions, p, selected by the two methods was different (for SCV, $p = 8$ and for GCV, $p = 7$). A question naturally arises: which rule should be used, GCV or SCV?

From a computational effort perspective, GCV is definitely preferred since GCV is computationally less intensive. This is because GCV has a simpler formula (5.22) compared to SCV. Our personal experiences revealed that the time required by SCV is often about 10 to 100 times greater than that required by GCV, depending on the sample size, the number of smoothing parameter candidates, and the computer's CPU.

However, from a statistical efficiency viewpoint, which rule is preferred, GCV or SCV? To answer this question, we conducted a small-scale simulation study. In Section 4.8 of Chapter 4, we designed a simulation model (4.78) and proposed two criteria ASER and ASEB as in (4.84) and (4.85) for assessing the performance of a smoother, say, Smoother B, against another, say, Smoother A. When the ASER is positive or the ASEB is larger than 50%, Smoother B performs better than Smoother A. Conversely, when ASER is negative or the ASEB is smaller than 50%, Smoother A performs better than Smoother B. In this simulation study, we adopted the same simulation model with these two criteria for assessing the performance of the NRS smoother (5.11). For simplicity, we used the quadratic truncated power basis (5.18) $\mathbf{\Phi}_p(t)$ with K (equivalently $p = K+k+1$ with $k = 2$ fixed) selected by GCV, named K_{gcv} against SCV, named K_{scv}. We also adopted all the 27 scenarios of simulation parameters used in the simulation studies of Section 4.8 but set $N = 100$ instead. That is, we repeated 100 times each scenario.

Table 5.1 shows the simulation results. In the first column, n and m are the number of subjects and the number of distinct design time points, respectively, while r_{miss} indicates the missing rate. The second column lists the minimum correlation coefficient $\rho_{\min}$. The ASER and ASEB together with their 95% standard deviation (2SD) errors are listed in the third and fourth columns, respectively. In the fifth and sixth columns, $\bar{K}_{gcv}$ and $\bar{K}_{scv}$ are listed. As an example, we explain the first scenario listed in the first row in detail. In this scenario, there are $n = 30$ subjects, $m = 20$ distinct design time points, the missing rate is $r_{miss} = 20\%$, and the minimum correlation coefficient is $\rho_{min} = .43$. For this scenario, GCV outperformed SCV in the sense of both ASER $= -6.5\% < 0$, and ASEB $= 21\% < 50\%$. The average number of knots selected by GCV and SCV was $\bar{K}_{gcv} = 3.03$ and $\bar{K}_{scv} = 4.57$, respectively.

Table 5.1 ASER, ASEB (in percentage), $\bar{K}_{gcv}$ and $\bar{K}_{scv}$ with their 95% standard deviations (2SD) (in parentheses) for the NRS fits of $\eta(t)$ using GCV against SCV, $N = 100$.

(n, m, r_{miss})	$\rho_{\min}$	ASER	ASEB	$\bar{K}_{gcv}$	$\bar{K}_{scv}$
(30,20,20%)	.43	−6.5(5.4)	21(8.1)	3.03(0.12)	4.57(0.46)
	.67	−0.1(6.5)	34(9.5)	2.51(0.15)	3.57(0.31)
	.79	6.5(4.6)	48(9.9)	2.15(0.15)	3.87(0.42)
(50,20,20%)	.43	−7.7(7.9)	18(7.7)	3.27(0.11)	4.51(0.40)
	.67	−0.2(6.4)	34(9.5)	3.05(0.14)	4.32(0.33)
	.79	0.6(3.8)	40(9.8)	2.66(0.14)	4.05(0.38)
(70,20,20%)	.43	−5.1(4.9)	25(8.7)	3.59(0.13)	4.86(0.43)
	.67	−4.0(5.4)	39(9.8)	3.16(0.09)	4.94(0.51)
	.79	0.9(4.3)	35(9.5)	2.92(0.12)	4.59(0.43)
(30,20,50%)	.43	−4.8(5.4)	12(6.5)	2.52(0.16)	3.35(0.39)
	.67	−6.5(9.0)	18(7.7)	2.32(0.21)	3.42(0.42)
	.79	−8.2(9.5)	20(8.0)	1.91(0.18)	3.03(0.49)
(50,20,50%)	.43	−10.5(8.4)	11(6.2)	3.18(0.17)	4.06(0.41)
	.67	0.2(4.3)	19(7.8)	2.61(0.23)	3.34(0.27)
	.79	−3.9(4.7)	20(8.0)	2.29(0.19)	3.22(0.33)
(70,20,50%)	.43	−7.0(6.9)	14(6.9)	3.21(0.17)	3.99(0.34)
	.67	−5.7(6.4)	12(6.5)	3.06(0.18)	3.96(0.46)
	.79	−4.8(8.0)	21(8.1)	2.44(0.16)	3.76(0.43)
(30,20,80%)	.43	−1.9(5.6)	11(6.2)	2.18(0.28)	2.69(0.40)
	.67	−13.4(14.2)	10(6.0)	2.27(0.34)	2.80(0.43)
	.79	−9.6(10.1)	8(5.4)	1.97(0.33)	2.47(0.47)
(50,20,80%)	.43	−10.1(12.3)	7(5.1)	2.87(0.42)	3.42(0.52)
	.67	−11.2(10.8)	5(4.3)	2.17(0.23)	2.81(0.45)
	.79	−19.5(13.3)	7(5.1)	2.02(0.32)	2.84(0.50)
(70,20,80%)	.43	−2.2(4.7)	5(4.3)	2.82(0.29)	3.07(0.35)
	.67	−8.0(7.8)	8(5.4)	2.40(0.23)	3.01(0.36)
	.79	−9.2(9.0)	6(4.7)	2.27(0.32)	2.79(0.42)

Some interesting results may be found with a careful inspection of Table 5.1:

- In terms of ASEB, GCV always outperformed SCV in the sense that all the ASEBs are less than 50%. Moreover, when $r_{miss} = 20\%, 50\%$ and 80%, the associated overall ASEBs are not larger than $48\%, 21\%$ and 11%, respectively.

- In terms of ASER, GCV outperformed SCV most of time in the sense that most of the ASERs are negative, with only one case showing a clearly positive ASER. This case (the third scenario with $n = 30, M = 20, r_{miss} = 20\%$ and $\rho_{\min} = .79$) represents a small sample size with a high level of correlation. Moreover, when $r_{miss} = 20\%, 50\%$ and 80%, the associated overall ASER's are not larger than $6.5\%, .2\%$ and -1.9%, respectively.

- The average number of knots selected by SCV and the associated standard deviations are always larger than those by GCV. This may provide a possible explanation why the GCV rule always outperformed the SCV rule.

The above results allow us to recommend using GCV other than SCV in NRS modeling of the NPM model (5.1) when data are sparse or less correlated. As a rule of thumb, when the missing rate is larger than 70%, GCV should generally be used to select the number of knots, K, for the truncated power basis (5.18). The lessons we learned from this simulation study are also applicable to selecting the number of basis number, p, for a truncated power basis or a B-spline basis.

5.3 GENERALIZED REGRESSION SPLINES

The naive regression spline (NRS) (5.11) method does not take the correlation of the longitudinal data (5.2) into account. Very often, properly taking the within-subject correlation is beneficial. Wang and Tayor (1995) and Zhang (1997) took the within-subject correlation into account via a regression spline approximation of the population mean function using a generalized least squares (GLS) method and specifying the covariance structure by a prespecified parametric model. We refer to this technique as a generalized regression spline (GRS) method.

5.3.1 The GRS Smoother

Let $\mathbf{V}_i$ denote the covariance matrix of $\mathbf{y}_i$ for $i = 1, 2, \cdots, n$. Since $\mathbf{y}_1, \mathbf{y}_2, \cdots, \mathbf{y}_n$ are independent, we have $\mathbf{V} = \text{Cov}(\mathbf{y}) = \text{diag}(\mathbf{V}_1, \cdots, \mathbf{V}_n)$. That is, $\mathbf{V}$ is a block diagonal matrix with diagonal matrices $\mathbf{V}_1, \cdots, \mathbf{V}_n$. By assuming $\mathbf{V}$ known, the GRS method accounts for the correlation of the longitudinal data via using the following GLS criterion:

$$(\mathbf{y} - \mathbf{X}\beta)^T\mathbf{V}^{-1}(\mathbf{y} - \mathbf{X}\beta), \tag{5.25}$$

where $\mathbf{X}, \mathbf{y}, \beta$ are defined as before, specifically $\mathbf{X}$ is defined in (5.6) and (5.8) based on a regression spline basis, $\mathbf{\Phi}_p(t)$. This criterion is a natural generalization of the OLS criterion (5.9). When $\mathbf{V} = \mathbf{I}_N$, the former reduces to the latter.

The GLS criterion (5.25) can be derived using the OLS criterion (5.9). In fact, when $\mathbf{V} \neq \mathbf{I}_N$, we can define $\tilde{\mathbf{y}} = \mathbf{V}^{-1/2}\mathbf{y}$, $\tilde{\mathbf{X}} = \mathbf{V}^{-1/2}\mathbf{X}$ and $\tilde{\mathbf{e}} = \mathbf{V}^{-1/2}\mathbf{e}$ so that the following linear model:

$$\tilde{\mathbf{y}} = \tilde{\mathbf{X}}\boldsymbol{\beta} + \tilde{\mathbf{e}} \tag{5.26}$$

satisfies the working independence assumption (5.15). Therefore, we can use the OLS criterion (5.9). That is $(\tilde{\mathbf{y}} - \tilde{\mathbf{X}})^T(\tilde{\mathbf{y}} - \tilde{\mathbf{X}})$ which is exactly equal to the GLS criterion (5.25).

Minimizing (5.25) with respect to $\boldsymbol{\beta}$ leads to

$$\hat{\boldsymbol{\beta}}_{grs} = (\mathbf{X}^T\mathbf{V}^{-1}\mathbf{X})^{-1}\mathbf{X}^T\mathbf{V}^{-1}\mathbf{y}. \tag{5.27}$$

Then the GRS smoother of $\eta(t)$ in the model (5.1) is

$$\hat{\eta}_{grs}(t) = \boldsymbol{\Phi}_p(t)^T(\mathbf{X}^T\mathbf{V}^{-1}\mathbf{X})^{-1}\mathbf{X}^T\mathbf{V}^{-1}\mathbf{y}. \tag{5.28}$$

Evaluating $\hat{\eta}_{grs}(t)$ at all the design time points leads to the fitted response vector :

$$\hat{\mathbf{y}}_{grs} = \mathbf{X}(\mathbf{X}^T\mathbf{V}^{-1}\mathbf{X})^{-1}\mathbf{X}^T\mathbf{V}^{-1}\mathbf{y} = \mathbf{A}_{grs}\mathbf{y}, \tag{5.29}$$

where

$$\mathbf{A}_{grs} = \mathbf{X}(\mathbf{X}^T\mathbf{V}^{-1}\mathbf{X})^{-1}\mathbf{X}^T\mathbf{V}^{-1} \tag{5.30}$$

is the associated GRS smoother matrix.

It is noteworthy that the GRS smoother matrix (5.30) is a generalization of the NRS smoother matrix (5.13), that is, when $\mathbf{V} = \mathbf{I}_N$, they are the same. However, the GRS smoother matrix (5.30) is no longer a projection matrix since it is not symmetric. But we still have $\mathbf{A}_{grs}^2 = \mathbf{A}_{grs}$ and $\text{tr}(\mathbf{A}_{grs}) = p$. Moreover, the GRS smoother (5.28) is still a linear estimator of $\eta(t)$ when $\mathbf{V}$ is known and we still have (5.14) but with the $\mathbf{A}_{nrs}$ replaced by the $\mathbf{A}_{grs}$.

5.3.2 Variability Band Construction

We can construct the pointwise SD band of $\eta(t)$ based on the GRS smoother (5.28). Since $\text{Cov}(\mathbf{y}) = \mathbf{V}$, by (5.27) and (5.28), we have

$$\begin{aligned} \text{Cov}(\hat{\boldsymbol{\beta}}_{grs}) &= (\mathbf{X}^T\mathbf{V}^{-1}\mathbf{X})^{-1}, \\ \text{Cov}(\hat{\eta}_{grs}(s), \hat{\eta}_{grs}(t)) &= \boldsymbol{\Phi}_p(s)^T(\mathbf{X}^T\mathbf{V}^{-1}\mathbf{X})^{-1}\boldsymbol{\Phi}_p(t). \end{aligned} \tag{5.31}$$

In particular,

$$\text{Var}(\hat{\eta}_{grs}(t)) = \boldsymbol{\Phi}_p(t)^T(\mathbf{X}^T\mathbf{V}^{-1}\mathbf{X})^{-1}\boldsymbol{\Phi}_p(t). \tag{5.32}$$

In practice, the $\mathbf{V}$ in the above expressions is usually unknown and should be replaced by its estimator $\hat{\mathbf{V}}$. Then the pointwise SD band of $\eta(t)$ can be constructed using (5.17) by replacing $\hat{\eta}_{nrs}(t)$ with $\hat{\eta}_{grs}(t)$.

5.3.3 Selection of the Number of Basis Functions

As for the NRS smoother (5.11), we need to choose the number of basis functions in $\mathbf{\Phi}_p(t)$ for the GRS smoother (5.28). Let us see how we can develop the GCV (5.22) for the GRS smoother. Notice that the model (5.26) transformed from the model (5.1) is a standard linear model with $\text{Cov}(\tilde{\mathbf{y}}) = \text{Cov}(\tilde{\mathbf{e}}) = \mathbf{I}_N$. It follows that $\hat{\beta}_{grs} = (\tilde{\mathbf{X}}^T\tilde{\mathbf{X}})^{-1}\tilde{\mathbf{X}}^T\tilde{\mathbf{y}}$ and $\hat{\tilde{\mathbf{y}}}_{nrs} = \tilde{\mathbf{A}}_{nrs}\tilde{\mathbf{y}}$ where the associated smoother matrix is $\tilde{\mathbf{A}}_{nrs} = \tilde{\mathbf{X}}(\tilde{\mathbf{X}}^T\tilde{\mathbf{X}})^{-1}\tilde{\mathbf{X}}^T$, which is a projection matrix with $\text{tr}(\tilde{\mathbf{A}}_{nrs}) = p$. Therefore, the GCV criterion for the standard linear model (5.26) can be written as

$$\begin{aligned} & \text{GCV}_{grs}(p) \\ = & \sum_{i=1}^{n}(\tilde{\mathbf{y}}_i - \hat{\tilde{\mathbf{y}}}_{nrs,i})^T\mathbf{W}_i(\tilde{\mathbf{y}}_i - \hat{\tilde{\mathbf{y}}}_{nrs,i})/(1-p/N)^2 \\ = & \frac{\sum_{i=1}^{n}(\mathbf{y}_i - \hat{\mathbf{y}}_{grs,i})^T\mathbf{V}_i^{-1/2}\mathbf{W}_i\mathbf{V}_i^{-1/2}(\mathbf{y}_i - \hat{\mathbf{y}}_{grs,i})}{(1-p/N)^2}, \end{aligned} \tag{5.33}$$

since actually we have

$$\begin{aligned} \tilde{\mathbf{A}}_{nrs} &= \mathbf{V}^{-1/2}\mathbf{X}(\mathbf{X}^T\mathbf{V}^{-1}\mathbf{X})^{-1}\mathbf{X}^T\mathbf{V}^{-1/2}, \\ \hat{\tilde{\mathbf{y}}}_{nrs} &= \mathbf{V}^{-1/2}\mathbf{X}(\mathbf{X}^T\mathbf{V}^{-1}\mathbf{X})^{-1}\mathbf{X}^T\mathbf{V}^{-1}\mathbf{y} = \mathbf{V}^{-1/2}\hat{\mathbf{y}}_{grs}. \end{aligned}$$

When $\mathbf{W}^{-1} = N^{-1}\mathbf{I}_N$, the GCV criterion (5.33) reduces to

$$\text{GCV}_{grs}(p) = \frac{N^{-1}\sum_{i=1}^{n}(\mathbf{y}_i - \hat{\mathbf{y}}_{grs,i})^T\mathbf{V}_i^{-1}(\mathbf{y}_i - \hat{\mathbf{y}}_{grs,i})}{(1-p/N)^2}.$$

It seems that there is no need to define the SCV for the GRS smoother (5.28) since when $\mathbf{V}$ is known, the model (5.26) is a standard linear model with i.i.d errors.

5.3.4 Estimating the Covariance Structure

In the above subsection, we assumed that $\mathbf{V}$ is known. However in practice, it is often not the case and $\mathbf{V}$ has to be estimated. In the literature, for simplicity, many authors (Wang and Taylor 1995, Wang 1998a, Zhang 1997 among others) assumed that $\mathbf{V}$ follows a known parametric model with some unknown parameter vector $\boldsymbol{\theta}$. We write $\mathbf{V}(\boldsymbol{\theta})$ to emphasize the dependence of $\mathbf{V}$ on $\boldsymbol{\theta}$.

Under the normality assumption, $\boldsymbol{\theta}$ can be estimated using the maximum likelihood (ML) method (Wang and Taylor 1995) or restricted maximum likelihood (REML) method (Diggle et al. 2002). The ML estimate of $\boldsymbol{\theta}$ is obtained via maximizing the following log-likelihood function (up to a constant) with respect to $\boldsymbol{\theta}$:

$$\text{Loglik}(\boldsymbol{\theta}) = -\frac{1}{2}\sum_{i=1}^{n}\left\{\log|\mathbf{V}_i(\boldsymbol{\theta})| + (\mathbf{y}_i - \hat{\mathbf{y}}_{grs,i})^T\mathbf{V}_i^{-1}(\boldsymbol{\theta})(\mathbf{y}_i - \hat{\mathbf{y}}_{grs,i})\right\}, \tag{5.34}$$

where $\hat{\mathbf{y}}_{grs} = [\hat{\mathbf{y}}_{grs,1}^T, \cdots, \hat{\mathbf{y}}_{grs,n}^T]^T$ is computed using (5.29) with the $\mathbf{V}(\boldsymbol{\theta})$. The REML estimate of $\hat{\boldsymbol{\theta}}$ is obtained via maximizing the following restricted log-likelihood function (up to a constant) with respect to $\boldsymbol{\theta}$:

$$\text{Loglik}^*(\boldsymbol{\theta}) = \text{Loglik}(\boldsymbol{\theta}) - \frac{1}{2}\sum_{i=1}^{n} \log |\mathbf{X}_i^T \mathbf{V}_i(\boldsymbol{\theta})^{-1} \mathbf{X}_i|. \tag{5.35}$$

The above maximization problems can be solved using the Newton-Raphson method, the Fisher-scoring method or a quasi-Newton method. Many parametric models for $\mathbf{V}(\boldsymbol{\theta})$ have been proposed in the literature; see Wang and Taylor (1995) and Diggle et al. (2002, Chapter 5), among others.

5.4 MIXED-EFFECTS REGRESSION SPLINES

In the previous sections, we described the NRS and GRS smoothers. These two smoothers aim to estimate the population mean function $\eta(t)$ in the NPM model (5.1). In many applications, one is also interested in estimating the subject-specific curves. For example, using the ACTG 388 data, a doctor may be particularly interested in predicting the CD4 counts of individual patients. In this case, we need to extend the NPM model (5.1) to the following nonparametric mixed-effects (NPME) model:

$$\begin{array}{c} y_{ij} = \eta(t_{ij}) + v_i(t_{ij}) + \epsilon_{ij}, \\ v_i(t) \sim \text{GP}(0, \gamma), \boldsymbol{\epsilon}_i = [\epsilon_{i1}, \cdots, \epsilon_{in_i}]^T \sim N(0, \mathbf{R}_i), \\ j = 1, 2, \cdots, n_i; \ i = 1, 2, \cdots, n, \end{array} \tag{5.36}$$

where $\eta(t)$ is the smooth population mean function of the longitudinal data as in the NPM model (5.1); $v_i(t), \ i = 1, 2, \cdots, n$ are the smooth subject-specific deviations from $\eta(t)$, which are assumed to be realizations of a Gaussian process with mean 0 and covariance function $\gamma(s,t) = \text{Cov}(v_i(s), v_i(t))$; and the vector of measurement errors $\boldsymbol{\epsilon}_i$ is assumed to be normal with covariance matrix $\mathbf{R}_i$. Often, one may assume $\mathbf{R}_i = \sigma^2 \mathbf{I}_{n_i}$ for simplicity. In the above model, no parametric forms are available for $\eta(t)$ and $v_i(t), \ i = 1, 2, \cdots, n$ instead $\eta(t)$ and $v_i(t), \ i = 1, 2, \cdots, n$ are merely assumed to be smooth.

Mixed-effects regression spline (MERS) methods have been studied by several authors including Shi, Taylor, and Wang (1996), and Rice and Wu (2001). The MERS methods aim to estimate $\eta(t)$ and $v_i(t), \ i = 1, 2, \cdots, n$ and hence the individual functions $s_i(t) = \eta(t) + v_i(t)$ using regression spline smoothing. Given two regression spline bases:

$$\boldsymbol{\Phi}_p(t) = [\phi_1(t), \cdots, \phi_p(t)]^T, \ \boldsymbol{\Psi}_q(t) = [\psi_1(t), \cdots, \psi_q(t)]^T,$$

a MERS method will transform the NPME model (5.36) into a standard LME model that can be solved by standard statistical software such as SAS and S-PLUS. The above bases can be constructed using any proper basis, e.g., the truncated power basis (5.18). Another possible basis is the B-spline basis (de Boor 1978).

Using the two bases $\boldsymbol{\Phi}_p(t)$ and $\boldsymbol{\Psi}_q(t)$, we may approximate $\eta(t)$ and $v_i(t)$ by

$$\eta(t) \approx \boldsymbol{\Phi}_p(t)^T \boldsymbol{\beta}, \ \ v_i(t) \approx \boldsymbol{\Psi}_q(t)^T \mathbf{b}_i, \ \ i = 1, 2, \cdots, n, \tag{5.37}$$

where $\beta = [\beta_1, \cdots, \beta_p]^T$ and $\mathbf{b}_i = [b_{i1}, \cdots, b_{iq}]^T$, $i = 1, 2, \cdots, n$ are the coefficient vectors. Since the basis $\mathbf{\Psi}_q(t)$ is not random, the randomness of $v_1(t), \cdots, v_n(t)$ is then shifted to that of the coefficient vectors $\mathbf{b}_1, \cdots, \mathbf{b}_n$.

For simplicity and as before, we set $\mathbf{x}_{ij} = \mathbf{\Phi}_p(t_{ij}), \mathbf{z}_{ij} = \mathbf{\Psi}_q(t_{ij})$. Then approximately the NPME model (5.36) can be written as

$$\begin{gathered} y_{ij} = \mathbf{x}_{ij}^T\beta + \mathbf{z}_{ij}^T\mathbf{b}_i + \epsilon_{ij}, \\ \mathbf{b}_i \sim N(0, \mathbf{D}), \epsilon_i \sim N(0, \mathbf{R}_i), \\ j = 1, 2, \cdots, n_i;\ i = 1, 2, \cdots, n, \end{gathered} \tag{5.38}$$

where $\mathbf{D} = \mathrm{Cov}(\mathbf{b}_i)$. By (5.37), a direct connection between γ and $\mathbf{D}$ is

$$\gamma(s, t) \approx \mathbf{\Psi}_q(s)^T\mathbf{D}\mathbf{\Psi}_q(t). \tag{5.39}$$

For the given two bases $\mathbf{\Phi}_p(t)$ and $\mathbf{\Psi}_q(t)$, the model (5.38) is a standard LME model. One may be more comfortable with its vector form or matrix form. To this end, we shall use all the notations defined in the previous sections for $\mathbf{y}_i, \mathbf{y}$ and $\mathbf{X}_i, \mathbf{X}$. In addition, define

$$\begin{array}{rclrcl} \mathbf{Z}_i & = & [\mathbf{z}_{i1}, \cdots, \mathbf{z}_{in_i}]^T, & \mathbf{Z} & = & \mathrm{diag}(\mathbf{Z}_1, \cdots, \mathbf{Z}_n), \\ \boldsymbol{\epsilon}_i & = & [\epsilon_{i1}, \cdots, \epsilon_{in_i}]^T, & \boldsymbol{\epsilon} & = & [\boldsymbol{\epsilon}_1^T, \cdots, \boldsymbol{\epsilon}_n^T]^T, \\ \tilde{\mathbf{D}} & = & \mathrm{diag}(\mathbf{D}, \mathbf{D}, \cdots, \mathbf{D}), & \mathbf{R} & = & \mathrm{diag}(\mathbf{R}_1, \cdots, \mathbf{R}_n), \\ \mathbf{b} & = & (\mathbf{b}_1^T, \mathbf{b}_2^T, \cdots, \mathbf{b}_n^T)^T. & & & \end{array}$$

Then, the model (5.38) can be expressed in a vector form as

$$\begin{array}{rcl} \mathbf{y}_i & = & \mathbf{X}_i\beta + \mathbf{Z}_i\mathbf{b}_i + \epsilon_i, \\ \mathbf{b}_i & \sim & N(\mathbf{0}, \mathbf{D}),\ \boldsymbol{\epsilon}_i \sim N(\mathbf{0}, \mathbf{R}_i), \\ & & i = 1, 2, \cdots, n, \end{array} \tag{5.40}$$

or in a matrix form as

$$\begin{array}{rcl} \mathbf{y} & = & \mathbf{X}\beta + \mathbf{Z}\mathbf{b} + \boldsymbol{\epsilon}, \\ \mathbf{b} & \sim & N(\mathbf{0}, \tilde{\mathbf{D}}),\ \boldsymbol{\epsilon} \sim N(\mathbf{0}, \mathbf{R}). \end{array} \tag{5.41}$$

Notice that $\mathbf{y}$ and $\boldsymbol{\epsilon}$ are vectors of size $N \times 1$, $\mathbf{b}$ is a vector of size $(nq) \times 1$, $\mathbf{X}$ is a $N \times p$ matrix, and $\mathbf{Z}$ is a block diagonal matrix of size $N \times (nq)$.

5.4.1 Fits and Smoother Matrices

The standard LME model (5.41) can be easily solved for $\hat{\beta}, \hat{\mathbf{b}}_i, \hat{\mathbf{D}}$ and $\hat{\mathbf{R}}_i$ by several approaches via estimating the variance components using the EM-algorithm (Laird and Ware 1982); see Chapter 2 for details. The associated codes are available in several popular statistical softwares such as the S-PLUS function *lme* or the SAS procedure *PROC MIXED*. Then the MERS estimators of $\eta(t), v_i(t)$, and $\gamma(s, t)$ can be constructed as

$$\begin{array}{rcl} \hat{\eta}(t) & = & \mathbf{\Phi}_p(t)^T\hat{\beta}, \\ \hat{v}_i(t) & = & \mathbf{\Psi}_q(t)^T\hat{\mathbf{b}}_i,\ i = 1, 2, \cdots, n, \\ \hat{\gamma}(s, t) & = & \mathbf{\Psi}_q(s)^T\hat{\mathbf{D}}\mathbf{\Psi}_q(t). \end{array} \tag{5.42}$$

Obviously, the estimator of γ above is obtained using the relationship (5.39). The natural MERS estimators of the individual functions $s_i(t),\ i = 1, 2, \cdots, n$ are then

$$\hat{s}_i(t) = \hat{\eta}(t) + \hat{v}_i(t),\ i = 1, 2, \cdots, n. \tag{5.43}$$

It is seen that an individual function estimator uses information from the entire data set and data from the individual. The individual function estimators are well defined even when some individuals have sparse data points that do not allow a direct individual function fit.

A direct relationship between the response vector $\mathbf{y}$ and its fitted vector is often useful. To establish such a relationship, we need the general formulas that connect $\hat{\beta}$ and $\hat{\mathbf{b}}$ to $\mathbf{y}$. When $\mathbf{D}$ and $\mathbf{R}_i$ are known, by (5.40) or (5.41), we have the following formulas

$$\begin{aligned} \hat{\beta} &= \left(\sum_{i=1}^n \mathbf{X}_i^T \mathbf{V}_i^{-1} \mathbf{X}_i\right)^{-1} \left(\sum_{i=1}^n \mathbf{X}_i \mathbf{V}_i^{-1} \mathbf{y}_i\right), \\ \hat{\mathbf{b}}_i &= \mathbf{D} \mathbf{Z}_i^T \mathbf{V}_i^{-1} \left(\mathbf{y}_i - \mathbf{X}_i \hat{\beta}\right),\ i = 1, 2, \cdots, n, \end{aligned} \tag{5.44}$$

or equivalently,

$$\begin{aligned} \hat{\beta} &= (\mathbf{X}^T \mathbf{V}^{-1} \mathbf{X})^{-1} (\mathbf{X}^T \mathbf{V}^{-1} \mathbf{y}), \\ \hat{\mathbf{b}} &= \tilde{\mathbf{D}} \mathbf{Z}^T \mathbf{V}^{-1} (\mathbf{y} - \mathbf{X}\hat{\beta}), \end{aligned} \tag{5.45}$$

where $\tilde{\mathbf{D}} = \text{diag}(\mathbf{D}, \cdots, \mathbf{D})$ is a block diagonal matrix with n blocks, and

$$\begin{aligned} \mathbf{V} &= \text{diag}(\mathbf{V}_1, \cdots, \mathbf{V}_n) = \text{Cov}(\mathbf{y}), \\ \mathbf{V}_i &= \mathbf{Z}_i \mathbf{D} \mathbf{Z}_i^T + \mathbf{R}_i = \text{Cov}(\mathbf{y}_i),\ i = 1, 2, \cdots, n. \end{aligned} \tag{5.46}$$

Then the fitted vector of $\eta(t)$ at all the design time points (5.3) can be expressed as

$$\hat{\eta} = \mathbf{X}\hat{\beta} = \mathbf{A}\mathbf{y}, \tag{5.47}$$

where

$$\mathbf{A} = \mathbf{X}(\mathbf{X}^T \mathbf{V}^{-1} \mathbf{X})^{-1} \mathbf{X}^T \mathbf{V}^{-1} \tag{5.48}$$

is the associated smoother matrix for estimating $\eta(t)$. When $\mathbf{V}$ is known, the expression (5.47) suggests that the estimator is a linear estimator of $\eta(t)$, evaluated at the design time points. Following Buja et al. (1989), we shall define the trace of $\mathbf{A}$ as the number of the degrees of freedom of the estimator (5.47), which quantifies the model complexity for estimating $\eta(t)$. It is easy to show that $\text{tr}(\mathbf{A}) = p$ for any $\mathbf{V}$ (see Theorem 5.1). This reflects the fact that we use p basis functions to approximate $\eta(t)$.

Notice that the MERS smoother (5.47) for $\eta(t)$ at all the design time points is the same as the GRS smoother (5.29) when the basis $\mathbf{\Phi}_p(t)$ and the covariance matrix $\mathbf{V}$ are the same for both smoothers. Moreover, in this case, the smoother matrices (5.30) and (5.48) are the same. The only difference between these two smoothers is how the covariance matrix $\mathbf{V}$ is specified. In the GRS smoother, $\mathbf{V}$ is assumed to be known or estimated using a parametric model, while in the MERS smoother, $\mathbf{V}$ is estimated nonparametrically.

The fits of $v_i(t),\ i = 1, 2, \cdots, n$ at the design time points (5.3) can be expressed as

$$\hat{\mathbf{v}} = \mathbf{Z}\hat{\mathbf{b}} = \mathbf{Z}\tilde{\mathbf{D}}\mathbf{Z}^T\mathbf{V}^{-1}(\mathbf{y} - \mathbf{X}\hat{\beta}) = \mathbf{A}_v\mathbf{y}, \tag{5.49}$$

where the associated smoother matrix for $\hat{v}_i(t),\ i = 1, 2, \cdots, n$ is

$$\mathbf{A}_v = \mathbf{Z}\tilde{\mathbf{D}}\mathbf{Z}^T\mathbf{V}^{-1}(\mathbf{I}_N - \mathbf{A}). \tag{5.50}$$

Notice that $\hat{s}_i(t) = \hat{\eta}(t) + \hat{v}_i(t),\ i = 1, 2, \cdots, n$. It follows that the smoother matrix of the individual curve estimators is $\mathbf{A} + \mathbf{A}_v$. Actually, we have the ANOVA decomposition for the response vector $\mathbf{y}$:

$$\mathbf{y} = \mathbf{A}\mathbf{y} + \mathbf{A}_v\mathbf{y} + (\mathbf{I}_N - \mathbf{A} - \mathbf{A}_v)\mathbf{y} = \hat{\eta} + \hat{\mathbf{v}} + (\mathbf{y} - \hat{\eta} - \hat{\mathbf{v}}). \tag{5.51}$$

Thus, the fitted response vector of the individual curves at all the design time points (5.3) can be expressed as $\hat{\eta} + \hat{\mathbf{v}}$.

5.4.2 Variability Band Construction

A pointwise SD band can give us a basic idea of how accurate the estimators are for the underlying target functions. Based on the formulas in (5.41) and (5.45), when $\mathbf{V}$ is fixed, it is easy to see that

$$\begin{array}{rcl} \text{Cov}(\hat{\beta}) & = & (\mathbf{X}^T\mathbf{V}^{-1}\mathbf{X})^{-1}, \\ \text{Cov}(\hat{\mathbf{b}} - \mathbf{b}) & = & \mathbf{H}\mathbf{V}\mathbf{H}^T - \mathbf{H}\mathbf{Z}\tilde{\mathbf{D}} - \tilde{\mathbf{D}}\mathbf{Z}^T\mathbf{H}^T + \tilde{\mathbf{D}}, \end{array} \tag{5.52}$$

where $\mathbf{H} = \tilde{\mathbf{D}}\mathbf{Z}^T\mathbf{V}^{-1}(\mathbf{I}_N - \mathbf{A})$. Here we take the variation of $\mathbf{b}$ into account since it is random. The above two formulas can be used to construct confidence intervals for $\hat{\beta}$ or $\hat{\mathbf{b}}$, and the pointwise SD bands for $\eta(t)$ and $v_i(t)$ as well. In fact, we have

$$\begin{array}{rcl} \text{Var}[\hat{\eta}(t)] & = & \mathbf{\Phi}_p(t)^T(\mathbf{X}^T\mathbf{V}^{-1}\mathbf{X})^{-1}\mathbf{\Phi}_p(t), \\ \text{Var}[\hat{v}_i(t) - v_i(t)] & \approx & \mathbf{\Psi}_q(t)^T\text{Cov}(\hat{\mathbf{b}}_i - \mathbf{b}_i)\mathbf{\Psi}_q(t),\ i = 1, 2, \cdots, n, \end{array} \tag{5.53}$$

where $\text{Cov}(\hat{\mathbf{b}}_i - \mathbf{b}_i)$ can be easily extracted from (5.52) and the "$\approx$" is used because $v_i(t) \approx \mathbf{\Psi}_q(t)^T\mathbf{b}_i$. Then the 95% pointwise SD band of $\eta(t)$ can be constructed as $\hat{\eta}(t) \pm 1.96\sqrt{\widehat{\text{Var}}[\hat{\eta}(t)]}$ where $\widehat{\text{Var}}[\hat{\eta}(t)]$ is given in (5.53) with $\mathbf{V}$ replaced by its estimate $\hat{\mathbf{V}}$. However, the 95% pointwise SD band for $v_i(t)$ should be constructed in a slightly different way since $v_i(t)$ are random. They should be constructed as

$$\hat{v}_i(t) \pm 1.96\sqrt{\widehat{\text{Var}}[\hat{v}_i(t) - v_i(t)]}. \tag{5.54}$$

A drawback is that the resulting 95% pointwise SD bands for $\eta(t)$ and $v_i(t)$ do not take the variation of $\hat{\mathbf{V}}$ into account.

It is often sufficient to construct the pointwise SD band of $\eta(t)$ or $v_i(t)$ at the design time points (5.3). For this purpose, we compute

$$\begin{array}{rcl} \text{Cov}(\hat{\mathbf{y}}) & = & \mathbf{A}\mathbf{V}\mathbf{A}^T, \\ \text{Cov}(\hat{\mathbf{v}}_i - \mathbf{v}_i) & \approx & \mathbf{Z}_i\text{Cov}(\hat{\mathbf{b}}_i - \mathbf{b}_i)\mathbf{Z}_i^T,\ i = 1, 2, \cdots, n, \end{array} \tag{5.55}$$

where $\mathbf{v}_i = [v_i(t_{i1}), \cdots, v_i(t_{in_i})]^T$. These two formulas allow us to construct pointwise SD bands for $\eta(t)$ and $v_i(t)$ at all the design time points using a similar procedure as described above.

It may be more common to construct the pointwise SD bands for the individual curves $s_i(t),\ i = 1, 2, \cdots, n$ than for the individual random-effects functions $v_i(t),\ i = 1, 2, \cdots, n$. For this purpose, we need to compute the covariance of β and $\mathbf{b}$. In what follows, we shall show how this can be done.

From (5.41), it is easy to verify that

$$\begin{pmatrix} \hat{\beta} \\ \hat{\mathbf{b}} \end{pmatrix} = \begin{pmatrix} \mathbf{X}^T\mathbf{R}^{-1}\mathbf{X} & \mathbf{X}^T\mathbf{R}^{-1}\mathbf{Z} \\ \mathbf{Z}^T\mathbf{R}^{-1}\mathbf{X} & \mathbf{Z}^T\mathbf{R}^{-1}\mathbf{Z} + \tilde{\mathbf{D}}^{-1} \end{pmatrix}^{-1} \begin{pmatrix} \mathbf{X}^T\mathbf{R}^{-1} \\ \mathbf{Z}^T\mathbf{R}^{-1} \end{pmatrix} \mathbf{y} \equiv \tilde{\mathbf{H}}\mathbf{y},$$

where $\tilde{\mathbf{H}}$ is obviously defined. It follows that

$$\text{Cov}\begin{pmatrix} \hat{\beta} - \beta \\ \hat{\mathbf{b}} - \mathbf{b} \end{pmatrix} = \tilde{\mathbf{H}}\mathbf{V}\tilde{\mathbf{H}} - \tilde{\mathbf{H}}\mathbf{Z}\tilde{\mathbf{D}} - \tilde{\mathbf{D}}\mathbf{Z}^T\tilde{\mathbf{H}} + \tilde{\mathbf{D}}. \tag{5.56}$$

Notice that

$$\hat{s}_i(t) = \mathbf{\Phi}_p(t)^T\hat{\beta} + \mathbf{\Psi}_q(t)^T\hat{\mathbf{b}}_i = \mathbf{\Phi}_p(t)^T\hat{\beta} + \mathbf{\Psi}_q(t)^T\mathbf{E}_i\hat{\mathbf{b}},$$

where $\mathbf{E}_i$ is a $q \times (nq)$ matrix such that $\mathbf{E}_i\hat{\mathbf{b}} = \hat{\mathbf{b}}_i$, and $s_i(t) \approx \mathbf{\Phi}_p(t)^T\beta + \mathbf{\Psi}_q(t)^T\mathbf{E}_i\mathbf{b}$. Therefore,

$$\hat{s}_i(t) - s_i(t) \approx [\mathbf{\Phi}_p(t)^T, \mathbf{\Psi}_q(t)^T\mathbf{E}_i]\begin{pmatrix} \hat{\beta} - \beta \\ \hat{\mathbf{b}} - \mathbf{b} \end{pmatrix} \equiv \mathbf{S}_i\begin{pmatrix} \hat{\beta} - \beta \\ \hat{\mathbf{b}} - \mathbf{b} \end{pmatrix}.$$

It follows that

$$\begin{aligned} \text{Var}[\hat{s}_i(t) - s_i(t)] &\approx \mathbf{S}_i\text{Cov}\begin{pmatrix} \hat{\beta} - \beta \\ \hat{\mathbf{b}} - \mathbf{b} \end{pmatrix}\mathbf{S}_i^T \\ &= \mathbf{S}_i\left[\tilde{\mathbf{H}}\mathbf{V}\tilde{\mathbf{H}} - \tilde{\mathbf{H}}\mathbf{Z}\tilde{\mathbf{D}} - \tilde{\mathbf{D}}\mathbf{Z}^T\tilde{\mathbf{H}} + \tilde{\mathbf{D}}\right]\mathbf{S}_i^T. \end{aligned} \tag{5.57}$$

Then we can construct the 95% pointwise SD band for $s_i(t)$ as

$$\hat{s}_i(t) \pm 1.96\sqrt{\widehat{\text{Var}}[\hat{s}_i(t) - s_i(t)]},$$

where $\widehat{\text{Var}}[\hat{s}_i(t) - s_i(t)]$ is given by (5.57) with $\mathbf{V}$ replaced by its estimate $\hat{\mathbf{V}}$.

5.4.3 No-Effect Test

Hypothesis testing about some characteristics in the NPME model (5.36) is also important. In the ACTG 388 study example, AIDS clinicians may want to know whether the effect of antiviral treatments on CD4 recovery is significant. To answer this question, we need to perform a hypothesis test with a null hypothesis of no-effect.

In general, testing hypotheses about the NPME model (5.36) can be approximately carried out by testing the associated hypotheses about the LME model (5.41). For example, when $\phi_1(t) \equiv 1$, conducting the following no-effect test: for some unknown constant c,

$$H_0 : \eta(t) \equiv c, \text{vs } H_1 : \eta(t) \neq c, \tag{5.58}$$

is approximately equivalent to conducting the following no-effect test:

$$H_0 : \mathbf{C}\boldsymbol{\beta} = 0, \text{vs } H_1 : \mathbf{C}\boldsymbol{\beta} \neq 0, \tag{5.59}$$

based on the LME model (5.41), where $\mathbf{C} = \text{diag}(0, 1, 1, \cdots, 1)$. The test (5.59) can be conducted using a Wald-type test or a χ^2-type test (Davidian and Giltinan 1995, Wu and Zhang 2002b). The χ^2-type test statistic is constructed using the fact that as $n \to \infty$, $\mathbf{C}\hat{\boldsymbol{\beta}} \sim \text{AN}(\mathbf{C}\boldsymbol{\beta}, \mathbf{C}(\mathbf{X}^T\mathbf{V}^{-1}\mathbf{X})^{-1}\mathbf{C}^T)$. It follows that asymptotically and under H_0, we have

$$T = \hat{\boldsymbol{\beta}}^T \mathbf{C}^T[\mathbf{C}(\mathbf{X}^T\mathbf{V}^{-1}\mathbf{X})^{-1}\mathbf{C}^T]^{-1}\mathbf{C}\hat{\boldsymbol{\beta}} \sim \chi^2_{p-1},$$

since the rank of $\mathbf{C}$ is $(p-1)$. In practice, T can be computed via replacing $\mathbf{V}$ and p with their estimates $\hat{\mathbf{V}}$ and $\hat{p}$. The $\hat{\mathbf{V}}$ can be obtained using the EM-algorithm while $\hat{p}$ can be selected by some smoothing parameter selector such as AIC or BIC which will be developed in the next subsection.

5.4.4 Choice of the Bases

For the MERS method, we need two bases $\boldsymbol{\Phi}_p(t)$ and $\boldsymbol{\Psi}_q(t)$. They can be any basis such as B-spline basis (de Boor 1978), truncated power basis (Wold 1974) or wavelet basis, among others. For example, Shi, Taylor and Wang (1996), and Rice and Wu (2001) employed the B-spline basis. We shall use the truncated power basis (5.18). We let $\boldsymbol{\Phi}_p(t)$ be the truncated power basis (5.18) of degree k with knots $\tau_1, \tau_2, \cdots, \tau_K$, and let $\boldsymbol{\Psi}_q(t)$ be the truncated power basis (5.18) of degree k_v with knots $\delta_1, \cdots, \delta_{K_v}$ where $p = K + k + 1$ and $q = K_v + k_v + 1$. When there is a need to locate the knots for $\boldsymbol{\Phi}_p(t)$ and $\boldsymbol{\Psi}_q(t)$, we employ one of the three methods that were outlined in Section 5.2.4. As for k and k_v, we often use a lower degree, say, 1, 2 and 3, for computational convenience. In practice, linear and quadratic truncated power bases are widely used.

When k and k_v are specified, choosing the number of knots, K and K_v, is equivalent to choosing the number of basis functions in $\boldsymbol{\Phi}_p(t)$ and $\boldsymbol{\Psi}_q(t)$. In the next subsection, we shall discuss how to choose p and q for the two bases $\boldsymbol{\Phi}_p(t)$ and $\boldsymbol{\Psi}_q(t)$.

5.4.5 Choice of the Number of Basis Functions

In previous discussions we fixed the the bases $\boldsymbol{\Phi}_p(t)$ and $\boldsymbol{\Psi}_q(t)$ in the MERS method. In practice, we need to choose these two bases carefully to produce good fits. When the construction rule for $\boldsymbol{\Phi}_p(t)$ and $\boldsymbol{\Psi}_q(t)$ is determined, it is equivalent to choosing the number of basis functions, p and q. Good choice of p and q in the MERS method should lead to a good trade-off between two aspects: goodness of fit and model complexity.

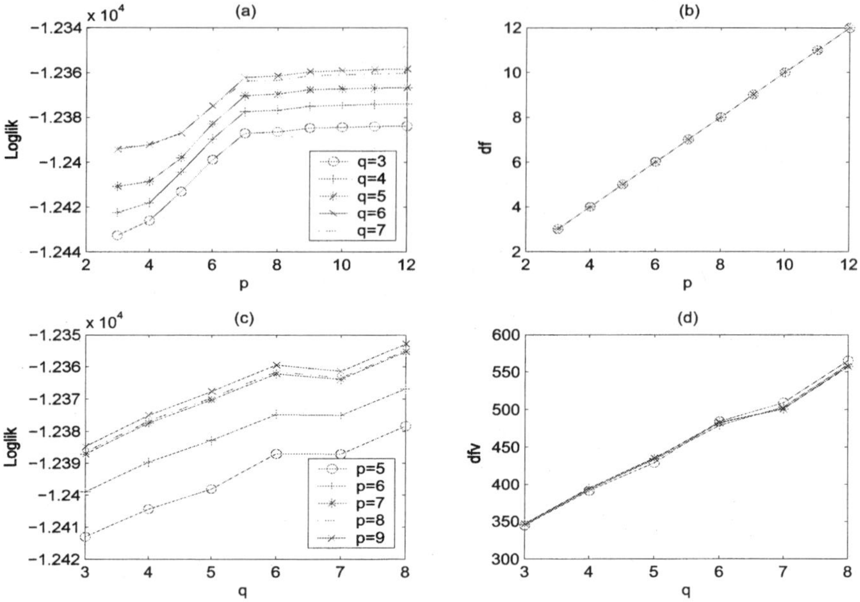

Fig. 5.3 Goodness of fit and model complexity of the MERS model fit to the ACTG 388 data. Profile curves of (a) Loglik (associated with q) vs p; (b) df (associated with q) vs p; (c) Loglik (associated with q) vs p; and (d) df_v (associated with p) vs q.

Goodness of Fit is a measure to quantify how closely a model fits the data. When the data are interpolated, the associated goodness of fit attains its optimal value. In the context of longitudinal data analysis, it is often measured by the likelihood or equivalently the log-likelihood of the resulting model. The larger the likelihood, the better the goodness of fit is.

Let $\boldsymbol{\eta}_i = [\eta(t_{i1}), \cdots, \eta(t_{in_i})]^T$ be the fixed-effects function $\eta(t)$ evaluated at the design time points of the i-th subject. Then by the NPME model (5.36), we have

$$\mathbf{y}_i \sim N(\boldsymbol{\eta}_i, \mathbf{V}_i), \; i = 1, 2, \cdots, n.$$

It follows that the log-likelihood, as a function of the number of basis functions, p and q, is (up to a constant):

$$\text{Loglik} = -\frac{1}{2}\sum_{i=1}^{n}\left\{(\mathbf{y}_i - \boldsymbol{\eta}_i)^T\mathbf{V}_i^{-1}(\mathbf{y}_i - \boldsymbol{\eta}_i) + \log|\mathbf{V}_i|\right\}. \tag{5.60}$$

In practice, the Loglik is computed via replacing the unknown $\boldsymbol{\eta}_i$ and $\mathbf{V}_i$ with their estimates. Under the LME model (5.41), $\boldsymbol{\eta}_i$ and $\mathbf{V}_i$ are estimated as

$$\hat{\boldsymbol{\eta}}_i = \mathbf{X}_i\hat{\boldsymbol{\beta}}, \quad \hat{\mathbf{V}}_i = \mathbf{Z}_i\hat{\mathbf{D}}\mathbf{Z}_i^T + \hat{\mathbf{R}}_i, \; i = 1, 2, \cdots, n. \tag{5.61}$$

It is expected that when p and q increase, the Loglik will generally increase due to increased number of covariates (basis functions) involved, so that the data are fitted more closely. Figure 5.3 (a) displays the Loglik as a function of p and q for the MERS model fitted to the ACTG 388 data. Five Loglik profile curves against p for $q = 3, 4, \cdots, 7$ are superimposed. Notice that the last two profile curves (corresponding to $q = 6, 7$) are nearly overlapping since they are too close to each other. It is seen that for a fixed q, the Loglik generally increases over p; and for a fixed p, the Loglik generally increases over q; see also Figure 5.3 (c) where five profile curves of Loglik against q for $p = 5, 6, 7, 8, 9$ are presented.

Model Complexity is generally used to quantify how complicated the model is. That is, it indicates approximately how many parameters are needed if an equivalent parametric model is used to fit the same data set. Following Buja et al. (1989), we measure model complexity by evaluating the trace of the associated smoother matrix.

For estimating $\eta(t)$ at all the design time points (5.3), the smoother matrix is $\mathbf{A}$ as defined in (5.48); for estimating $v_i(t),\ i = 1, 2, \cdots, n$ at the design time points, the smoother matrix is $\mathbf{A}_v$ as defined in (5.50). It follows that the degrees of freedom of MERS modeling, as functions of p and q, are, respectively:

$$\text{df} = \text{tr}(\mathbf{A}),\ \text{df}_v = \text{tr}(\mathbf{A}_v). \tag{5.62}$$

Theorem 5.1 *Assume* $\mathbf{X}$ *is of full rank. Then we have*

$$df = p,\quad 0 < df_v < n\min(q, N/n), \tag{5.63}$$

where n *and* N *are the number of subjects and the number of the total measurements for all the subjects, respectively.*

Proof: see Appendix.

Theorem 5.1 provides the ranges of the model complexity for estimating the mean function and the random-effects functions. It shows that the df is the same as p, and is independent of q. This indicates that the basis $\mathbf{\Psi}_q(t)$ has little effect on the model complexity for estimating $\eta(t)$. Since p is the number of basis functions in $\mathbf{\Phi}_p(t)$, in order to make the design matrix $\mathbf{X}$ full rank, p can not be larger than M, the number of distinct design time points. It follows that the df can not be larger than M. Notice that N/n represents the average number of measurements per subject. When $q < N/n$, we have $\text{df}_v < nq$. This suggests that it is sufficient to assume $1 \le q < N/n$. When N/n is small, q should be small.

Figure 5.3 (b) presents five profile curves of df against p for $q = 3, 4, \cdots, 7$ for the ACTG 388 data. The profile curves are tied together, indicating no effect of q [or equivalently $\mathbf{\Psi}_q(t)$] on df as expected. It is also seen that the df is a linear increasing function of p. Figure 5.3 (d) presents five nearly overlapping profile curves of df_v against q for $p = 5, 6, 7, 8, 9$. It is seen that the df_v generally increases as q increases and the effect of p on df_v is very small. Therefore, the df is mainly affected by p, and the df_v by q. It follows that the sum of df and df_v will generally increase when either p or q or both increase.

AIC and BIC. From the above discussions, we can see that the goodness of fit, as indicated by Loglik, will increase as either p or q or both increase, while at the

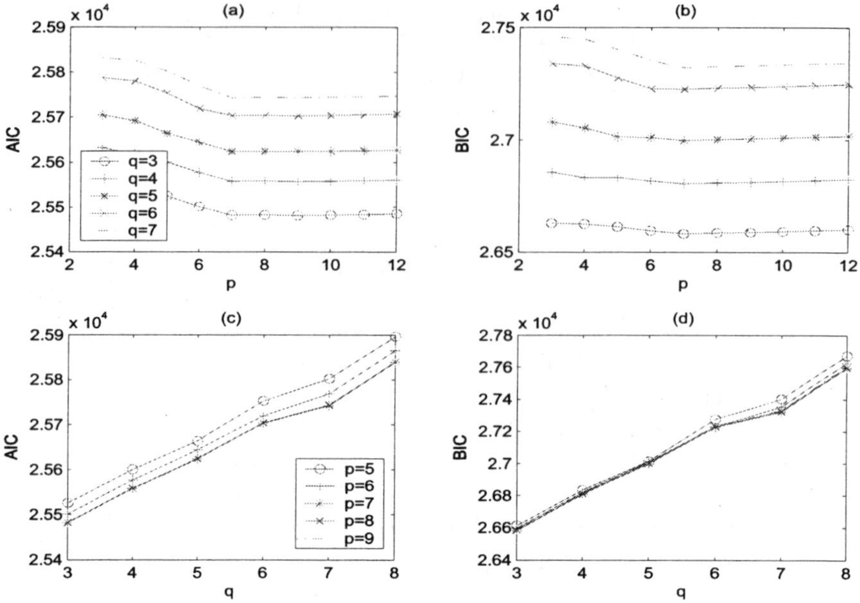

Fig. 5.4 Five profile curves of AIC and BIC for the MERS model fitted to the ACTG 388 data. (a) and (b): against p for $q = 3, 4, 5, \cdots, 7$; (c) and (d): against q for $p = 5, 6, 7, 8, 9$.

same time, the model complexity, as indicated by $\mathrm{df} + \mathrm{df}_v$ (the model complexity for estimating both $\eta(t)$ and $v_i(t),\ i = 1, 2, \cdots, n$), will increase too. To trade-off these two components, we may adopt the Akaike information criterion (AIC) (Akaike 1973) and the Bayesian information criterion (BIC) (Schwarz 1978, so BIC is also known as SIC) which have been extended for standard LME and NLME (nonlinear mixed-effects) models (Davidian and Giltinan 1995).

For MERS modeling, we can define the AIC and BIC as follows:

$$\begin{aligned} \mathrm{AIC} &= -2\mathrm{Loglik} + 2[\mathrm{df} + \mathrm{df}_v], \\ \mathrm{BIC} &= -2\mathrm{Loglik} + \log(n)[\mathrm{df} + \mathrm{df}_v], \end{aligned} \tag{5.64}$$

so that the twice negative log-likelihood function is penalized by the degrees of freedom for MERS modeling.

Figure 5.4 (a) presents five profile curves of the AIC against p for $q = 3, 4, 5, 6, 7$. Numerical outputs indicated that for each q, the AIC was always minimized at $p = 7$; and for each p, the AIC was always minimized at $q = 3$. Therefore, the optimal pair of p and q was $[p, q] = [7, 3]$. This is also seen from Figure 5.4 (c) where five profile curves of the AIC against q for $p = 5, 6, 7, 8, 9$ are presented. Similarly, numerical outputs also indicated that when the BIC was used, the best pair of p and q was also $[p, q] = [7, 3]$. See also Figures 5.4 (b) and (d). In this ACTG 388 data example, we actually used the quadratic truncated power basis (5.18) for $\mathbf{\Phi}_p(t)$ and $\mathbf{\Psi}_q(t)$. The above results indicate that both the AIC and BIC favored $K = 4$ knots (the same as the K selected by the GCV rule (5.22) for the NRS fit to the ACTG 388 data; see Figure 5.1) for estimating $\eta(t)$ and no knots for estimating $v_i(t),\ i = 1, 2, \cdots, n$. This implies that a quadratic polynomial model may be proper for the random-effects functions so that a semiparametric model may be proper for the ACTG 388 data. See Section 8.3.6 of Chapter 8 for a semiparametric model fit to the ACTG 388 data.

When the number of possible pairs of $[p, q]$ is large, direct minimization of AIC or BIC over $[p, q]$ is often computationally intensive. A simple trick for reducing the computational effort is to minimize the AIC or BIC on just one argument while keeping the other fixed. For example, letting $q = q_0$ for some given $q_0 > 0$, we may first find p^* to minimize the AIC. Then letting $p = p^*$, we find q^* to minimize the AIC. The pair $[p^*, q^*]$ obtained in this way may not minimize the AIC globally but it is often satisfactory for practical applications or simulation studies. Otherwise, one may repeat the above process a number of times.

5.4.6 Example and Model Checking

In Section 5.2, we present the NRS fit (5.11) to the ACTG 388 data. In this subsection, we present the MERS fit. We used the quadratic truncated power basis (5.18) with 4 knots ($p = 7$) for estimating $\eta(t)$, while another quadratic truncated power basis with no knots ($q = 3$) for estimating $v_i(t),\ i = 1, 2, \cdots, n$. The knots were spaced using the "equally spaced sample quantiles as knots" method as described in Section 5.2.4. The numbers of basis functions $[p, q] = [7, 3]$ were selected by the AIC rule (5.64); see Figure 5.4 (b).

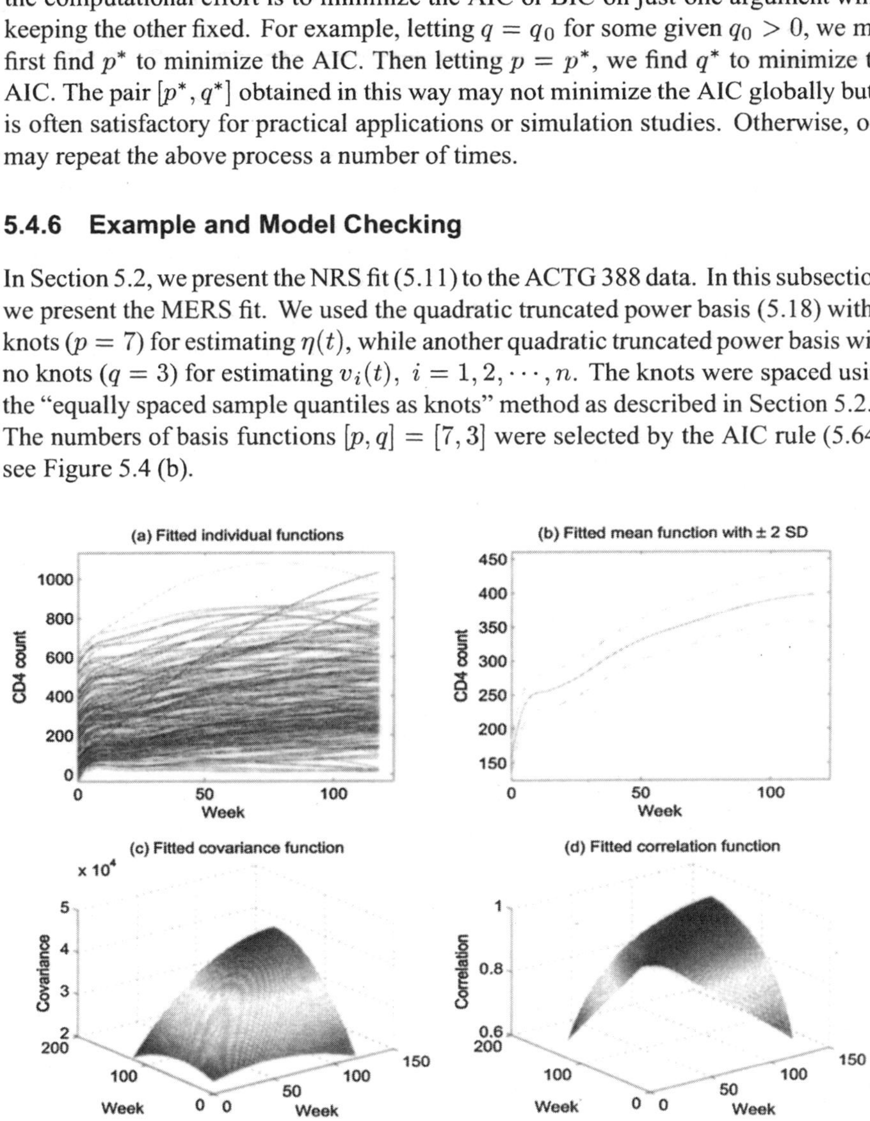

Fig. 5.5 Overall MERS fits to the ACTG 388 data.

Overall Fits. Figure 5.5 presents the overall MERS fits to the ACTG 388 data. Figure 5.5 (a) plots all the estimated individual curves. It is seen that the variation of the curves seems small at the beginning of the study, while increasing over time. This observation is consistent with the conclusion obtained by observing the surface of the estimated covariance function shown in Figure 5.5 (c). It indicates that the overall variation of the individual curves increased over time.

Figure 5.5 (b) shows the estimated population mean function (solid curve) with the 95% pointwise SD band (dot-dashed curves). The SD band indicates how accurately the population mean function was estimated based on the given data. The estimated population mean function increased dramatically during the early weeks and continued to increase until the end of the study where it decreased slightly. Thus, overall, the antiviral drug was effective, especially within the first few weeks after the antiviral drug was initiated. Although the estimated overall mean function indicates that the drug was effective up to at least 110 weeks, the associated variations also increased with time, as indicated by the increasingly wider SD band. Figure 5.5 (d) presents the estimated correlation function. It seems that the within-subject correlation was higher than .6 in general, and higher than .8 when measurements were taken in close proximity. This means that the observations within a subject were strongly correlated and it is necessary to take the correlation into account to obtain efficient estimates for the underlying mean function, and the covariance function as well.

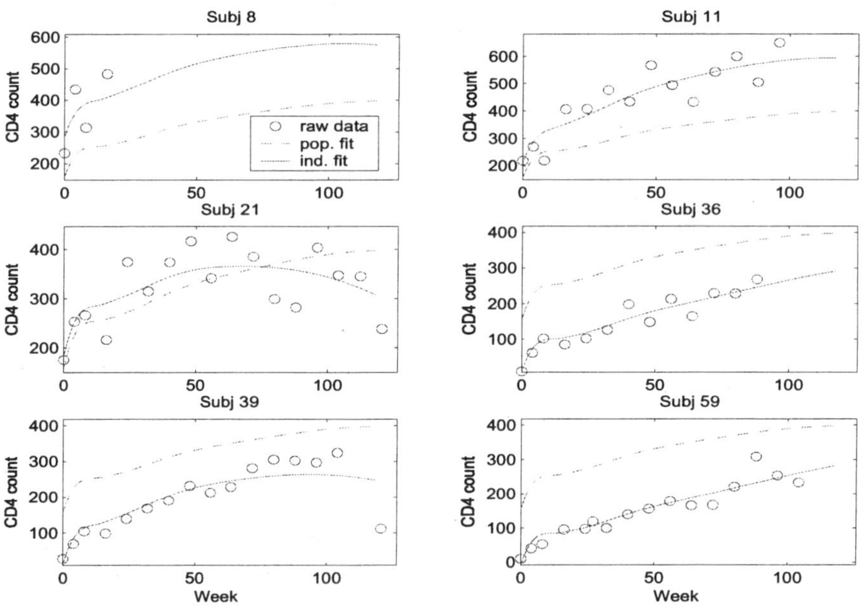

Fig. 5.6 Individual fits for six randomly selected subjects from the ACTG 388 data.

Individual Fits. Figure 5.6 presents the individual fits for six randomly selected subjects from the ACTG 388 data. In each panel, the raw observations (circles), the individual curve fit (solid curve) and the estimated mean function (dashed curve) are plotted. It is seen that the individual curves fit the data quite well for almost all of the subjects. The estimated population mean function sometimes was below, above or overlapping with the individual curves, as expected.

Note that Subject 8 had only four observations, which generally does not allow a sensible nonparametric regression fit, but the associated individual fit adapted to the data impressively well. This is due to the fact that MERS modeling is able to pull information from all the subjects as well as from the particular individual.

Also note that Subject 39 had an unusual observation at the right-lower corner of the panel. It is hard to determine if this observation was an outlier when only the observations of Subject 39 were considered. However, the appearance of a similar observation for Subject 21 and the Subject 21 individual fit indicated that this unusual observation was likely not an outlier.

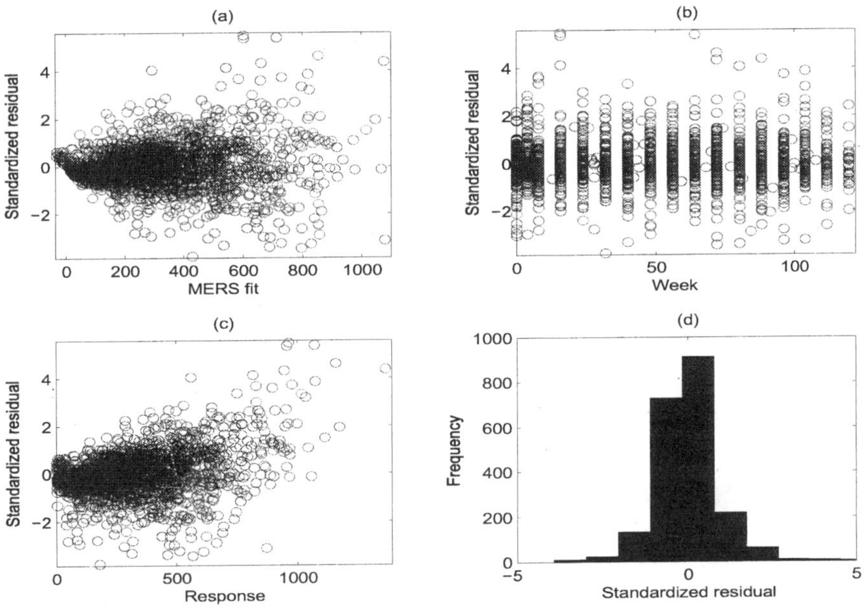

Fig. 5.7 Residual analysis based on the MERS fits to the ACTG 388 data.

Model Checking. To assess the adequacy of the MERS fits to the ACTG 388 data, we conducted a residual analysis. Notice that here the MERS fits are the individual fits evaluated at all the design time points (5.3), and the MERS residuals equal the responses subtracting the MERS fits. Figure 5.7 presents plots for the residual analysis. The plots of the standardized residual versus the MERS fit, time, and the response were displayed, respectively in Figures 5.7 (a), (b), and (c). It is seen that the MERS fit is satisfactory since the residuals are symmetrically scattered around the zero horizontal line; in particular, the undesirable linear trend in Figure 5.2 (c) no longer appears in Figure 5.7 (c), indicating that the subject-specific effects had been accounted for. Unfortunately, the residual plots in Figures 5.7 (a) and (c) also revealed that the underlying noise may not be homogeneous. Some data transformation may be needed to resolve the problem. Figure 5.7 (d) shows the histogram of the standardized residuals. It seems that the standardized residuals are symmetric and bell-shaped, as desired.

5.5 COMPARING MERS AGAINST NRS

We know that the MERS method takes the within-subject correlation into account, and the NRS method does not. In this section, we compare the MERS method against the NRS method via their applications to the ACTG 388 data and a simulation study.

5.5.1 Comparison via the ACTG 388 Data

We applied the NRS and MERS methods to the ACTG 388 data, respectively. For simplicity, we adopted the quadratic truncated power basis (5.18) for $\mathbf{\Phi}_p(t)$ and $\mathbf{\Psi}_q(t)$. For the NRS method, $p = 7$ was chosen by the GCV rule (5.22) and for the MERS method, $[p, q] = [7, 3]$ was chosen by the AIC rule (5.64).

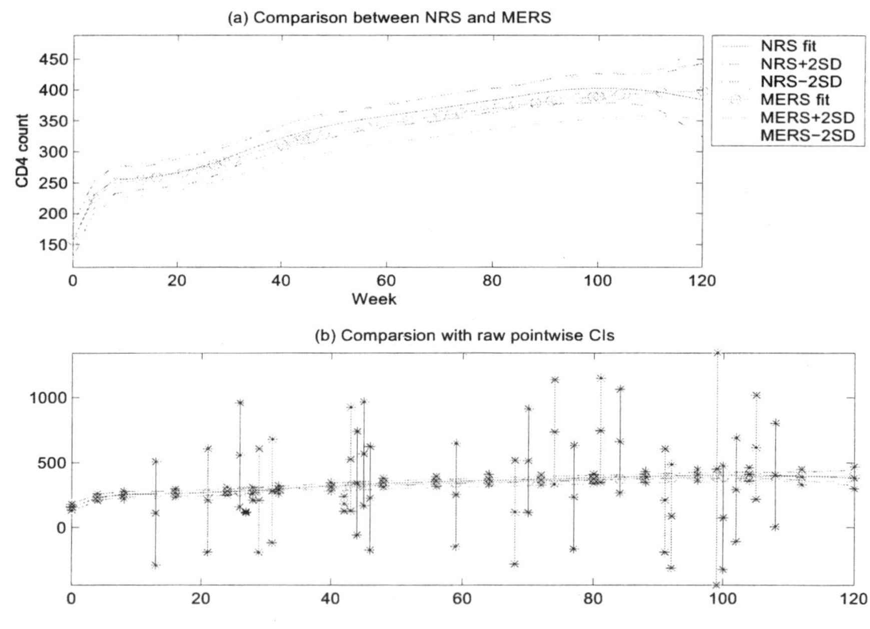

Fig. 5.8 Comparison of the NRS and MERS fits of the population mean function for the ACTG 388 data. (a) NRS and MERS fits and their 95% pointwise SD bands; and (b) NRS and MERS fits and their 95% pointwise SD bands, superimposed by the raw pointwise 95% CIs (vertical intervals with stars at the ends and the center).

Figure 5.8 (a) shows the NRS fit (solid curve) and its 95% pointwise SD band (dot-dashed curves) superimposed with the MERS fit (solid-circled curve) and its 95% pointwise SD band (dashed curves). It is seen that the NRS fit and the MERS fit are comparable. The NRS fit is within the 95% pointwise SD band of the MERS fit, while the MERS fit is within the 95% pointwise SD band of the NRS fit, indicating that both the NRS fit and the MERS fit are reasonable. It is also seen that the 95% pointwise SD band of the MERS fit is wider than that of the NRS fit. This is expected because the MERS fit took the correlation of the data into account while the NRS fit assumed working independence. Notice that the 95% pointwise SD band of the NRS fit at the right end is much wider than that of the MERS fit. This is possibly due to the boundary effect of the NRS fit since there were fewer measurements at the right end. Such a boundary effect seems not to appear in the MERS fit, indicating that the MERS fit is more robust.

In Figure 5.8 (b), we superimposed the NRS and MERS fits, their 95% pointwise SD bands and the raw pointwise 95% confidence intervals (CIs) (vertical intervals with stars at the ends and the center). It is seen that at time points where there are enough data, the raw pointwise CI is comparable to the associated NRS or MERS SD interval; while at time points where the data are sparse, the raw pointwise CI is much wider than the associated NRS or MERS interval. This is reasonable since the raw pointwise CI used just the data at the particular time point, while the NRS and MERS SD intervals used the data nearby. Moreover, the raw mean curve obtained by connecting the centers of the raw pointwise CIs is very rough. This is expected since no roughness was accounted for in the construction of the raw pointwise CIs.

5.5.2 Comparison via Simulations

For convenience, we adopted the simulation model and the parameter setup used for comparing the GCV and SCV rules in Section 5.2. As in the ACTG 388 data example, we adopted the quadratic truncated power basis (5.18) for $\mathbf{\Phi}_p(t)$ and $\mathbf{\Psi}_q(t)$. Since $k = k_v = 2$, choosing the number of basis functions, p and q is equivalent to choosing the number of knots, K and K_v. For the NRS method, we employed the GCV rule (5.22) to select K and denoted the resulting K as K_{gcv}, while for the MERS method, we used the BIC rule (5.64) to select K and K_v and denoted the resulting K as K_{bic}.

Table 5.2 presents the simulation results for comparing the MERS fits of the population mean function $\eta(t)$ against the associated NRS fits. It can be read in a similar way as Table 5.1. In particular, when the ASER is positive, the MERS fit outperforms the associated NRS fit, and when the ASEB is more than 50%, the MERS fit outperforms the associated NRS fit. It is seen that when the missing rate r_{miss} is 20% and 50%, the MERS fits outperform the NRS fits since all the associated ASERs are positive and all the associated ASEBs are more than 50% (some even more than 80%). Furthermore, when the associated minimum correlation coefficients $\rho_{\min}$ are bigger, the associated ASERs and ASEBs are generally larger. This indicates that the MERS method is preferred when the data missing rate is low and moderate, or the underlying correlation is large.

Table 5.2 ASER, ASEB (in percentage), $\bar{K}_{gcv}$ and $\bar{K}_{bic}$ with 2 standard deviations (in parentheses) for comparing the MERS fits of $\eta(t)$ against the associated NRS fits, $N = 100$.

$(n,\ m,\ r_{miss})$	$\rho_{\min}$	ASER	ASEB	$\bar{K}_{gcv}$	$\bar{K}_{bic}$
(30,20,20%)	.43	0.3(5.9)	58(9.9)	3.04(0.12)	3.24(0.14)
	.67	10.6(5.0)	74(8.8)	2.56(0.15)	3.11(0.10)
	.79	21.0(5.7)	76(8.5)	2.12(0.16)	3.22(0.17)
(50,20,20%)	.43	12.3(4.2)	76(8.5)	3.32(0.11)	3.43(0.11)
	.67	17.4(5.4)	77(8.4)	3.01(0.12)	3.45(0.13)
	.79	17.5(6.2)	75(8.7)	2.67(0.15)	3.41(0.12)
(70,20,20%)	.43	4.0(5.6)	66(9.5)	3.58(0.14)	3.55(0.12)
	.67	9.0(4.7)	70(9.2)	3.26(0.09)	3.51(0.13)
	.79	15.9(6.8)	74(8.8)	2.97(0.12)	3.39(0.11)
(30,20,50%)	.43	10.5(8.1)	71(9.1)	2.52(0.17)	2.75(0.15)
	.67	23.3(5.7)	83(7.5)	2.27(0.20)	2.64(0.15)
	.79	23.9(7.3)	72(9.0)	1.95(0.19)	2.46(0.15)
(50,20,50%)	.43	12.3(7.1)	71(9.1)	3.19(0.19)	2.96(0.11)
	.67	6.4(20.4)	71(9.1)	2.72(0.22)	2.75(0.15)
	.79	21.3(6.5)	80(8.0)	2.25(0.21)	2.78(0.11)
(70,20,50%)	.43	13.9(5.7)	71(9.1)	3.52(0.19)	3.16(0.11)
	.67	16.3(8.5)	80(8.0)	2.71(0.18)	3.14(0.09)
	.79	21.9(7.4)	79(8.1)	2.47(0.18)	2.91(0.09)
(30,20,80%)	.43	−2019.6(290.5)	0(0)	2.23(0.29)	0(0)
	.67	−1281.7(236.4)	0(0)	2.22(0.34)	0(0)
	.79	−829.8(140.2)	0(0)	1.96(0.33)	0(0)
(50,20,80%)	.43	−3049.4(562.3)	0(0)	2.43(0.24)	0(0)
	.67	−1784.6(261.1)	0(0)	2.42(0.37)	0(0)
	.79	−1445.6(319.1)	0(0)	2.27(0.33)	0(0)
(70,20,80%)	.43	−3972.7(419.5)	0(0)	2.74(0.26)	0(0)
	.67	−2604.0(419.5)	0(0)	2.36(0.27)	0(0)
	.79	−1691.8(237.5)	0(0)	2.05(0.31)	0(0)

Table 5.3 ASER, ASEB (in percentage), $\bar{K}_{gcv}$ and $\bar{K}_{bic}$ with 2 standard deviations (in parentheses) for comparing the MERS fits of $\eta(t)$ against the associated NRS fits when $r_{miss} =$ 70% and $m = 40$, $N = 100$.

$(n,\ m,\ r_{miss})$	$\rho_{\min}$	ASER	ASEB	$\bar{K}_{gcv}$	$\bar{K}_{bic}$
(30,40,70%)	.43	20.5(6.7)	78(8.3)	2.79(0.23)	2.89(0.16)
	.67	23.8(8.2)	80(8.0)	2.32(0.23)	2.90(0.18)
	.79	28.3(7.7)	79(8.1)	2.18(0.23)	2.95(0.18)
(50,40,70%)	.43	13.9(6.9)	72(9.0)	3.63(0.35)	3.34(0.15)
	.67	34.0(6.7)	84(7.3)	2.73(0.19)	3.24(0.13)
	.79	36.0(7.2)	84(7.3)	2.37(0.23)	3.18(0.16)
(70,40,70%)	.43	15.0(9.4)	74(8.8)	3.71(0.39)	3.42(0.15)
	.67	27.6(7.3)	84(7.3)	3.22(0.27)	3.27(0.14)
	.79	35.8(6.9)	86(6.9)	2.85(0.29)	3.28(0.14)

When we compare the two methods for the cases when the missing rate $r_{miss} = 80\%$, the NRS fits outperform the MERS fits as shown by the associated ASERs (they all are negative) and ASEBs (they all are 0). Notice that in these cases, the average number of measurements per subject is about $(1 - 80\%) \times 20 = 4$. In these cases, the more complicated MERS model (which has more than 4 free parameters) was not allowed to be fitted. Therefore, when longitudinal data do not allow the MERS model to be fitted, the NRS model is a good alternative.

It is also interesting to compare the average number of the knots selected by GCV for the NRS fit for a scenario, namely $\bar{K}_{gcv}$ against that selected by BIC for the MERS fit, namely $\bar{K}_{bic}$ listed in the last two columns of Table 5.2. For the first 18 scenarios (when $r_{miss} = 20\%$ and 50%), we have the following interesting conclusions:

- In most of the scenarios, $\bar{K}_{gcv} < \bar{K}_{bic} < \bar{K}_{scv}$ where $\bar{K}_{scv}$ was reported in Table 5.1.

- With n, m and r_{miss} fixed, $\bar{K}_{gcv}$ decreases as $\rho_{\min}$ increases. However, this is not the case for $\bar{K}_{bic}$. This implies that the NRS fit attempts to smooth the data more when the correlation is larger while the MERS fit does not since it already accounts for the correlation directly. This is possibly one of the reasons why the MERS fit outperforms the NRS fit when the correlation is large.

- With n, m and $\rho_{\min}$ fixed, both $\bar{K}_{gcv}$ and $\bar{K}_{bic}$ decrease as r_{miss} increases. This is reasonable since fewer measurements are available for the higher missing rate cases that can only support fewer number of basis functions.

- With m, r_{miss} and $\rho_{\min}$ fixed, both $\bar{K}_{gcv}$ and $\bar{K}_{bic}$ generally increase as n increases since more observations are available to support more basis functions.

However, for the scenarios with $r_{miss} = 80\%$, GCV can choose proper K_{gcv}'s for the NRS fits but BIC cannot get proper K_{bic}'s for the MERS fits as shown in Table 5.2.

What would happen if we modify the simulation parameter setup for the last 8 scenarios in Table 5.2 so that BIC can properly select K_{bic}'s for the MERS fits? To answer this question, we conducted another simulation study. We increased m from 20 to 40 and reduced r_{miss} from 80% to 70%. The simulation results were reported in Table 5.3. It is seen that BIC was indeed able to choose proper K_{bic}'s for the MERS fits as shown in the last column of the table. In most of the scenarios, the $\bar{K}_{bic}$'s were larger than the associated $\bar{K}_{gcv}$'s. It is seen that the MERS fits consistently outperformed the NRS fits (all the ASERs are larger than 13%, and all the ASEBs are greater than 70%).

5.6 SUMMARY AND BIBLIOGRAPHICAL NOTES

In this chapter, we discussed the NRS, GRS and MERS methods for longitudinal data. When longitudinal data are weakly correlated, the NRS method is useful, and when the data are strongly correlated, the GRS or MERS methods are better since the correlation needs to be accounted for. Application of the GRS method assumes a known parametric correlation structure while no such assumption is needed for the MERS method. Moreover, the MERS method not only produces a good fit of the population mean function, but also gives the estimates of the individual functions. Simulation studies show that the MERS method outperforms the NRS method especially when the data are strongly correlated and when the sample size is large enough to support a MERS fit.

We recommend the GCV method to select the number of basis functions for the NRS method since we found that GCV outperforms SCV most of time in simulation studies when the data are weakly correlated. We also defined the "model complexity" for the MERS method and hence generalized the AIC and BIC for selecting the number of basis functions for the MERS method.

The early discussions of regression splines for data analysis include Wold (1974), Smith (1979) and Wegman and Wright (1983), among others. An interesting account about regression splines can be found in Eubank (1988, 1999).

Discussions on the NRS method can be found in Heuer (1997), Besse, Cardot, and Ferraty (1997), and Huggins and Loesch (1998), among others. Recently, Huang, Wu and Zhou (2002) extended this technique to time-varying coefficient models for longitudinal data. The NRS method is appropriate if the longitudinal data are weakly correlated or highly sparse since in these cases, a working independence assumption for longitudinal data may be valid. When a parametric model for the within-subject correlation is available, this correlation can be accounted for properly via the generalized least-squares method using regression splines. Several authors have investigated the GRS method, including Wang and Taylor (1995), Zhang (1997, 1999), among others. In both the NRS and GRS methods, only the population mean function is estimated. The MERS method proposed and investigated in Shi, Taylor and Wang (1996) and Rice and Wu (2001) allows both population mean function and individual functions to be estimated.

For multivariate adaptive splines and their applications in longitudinal data analysis, see Friedman (1991), Zhang (1994, 1997, 1999), among others.

5.7 APPENDIX: PROOFS

Proof of Theorem 5.1 The first equation in (5.63) is easily verified as follows:

$$\begin{aligned}\text{df} &= \text{tr}(\mathbf{A}) = \text{tr}\left[\mathbf{X}(\mathbf{X}^T\mathbf{V}^{-1}\mathbf{X})^{-1}\mathbf{X}^T\mathbf{V}^{-1}\right] \\ &= \text{tr}\left[(\mathbf{X}^T\mathbf{V}^{-1}\mathbf{X})^{-1}\mathbf{X}^T\mathbf{V}^{-1}\mathbf{X}\right] \\ &= p.\end{aligned}$$

The above verification is valid even when $\mathbf{V}$ is replaced by its estimate $\hat{\mathbf{V}}$.

To verify the second equation in (5.63), by (5.50), we have

$$\begin{aligned}\text{df}_v &= \text{tr}(\mathbf{A}_v) \\ &= \text{tr}\left[\mathbf{Z}\tilde{\mathbf{D}}\mathbf{Z}^T\mathbf{V}^{-1} - \mathbf{Z}\tilde{\mathbf{D}}\mathbf{Z}^T\mathbf{V}^{-1}\mathbf{X}(\mathbf{X}^T\mathbf{V}^{-1}\mathbf{X})^{-1}\mathbf{X}^T\mathbf{V}^{-1}\right] \\ &= \text{tr}[\mathbf{Z}\tilde{\mathbf{D}}\mathbf{Z}^T\mathbf{V}^{-1}] - \text{tr}[\tilde{\mathbf{D}}^{1/2}\mathbf{Z}^T\mathbf{V}^{-1}\mathbf{X}(\mathbf{X}^T\mathbf{V}^{-1}\mathbf{X})^{-1}\mathbf{X}^T\mathbf{V}^{-1}\mathbf{Z}\tilde{\mathbf{D}}^{1/2}]\end{aligned}$$

$$
\begin{aligned}
&\leq \text{tr}[\mathbf{HH}^T(\mathbf{HH}^T + \mathbf{I}_N)^{-1}] \\
&= N - \text{tr}[(\mathbf{HH}^T + \mathbf{I}_N)^{-1}],
\end{aligned}
$$

where $\mathbf{H} = \mathbf{R}^{-1/2}\mathbf{Z}\tilde{\mathbf{D}}^{1/2}$, which is a matrix of size $N \times (nq)$. It follows that $0 < \text{df}_v < N$. Using the above derivation, on the other hand, we have

$$
\begin{aligned}
\text{df}_v = \text{tr}(\mathbf{A}_v) &\leq \text{tr}[\mathbf{HH}^T(\mathbf{HH}^T + \mathbf{I}_N)^{-1}] \\
&= \text{tr}[\mathbf{H}^T(\mathbf{I}_N + \mathbf{HH}^T)^{-1}\mathbf{H}].
\end{aligned}
$$

Applying Lemma 2.1 of Chapter 2,

$$
(\mathbf{I}_N + \mathbf{HH}^T)^{-1} = \mathbf{I}_N - \mathbf{H}(\mathbf{I}_{nq} + \mathbf{H}^T\mathbf{H})^{-1}\mathbf{H}^T.
$$

Therefore,

$$
\begin{aligned}
\mathbf{H}^T(\mathbf{I}_N + \mathbf{HH}^T)^{-1}\mathbf{H} &= \mathbf{H}^T\mathbf{H} - \mathbf{H}^T\mathbf{H}(\mathbf{I}_{nq} + \mathbf{H}^T\mathbf{H})^{-1}\mathbf{H}^T\mathbf{H} \\
&= \mathbf{H}^T\mathbf{H}(\mathbf{I}_{nq} + \mathbf{H}^T\mathbf{H})^{-1}.
\end{aligned}
$$

It follows that

$$
\begin{aligned}
\text{df}_v = \text{tr}(\mathbf{A}_v) &\leq \text{tr}[\mathbf{H}^T\mathbf{H}(\mathbf{I}_{nq} + \mathbf{H}^T\mathbf{H})^{-1}] \\
&= nq - \text{tr}[(\mathbf{I}_{nq} + \mathbf{H}^T\mathbf{H})^{-1}].
\end{aligned}
$$

Thus, $0 < \text{df}_v < nq$. The second equation in (5.63) then follows. The theorem is proved.

6

Smoothing Splines Methods

6.1 INTRODUCTION

In the previous chapter, we investigated regression spline methods for longitudinal data modeling and analysis. Smoothing splines are a powerful technique for smoothing uncorrelated data, as demonstrated in Section 3.4 of Chapter 3 and in Wahba (1990), Green and Silverman (1994), Eubank (1988, 1999) and references therein. In this chapter, we shall focus on smoothing spline methods for longitudinal data analysis. These include naive smoothing splines (Rice and Silverman 1991, Hoover et al. 1998), generalized smoothing splines (Wang 1998a), extended smoothing splines (Brumback and Rice 1998, Wang 1998a, Zhang et al. 1998, Guo 2002a, among others), and mixed-effects smoothing splines. The first two methods mainly consider estimation of the population mean function of longitudinal data, while the latter two methods also consider estimation of the subject-specified individual functions.

6.2 NAIVE SMOOTHING SPLINES

Rice and Silverman (1991) directly applied smoothing splines to model the mean function of functional data while Hoover et al. (1998) did similarly for longitudinal data. In both cases, the methodology does not take the within-subject correlation of functional data or longitudinal data into account. That is, it treats the functional data or the longitudinal data as if they were independent although in reality they may be dependent. For convenience, we refer to this technique as a naive smoothing spline (NSS) method.

6.2.1 The NSS Estimator

Consider the following nonparametric population mean (NPM) model:

$$y_{ij} = \eta(t_{ij}) + e_{ij},\ j = 1, 2, \cdots, n_i;\ i = 1, 2, \cdots, n, \tag{6.1}$$

with the design time points for all the subjects:

$$t_{ij},\ j = 1, 2, \cdots, n_i;\ i = 1, 2, \cdots, n. \tag{6.2}$$

The NSS estimator, $\hat{\eta}(t)$, of $\eta(t)$ is defined as the minimizer of the following penalized least squares (PLS) criterion:

$$\sum_{i=1}^{n}\sum_{j=1}^{n_i}[y_{ij} - \eta(t_{ij})]^2 + \lambda \int_a^b \left[\eta^{(k)}(t)\right]^2 dt, \tag{6.3}$$

over the k-th order Sobolev space $\mathcal{W}_2^k[a, b]$ defined in (3.32) where as before, λ is a smoothing parameter controlling the tradeoff between the goodness of fit to the data (the first term) and the roughness of the estimated curve (the second term). It is easy to show that the estimator $\hat{\eta}(t)$ is a smoothing spline of degree $(2k - 1)$ with knots at all the distinct design time points:

$$\tau_1, \tau_2, \cdots, \tau_M, \tag{6.4}$$

for all the subjects. First we need to find out the explicit expression of the NSS estimator $\hat{\eta}(t)$.

6.2.2 Cubic NSS Estimator

We consider constructing the cubic NSS smoother when $k = 2$. The construction of general degree NSS smoothers will be deferred until Section 6.6. The cubic NSS smoother is widely used due to its simplicity of construction (Green and Silverman 1994). By (3.30), we can easily express the roughness term in (6.3) as

$$\int_a^b [\eta''(t)]^2 dt = \boldsymbol{\eta}^T \mathbf{G} \boldsymbol{\eta}, \tag{6.5}$$

where $\boldsymbol{\eta} = [\eta(\tau_1), \cdots, \eta(\tau_M)]^T$ is the vector obtained via evaluating the mean function $\eta(t)$ at all the distinct design time points (6.4), and $\mathbf{G}$ is the associated roughness matrix (3.33).

We can express $\eta(t_{ij})$, the value of $\eta(t)$ at any design time point, t_{ij}, using $\boldsymbol{\eta}$:

$$\begin{aligned} \eta(t_{ij}) &= \mathbf{x}_{ij}^T \boldsymbol{\eta},\ \mathbf{x}_{ij} = [x_{ij1}, \cdots, x_{ijM}]^T, \\ x_{ijr} &= 1,\ \text{if } t_{ij} = \tau_r \text{ and } 0, \text{ otherwise for each } r = 1, \cdots, M. \end{aligned} \tag{6.6}$$

That is, $\mathbf{x}_{ij}$ is an indicator vector indicating which value among $\tau_1, \cdots, \tau_M$ equals t_{ij}. Using (6.5) and (6.6), we can write the expression (6.3) as

$$\sum_{i=1}^{n}\sum_{j=1}^{n_i}\left[y_{ij} - \mathbf{x}_{ij}^T \boldsymbol{\eta}\right]^2 + \lambda \boldsymbol{\eta}^T \mathbf{G} \boldsymbol{\eta}. \tag{6.7}$$

Set

$$\begin{aligned} \mathbf{X}_i &= [\mathbf{x}_{i1}, \cdots, \mathbf{x}_{in_i}]^T, & \mathbf{X} &= [\mathbf{X}_1^T, \cdots, \mathbf{X}_n^T]^T, \\ \mathbf{y}_i &= [y_{i1}, \cdots, y_{in_i}]^T, & \mathbf{y} &= [\mathbf{y}_1^T, \cdots, \mathbf{y}_n^T]^T. \end{aligned} \tag{6.8}$$

Then the expression (6.7) can be further expressed as

$$\sum_{i=1}^{n} \|\mathbf{y}_i - \mathbf{X}_i \boldsymbol{\eta}\|^2 + \lambda \boldsymbol{\eta}^T \mathbf{G} \boldsymbol{\eta},$$

or

$$\|\mathbf{y} - \mathbf{X}\boldsymbol{\eta}\|^2 + \lambda \boldsymbol{\eta}^T \mathbf{G} \boldsymbol{\eta}, \tag{6.9}$$

where $\|\cdot\|$ denotes the usual L_2-norm. It follows that the minimizer of (6.7) is

$$\hat{\boldsymbol{\eta}}_{nss} = (\mathbf{X}^T\mathbf{X} + \lambda \mathbf{G})^{-1}\mathbf{X}^T\mathbf{y}. \tag{6.10}$$

It follows that the values of $\hat{\eta}(t)$ evaluated at all the design time points (6.2) are

$$\hat{\mathbf{y}}_{nss} = \mathbf{X}\hat{\boldsymbol{\eta}}_{nss} = \mathbf{X}(\mathbf{X}^T\mathbf{X} + \lambda \mathbf{G})^{-1}\mathbf{X}^T\mathbf{y} = \mathbf{A}_{nss}\mathbf{y}, \tag{6.11}$$

where $\mathbf{A}_{nss} = \mathbf{X}(\mathbf{X}^T\mathbf{X}+\lambda\mathbf{G})^{-1}\mathbf{X}^T$ is the associated NSS smoother matrix. From

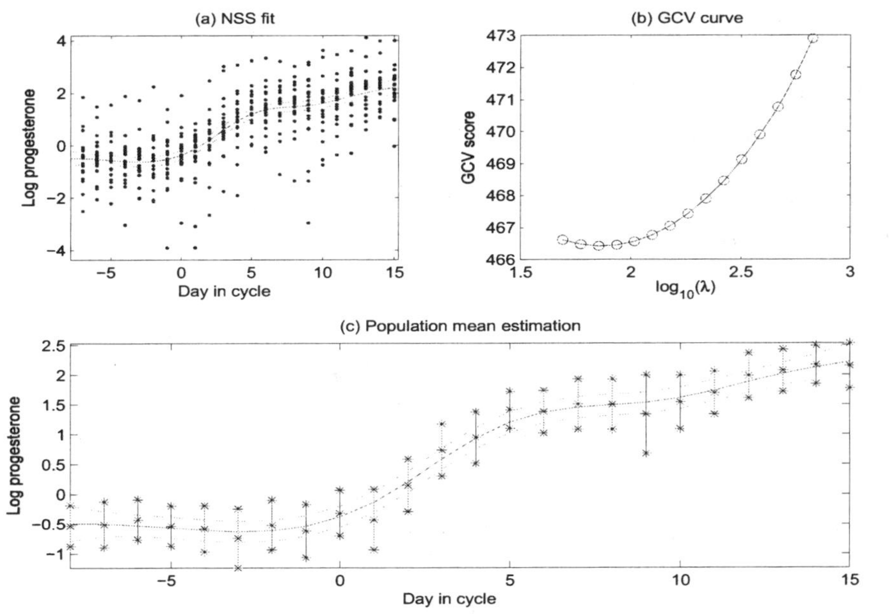

Fig. 6.1 NSS fit to the conceptive progesterone data. (a) NSS fit (solid curve) with 95% pointwise SD band (dashed curves); (b) GCV curve against the smoothing parameter in $\log_{10}$-scale; and (c) NSS fit (solid curve) with 95% pointwise SD band (dashed curves), raw 95% pointwise CIs (vertical intervals with stars at the ends and the center) at the distinct time points.

the above expression, it is seen that the NSS estimator is a linear smoother of $\eta(t)$.

Although (6.10) just gives the values of the cubic NSS estimator $\hat{\eta}(t)$ at all the distinct design time points, its value at any other time point t can be obtained via a simple formula given in Green and Silverman (1994) or via a simple interpolation.

As an example, we applied the NSS fit to the conceptive progesterone data introduced in Section 1.1.1 of Chapter 1. Figure 6.1 (a) shows the NSS fit (solid curve), the conceptive progesterone data (dots), and the associated 95% pointwise standard deviation (SD) band (dashed curves). It seems that the NSS fit catches the main tendency of the conceptive progesterone curves quite well. Figure 6.1 (b) shows the GCV curve against the smoothing parameter λ in $\log_{10}$-scale. The GCV rule will be introduced in Section 6.2.5. Figure 6.1 (c) displays the NSS fit (solid curve), the associated 95% pointwise SD band (dashed curves), and the raw 95% pointwise confidence intervals (CIs) (vertical intervals with stars at the ends and the center) at the distinct design time points (6.4). It is seen that the NSS method provides a substantially better fit of the data compared to the raw pointwise means (the centers of the raw 95% pointwise CIs).

6.2.3 Cubic NSS Estimator for Panel Data

When each subject has the same design time points, the associated longitudinal data are known as panel data. That is, for panel data, $t_{ij} = \tau_j, j = 1, \cdots, M$ and $n_i = M$ for $i = 1, \cdots, n$. Rice and Silverman (1991) noticed that in this case, the cubic NSS estimator $\hat{\eta}(t)$ is equivalent to the cubic smoothing spline estimator obtained via smoothing the pointwise means of the responses at all the distinct design time points $\tau_1, \cdots, \tau_M$.

To verify this statement via the expression (6.10), notice that for panel data, $\mathbf{X}_i = \mathbf{I}_M$ after a possible reordering of the measurements. It follows from the first equation of (6.10) that

$$
\begin{aligned}
\hat{\eta}_{nss} &= \left(\sum_{i=1}^{n} \mathbf{X}_i^T \mathbf{X}_i + \lambda \mathbf{G}\right)^{-1} \sum_{i=1}^{n} \mathbf{X}_i^T \mathbf{y}_i \\
&= (n\mathbf{I}_M + \lambda \mathbf{G})^{-1} \sum_{i=1}^{n} \mathbf{y}_i \\
&= (\mathbf{I}_M + n^{-1}\lambda \mathbf{G})^{-1} \bar{\mathbf{y}},
\end{aligned}
$$

where $\bar{\mathbf{y}} = n^{-1} \sum_{i=1}^{n} \mathbf{y}_i$ denotes the pointwise means of the responses $\mathbf{y}_i$, $i = 1, 2, \cdots, n$ at all the distinct design time points. Therefore, the cubic NSS estimator $\hat{\eta}(t)$ is indeed equivalent to the cubic smoothing spline obtained via smoothing the pointwise means of the responses. Verification for other degree NSS estimators can be done similarly.

6.2.4 Variability Band Construction

We can construct the pointwise SD band for $\eta(t)$ based on the NSS smoother (6.10) under the following working independence assumption for the errors e_{ij}:

$$e_{ij} \sim N(0, \sigma^2),\ j = 1, 2, \cdots, n_i;\ i = 1, 2, \cdots, n. \tag{6.12}$$

It follows that $\mathbf{V} = \text{Cov}(\mathbf{y}) = \sigma^2 \mathbf{I}_N$. Then by (6.10), we have

$$\text{Cov}(\hat{\boldsymbol{\eta}}_{nss}) = \sigma^2 (\mathbf{X}^T\mathbf{X} + \lambda \mathbf{G})^{-1} \mathbf{X}^T\mathbf{X} (\mathbf{X}^T\mathbf{X} + \lambda \mathbf{G})^{-1}. \tag{6.13}$$

It follows that the working variance of $\hat{\eta}_{nss}(\tau_r)$ is the r-th diagonal entry of the above matrix $\text{Cov}(\hat{\boldsymbol{\eta}}_{nss})$. That is,

$$\text{Var}\left(\hat{\eta}_{nss}(\tau_r)\right) = \mathbf{e}_r^T \text{Cov}(\hat{\boldsymbol{\eta}}_{nss}) \mathbf{e}_r, \tag{6.14}$$

where $\mathbf{e}_r$ is a M-dimensional unit vector with its r-th entry 1 and others 0. In practice, the working variance σ^2 in the above expressions is usually unknown and should be replaced by its estimator $\hat{\sigma}^2$. Under the working independence assumption, the working variance σ^2 may be estimated by

$$\hat{\sigma}^2 = \|\mathbf{y} - \hat{\mathbf{y}}\|^2 / [N - \text{tr}(\mathbf{A}_{nss})]^2.$$

Then the pointwise SD band of $\eta(t)$ can be constructed approximately. For example, the 95% pointwise standard deviation interval of $\eta(t)$ at τ_r can be constructed as

$$\hat{\eta}_{nss}(\tau_r) \pm 1.96 \sqrt{\widehat{\text{Var}}(\hat{\eta}_{nss}(\tau_r))}, \tag{6.15}$$

where $\widehat{\text{Var}}(\hat{\eta}_{nss}(\tau_r))$ is obtained via replacing the σ^2 in (6.14) by its estimator $\hat{\sigma}^2$.

Examples of the variability band construction using the above method were given in Figures 6.1 (a) and (c).

6.2.5 Choice of the Smoothing Parameter

The smoothing parameter λ needs to be carefully selected. Rice and Silverman (1991) suggested a "leave-one-subject-out" cross-validation (SCV) method. Hoover et al. (1998) advocated the SCV method to select the smoothing parameter for smoothing splines and local polynomial estimators for time-varying coefficient models. For the cubic NSS estimator, the SCV score is defined as

$$\text{SCV}_{nss}(\lambda) = \sum_{i=1}^{n} \left(\mathbf{y}_i - \hat{\boldsymbol{\eta}}_{nss,i}^{(-i)}\right)^T \mathbf{W}_i \left(\mathbf{y}_i - \hat{\boldsymbol{\eta}}_{nss,i}^{(-i)}\right), \tag{6.16}$$

where $\mathbf{W} = \text{diag}(\mathbf{W}_1, \cdots, \mathbf{W}_n)$ is a matrix of weights, and $\hat{\boldsymbol{\eta}}_{nss,i}^{(-i)} = \mathbf{X}_i \hat{\boldsymbol{\eta}}_{nss}^{(-i)}$ with $\hat{\boldsymbol{\eta}}_{nss}^{(-i)}$ being the cubic NSS estimator (6.10) computed using all the data except the data from subject i for each $i = 1, \cdots, n$. The weight matrix $\mathbf{W}$ can be specified by one of the following three possible methods:

Method 1. Take $\mathbf{W}_i = N^{-1}\mathbf{I}_{n_i}$ so that each of the measurements is treated equally.

Method 2. Take $\mathbf{W}_i = (nn_i)^{-1}\mathbf{I}_{n_i}$ so that each of the measurements within a subject is treated equally but the measurements from different subjects are weighted differently based on the number of measurements for each subject.

Method 3. Take $\mathbf{W}_i = \mathbf{V}_i^{-1}$ where $\mathbf{V}_i = \text{Cov}(\mathbf{y}_i)$ so that the within-subject correlation is taken into account.

For the NSS estimator, it is natural to use Method 1 or 2 for specifying $\mathbf{W}$, but for the generalized smoothing spline (GSS) estimator, which will be discussed in the next section, Method 3 is more appropriate.

It is seen that for each given λ, we need to compute the cubic NSS estimator (6.10) n times to obtain $\text{SCV}_{nss}(\lambda)$. Therefore, the SCV_{nss} is computationally intensive, especially when n is large. A trick for reducing some of the computational effort is suggested by Rice and Silverman (1991) and Hoover et al. (1998).

It is easy to check that the NSS estimator $\hat{\eta}(t)$ is a linear smoother of $\eta(t)$. For the cubic NSS estimator, this is demonstrated by the expression (6.11). For this reason, the "leave-one-point-out" cross-validation (PCV) is natural to construct, as well as the associated generalized cross-validation (GCV). For the cubic NSS estimator, the PCV score may be defined as

$$\text{PCV}_{nss}(\lambda) = \sum_{i=1}^{n} \left(\mathbf{y}_i - \hat{\boldsymbol{\eta}}_{nss,i}^{(-pt)}\right)^T \mathbf{W}_i \left(\mathbf{y}_i - \hat{\boldsymbol{\eta}}_{nss,i}^{(-pt)}\right), \tag{6.17}$$

where $\hat{\boldsymbol{\eta}}_{nss,i}^{(-pt)} = \left[\hat{\eta}_{nss,i1}^{(-i1)}, \cdots, \hat{\eta}_{nss,in_i}^{(-in_i)}\right]^T$ and $\hat{\eta}_{nss,ij}^{(-ij)} = \mathbf{x}_{ij}^T\hat{\boldsymbol{\eta}}_{nss}^{(-ij)}$ with $\hat{\boldsymbol{\eta}}_{nss}^{(-ij)}$ denoting the cubic NSS estimator (6.10) computed using all the data except the datum point (t_{ij}, y_{ij}). From (6.17), it seems that for each λ, it requires $N = \sum_{i=1}^n n_i$ computations of the cubic NSS estimator (6.10) using essentially all the data. However, it is easy to show that the PCV score (6.17) can be rewritten as

$$\text{PCV}_{nss}(\lambda) = \sum_{i=1}^{n} (\mathbf{y}_i - \mathbf{X}_i\hat{\boldsymbol{\eta}}_{nss})^T \mathbf{S}_i^{-1}\mathbf{W}_i\mathbf{S}_i^{-1}(\mathbf{y}_i - \mathbf{X}_i\hat{\boldsymbol{\eta}}_{nss}), \tag{6.18}$$

where $\mathbf{S}_i = \mathbf{I}_{n_i} - \text{diag}(a_{i1,i1}, \cdots, a_{in_i,in_i})$ with $a_{ij,ij}$ being the (ij)-th diagonal entry of the smoother matrix $\mathbf{A}_{nss}$ defined in (6.11). This indicates that for each λ, it only requires one computation of the cubic NSS estimator (6.10), resulting in a great saving of computational effort. To further simplify the computation, the diagonal entries $a_{ij,ij}$ may be replaced with their average $N^{-1}\sum_{i=1}^n \sum_{j=1}^{n_i} a_{ij,ij} = \text{tr}(\mathbf{A}_{nss})/N$, resulting in the so called GCV score:

$$\text{GCV}_{nss}(\lambda) = \frac{\sum_{i=1}^n (\mathbf{y}_i - \mathbf{X}_i\hat{\boldsymbol{\eta}}_{nss})^T \mathbf{W}_i(\mathbf{y}_i - \mathbf{X}_i\hat{\boldsymbol{\eta}}_{nss})}{[1 - \text{tr}(\mathbf{A}_{nss})/N]^2}. \tag{6.19}$$

Figure 6.1 (b) displays an example of GCV scores plotted against the smoothing parameter λ in $\log_{10}$-scale. The minimizer of the GCV curve gives the best value of λ for the NSS fit to the conceptive progesterone data as shown in Figure 6.1 (a). It seems that the GCV rule (6.19) yielded a good smoothing parameter in this example.

6.2.6 NSS Fit as BLUP of a LME Model

Following Brumback and Rice (1998), we can express the NSS fit (6.10) as the BLUP of a LME model. As the roughness matrix of cubic smoothing splines with knots at all the distinct design time points (6.4), $\mathbf{G}$ has two zero and $(M-2)$ nonzero eigenvalues. Following Section 3.4 of Chapter 3, we denote the singular value decomposition (SVD) of $\mathbf{G}$ as

$$\mathbf{G} = \mathbf{U}\text{diag}(\mathbf{0}, \mathbf{\Lambda})\mathbf{U}^T, \tag{6.20}$$

where $\mathbf{U}$ is an orthonormal matrix whose column vectors are the eigenvectors of $\mathbf{G}$, $\mathbf{0}$ is a 2×2 zero matrix, and $\mathbf{\Lambda}$ is a $(M-2) \times (M-2)$ diagonal matrix whose diagonal entries are the $(M-2)$ nonzero eigenvalues of $\mathbf{G}$. Let $\mathbf{U}_1$ and $\mathbf{U}_2$ be the matrices consisting of the first 2 columns and the remaining columns of $\mathbf{U}$, respectively.

Using the above SVD, we can reparameterize the expression (6.7) so that it can be easily connected to a LME model. First, we define

$$\beta = \mathbf{U}_1^T\eta, \quad \mathbf{b} = \mathbf{\Lambda}^{1/2}\mathbf{U}_2^T\eta. \tag{6.21}$$

Notice that β is a vector of dimension 2, and $\mathbf{b}$ is a vector of dimension $(M-2)$. Then

$$[\beta^T, \mathbf{b}^T]^T = \text{diag}(\mathbf{I}_2, \mathbf{\Lambda}^{1/2})\mathbf{U}^T\eta, \tag{6.22}$$

and hence

$$\eta = \mathbf{U}\text{diag}(\mathbf{I}_2, \mathbf{\Lambda}^{-1/2})[\beta^T, \mathbf{b}^T]^T. \tag{6.23}$$

It is easy to show that

$$\eta^T\mathbf{G}\eta = \mathbf{b}^T\mathbf{b}. \tag{6.24}$$

Define

$$\tilde{\mathbf{X}} = \mathbf{X}\mathbf{U}_1, \quad \tilde{\mathbf{Z}} = \mathbf{X}\mathbf{U}_2\mathbf{\Lambda}^{-1/2}. \tag{6.25}$$

Using this and (6.23), we have $\mathbf{X}\eta = \tilde{\mathbf{X}}\beta + \tilde{\mathbf{Z}}\mathbf{b}$. We then reparameterize (6.9) as

$$\|\mathbf{y} - \tilde{\mathbf{X}}\beta - \tilde{\mathbf{Z}}\mathbf{b}\|^2 + \lambda\mathbf{b}^T\mathbf{b}. \tag{6.26}$$

Under the working independence assumption (6.12), we can connect (6.26) to the following LME model:

$$\begin{array}{c} \mathbf{y} = \tilde{\mathbf{X}}\beta + \tilde{\mathbf{Z}}\mathbf{b} + \epsilon, \\ \mathbf{b} \sim N(0, \sigma^2/\lambda\mathbf{I}_{M-2}), \quad \epsilon \sim N(0, \sigma^2\mathbf{I}_N), \end{array} \tag{6.27}$$

in the sense that the minimizers $\hat{\beta}$ and $\hat{\mathbf{b}}$ of (6.26) are the BLUPs of the above LME model. Notice that there are 2 free variance components in the LME model (6.27), σ^2 and $\sigma_\lambda^2 = \sigma^2/\lambda$ for the measurement errors and the population mean function $\eta(t)$, respectively.

The LME model (6.27) can be solved using the S-PLUS function *lme* or the SAS procedure *PROC MIXED* to obtain the BLUPs $\hat{\beta}, \hat{\mathbf{b}}$ and the variance components $\hat{\sigma}^2$ and $\hat{\sigma}_\lambda^2$ of the LME model (6.27). Then the fits of $\eta(t)$ at all the design time points (6.2) can be summarized as $\hat{\eta}^* = \tilde{\mathbf{X}}\hat{\beta} + \tilde{\mathbf{Z}}\hat{\mathbf{b}}$. The standard deviations of $\hat{\eta}^*$

can also be obtained via using the covariance matrices of $\hat{\beta}$ and $\hat{\mathbf{b}}$. Moreover, using the definition of σ_λ^2, a proper smoothing parameter can be obtained as $\lambda = \hat{\sigma}^2/\hat{\sigma}_\lambda^2$. Similar to Section 3.4.3, the connection between the NSS fit and the BLUPs of a LME model is just for computational convenience. Here the LME model is different from the subject-specific LME model described in Section 2.2 of Chapter 2.

6.2.7 Model Checking

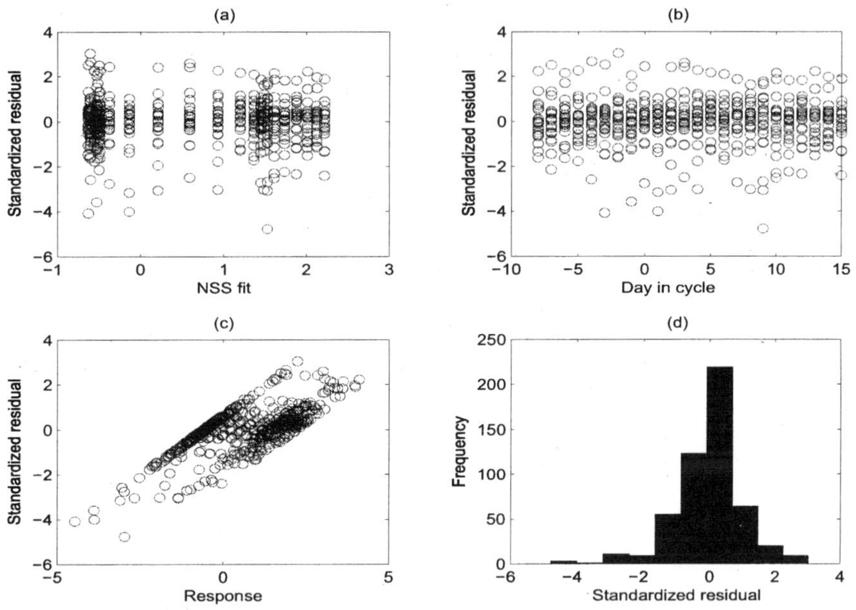

Fig. 6.2 Residual analysis of the NSS fit to the conceptive progesterone data.

Standard residual analysis may be performed to check the adequacy of the NSS fit. For example, standard residual plots for the NSS fit to the conceptive progesterone data are given in Figure 6.2. Figures 6.2 (a), (b) and (c) display the plots of the standardized residual against NSS fit $\hat{\eta}_{ij}$, time t_{ij}, and response y_{ij}, respectively. From Figures 6.2 (a) and (b), it is seen that the population mean function was adequately fitted. However, Figure 6.2 (c) shows a strong linear tendency against the response: the larger the response, the larger the standardized residual. This implies that the NPM model (6.1) is not adequate to the data. Figure 6.2 (d) shows the histogram of the standardized residual. It indicates that the standardized residuals are approximately symmetric and bell shaped.

6.3 GENERALIZED SMOOTHING SPLINES

The naive smoothing spline (NSS) smoother (6.11) is constructed without taking the correlation structure of longitudinal data into account (Rice and Silverman 1991, Hoover et al. 1998). It should generally perform well when the longitudinal data are not strongly correlated. Otherwise, it is efficient to take the correlation into account. By doing so, Wang (1998b) proposed and studied a generalized smoothing spline (GSS) method.

6.3.1 Constructing a Cubic GSS Estimator

We shall focus on constructing a cubic GSS estimator due to its simplicity and wide application. A general degree GSS estimator can be constructed similarly. We defer this to Section 6.6.

The cubic GSS estimator of $\eta(t)$ is defined as the minimizer $\hat{\eta}(t)$ of the penalized generalized least squares (PGLS) criterion:

$$\begin{gathered}\sum_{i=1}^n (\mathbf{y}_i - \boldsymbol{\eta}_i)^T \mathbf{V}_i^{-1} (\mathbf{y}_i - \boldsymbol{\eta}_i) + \lambda \int_a^b [\eta''(t)]^2\, dt, \\ \text{where } \boldsymbol{\eta}_i = [\eta(t_{i1}), \cdots, \eta(t_{in_i})]^T, \quad \mathbf{V}_i = \text{Cov}(\mathbf{y}_i),\end{gathered} \tag{6.28}$$

with respect to $\eta(t)$ over the second order Sobolev space $\mathcal{W}_2^2[a,b]$ as defined in (3.32) with $k = 2$. Unlike the criterion (6.3), the above criterion takes the within-subject correlation of $\mathbf{y}_i$ into account.

When all the $\mathbf{V}_i$ are known, it is easy to minimize the PGLS criterion (6.28). Wang (1998b) solved this problem using the smoothing spline based on the reproducing kernel Hilbert space bases developed in Wahba (1990). We are more familiar with the smoothing splines implemented in Green and Silverman (1994). This is the reason why we adopt the implementation of Green and Silverman (1994), although there is little difference in the essential ideas of both methods.

Using the definitions of $\mathbf{X}_i$ and η given in the previous section and the roughness expression (6.5), the PGLS criterion (6.28) can be re-expressed as

$$\sum_{i=1}^n (\mathbf{y}_i - \mathbf{X}_i\eta)^T \mathbf{V}_i^{-1} (\mathbf{y}_i - \mathbf{X}_i\eta) + \lambda \eta^T \mathbf{G} \eta,$$

or

$$(\mathbf{y} - \mathbf{X}\eta)^T \mathbf{V}^{-1} (\mathbf{y} - \mathbf{X}\eta) + \lambda \eta^T \mathbf{G} \eta, \tag{6.29}$$

where $\mathbf{V} = \text{diag}(\mathbf{V}_1, \cdots, \mathbf{V}_n)$ and $\mathbf{X} = [\mathbf{X}_1^T, \cdots, \mathbf{X}_n^T]^T$ as defined in (6.8). Simple algebra leads to

$$\hat{\eta}_{gss} = (\mathbf{X}^T \mathbf{V}^{-1} \mathbf{X} + \lambda \mathbf{G})^{-1} \mathbf{X}^T \mathbf{V}^{-1} \mathbf{y}. \tag{6.30}$$

It follows that the values of $\hat{\eta}(t)$ at all the design time points (6.2) form a vector:

$$\hat{\mathbf{y}}_{gss} = \mathbf{X}(\mathbf{X}^T \mathbf{V}^{-1} \mathbf{X} + \lambda \mathbf{G})^{-1} \mathbf{X}^T \mathbf{V}^{-1} \mathbf{y} = \mathbf{A}_{gss} \mathbf{y}, \tag{6.31}$$

where the associated GSS estimator matrix is

$$\mathbf{A}_{gss} = \mathbf{X}(\mathbf{X}^T \mathbf{V}^{-1} \mathbf{X} + \lambda \mathbf{G})^{-1} \mathbf{X}^T \mathbf{V}^{-1}. \tag{6.32}$$

It is noteworthy that the GSS estimator matrix (6.32) is a generalization of the NSS smoother matrix defined in (6.11). When $\mathbf{V} = \mathbf{I}_N$, the GSS estimator matrix (6.32) reduces to the NSS smoother matrix in (6.11). However, the GSS estimator matrix (6.32) is no longer a projection matrix since it is not symmetric.

6.3.2 Variability Band Construction

As we did for the NSS smoother in the previous section, we can construct the pointwise SD band for $\eta(t)$ based on the GSS estimator (6.30). Since $\text{Cov}(\mathbf{y}) = \mathbf{V}$, by (6.30), we have

$$\text{Cov}(\hat{\boldsymbol{\eta}}_{gss}) = (\mathbf{X}^T\mathbf{V}^{-1}\mathbf{X} + \lambda\mathbf{G})^{-1}\mathbf{X}^T\mathbf{V}^{-1}\mathbf{X}(\mathbf{X}^T\mathbf{V}^{-1}\mathbf{X} + \lambda\mathbf{G})^{-1}. \quad (6.33)$$

When we replace the $\mathbf{V}$ by its estimator, we can obtain an estimate of $\text{Cov}(\hat{\boldsymbol{\eta}}_{gss})$. Based on this, we can easily construct the approximate pointwise SD band, which is similar to that of the NSS smoother, described in Section 6.2.4.

6.3.3 Choice of the Smoothing Parameter

For the cubic NSS smoother (6.10), we proposed to select the smoothing parameter, λ, using the SCV, PCV or GCV rule. These rules may be extended for the cubic GSS estimator (6.30). As an example, let us see how we can extend the GCV criterion (6.19).

Set $\tilde{\mathbf{y}} = \mathbf{V}^{-1/2}\mathbf{y}$ and $\tilde{\mathbf{X}} = \mathbf{V}^{-1/2}\mathbf{X}$. Then we have $\text{Cov}(\tilde{\mathbf{y}}) = \mathbf{I}_N$. This means that the entries of $\tilde{\mathbf{y}}$ are i.i.d. Therefore the GCV rule (6.19) should apply for the linear smoother based on $\tilde{\mathbf{y}}$ and $\tilde{\mathbf{X}}$. From (6.30), we have

$$\hat{\boldsymbol{\eta}}_{gss} = (\tilde{\mathbf{X}}^T\tilde{\mathbf{X}} + \lambda\mathbf{G})^{-1}\tilde{\mathbf{X}}^T\tilde{\mathbf{y}}. \quad (6.34)$$

It follows that

$$\hat{\tilde{\mathbf{y}}}_{nss} = \tilde{\mathbf{X}}\hat{\boldsymbol{\eta}}_{gss} = \tilde{\mathbf{X}}(\tilde{\mathbf{X}}^T\tilde{\mathbf{X}} + \lambda\mathbf{G})^{-1}\tilde{\mathbf{X}}^T\tilde{\mathbf{y}} = \tilde{\mathbf{A}}_{nss}\tilde{\mathbf{y}},$$

where $\tilde{\mathbf{A}}_{nss}$ is the associated smoother matrix for regressing $\tilde{\mathbf{y}}$ against $\tilde{\mathbf{X}}$ using a NSS smoother. Applying the GCV rule (6.19), we have

$$\text{GCV}_{gss}(\lambda) = \sum_{i=1}^{n}(\tilde{\mathbf{y}}_i - \tilde{\mathbf{X}}_i\hat{\boldsymbol{\eta}}_{gss})^T\mathbf{W}_i(\tilde{\mathbf{y}}_i - \tilde{\mathbf{X}}_i\hat{\boldsymbol{\eta}}_{gss})/[1 - \text{tr}(\tilde{\mathbf{A}}_{nss})/N]^2.$$

Notice that

$$(\tilde{\mathbf{y}}_i - \tilde{\mathbf{X}}_i\hat{\boldsymbol{\eta}}_{gss})^T\mathbf{W}_i(\tilde{\mathbf{y}}_i - \tilde{\mathbf{X}}_i\hat{\boldsymbol{\eta}}_{gss}) = (\mathbf{y}_i - \mathbf{X}_i\hat{\boldsymbol{\eta}}_{gss})^T\mathbf{V}_i^{-1/2}\mathbf{W}_i\mathbf{V}_i^{-1/2}(\mathbf{y}_i - \mathbf{X}_i\hat{\boldsymbol{\eta}}_{gss}),$$

and

$$\text{tr}(\tilde{\mathbf{A}}_{nss}) = \text{tr}((\mathbf{X}^T\mathbf{V}^{-1}\mathbf{X} + \lambda\mathbf{G})^{-1}\mathbf{X}^T\mathbf{V}^{-1}\mathbf{X}) = \text{tr}(\mathbf{A}_{gss}),$$

we have

$$\text{GCV}_{gss}(\lambda) = \frac{\sum_{i=1}^{n}(\mathbf{y}_i - \mathbf{X}_i\hat{\boldsymbol{\eta}}_{gss})^T\mathbf{V}_i^{-1/2}\mathbf{W}_i\mathbf{V}_i^{-1/2}(\mathbf{y}_i - \mathbf{X}_i\hat{\boldsymbol{\eta}}_{gss})}{[1 - \text{tr}(\mathbf{A}_{gss})/N]^2}. \quad (6.35)$$

In practice, we can take $\mathbf{W}_i = (nn_i)^{-1}\mathbf{I}_{n_i}$ so that $\mathbf{V}_i^{-1/2}\mathbf{W}_i\mathbf{V}_i^{-1/2} = (nn_i)^{-1}\mathbf{V}_i^{-1}$.

6.3.4 Covariance Matrix Estimation

In the above subsections, the covariance matrix $\mathbf{V}$ is assumed to be known. In practice, it has to be estimated. For the GSS estimators, it is assumed that $\mathbf{V}$ follows a known parametric model with some unknown parameter vector $\boldsymbol{\theta}$. Wang (1998a) suggested to estimate $\boldsymbol{\theta}$ and the smoothing parameter λ at the same time using the restricted maximum likelihood (REML) method, i.e., the generalized maximum likelihood (GML) method of Wahba (1990). See also Wang and Taylor (1995). Some parametric models for the covariance structure are given in Diggle et al. (2002).

6.3.5 GSS Fit as BLUP of a LME Model

In Section 6.2.6, we demonstrated that the NSS fit (6.10) can be expressed as the BLUP of a LME model. In this section, we can easily demonstrate that the GSS fit (6.30) can also be expressed as the BLUP of a LME model. For this purpose, it is sufficient to show that the expression (6.29) can be easily transformed into the expression (6.9).

In fact, when $\mathbf{V}$ is known, we can set $\tilde{\mathbf{y}} = \mathbf{V}^{-1/2}\mathbf{y}$ and $\tilde{\mathbf{X}} = \mathbf{V}^{-1/2}\mathbf{X}$. Then (6.29) can be written as

$$\|\tilde{\mathbf{y}} - \tilde{\mathbf{X}}\boldsymbol{\eta}\|^2 + \lambda\boldsymbol{\eta}^T\mathbf{G}\boldsymbol{\eta}, \tag{6.36}$$

which is the same form as that of the expression (6.9) as desired.

6.4 EXTENDED SMOOTHING SPLINES

The NSS and GSS estimators aim to estimate the population mean function only. Sometimes, the individual curves, i.e., the subject-specific curves are also of interest. In this section, we extend smoothing splines to fit subject-specific curves. For simplicity, we call the resulting smoothers as extended smoothing splines.

6.4.1 Subject-Specific Curve Fitting

For a longitudinal data set, the subject-specific curve fitting (SSCF) model can be written as

$$\begin{array}{c} y_{ij} = \eta(t_{ij}) + v_i(t_{ij}) + \epsilon_{ij}, \\ j = 1, 2, \cdots, n_i;\ i = 1, 2, \cdots, n, \\ \text{Subject to } \sum_{i=1}^n v_i(t) = 0, \end{array} \tag{6.37}$$

where $\eta(t)$ is the smooth population mean curve, and $v_i(t)$ are the smooth subject-specific deviation curves of the smooth individual curves $s_i(t) = \eta(t) + v_i(t),\ i = 1, 2, \cdots, n$ from the population mean curve. The model (6.37) may be obtained via replacing e_{ij} of the NPM model (6.1) by $v_i(t_{ij}) + \epsilon_{ij}$ and adding an identifiability condition $\sum_{i=1}^n v_i(t) = 0$. This allows modeling the population mean and subject-specific deviation curves of longitudinal data at the same time.

Recall that $\boldsymbol{\eta} = [\eta(\tau_1), \cdots, \eta(\tau_M)]^T$ contains the values of $\eta(t)$ at all the distinct design time points (6.4). Define similarly $\mathbf{v}_i = [v_i(\tau_1), \cdots, v_i(\tau_M)]^T$. Using the definition of $\mathbf{x}_{ij}$ given in (6.6), we have $\eta(t_{ij}) = \mathbf{x}_{ij}^T\boldsymbol{\eta}$ and $v_i(t_{ij}) = \mathbf{x}_{ij}^T\mathbf{v}_i$. Therefore, the model (6.37) can be further written as

$$\begin{array}{c} \mathbf{y}_i = \mathbf{X}_i\boldsymbol{\eta} + \mathbf{X}_i\mathbf{v}_i + \boldsymbol{\epsilon}_i, \quad i = 1, 2, \cdots, n, \\ \text{Subject to } \sum_{i=1}^n v_i(t) = 0, \end{array} \tag{6.38}$$

where $\mathbf{y}_i$ and $\mathbf{X}_i$ are defined as before and $\boldsymbol{\epsilon}_i = [\epsilon_{i1}, \cdots, \epsilon_{in_i}]^T$.

Denote $\mathbf{R}_i = \text{Cov}(\boldsymbol{\epsilon}_i)$. Since $\eta(t)$ and $v_i(t),\ i = 1, 2, \cdots, n$ are assumed to be smooth, following Brumback and Rice (1998), we define their estimators $\hat{\eta}(t)$ and $\hat{v}_i(t),\ i = 1, 2, \cdots, n$ as the minimizers of the following constrained penalized least squares (PLS) criterion:

$$\begin{array}{c} \sum_{i=1}^n \left\{(\mathbf{y}_i - \mathbf{X}_i\boldsymbol{\eta} - \mathbf{X}_i\mathbf{v}_i)^T\mathbf{R}_i^{-1}(\mathbf{y}_i - \mathbf{X}_i\boldsymbol{\eta} - \mathbf{X}_i\mathbf{v}_i)\right\} \\ + \lambda_v \sum_{i=1}^n \left\{ \int_a^b [v_i''(t)]^2 dt \right\} + \lambda \int_a^b [\eta''(t)]^2 dt, \\ \text{Subject to } \sum_{i=1}^n v_i(t) = 0, \end{array} \tag{6.39}$$

over the second order Sobolev space $\mathcal{W}_2^2[a, b]$ of functions with continuous second derivatives; see (3.32) for a general definition of the Sobolev space $\mathcal{W}_2^k[a, b]$ where $[a, b]$ is an interval that contains all the design time points. In the expression (6.39), the first term is the weighted sum of squared residuals, representing the goodness of fit of SSCF modeling, the second term is the sum of the roughness of all the smooth subject-specific deviation curves $v_i(t)$ multiplied by a smoothing parameter λ_v, and the last term is the roughness of $\eta(t)$ multiplied by another smoothing parameter λ. Therefore, the smoothing parameters λ_v and λ are used to tradeoff the goodness of fit with the roughness of $v_i(t)$ and $\eta(t)$. We may show that the estimators $\hat{\eta}(t)$ and $\hat{v}_i(t),\ i = 1, 2, \cdots, n$ are natural cubic smoothing splines with knots at all the distinct design time points (6.4).

6.4.2 The ESS Estimators

To solve the above minimization problem (6.39), we need to compute the two roughness terms. Using (3.30) and (6.5) again, we have

$$\int_a^b [v_i''(t)]^2 dt = \mathbf{v}_i^T\mathbf{G}\mathbf{v}_i, \quad \int_a^b [\eta''(t)]^2 dt = \boldsymbol{\eta}^T\mathbf{G}\boldsymbol{\eta}, \tag{6.40}$$

where $\mathbf{G}$ is the associated roughness matrix computed in (3.33) using the method of Green and Silverman (1994). It follows that the PLS criterion (6.39) can be written as

$$\begin{array}{c} \sum_{i=1}^n \left\{(\mathbf{y}_i - \mathbf{X}_i\boldsymbol{\eta} - \mathbf{X}_i\mathbf{v}_i)^T\mathbf{R}_i^{-1}(\mathbf{y}_i - \mathbf{X}_i\boldsymbol{\eta} - \mathbf{X}_i\mathbf{v}_i)\right\} \\ + \lambda_v \sum_{i=1}^n \mathbf{v}_i^T\mathbf{G}\mathbf{v}_i + \lambda\boldsymbol{\eta}^T\mathbf{G}\boldsymbol{\eta}, \\ \text{Subject to } \sum_{i=1}^n \mathbf{v}_i = 0. \end{array} \tag{6.41}$$

Note that the identifiability condition $\sum_{i=1}^n \mathbf{v}_i = 0$ is derived from the identifiability condition $\sum_{i=1}^n v_i(t) = 0$ of (6.37) since the latter implies that $\sum_{i=1}^n v_i(\tau_l) = 0, l =$

$1, 2, \cdots, M$. Set

$$\begin{array}{llll} \mathbf{y} & = [\mathbf{y}_1^T, \cdots, \mathbf{y}_n^T]^T, & \mathbf{X} = [\mathbf{X}_1, \cdots, \mathbf{X}_n]^T, & \\ \mathbf{v} & = [\mathbf{v}_1^T, \cdots, \mathbf{v}_n^T]^T, & \mathbf{Z} = \mathrm{diag}(\mathbf{X}_1, \cdots, \mathbf{X}_n), & (6.42) \\ \mathbf{R} & = \mathrm{diag}(\mathbf{R}_1, \cdots, \mathbf{R}_n), & \tilde{\mathbf{G}} = \mathrm{diag}(\mathbf{G}, \cdots, \mathbf{G}). & \end{array}$$

Then we can further write (6.41) as

$$\begin{array}{c} (\mathbf{y} - \mathbf{X}\eta - \mathbf{Z}\mathbf{v})^T \mathbf{R}^{-1} (\mathbf{y} - \mathbf{X}\eta - \mathbf{Z}\mathbf{v}) + \lambda_v \mathbf{v}^T \tilde{\mathbf{G}} \mathbf{v} + \lambda \eta^T \mathbf{G} \eta, \\ \mathbf{H}\mathbf{v} = 0, \end{array} \tag{6.43}$$

where $\mathbf{H} = [\mathbf{I}_M, \mathbf{I}_M, \cdots, \mathbf{I}_M]$, a block matrix with n blocks of $\mathbf{I}_M$. Due to the constraint $\mathbf{H}\mathbf{v} = 0$, it is not easy to write down the closed-form formulas of the minimizers of the constrained PLS criterion (6.43).

6.4.3 ESS Fits as BLUPs of a LME Model

It is challenging to solve (6.39) since it is a constrained PLS problem. Moreover, it has $n + 1$ functions, $\eta(t)$ and $v_i(t)$, $i = 1, 2, \cdots, n$, need to be estimated and two smoothing parameters, λ and λ_v, need to be selected. Brumback and Rice (1998), Wang (1998a), and Zhang et al. (1998) overcame these difficulties via connecting the SSCF problem (6.39), or equivalently, the SSCF problem (6.41), to a LME model that can be solved using the S-PLUS function *lme* or the SAS procedure *PROC MIXED*.

This can be done in a similar way as described in Section 6.2.6, but now we have to deal with the population mean function $\eta(t)$ and the n individual curves $s_i(t)$, $i = 1, 2, \cdots, n$. The key again is to use the singular value decomposition (SVD) of the roughness matrix $\mathbf{G}$ as defined in (6.20) to reparameterize the PLS criterion (6.41) so that the SSCF model can be easily connected to a LME model. First, we define

$$\begin{array}{rcl} \beta_0 & = & \mathbf{U}_1^T \eta, \;\; \mathbf{b}_0 = \mathbf{\Lambda}^{1/2} \mathbf{U}_2^T \eta, \\ \beta_i & = & \mathbf{U}_1^T \mathbf{v}_i, \;\; \mathbf{b}_i = \mathbf{\Lambda}^{1/2} \mathbf{U}_2^T \mathbf{v}_i. \end{array} \tag{6.44}$$

The identifiability condition $\sum_{i=1}^n \mathbf{v}_i = 0$ implies that

$$\sum_{i=1}^n \beta_i = 0, \;\; \sum_{i=1}^n \mathbf{b}_i = 0.$$

Notice that $\beta_0, \beta_1, \cdots, \beta_n$ are vectors of dimension 2, and $\mathbf{b}_0, \mathbf{b}_1, \cdots, \mathbf{b}_n$ are vectors of dimension $(M - 2)$. By (6.44), we have

$$\begin{array}{rcl} [\beta_0^T, \mathbf{b}_0^T]^T & = & \mathrm{diag}(\mathbf{I}_2, \mathbf{\Lambda}^{1/2}) \mathbf{U}^T \eta, \\ [\beta_i^T, \mathbf{b}_i^T]^T & = & \mathrm{diag}(\mathbf{I}_2, \mathbf{\Lambda}^{1/2}) \mathbf{U}^T \mathbf{v}_i, \end{array} \tag{6.45}$$

and hence

$$\begin{array}{rcl} \eta & = & \mathbf{U}\mathrm{diag}(\mathbf{I}_2, \mathbf{\Lambda}^{-1/2}) [\beta_0^T, \mathbf{b}_0^T]^T = \mathbf{U}_1 \beta_0 + \mathbf{U}_2 \mathbf{\Lambda}^{-1/2} \mathbf{b}_0, \\ \mathbf{v}_i & = & \mathbf{U}\mathrm{diag}(\mathbf{I}_2, \mathbf{\Lambda}^{-1/2}) [\beta_i^T, \mathbf{b}_i^T]^T = \mathbf{U}_1 \beta_i + \mathbf{U}_2 \mathbf{\Lambda}^{-1/2} \mathbf{b}_i. \end{array} \tag{6.46}$$

Therefore

$$\eta^T \mathbf{G}\eta = \mathbf{b}_0^T \mathbf{b}_0, \quad \mathbf{v}_i^T \mathbf{G}\mathbf{v}_i = \mathbf{b}_i^T \mathbf{b}_i. \tag{6.47}$$

Let $\tilde{\mathbf{X}}_i = \mathbf{X}_i \mathbf{U}_1$ and $\tilde{\mathbf{Z}}_i = \mathbf{X}_i \mathbf{U}_2 \mathbf{\Lambda}^{-1/2}$. Then by (6.46), we have

$$\mathbf{X}_i \eta = \tilde{\mathbf{X}}_i \boldsymbol{\beta}_0 + \tilde{\mathbf{Z}}_i \mathbf{b}_0, \quad \mathbf{X}_i \mathbf{v}_i = \tilde{\mathbf{X}}_i \boldsymbol{\beta}_i + \tilde{\mathbf{Z}}_i \mathbf{b}_i. \tag{6.48}$$

We then reparameterize (6.41) as

$$\begin{array}{c} \sum_{i=1}^n \Big\{ [\mathbf{y}_i - (\tilde{\mathbf{X}}_i \boldsymbol{\beta}_0 + \tilde{\mathbf{Z}}_i \mathbf{b}_0) - (\tilde{\mathbf{X}}_i \boldsymbol{\beta}_i + \tilde{\mathbf{X}}_i \mathbf{b}_i)]^T \mathbf{R}_i^{-1} \\ \times [\mathbf{y}_i - (\tilde{\mathbf{X}}_i \boldsymbol{\beta}_0 + \tilde{\mathbf{Z}}_i \mathbf{b}_0) - (\tilde{\mathbf{X}}_i \boldsymbol{\beta}_i + \tilde{\mathbf{X}}_i \mathbf{b}_i)] \Big\} \\ + \lambda_v \sum_{i=1}^n \mathbf{b}_i^T \mathbf{b}_i + \lambda \mathbf{b}_0^T \mathbf{b}_0, \\ \text{Subject to } \sum_{i=1}^n \boldsymbol{\beta}_i = 0, \ \sum_{i=1}^n \mathbf{b}_i = 0. \end{array} \tag{6.49}$$

Combining the parameter vectors, we get

$$\boldsymbol{\beta} = [\boldsymbol{\beta}_0^T, \boldsymbol{\beta}_1^T, \cdots, \boldsymbol{\beta}_n^T]^T, \quad \mathbf{b} = [\mathbf{b}_0^T, \mathbf{b}_1^T, \cdots, \mathbf{b}_n^T]^T,$$

and

$$\mathbf{D} = \text{diag}[\mathbf{I}_{M-2}/\lambda, \mathbf{I}_{M-2}/\lambda_v, \cdots, \mathbf{I}_{M-2}/\lambda_v],$$

where $\boldsymbol{\beta}$ is a $[2(n+1)]$ dimensional vector obtained by stacking $\boldsymbol{\beta}_0, \cdots, \boldsymbol{\beta}_n$ together, b is a $[(M-2)(n+1)]$ dimensional vector obtained by stacking $\mathbf{b}_0, \cdots, \mathbf{b}_n$ together, and $\mathbf{D}$ is a diagonal matrix of size $[(M-2)(n+1)] \times [(M-2)(n+1)]$. Then (6.49) can be further written as

$$\begin{array}{c} (\mathbf{y} - \tilde{\mathbf{X}}\boldsymbol{\beta} - \tilde{\mathbf{Z}}\mathbf{b})^T \mathbf{R}^{-1} (\mathbf{y} - \tilde{\mathbf{X}}\boldsymbol{\beta} - \tilde{\mathbf{Z}}\mathbf{b}) + \mathbf{b}^T \mathbf{D}^{-1} \mathbf{b}, \\ \text{Subject to } \mathbf{H}_0 \boldsymbol{\beta} = 0, \ \mathbf{H}_1 \mathbf{b} = 0, \end{array} \tag{6.50}$$

where $\mathbf{H}_0 = [\mathbf{0}_{2\times 2}, \mathbf{I}_2, \cdots, \mathbf{I}_2]$, $\mathbf{H}_1 = [\mathbf{0}_{(M-2)\times(M-2)}, \mathbf{I}_{M-2}, \cdots, \mathbf{I}_{M-2}]$ are two block matrices so that $\mathbf{H}_0 \boldsymbol{\beta} = 0$ and $\mathbf{H}_1 \mathbf{b} = 0$ are equivalent to $\sum_{i=1}^n \boldsymbol{\beta}_i = 0$ and $\sum_{i=1}^n \mathbf{b}_i = 0$, respectively, and $\tilde{\mathbf{X}}$ and $\tilde{\mathbf{Z}}$ are matrices of size $N \times [2(n+1)]$ and $N \times [(M-2)(n+1)]$, respectively. The $\tilde{\mathbf{X}}$ is defined as

$$\tilde{\mathbf{X}} = \begin{pmatrix} \tilde{\mathbf{X}}_1, & \tilde{\mathbf{X}}_1, & \mathbf{0}, & \cdots, & \mathbf{0}, & \mathbf{0} \\ \tilde{\mathbf{X}}_2, & \mathbf{0}, & \tilde{\mathbf{X}}_2, & \cdots, & \mathbf{0}, & \mathbf{0} \\ \vdots, & \vdots, & \vdots, & \vdots, & \vdots, & \vdots \\ \tilde{\mathbf{X}}_{n-1}, & \mathbf{0}, & \mathbf{0}, & \cdots, & \tilde{\mathbf{X}}_{n-1}, & \mathbf{0} \\ \tilde{\mathbf{X}}_n, & \mathbf{0}, & \mathbf{0}, & \cdots, & \mathbf{0}, & \tilde{\mathbf{X}}_n \end{pmatrix},$$

and $\tilde{\mathbf{Z}}$ is defined similarly but replacing all the $\tilde{\mathbf{X}}_i$ with $\tilde{\mathbf{Z}}_i$ in the above definition. We then connect (6.50) to the following LME model:

$$\begin{array}{c} \mathbf{y} = \tilde{\mathbf{X}}\boldsymbol{\beta} + \tilde{\mathbf{Z}}\mathbf{b} + \boldsymbol{\epsilon}, \\ \mathbf{b} \sim N(0, \mathbf{D}), \quad \boldsymbol{\epsilon} \sim N(0, \mathbf{R}), \\ \text{Subject to } \mathbf{H}_0 \boldsymbol{\beta} = 0, \end{array} \tag{6.51}$$

in the sense that the minimizers $\hat{\beta}$ and $\hat{\mathbf{b}}$ of (6.50) are the BLUPs of the above LME model. In the above LME model, we drop the identifiability condition $\mathbf{H}_1\mathbf{b} = \sum_{i=1}^n \mathbf{b}_i = 0$ since we assume that $\mathrm{E}\mathbf{b} = 0$ so that the condition is automatically satisfied at least asymptotically. Notice that the above LME model has $2(n+1)$ fixed-effects parameters in β for the population mean curve $\eta(t)$ and the n subject-specific deviation curves $v_i(t),\ i = 1, 2, \cdots, n$. The population mean curve contributes β_0, the first level fixed-effects; and the subject-specific deviation curves contribute $\beta_1, \cdots, \beta_n$, the second level fixed-effects. Usually, the second level fixed-effects are not identifiable and a natural constraint is to make them sum up to 0. Therefore, we still need to keep the identifiability condition $\mathbf{H}_0\beta = \sum_{i=1}^n \beta_i = 0$ (Brumback and Rice 1998, Ruppert, Wand and Carroll 2003).

Notice that when we assume $\mathbf{R} = \sigma^2\mathbf{I}_N$, as we usually do, there are 3 free variance components in the LME model (6.51):

$$\sigma^2, \quad \sigma_\lambda^2 = 1/\lambda, \quad \text{and} \quad \sigma_{\lambda_v}^2 = 1/\lambda_v, \tag{6.52}$$

for the measurement errors, the population mean curve $\eta(t)$ and the subject-specific deviation curves $v_i(t)$, respectively.

The above connection is a generalization of the one presented in Section 3.4 of Chapter 3 where only one smoothing spline is connected with the BLUPs of a LME model, while here a group of smoothing splines are connected to the BLUPs of a LME model. This connection can be used to make inferences about the SSCF model (6.37) using the BLUPs of the LME model (6.51).

The LME model (6.51) is a standard constrained LME model and it can be fitted using the S-PLUS function *lme* or the SAS procedure *PROC MIXED*. Due to its special structure (i.e., the covariance matrix is diagonal), Brumback and Rice (1998) also discussed some strategies for reducing its computation effort. Let us assume the BLUPs, $\hat{\beta}$ and $\hat{\mathbf{b}}_i$, of the constrained LME model (6.51) are available and the variance components estimated as $\hat{\sigma}^2, \hat{\sigma}_\lambda^2$ and $\hat{\sigma}_{\lambda_v}^2$ using a ML-based or REML-based EM-algorithm. Then using (6.46), we can obtain the estimates of $\eta(t)$ and $v_i(t)$ at all the distinct design time points (6.4):

$$\begin{aligned} \hat{\eta} &= \mathbf{U}\mathrm{diag}[\mathbf{I}_2, \mathbf{\Lambda}^{-1/2}][\hat{\beta}_0^T, \hat{\mathbf{b}}_0^T]^T, \\ \hat{\mathbf{v}}_i &= \mathbf{U}\mathrm{diag}[\mathbf{I}_2, \mathbf{\Lambda}^{-1/2}][\hat{\beta}_i^T, \hat{\mathbf{b}}_i^T]^T,\ i = 1, 2, \cdots, n. \end{aligned} \tag{6.53}$$

Using (6.52), we obtain the associated smoothing parameters $\lambda = 1/\hat{\sigma}_\lambda^2$ and $\lambda_v = 1/\hat{\sigma}_{\lambda_v}^2$. This implies that the EM-algorithm can be used as an automatic smoothing parameter selector for selecting λ and λ_v.

The S-PLUS function *lme* or the SAS procedure *PROC MIXED* also provide the standard deviations of $\hat{\beta}$ and $\hat{\mathbf{b}}$. Using (6.53), we can also compute the pointwise standard deviations of $\hat{\eta}$ and hence construct the pointwise SD band for $\eta(t)$.

The technique using the connection between a smoothing spline and a LME model (Speed 1991) is widely used in the recent smoothing spline literature (Brumback and Rice 1998, Wang 1998a, Zhang et al. 1998, Guo 2002a). It has several advantages under the SSCF model:

(1). The SSCF problem (6.39) is connected to a standard constrained LME model, which can be solved by some existing software routines;

(2). The smoothing parameters involved in (6.41) now become part of the variance components of the constrained LME model (6.51) and can be computed via the powerful EM-algorithm, which serves as a smoothing parameter selector. This is quite different from the traditional smoothing parameter selector such as CV or GCV; and

(3). The resulting estimates are showed to be the BLUPs of the constrained LME model.

However, this technique has some limitations too. For example, it is computationally expensive since the variance matrix $\mathbf{V} = \tilde{\mathbf{Z}}\mathbf{D}\tilde{\mathbf{Z}}^T + \mathbf{R}$ is a $N \times N$ matrix which is not block diagonal (since $\tilde{\mathbf{Z}}$ is not block diagonal). It requires a large storage capacity and a long time to inverse $\mathbf{V}$ when N is large. For the progesterone data that we have introduced in Chapter 1, we have $M = 24$ distinct design time points. The number of conceptive and nonconceptive curves is 91 so that for the whole data set, N is about $24 \times 91 = 2184$. For such a big $\mathbf{V}$, Brumback and Rice (1998) reported that the $\mathbf{V}$ required approximately 32 MB of storage, and using a Sparc 2000 machine, just one inversion of such a matrix took about 15 minutes, not mentioning the implementation of the EM-algorithm for estimating the variance components.

6.4.4 Reduction of the Number of Fixed-Effects Parameters

In the constrained LME model (6.51), there are $(2n + 2)$ fixed-effects parameters in $\beta = [\beta_0^T, \beta_1^T, \cdots, \beta_n^T]^T$, three free variance components parameters $\sigma^2, \sigma_\lambda^2$ and $\sigma_{\lambda_v}^2$ and a constraint $\mathbf{H}_0\beta = 0$. It is often challenging to estimate so many fixed-effects parameters in a constrained LME model. To reduce the number of fixed-effects parameters and to remove the constraint, Guo (2002a) and Ruppert et al. (2003) assumed that the second level fixed-effects parameters $\beta_1, \beta_2, \cdots, \beta_n$ are also random, following a multivariate normal distribution with mean 0 and covariance matrix $\mathbf{D}_\beta$ of size 2×2 so that the constraint $\mathbf{H}_0\beta = \sum_{i=1}^n \beta_i = 0$ is automatically satisfied at least asymptotically and hence can be removed. There are 3 free parameters in the $\mathbf{D}_\beta$. The resulting (unconstrained) LME model then has only 2 fixed-effects parameters and 6 free variance components parameters. It is much easier to fit the new (unconstrained) LME model than to fit the original constrained LME model (6.51).

6.5 MIXED-EFFECTS SMOOTHING SPLINES

In the SSCF model (6.37), the subject-specific deviation curves $v_i(t),\ i = 1, 2, \cdots, n$ were assumed to be fixed. This contradicts with the fact that these deviation curves are different from one subject to another. To overcome this difficulty, we assume these deviation curves are random. Formally, we assume $v_i(t),\ i = 1, 2, \cdots, n$ are i.i.d realizations of an underlying stochastic process. For example, we may assume

$v_i(t) \sim \text{GP}(0, \gamma)$, a Gaussian stochastic process with mean 0 and covariance function $\gamma(s,t)$. In this case, the resulting model is known as an nonparametric mixed-effects (NPME) model (5.36) as described in Chapter 5. For reference convenience, we rewrite it here:

$$\begin{array}{c} y_{ij} = \eta(t_{ij}) + v_i(t_{ij}) + \epsilon_{ij}, \\ v_i(t) \sim \text{GP}(0, \gamma),\ \epsilon_i = [\epsilon_{i1}, \cdots, \epsilon_{in_i}]^T \sim N(0, \mathbf{R}_i), \\ j = 1, 2, \cdots, n_i;\ i = 1, 2, \cdots, n, \end{array} \tag{6.54}$$

where $\eta(t)$ and $v_i(t)$ are smooth functions over $[a, b]$ for some a and b such that $-\infty < a < b < \infty$. In the framework of the NPME model, the population mean function, $\eta(t)$, is often referred to as a nonparametric fixed-effect function, and the subject-specific deviation functions, $v_i(t)$, are often referred to as nonparametric random-effect functions.

In Chapter 5, a regression spline-based technique was developed to fit the NPME model (6.54). In this section, we shall develop a smoothing spline-based technique to fit it. For simplicity, we refer to the resulting estimators as mixed-effects smoothing splines (MESS) estimators.

6.5.1 The Cubic MESS Estimators

Using the notations $\boldsymbol{\eta}, \mathbf{v}_i, \mathbf{X}_i$ and $\mathbf{y}_i$ defined in Section 6.4, we can re-write the model (6.54) as

$$\begin{array}{c} \mathbf{y}_i = \mathbf{X}_i\boldsymbol{\eta} + \mathbf{X}_i\mathbf{v}_i + \boldsymbol{\epsilon}_i, \\ \mathbf{v}_i \sim N(\mathbf{0}, \mathbf{D}),\ \ \boldsymbol{\epsilon}_i \sim N(\mathbf{0}, \mathbf{R}_i), \\ i = 1, 2, \cdots, n, \end{array} \tag{6.55}$$

where $\boldsymbol{\eta}$ and $\mathbf{v}_i$ are the values of $\eta(t)$ and $v_i(t)$ at all the distinct design time points (6.4), and the entries of $\mathbf{D}$ are

$$\mathbf{D}(r, s) = \gamma(\tau_r, \tau_s),\ r, s = 1, 2, \cdots, M. \tag{6.56}$$

Then given $\mathbf{D}$ and $\mathbf{R}_i$, $i = 1, 2, \cdots, n$, the cubic MESS estimators of $\eta(t), v_i(t)$, $i = 1, 2, \cdots, n$ are defined as the minimizers of the following penalized generalized log-likelihood (PGLL) criterion:

$$\begin{aligned} \sum_{i=1}^{n} \Big\{ & (\mathbf{y}_i - \mathbf{X}_i\boldsymbol{\eta} - \mathbf{X}_i\mathbf{v}_i)^T\mathbf{R}_i^{-1}(\mathbf{y}_i - \mathbf{X}_i\boldsymbol{\eta} - \mathbf{X}_i\mathbf{v}_i) \\ & + \log|\mathbf{D}| + \mathbf{v}_i^T\mathbf{D}^{-1}\mathbf{v}_i + \log|\mathbf{R}_i| \Big\} \\ & + \lambda_v \sum_{i=1}^{n} \Big\{ \int_a^b [v_i''(t)]^2 dt \Big\} + \lambda \int_a^b [\eta''(t)]^2 dt. \end{aligned} \tag{6.57}$$

In the above expression, the first term is the twice negative logarithm of the generalized likelihood of $\{\boldsymbol{\eta}, \mathbf{v}_i,\ i = 1, 2, \cdots, n\}$ (the joint probability density function of $\{\mathbf{y}_i, \mathbf{v}_i,\ i = 1, 2, \cdots, n\}$); the second term is the sum of the roughness of the random-effect functions $v_i(t),\ i = 1, 2, \cdots, n$ multiplied by a common smoothing parameter λ_v; and the last term is the roughness of the fixed-effect function $\eta(t)$ multiplied by another smoothing parameter λ. As in (6.39), λ is used to tradeoff between the goodness of fit and the roughness of $\eta(t)$, and λ_v is used to tradeoff between the goodness of fit and the roughness of $v_i(t),\ i = 1, 2, \cdots, n$. Notice that the main difference between (6.39) and (6.57) is that the latter takes the within-subject correlation [represented by $\gamma(s, t)$ or equivalently by $\mathbf{D}$] into account.

In (6.57), as in (6.39), the same smoothing parameter λ_v is used for all the random-effect functions $v_i(t),\ i = 1, 2, \cdots, n$. This is reasonable since the random-effect functions $v_i(t),\ i = 1, 2, \cdots, n$ are assumed to be the realizations of the same underlying zero-mean Gaussian process. Conceptually, however, different smoothing parameters may be used for different $v_i(t)$ but selecting these smoothing parameters requires a much greater computational effort.

Using the formulas in (6.40), the expression (6.57) can be re-written as

$$\begin{aligned} &\sum_{i=1}^{n}\Big\{(\mathbf{y}_i - \mathbf{X}_i\boldsymbol{\eta} - \mathbf{X}_i\mathbf{v}_i)^T\mathbf{R}_i^{-1}(\mathbf{y}_i - \mathbf{X}_i\boldsymbol{\eta} - \mathbf{X}_i\mathbf{v}_i) \\ &\qquad + \log|\mathbf{D}| + \mathbf{v}_i^T\mathbf{D}^{-1}\mathbf{v}_i + \log|\mathbf{R}_i|\Big\} \\ &\qquad + \lambda_v\sum_{i=1}^{n}\Big\{\mathbf{v}_i^T\mathbf{G}\mathbf{v}_i\Big\} + \lambda\boldsymbol{\eta}^T\mathbf{G}\boldsymbol{\eta}. \end{aligned} \tag{6.58}$$

In the above expression, it is easy to notice that

$$\mathbf{v}_i^T\mathbf{D}^{-1}\mathbf{v}_i + \lambda_v\mathbf{v}_i^T\mathbf{G}\mathbf{v}_i = \mathbf{v}_i^T\mathbf{D}_v^{-1}\mathbf{v}_i,\ i = 1, 2, \cdots, n,$$

where

$$\mathbf{D}_v = (\mathbf{D}^{-1} + \lambda_v\mathbf{G})^{-1}. \tag{6.59}$$

This motivates us to call $\mathbf{D}_v$ as a "regularized" covariance matrix of the random-effects $\mathbf{v}_i,\ i = 1, 2, \cdots, n$ since it is obtained by regularizing $\mathbf{D}$ with the roughness matrix $\mathbf{G}$ and the smoothing parameter λ_v. Consequently, we call

$$\tilde{\mathbf{V}}_i = \mathbf{X}_i\mathbf{D}_v\mathbf{X}_i^T + \mathbf{R}_i,\ i = 1, 2, \cdots, n, \tag{6.60}$$

as "regularized" covariance matrices of $\mathbf{y}_i,\ i = 1, 2, \cdots, n$. Therefore, the regularized variations of $\mathbf{v}_i$ and $\mathbf{y}_i$ come from two sources: their original variations and the roughness of $v_i(t),\ i = 1, 2, \cdots, n$. These two components join together in a natural manner.

Theorem 6.1 *Assume the variance components* $\mathbf{D},\ \mathbf{R}_i,\ i = 1, 2, \cdots, n$, *and the smoothing parameters* λ *and* λ_v *are known. Then the minimizers of the PGLL criterion (6.58) are*

$$\begin{aligned} \hat{\boldsymbol{\eta}} &= \left(\textstyle\sum_{i=1}^{n}\mathbf{X}_i^T\tilde{\mathbf{V}}_i^{-1}\mathbf{X}_i + \lambda\mathbf{G}\right)^{-1}\left(\textstyle\sum_{i=1}^{n}\mathbf{X}_i^T\tilde{\mathbf{V}}_i^{-1}\mathbf{y}_i\right), \\ \hat{\mathbf{v}}_i &= \mathbf{D}_v\mathbf{X}_i^T\tilde{\mathbf{V}}_i^{-1}(\mathbf{y}_i - \mathbf{X}_i\hat{\boldsymbol{\eta}}),\ i = 1, 2, \cdots, n. \end{aligned} \tag{6.61}$$

Proof: see Appendix.

In a matrix form, the minimizers (6.61) in the above theorem can be written as

$$\begin{aligned} \hat{\eta} &= \left(\mathbf{X}^T\tilde{\mathbf{V}}^{-1}\mathbf{X} + \lambda\mathbf{G}\right)^{-1}\mathbf{X}^T\tilde{\mathbf{V}}^{-1}\mathbf{y}, \\ \hat{\mathbf{v}} &= \tilde{\mathbf{D}}_v\mathbf{Z}^T\tilde{\mathbf{V}}^{-1}(\mathbf{y} - \mathbf{X}\hat{\eta}), \end{aligned} \tag{6.62}$$

where

$$\begin{aligned} \mathbf{y} &= [\mathbf{y}_1^T, \cdots, \mathbf{y}_n^T]^T, & \mathbf{X} &= [\mathbf{X}_1^T, \cdots, \mathbf{X}_n^T]^T, \\ \mathbf{Z} &= \mathrm{diag}(\mathbf{X}_1, \cdots, \mathbf{X}_n), & \tilde{\mathbf{V}} &= \mathrm{diag}(\tilde{\mathbf{V}}_1, \tilde{\mathbf{V}}_2, \cdots, \tilde{\mathbf{V}}_n), \\ \tilde{\mathbf{D}}_v &= \mathrm{diag}(\mathbf{D}_v, \mathbf{D}_v, \cdots, \mathbf{D}_v). \end{aligned} \tag{6.63}$$

6.5.2 Bayesian Interpretation

The cubic MESS estimators (6.62) can be interpreted as the posterior expectations of the parameters of a Bayesian problem. This Bayesian interpretation is closely related to the EM-algorithm that we shall develop later.

Using the notations in (6.63), we can define the Bayesian problem as follows:

$$\mathbf{y}|\eta, \mathbf{v} \sim N(\mathbf{X}\eta + \mathbf{Z}\mathbf{v}, \mathbf{R}), \tag{6.64}$$

where the parameters η and $\mathbf{v}$ are independent, and they have the following prior distributions:

$$\eta \sim N(0, \lambda^{-1}\mathbf{G}^-), \quad \mathbf{v} \sim N(0, \tilde{\mathbf{D}}_v). \tag{6.65}$$

Since the roughness matrix $\mathbf{G}$ is not of full rank, in the above expression, we use $\mathbf{G}^-$ to denote one of its generalized inverses. Therefore, the prior distribution for η is improper. Nevertheless, the prior distribution for $\mathbf{v}$ is proper.

Notice that we can change the prior of η slightly so that the prior is proper. The new prior of η is $\eta \sim N\left[0, (\lambda_0\mathbf{I}_M + \lambda\mathbf{G})^{-1}\right]$. Using limit theory arguments, it is easy to show that the limit posterior distributions (as $\lambda_0 \to 0$) are the same as the posterior distributions under the Bayesian framework (6.64) and (6.65).

We have the following theorem.

Theorem 6.2 *The MESS estimators (6.62) that minimize the PGLL criterion (6.58) are the same as the posterior expectations of the parameters of the Bayesian problem defined in (6.64) and (6.65). That is,*

$$\hat{\eta} = E(\eta|\mathbf{y}), \quad \hat{\mathbf{v}} = E(\mathbf{v}|\mathbf{y}). \tag{6.66}$$

Moreover, the posterior distributions of $\eta, \mathbf{v}$ and $\epsilon = \mathbf{y} - \mathbf{X}\eta - \mathbf{Z}\mathbf{v}$ are, respectively, given by

$$\eta|\mathbf{y} \sim N\left[\hat{\eta}, (\lambda\mathbf{G} + \mathbf{X}^T\tilde{\mathbf{V}}^{-1}\mathbf{X})^{-1}\right], \tag{6.67}$$

$$\mathbf{v}|\mathbf{y} \sim N\left[\hat{\mathbf{v}}, \tilde{\mathbf{D}}_v - \tilde{\mathbf{D}}_v\mathbf{Z}^T\mathbf{W}\mathbf{Z}\tilde{\mathbf{D}}_v\right], \tag{6.68}$$

$$\epsilon|\mathbf{y} \sim N\left[\hat{\epsilon}, \mathbf{R} - \mathbf{R}\mathbf{W}\mathbf{R}\right], \tag{6.69}$$

where $\hat{\epsilon} = \mathbf{y} - \mathbf{X}\hat{\eta} - \mathbf{Z}\hat{\mathbf{v}}$ *and* $\mathbf{W} = \tilde{\mathbf{V}}^{-1} - \tilde{\mathbf{V}}^{-1}\mathbf{X}(\lambda\mathbf{G} + \mathbf{X}^T\tilde{\mathbf{V}}^{-1}\mathbf{X})^{-1}\mathbf{X}^T\tilde{\mathbf{V}}^{-1}$.

Proof: see Appendix.

Let $\mathbf{W}_i = \tilde{\mathbf{V}}_i^{-1} - \tilde{\mathbf{V}}_i^{-1}\mathbf{X}_i(\lambda\mathbf{G} + \mathbf{X}^T\tilde{\mathbf{V}}^{-1}\mathbf{X})^{-1}\mathbf{X}_i^T\tilde{\mathbf{V}}_i^{-1}$, $i = 1, 2, \cdots, n$. Then it is easy to show that

$$\begin{array}{ll} \mathrm{E}(\mathbf{v}_i|\mathbf{y}) = \hat{\mathbf{v}}_i, & \mathrm{Cov}(\mathbf{v}_i|\mathbf{y}) = \mathbf{D}_v - \mathbf{D}_v\mathbf{X}_i^T\mathbf{W}_i\mathbf{X}_i\mathbf{D}_v, \\ \mathrm{E}(\epsilon_i|\mathbf{y}) = \hat{\epsilon}_i, & \mathrm{Cov}(\epsilon_i|\mathbf{y}) = \mathbf{R}_i - \mathbf{R}_i\mathbf{W}_i\mathbf{R}_i. \end{array} \tag{6.70}$$

In next section, we shall discuss REML-based and ML-based EM-algorithms for estimating the variance components $\mathbf{D}$ and $\mathbf{R}_i$, $i = 1, 2, \cdots, n$. The posterior means and covariances of $\mathbf{v}_i$ and ϵ_i given in (6.70) are closely related to the REML-based EM-algorithm for the variance component estimation. For the ML-based EM-algorithm, the posterior means and covariances of $\mathbf{v}_i$ and ϵ_i, given $\eta = \hat{\eta}$ in (6.72), are more related.

Theorem 6.3 *Under the Bayesian framework defined in (6.64) and (6.65), we have*

$$\begin{array}{rcl} \mathbf{v}|\mathbf{y}, \eta & \sim & N\left[\tilde{\mathbf{D}}_v\mathbf{Z}^T\tilde{\mathbf{V}}^{-1}(\mathbf{y} - \mathbf{X}\eta), \tilde{\mathbf{D}}_v - \tilde{\mathbf{D}}_v\mathbf{Z}^T\tilde{\mathbf{V}}^{-1}\mathbf{Z}\tilde{\mathbf{D}}_v\right], \\ \epsilon|\mathbf{y}, \eta & \sim & N\left[\mathbf{R}\tilde{\mathbf{V}}^{-1}(\mathbf{y} - \mathbf{X}\eta), \mathbf{R} - \mathbf{R}\tilde{\mathbf{V}}^{-1}\mathbf{R}\right]. \end{array} \tag{6.71}$$

Proof: see Appendix.

From the above theorem, we actually have

$$\begin{array}{rcl} \mathrm{E}(\mathbf{v}|\mathbf{y}, \eta = \hat{\eta}) & = & \tilde{\mathbf{D}}_v\mathbf{Z}^T\tilde{\mathbf{V}}^{-1}(\mathbf{y} - \mathbf{X}\hat{\eta}) = \hat{\mathbf{v}} = \mathrm{E}(\mathbf{v}|\mathbf{y}), \\ \mathrm{E}(\epsilon|\mathbf{y}, \eta = \hat{\eta}) & = & \mathbf{R}\tilde{\mathbf{V}}^{-1}(\mathbf{y} - \mathbf{X}\hat{\eta}) = \hat{\epsilon} = \mathrm{E}(\epsilon|\mathbf{y}). \end{array}$$

Based on this and by Theorem 6.3, we have

$$\begin{array}{rclrcl} \mathrm{E}(\mathbf{v}_i|\mathbf{y}, \eta = \hat{\eta}) & = & \hat{\mathbf{v}}_i, & \mathrm{Cov}(\mathbf{v}_i|\mathbf{y}, \eta = \hat{\eta}) & = & \mathbf{D}_v - \mathbf{D}_v\mathbf{X}_i^T\tilde{\mathbf{V}}_i^{-1}\mathbf{X}_i\mathbf{D}_v, \\ \mathrm{E}(\epsilon_i|\mathbf{y}, \eta = \hat{\eta}) & = & \hat{\epsilon}_i, & \mathrm{Cov}(\epsilon_i|\mathbf{y}, \eta = \hat{\eta}) & = & \tilde{\mathbf{R}}_i - \mathbf{R}_i\tilde{\mathbf{V}}_i^{-1}\mathbf{R}_i. \end{array} \tag{6.72}$$

Comparing the above expressions to those in (6.70), we see that $\mathrm{Cov}(\mathbf{v}_i|\mathbf{y}, \eta = \hat{\eta})$ and $\mathrm{Cov}(\epsilon_i|\mathbf{y}, \eta = \hat{\eta})$ can be obtained, respectively, from $\mathrm{Cov}(\mathbf{v}_i|\mathbf{y})$ and $\mathrm{Cov}(\epsilon_i|\mathbf{y})$ by replacing $\mathbf{W}_i$ with $\tilde{\mathbf{V}}_i^{-1}$.

6.5.3 Variance Components Estimation

In the expression (6.61), we assumed that the variance components $\mathbf{D}$ and

$$\mathbf{R}_i = \sigma^2\mathbf{I}_{n_i}, \quad i = 1, 2, \cdots, n, \tag{6.73}$$

are known. In practice, both $\mathbf{D}$ and σ^2 need to be estimated. In the general LME model, the variance components $\mathbf{D}$ and σ^2 are estimated by a ML-based or REML-based EM-algorithm (Laird, Lange and Stram 1987). The EM-algorithm can not be directly applied to the variance component estimation of the MESS model. In this section, we develop a new EM-algorithm for the cubic MESS model.

Notice that under the model (6.54), both $\mathbf{v}_i$ and $\boldsymbol{\epsilon}_i$ are normal. Therefore when $\mathbf{v}_i$ and $\boldsymbol{\epsilon}_i$ were known, the maximum likelihood estimates (MLEs) of σ^2 and $\mathbf{D}$ would be

$$\hat{\sigma}^2 = N^{-1}\sum_{i=1}^{n} \boldsymbol{\epsilon}_i^T \boldsymbol{\epsilon}_i, \quad \hat{\mathbf{D}} = n^{-1}\sum_{i=1}^{n} \mathbf{v}_i \mathbf{v}_i^T. \tag{6.74}$$

Because $\boldsymbol{\epsilon}_i$ and $\mathbf{v}_i$ are actually unknown, the above MLEs are not computable. There are two methods that may be used to overcome this difficulty:

1. Replace $\hat{\sigma}^2$ and $\hat{\mathbf{D}}$ by their posterior expectations $\mathrm{E}(\hat{\sigma}^2|\mathbf{y})$ and $\mathrm{E}(\hat{\mathbf{D}}|\mathbf{y})$, respectively.

2. Replace $\hat{\sigma}^2$ and $\hat{\mathbf{D}}$ by their posterior expectations $\mathrm{E}(\hat{\sigma}^2|\mathbf{y}, \boldsymbol{\eta} = \hat{\boldsymbol{\eta}})$ and $\mathrm{E}(\hat{\mathbf{D}}|\mathbf{y}, \boldsymbol{\eta} = \hat{\boldsymbol{\eta}})$, respectively.

Based on either method, an EM-algorithm can be used to estimate σ^2 and $\mathbf{D}$. The maximum likelihood step (M-step) is based on (6.74) while the expectation step (E-step) is based on either Method 1 or Method 2 above. If Method 1 is used, the EM-algorithm is the REML-based EM-algorithm. If Method 2 is used, the EM-algorithm is the ML-based EM-algorithm. Using Theorems 6.2 and 6.3, we can show the following theorem.

Theorem 6.4 *Assume that $\mathbf{R}_i$, $i = 1, 2, \cdots, n$ satisfy (6.73). Then under the Bayesian framework defined in (6.64) and (6.65), the posterior expectations of $\hat{\sigma}^2$ and $\hat{\mathbf{D}}$ given $\mathbf{y}$ are*

$$E(\hat{\sigma}^2|\mathbf{y}) = N^{-1}\sum_{i=1}^{n}\left\{\hat{\boldsymbol{\epsilon}}_i^T\hat{\boldsymbol{\epsilon}}_i + \sigma^2\left[n_i - \sigma^2 tr(\mathbf{W}_i)\right]\right\}, \tag{6.75}$$

$$E(\hat{\mathbf{D}}|\mathbf{y}) = n^{-1}\sum_{i=1}^{n}\left\{\hat{\mathbf{v}}_i\hat{\mathbf{v}}_i^T + [\mathbf{D}_v - \mathbf{D}_v\mathbf{X}_i^T\mathbf{W}_i\mathbf{X}_i\mathbf{D}_v]\right\}, \tag{6.76}$$

and the posterior expectations $E[\hat{\sigma}^2|\mathbf{y}, \boldsymbol{\eta} = \hat{\boldsymbol{\eta}}]$ and $E[\hat{\mathbf{D}}|\mathbf{y}, \boldsymbol{\eta} = \hat{\boldsymbol{\eta}}]$ are the same as (6.75) and (6.76), respectively, after the $\mathbf{W}_i$ are replaced by the $\tilde{\mathbf{V}}_i^{-1}$.

Proof: see Appendix.

The above ideas for the EM-algorithms can be iteratively operated for some given initial values. For simplicity, the initial values can be taken as $\hat{\mathbf{D}} = \hat{\sigma}^2\mathbf{I}_M$ and $\hat{\sigma}^2 = 1$ where M is the number of distinct design time points. As an illustration, we state the REML-based EM-algorithm in details below. Notice that the smoothing parameters λ, λ_v and the roughness matrix $\mathbf{G}$ do not change during the whole iteration process. Moreover, the ML-based EM-algorithm can be obtained from the REML-based EM-algorithm by replacing the $\mathbf{W}_{i(\cdot)}$ with the $\tilde{\mathbf{V}}_{i(\cdot)}^{-1}$.

REML-Based EM-Algorithm

Step 0. Set $r = 0$; Initializing $\hat{\sigma}^2_{(r)} = 1$, and $\hat{\mathbf{D}}_{(r)} = \hat{\sigma}^2_{(r)}\mathbf{I}_M$.

Step 1. Set $r = r + 1$. Compute

$$\begin{aligned}\hat{\mathbf{D}}_{v(r-1)} &= \left(\hat{\mathbf{D}}_{(r-1)}^{-1} + \lambda_v \mathbf{G}\right)^{-1}, \\ \tilde{\mathbf{V}}_{i(r-1)} &= \mathbf{X}_i \hat{\mathbf{D}}_{v(r-1)} \mathbf{X}_i^T + \hat{\sigma}^2_{(r-1)} \mathbf{I}_{n_i}, \; i = 1, 2, \cdots, n.\end{aligned}$$

Then update $\hat{\boldsymbol{\eta}}_{(r)}$ and $\hat{\mathbf{v}}_{i(r)}$ using

$$\begin{aligned}\hat{\boldsymbol{\eta}}_{(r)} &= \left(\sum_{i=1}^n \mathbf{X}_i^T \tilde{\mathbf{V}}_{i(r-1)}^{-1} \mathbf{X}_i + \lambda \mathbf{G}\right)^{-1} \left(\sum_{i=1}^n \mathbf{X}_i^T \tilde{\mathbf{V}}_{i(r-1)}^{-1} \mathbf{y}_i\right), \\ \hat{\mathbf{v}}_{i(r)} &= \hat{\mathbf{D}}_{v(r-1)} \mathbf{X}_i^T \tilde{\mathbf{V}}_{i(r-1)}^{-1} \left(\mathbf{y}_i - \mathbf{X}_i \hat{\boldsymbol{\eta}}_{(r)}\right), \; i = 1, 2, \cdots, n.\end{aligned}$$

Step 2. Compute the residuals $\hat{\boldsymbol{\epsilon}}_{i(r)} = \mathbf{y}_i - \mathbf{X}_i \hat{\boldsymbol{\eta}}_{(r)} - \mathbf{X}_i \hat{\mathbf{v}}_{i(r)}$ and the matrices $\mathbf{W}_{i(r-1)} = \tilde{\mathbf{V}}_{i(r-1)}^{-1} - \tilde{\mathbf{V}}_{i(r-1)}^{-1} \mathbf{X}_i (\lambda \mathbf{G} + \sum_{k=1}^n \mathbf{X}_k^T \tilde{\mathbf{V}}_{k(r-1)}^{-1} \mathbf{X}_k)^{-1} \mathbf{X}_i^T \tilde{\mathbf{V}}_{i(r-1)}^{-1}$. Then update $\hat{\sigma}^2_{(r)}$ and $\hat{\mathbf{D}}_{(r)}$ using

$$\begin{aligned}\hat{\sigma}^2_{(r)} &= N^{-1} \sum_{i=1}^n \left\{\hat{\boldsymbol{\epsilon}}_{i(r)}^T \hat{\boldsymbol{\epsilon}}_{i(r)} + \hat{\sigma}^2_{(r-1)} [n_i - \hat{\sigma}^2_{(r-1)} \text{tr}(\mathbf{W}_{i(r-1)})]\right\}, \\ \hat{\mathbf{D}}_{(r)} &= n^{-1} \sum_{i=1}^n \Big\{\hat{\mathbf{v}}_{i(r)} \hat{\mathbf{v}}_{i(r)}^T \\ &\quad + [\hat{\mathbf{D}}_{v(r-1)} - \hat{\mathbf{D}}_{v(r-1)} \mathbf{X}_i^T \mathbf{W}_{i(r-1)} \mathbf{X}_i \hat{\mathbf{D}}_{v(r-1)}]\Big\}.\end{aligned}$$

Step 3. When some convergence condition is satisfied, stop; otherwise, go back to Step 1.

6.5.4 Fits and Smoother Matrices

Estimates of $\eta(t), v_i(t)$ and $\gamma(s,t)$, among others, of the NPME model (6.54) can be obtained via the cubic MESS estimators (6.61). First of all, let us focus on the estimation of these quantities at the M distinct design time points (6.4). For other time points within the range of interest, estimates can be obtained via some simple interpolation.

Let $\mathbf{e}_r$ denote a M-dimensional vector with the r-th entry being 1 and otherwise 0. Then the estimates of $\eta(t)$ and $v_i(t)$ at τ_r are

$$\hat{\eta}(\tau_r) = \mathbf{e}_r^T \hat{\boldsymbol{\eta}}, \quad \hat{v}_i(\tau_r) = \mathbf{e}_r^T \hat{\mathbf{v}}_i, \; i = 1, 2, \cdots, n, \tag{6.77}$$

and the estimate of $\gamma(\tau_r, \tau_s)$ is

$$\hat{\gamma}(\tau_r, \tau_s) = \hat{\mathbf{D}}(r, s), \; r, \; s = 1, 2, \cdots, M. \tag{6.78}$$

The natural estimates of the individual functions $s_i(t),\ i = 1, 2, \cdots, n$ at the M distinct design time points (6.4) are then

$$\hat{s}_i(\tau_r) = \hat{\eta}(\tau_r) + \hat{v}_i(\tau_r),\ r = 1, 2, \cdots, M;\ \ i = 1, 2, \cdots, n.$$

We now consider fits of $\eta(t)$, and $v_i(t),\ i = 1, 2, \cdots, n$ at all the design time points (6.2). It can be shown that the fits of $\eta(t)$ and $v_i(t),\ i = 1, 2, \cdots, n$ at all the design time points can be expressed as the product of a smoother matrix and the response vector $\mathbf{y}$.

Recall that $\hat{\eta}$ contains the fits of $\eta(t)$ at all the distinct design time points (6.4). Using the formulas in (6.62), the fits of $\eta(t)$ at all the design time points (6.2) can be expressed as

$$\hat{\eta}^* = \mathbf{X}\hat{\eta} = \mathbf{X}(\mathbf{X}^T\tilde{\mathbf{V}}^{-1}\mathbf{X} + \lambda\mathbf{G})^{-1}\mathbf{X}^T\tilde{\mathbf{V}}^{-1}\mathbf{y} \equiv \mathbf{A}\mathbf{y}, \tag{6.79}$$

where $\mathbf{A}$ is the associated MESS smoother matrix for estimating $\eta(t)$ at the design time points. When $\tilde{\mathbf{V}}$ is known, the matrix $\mathbf{A}$ does not depend on $\mathbf{y}$. In this case, $\hat{\eta}^*$ is a linear smoother of $\eta(t)$. Following Buja et al. (1989), we use the trace of $\mathbf{A}$ to quantify how complicated the MESS model is.

Also notice that $\hat{\mathbf{v}}$ contains the fits of $v_i(t),\ i = 1, 2, \cdots, n$ at all the distinct design time points (6.4). Then the fits of $v_i(t),\ i = 1, 2, \cdots, n$ at all the design time points can be expressed as

$$\hat{\mathbf{v}}^* = \mathbf{Z}\hat{\mathbf{v}} = \mathbf{Z}\tilde{\mathbf{D}}_v\mathbf{Z}^T\tilde{\mathbf{V}}^{-1}(\mathbf{I}_N - \mathbf{A})\mathbf{y} \equiv \mathbf{A}_v\mathbf{y}, \tag{6.80}$$

where $\mathbf{A}_v$ is the associated MESS smoother matrix for estimating $v_i(t),\ i = 1, 2, \cdots, n$ at all the design time points. It may be worthwhile to emphasize again that $\hat{\eta}^*$ and $\hat{\mathbf{v}}^*$ collect the fits of $\eta(t)$ and $v_i(t),\ i = 1, 2, \cdots, n$ at all the design time points (6.2) for all the subjects, while $\hat{\eta}$ and $\hat{\mathbf{v}}$ collect the fits only at the M distinct design time points (6.4).

6.5.5 Variability Band Construction

The standard deviation (SD) band of $\eta(t)$ can be constructed based on the covariance of $\hat{\eta}$. Based on (6.62), it is easy to see that

$$\text{Cov}(\hat{\eta}) = (\mathbf{X}^T\tilde{\mathbf{V}}^{-1}\mathbf{X} + \lambda\mathbf{G})^{-1}\mathbf{X}^T\tilde{\mathbf{V}}^{-1}\mathbf{V}\tilde{\mathbf{V}}^{-1}\mathbf{X}(\mathbf{X}^T\tilde{\mathbf{V}}^{-1}\mathbf{X} + \lambda\mathbf{G})^{-1}, \tag{6.81}$$

where $\mathbf{V} = \text{Cov}(\mathbf{y}) = \mathbf{Z}\tilde{\mathbf{D}}\mathbf{Z}^T + \mathbf{R}$ with $\mathbf{Z}$ and $\mathbf{R}$ defined in (6.63) and $\tilde{\mathbf{D}} = \text{diag}(\mathbf{D}, \cdots, \mathbf{D})$. In the above expression, we can replace $\mathbf{V}$ with $\tilde{\mathbf{V}}$ to account for the roughness of the random-effect functions.

Using (6.81), it is easy to construct the pointwise SD band for $\eta(t)$ at all the distinct design time points (6.4) provided we follow the procedure for constructing the pointwise SD band for the NSS smoother as described in Section 6.2.

Wahba (1983) suggested to construct Bayesian confidence intervals for a smoothing spline using its Bayesian interpretation. Similarly, we can construct a pointwise

Bayesian confidence band for $\eta(t)$ based on the Bayesian interpretation of $\hat{\eta}$ given in Theorem 6.2. That is, we simply repeat the above process that constructs the pointwise SD band for $\eta(t)$ via replacing $\text{Cov}(\hat{\eta})$ by

$$\text{Cov}(\hat{\eta}|\mathbf{y}) = (\mathbf{X}^T\tilde{\mathbf{V}}^{-1}\mathbf{X} + \lambda\mathbf{G})^{-1},$$

which was given in Theorem 6.2.

6.5.6 Choice of the Smoothing Parameters

In Section 5.2.5 of Chapter 5, we extended the classic AIC (Akaike 1973) and BIC (Schwarz 1978) rules for the MERS model. Here we can extend these rules for the MESS model. The resulting AIC and BIC rules should be able to tradeoff between the "goodness of fit" and "model complexity" of the MESS model. The former can be quantified by the log-likelihood of the MESS model while the latter can be measured by the traces of the associated smoother matrices.

As before, let $\boldsymbol{\eta}_i = [\eta(t_{i1}), \cdots, \eta(t_{in_i})]^T$ denote the fixed-effects vector at the design time points for the i-th subject, and let $\mathbf{V}_i = \text{Cov}(\mathbf{y}_i)$. Then we have $\mathbf{y}_i \sim N(\boldsymbol{\eta}_i, \mathbf{V}_i),\ i = 1, 2, \cdots, n$. It follows that the log-likelihood of the MESS model can be expressed as follows:

$$\text{Loglik} = -\frac{1}{2}\sum_{i=1}^{n}\{(\mathbf{y}_i - \boldsymbol{\eta}_i)^T\mathbf{V}_i^{-1}(\mathbf{y}_i - \boldsymbol{\eta}_i) + \log(|\mathbf{V}_i|)\} - \frac{N}{2}\log(2\pi), \quad (6.82)$$

via replacing the unknown $\boldsymbol{\eta}_i$ and $\mathbf{V}_i$ with their estimates under the MESS model:

$$\hat{\boldsymbol{\eta}}_i = \mathbf{X}_i\hat{\boldsymbol{\eta}}, \quad \hat{\hat{\mathbf{V}}}_i = \mathbf{X}_i\hat{\mathbf{D}}_v\mathbf{X}_i^T + \hat{\mathbf{R}}_i,\ i = 1, 2, \cdots, n, \quad (6.83)$$

for the given smoothing parameters λ and λ_v. When the smoothing parameter λ (or λ_v) increases, more penalty is imposed for the roughness of $\eta(t)$ (or $v_i(t),\ i = 1, 2, \cdots, n$), implying that the model's goodness of fit is more poor. Therefore, it is expected that the Loglik decreases as λ (or λ_v) increases. This can be seen from Figures 6.3 (a) and (c). In Figure 6.3 (a) five profile curves of Loglik against $\log_{10}(\lambda)$ for five selected values of λ_v are presented, while in Figure 6.3 (c) five profile curves of Loglik against $\log_{10}(\lambda_v)$ for five selected values of λ are displayed. It is seen that for each fixed λ_v, the Loglik decreases as λ increases; for each fixed λ, the Loglik decreases as λ_v increases.

It is natural to define the model complexity for estimating $\eta(t)$ and $v_i(t),\ i = 1, 2, \cdots, n$, separately. The model complexity or the number of degrees of freedom (df) for estimating $\eta(t)$ using the cubic MESS model can be defined as the trace of the smoother matrix $\mathbf{A}$ (6.79):

$$\text{df} = \text{tr}(\mathbf{A}) = \text{tr}\left[(\mathbf{X}^T\tilde{\mathbf{V}}^{-1}\mathbf{X} + \lambda\mathbf{G})^{-1}\mathbf{X}^T\tilde{\mathbf{V}}^{-1}\mathbf{X}\right]. \quad (6.84)$$

We have the following lemma.

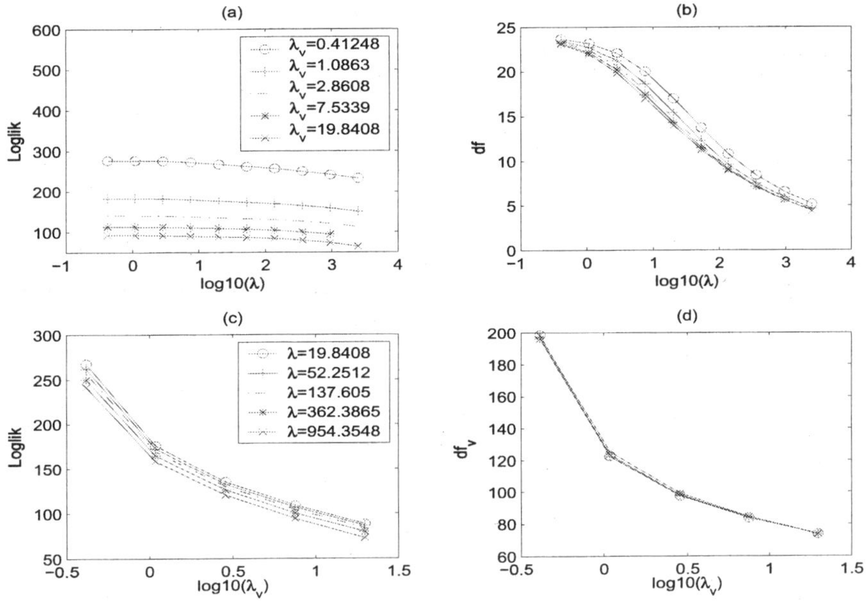

Fig. 6.3 Examples showing profile curves of Loglik, df and df_v against λ and λ_v in $\log_{10}$-scale.

Lemma 6.1 *For a fixed λ_v, df is a monotonically decreasing function of λ.*

Proof: see Appendix.

The above lemma states that when λ increases, i.e., when more smoothing is imposed for estimating $\eta(t)$, the cubic MESS model for estimating $\eta(t)$ becomes simpler regardless of the choice of λ_v. From the proof of Lemma 6.1 given in the Appendix of this chapter, it is known that for the cubic MESS model, $\mathbf{X}$ is of full rank, and

$$\text{df} = 2 + \sum_{l=3}^{M} (1 + \lambda w_l)^{-1},$$

where $w_l, l = 3, 4, \cdots, M$ are some positive numbers. From the above expression, it is seen that the model complexity, df, is mainly dominated by λ. This can also be seen from Figure 6.3 (b) in which five profile curves of df against $\log_{10}(\lambda)$ for five selected values of λ_v are presented. It is seen that the df indeed decreases as λ increases but it changes little as λ_v increases.

The model complexity for estimating $v_i(t),\ i = 1, 2, \cdots, n$ using the cubic MESS model can be defined as the trace of the smoother matrix $\mathbf{A}_v$ (6.80):

$$\text{df}_v = \text{tr}(\mathbf{A}_v) = \text{tr}\left[\mathbf{Z}\tilde{\mathbf{D}}_v\mathbf{Z}^T\tilde{\mathbf{V}}^{-1}(\mathbf{I}_N - \mathbf{A})\right]. \tag{6.85}$$

As expected, while the df decreases with λ and is mainly dominated by λ, the df_v decreases with λ_v and is mainly dominated by λ_v. This can be seen from the proof of Theorem 6.5 given in the Appendix of this chapter. It can also be observed from Figure 6.3 (d) in which five profile curves of df_v against $\log_{10}(\lambda_v)$ for five selected values of λ are displayed. It is seen that the df_v decreases as λ_v increases but it changes little as λ increases.

We state the ranges for df and df_v in the following theorem.

Theorem 6.5 *For the cubic MESS model,*

$$2 < df \leq M, \quad 0 < df_v < n\min(M, N/n), \tag{6.86}$$

where M is the number of all the distinct design time points and N is the number of all the design time points.

Proof: see Appendix.

We are now ready to define the AIC and BIC rules. Since in the cubic MESS model, both $\eta(t)$ and $v_i(t),\ i = 1, 2, \cdots, n$ are fitted, the cubic MESS model complexity is then defined as the sum of df and df_v. Then following the classical AIC rule (Akaike 1973), we can define the AIC rule as follows:

$$\mathrm{AIC}(\lambda, \lambda_v) = -2\mathrm{Loglik} + 2(\mathrm{df} + \mathrm{df}_v). \tag{6.87}$$

The minimizers of the above AIC rule may be referred to as the AIC smoothing parameters.

Similarly, following the classical BIC rule (Schwarz 1978), we can define the BIC rule for the cubic MESS model as

$$\mathrm{BIC}(\lambda, \lambda_v) = -2\mathrm{Loglik} + \log(n)(\mathrm{df} + \mathrm{df}_v), \tag{6.88}$$

where n is the number of subjects of the longitudinal data set. The minimizers of the above BIC rule may be referred to as the BIC smoothing parameters. When $n \geq 8$, we have $\log(n) > 2$. It follows that the BIC rule penalizes more the model complexity. Therefore, the BIC rule is more conservative than the AIC rule.

In practice, to minimize the AIC (6.87), we can minimize the AIC with respect to λ with a fixed λ_v, then minimize the AIC with respect to λ_v with λ fixed at the previous estimate. This process can be repeated until satisfactory results are obtained. We can use the BIC rule similarly.

Figures 6.4 (a) and (b) show the profile curves (associated with λ_v) of AIC and BIC against $\log_{10}(\lambda)$, and Figures 6.4 (c) and (d) show the profile curves (associated with λ) of AIC and BIC against $\log_{10}(\lambda_v)$. It seems that by the AIC rule, the best pair of (λ, λ_v) is $(2.8608, .4125)$, while by the BIC rule, the best pair of (λ, λ_v) is $(137.605, 1.0863)$.

6.5.7 Application to the Conceptive Progesterone Data

As an illustration, we applied the cubic MESS model to the conceptive progesterone data. The smoothing parameters $(\lambda, \lambda_v) = (137.605, 1.0863)$ were selected by the BIC rule (6.88). Figure 6.5 (a) shows the individual fits. Figure 6.5 (b) displays the estimated fixed-effect function (solid curve), together with the 95% standard deviation

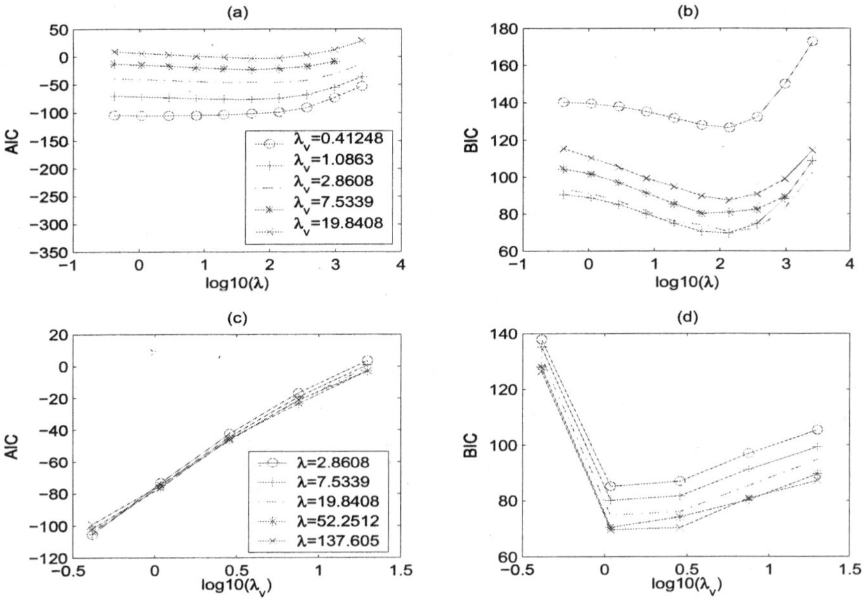

Fig. 6.4 Examples showing profile curves of AIC and BIC against λ or λ_v.

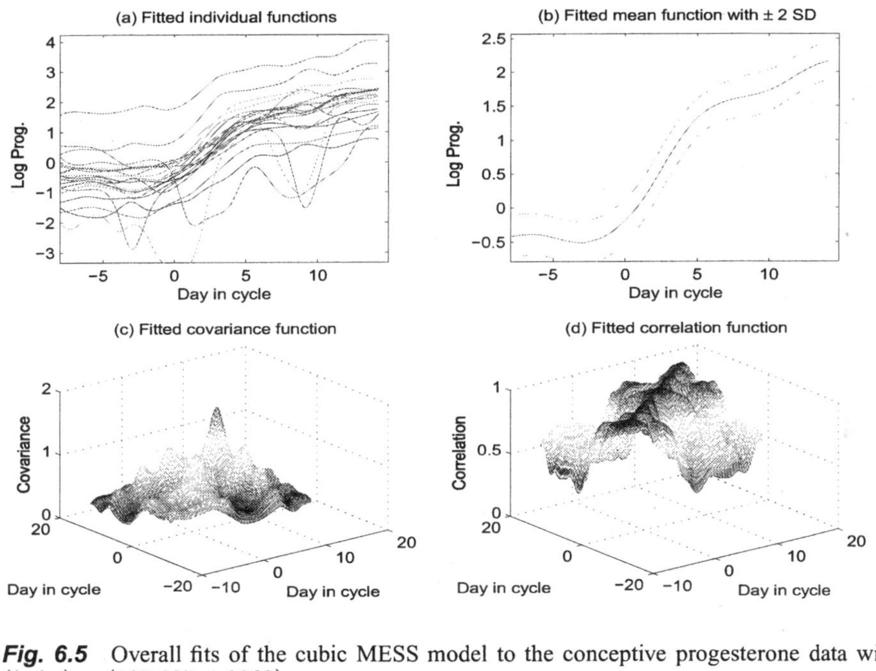

Fig. 6.5 Overall fits of the cubic MESS model to the conceptive progesterone data with $(\lambda, \lambda_v) = (137.605, 1.0863)$.

band (dot-dashed curves). Figures 6.5 (c) and (d) display the estimated covariance and correlation functions, respectively. It seems that the BIC rule chose proper amount of smoothing for both the fixed-effect function and the random-effect functions.

Figure 6.6 displays six randomly selected individual fits of the cubic MESS model to the conceptive progesterone data. In each panel, we superimposed the raw data (circles), the individual fit (solid curve) and the fixed-effect function (population) fit (dashed curve). It is seen that each of the six estimated individual curves fits the individual data quite well.

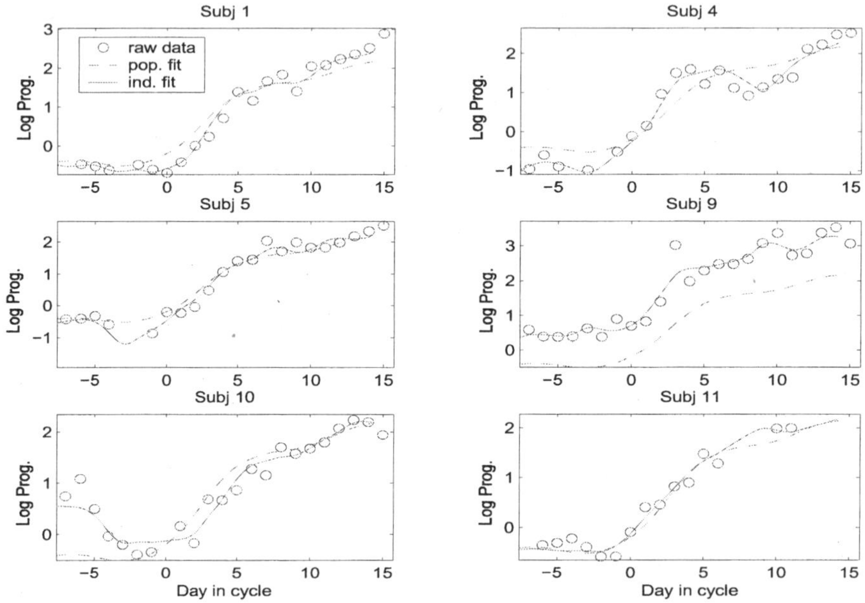

Fig. 6.6 Six randomly selected individual fits (solid curve) of the cubic MESS model to the conceptive progesterone data (circles) with $(\lambda, \lambda_v) = (137.605, 1.0863)$, superimposed with the fixed-effect function (dashed curve).

In Section 6.2, we used a residual analysis of the NSS fit to the conceptive progesterone data to conclude that the NPM model (6.1) does not adequately fit the data. We now perform a similar residual analysis to see if the cubic MESS fit to the conceptive progesterone data is adequate. Figures 6.7 (a), (b), and (c) display the plots of the standardized residuals of the cubic MESS fit against the individual fits $\hat{s}(t_{ij}) = \hat{\eta}(t_{ij}) + \hat{v}_i(t_{ij})$, time t_{ij}, and response y_{ij}, respectively. It is seen that almost all the residuals are within the 95% confidence interval $[-2, 2]$ and the linear trend appearing in Figure 6.2 (c) no longer exists in Figure 6.7 (c). This means that the NPME model (6.54) adequately fits the conceptive progesterone data. Figure 6.7 (d) shows that the standardized residuals are quite normal. In summary, the cubic MESS fit does improve the cubic NSS fit substantially.

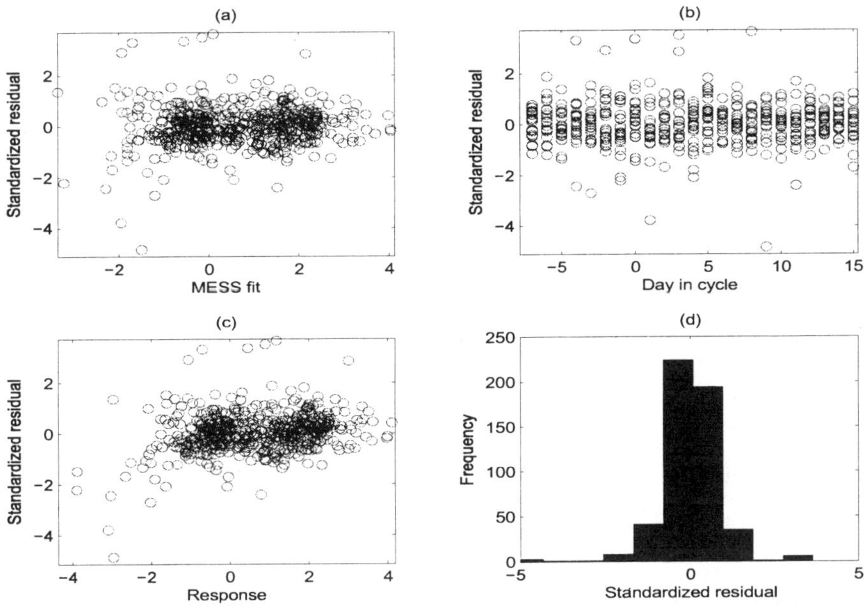

Fig. 6.7 Residual analysis of the cubic MESS model to the conceptive progesterone data.

6.6 GENERAL DEGREE SMOOTHING SPLINES

So far, we have discussed the cubic NSS, GSS, ESS, and MESS models for longitudinal data. A drawback of these models is that it is expensive to compute when M, the number of distinct design time points, is too large. Moreover, it is not easy to extend these models to higher degree MESS models if desired. This is because higher degree smoothing splines require more computational effort to handle the roughness which requires integration of higher order derivatives; see (6.3) for example. In this section, we show how to deal with general degree smoothing splines for the NSS, GSS, ESS and MESS models, respectively.

6.6.1 General Degree NSS

The key task for the general degree NSS model is how to compute the roughness term in the expression (6.3) for a general positive integer k so that it becomes a form like (6.7) using a given regression spline basis. The idea is along the same lines as that for computing the roughness matrix of a smoothing spline described in Section 3.4 of Chapter 3.

Assume $\mathbf{\Phi}_p(t) = [\phi_1(t), \cdots, \phi_p(t)]^T$ to be the given basis for $\eta(t)$. Then approximately, we have

$$\eta(t) = \mathbf{\Phi}_p(t)^T \beta, \quad (6.89)$$

where $\beta = [\beta_1, \cdots, \beta_p]^T$ is a vector of parameters (coefficients) that plays a similar role as the η in the expression (6.7). It follows that the roughness of $\eta(t)$ can be approximately computed as

$$\int_a^b \left[\eta^{(k)}(t)\right]^2 dt = \beta^T \mathbf{G}\beta, \tag{6.90}$$

where

$$\mathbf{G} = \int_a^b \mathbf{\Phi}_p^{(k)}(t)\mathbf{\Phi}_p^{(k)}(t)^T dt, \tag{6.91}$$

is the roughness matrix of $\eta(t)$ based on $\mathbf{\Phi}_p(t)$.

Define $\mathbf{x}_{ij} = \mathbf{\Phi}_p(t_{ij})$. Then the PLS criterion (6.3) becomes

$$\sum_{i=1}^n \sum_{j=1}^{n_i} \left[y_{ij} - \mathbf{x}_{ij}^T\beta\right]^2 + \lambda\beta^T\mathbf{G}\beta,$$

which is essentially the same as the PLS criterion (6.7). Therefore, the remaining development of the general degree NSS model will be the same as that of the cubic NSS model.

6.6.2 General Degree GSS

The general degree GSS can be defined, following (6.28), as the minimizer of the following PGLS criterion:

$$\begin{array}{c}\sum_{i=1}^n (\mathbf{y}_i - \boldsymbol{\eta}_i)^T \mathbf{V}_i^{-1}(\mathbf{y}_i - \boldsymbol{\eta}_i) + \lambda \int_a^b \left[\eta^{(k)}(t)\right]^2 dt, \\ \boldsymbol{\eta}_i = [\eta(t_{i1}), \cdots, \eta(t_{in_i})]^T, \quad \mathbf{V}_i = \text{Cov}(\mathbf{y}_i),\end{array} \tag{6.92}$$

with respect to $\eta(t)$ over the k-th order Sobolev space $\mathcal{W}_2^k[a,b]$ as defined in (3.32).

Assume again that $\mathbf{\Phi}_p(t)$ is the given basis for $\eta(t)$. Then the roughness of $\eta(t)$ can be computed approximately using (6.90). Moreover, we have $\boldsymbol{\eta}_i = \mathbf{X}_i\beta, \quad i = 1, 2, \cdots, n$, where $\mathbf{X}_i = [\mathbf{x}_{i1}, \cdots, \mathbf{x}_{in_i}]^T$. Then the PGLS criterion (6.92) can be written as

$$\sum_{i=1}^n (\mathbf{y}_i - \mathbf{X}_i\beta)^T \mathbf{V}_i^{-1}(\mathbf{y}_i - \mathbf{X}_i\beta) + \lambda\beta^T\mathbf{G}\beta,$$

which is similar to the PGLS criterion (6.29). The remaining development of the general degree GSS model will be the same as that of the cubic GSS model.

6.6.3 General Degree ESS

Similar to the cubic ESS model that starts from (6.39), the general degree ESS model starts from the following constrained PLS criterion:

$$\begin{array}{c}\sum_{i=1}^n \left\{(\mathbf{y}_i - \boldsymbol{\eta}_i - \mathbf{v}_i)^T \mathbf{R}_i^{-1}(\mathbf{y}_i - \boldsymbol{\eta}_i - \mathbf{v}_i)\right\} \\ + \lambda_v \sum_{i=1}^n \left\{\int_a^b \left[v_i^{(k)}(t)\right]^2 dt\right\} + \lambda \int_a^b \left[\eta^{(k)}(t)\right]^2 dt, \\ \text{Subject to } \sum_{i=1}^n v_i(t) = 0,\end{array} \tag{6.93}$$

where $\mathbf{v}_i = [v_i(t_{i1}), \cdots, v_i(t_{in_i})]^T$ denotes the vector of the values of $v_i(t)$ at the design time points $t_{ij}, j = 1, 2, \cdots, n_i$ for subject i, and $\sum_{i=1}^n v_i(t) = 0$ is an identifiability condition for the subject-specific deviation functions $v_i(t), \ i = 1, 2, \cdots, n$. For the above model, we need two bases: $\mathbf{\Phi}_p(t)$ for computing the roughness of $\eta(t)$, and $\mathbf{\Psi}_q(t) = [\psi_1(t), \cdots, \psi_q(t)]^T$ for computing the roughness of $v_i(t), \ i = 1, 2, \cdots, n$.

From the previous subsections, we approximately have $\eta(t) = \mathbf{\Phi}_p(t)^T\beta, \boldsymbol{\eta}_i = \mathbf{X}_i\beta$ and $\int_a^b \left[\eta^{(k)}(t)\right]^2 dt = \beta^T\mathbf{G}\beta$ where $\beta, \mathbf{X}_i, \mathbf{G}$ were defined in the previous subsections. Similarly, the $v_i(t), \ i = 1, 2, \cdots, n$ can be approximately expressed as

$$v_i(t) = \mathbf{\Psi}_q(t)^T\mathbf{b}_i, \ i = 1, 2, \cdots, n, \tag{6.94}$$

where $\mathbf{b}_i, \ i = 1, 2, \cdots, n$ are the vectors of the associated coefficients. In addition, we approximately have

$$\int_a^b \left[v_i^{(k)}(t)\right]^2 dt = \mathbf{b}_i^T\mathbf{G}_v\mathbf{b}_i, \ i = 1, 2, \cdots, n, \tag{6.95}$$

where the roughness matrix $\mathbf{G}_v$ is defined similarly as $\mathbf{G}$ (6.91), but is now based on the basis $\mathbf{\Psi}_q(t)$.

Define $\mathbf{z}_{ij} = \mathbf{\Psi}_q(t_{ij})$ and $\mathbf{Z}_i = [\mathbf{z}_{i1}, \cdots, \mathbf{z}_{in_i}]^T$. Then $v_i(t_{ij}) = \mathbf{z}_{ij}^T\mathbf{b}_i$ and $\mathbf{v}_i = \mathbf{Z}_i\mathbf{b}_i$. The constrained PLS criterion (6.93) can be written as

$$\begin{aligned} \textstyle\sum_{i=1}^n \Big\{(\mathbf{y}_i - \mathbf{X}_i\beta - \mathbf{Z}_i\mathbf{b}_i)^T\mathbf{R}_i^{-1}(\mathbf{y}_i - \mathbf{X}_i\beta - \mathbf{Z}_i\mathbf{b}_i)\Big\} \\ + \lambda_v \textstyle\sum_{i=1}^n \Big\{\mathbf{b}_i^T\mathbf{G}_v\mathbf{b}_i\Big\} + \lambda\beta^T\mathbf{G}\beta, \\ \text{Subject to } \textstyle\sum_{i=1}^n \mathbf{b}_i = 0, \end{aligned} \tag{6.96}$$

where the identifiability condition $\sum_{i=1}^n \mathbf{b}_i = 0$ is derived from the original identifiability condition $\sum_{i=1}^n v_i(t) = 0$. Then starting from the above expression, we can describe the general degree ESS model. The general degree ESS model differs from the cubic ESS model in the PLS criterion (6.39) with different covariate matrices $\mathbf{X}_i$ and $\mathbf{Z}_i$, and with different roughness matrices $\mathbf{G}$ and $\mathbf{G}_v$. We shall continue using the notations in (6.42) and re-define

$$\begin{aligned} \mathbf{b} = [\mathbf{b}_1^T, \cdots, \mathbf{b}_n^T]^T, \ \ \mathbf{Z} = \text{diag}(\mathbf{Z}_1, \cdots, \mathbf{Z}_n), \\ \tilde{\mathbf{G}}_v = \text{diag}(\mathbf{G}_v, \cdots, \mathbf{G}_v). \end{aligned} \tag{6.97}$$

Then we can further write (6.96) as

$$\begin{aligned} (\mathbf{y} - \mathbf{X}\beta - \mathbf{Z}\mathbf{b})^T\mathbf{R}^{-1}(\mathbf{y} - \mathbf{X}\beta - \mathbf{Z}\mathbf{b}) + \lambda_v\mathbf{b}^T\tilde{\mathbf{G}}_v\mathbf{b} + \lambda\beta^T\mathbf{G}\beta, \\ \text{Subject to } \mathbf{H}\mathbf{b} = 0, \end{aligned} \tag{6.98}$$

where $\mathbf{H} = [\mathbf{I}_q, \mathbf{I}_q, \cdots, \mathbf{I}_q]$, a block matrix with n blocks of $\mathbf{I}_q$ so that $\mathbf{H}\mathbf{b} = 0$ is equivalent to $\sum_{i=1}^n \mathbf{b}_i = 0$.

To establish the LME model associated with the minimizers of the constrained PLS criterion (6.98), we need to reparameterize the constrained PLS criterion (6.96).

Using the method described from (6.45) to (6.48) in Section 6.4.3, we can similarly use the singular value decompositions of $\mathbf{G}$ and $\mathbf{G}_v$ to obtain

$$\begin{aligned} \mathbf{X}_i\beta &= \tilde{\mathbf{X}}_{i0}\tilde{\beta}_0 + \tilde{\mathbf{Z}}_{i0}\tilde{\mathbf{b}}_0, \quad \beta^T\mathbf{G}\beta = \tilde{\mathbf{b}}_0^T\tilde{\mathbf{b}}_0, \\ \mathbf{Z}_i\mathbf{b}_i &= \tilde{\mathbf{X}}_{i1}\tilde{\beta}_i + \tilde{\mathbf{Z}}_{i1}\tilde{\mathbf{b}}_i, \quad \mathbf{b}_i^T\mathbf{G}_v\mathbf{b}_i = \tilde{\mathbf{b}}_i^T\tilde{\mathbf{b}}_i, \end{aligned}$$

where the new covariate matrices $\tilde{\mathbf{X}}_{i0}, \tilde{\mathbf{Z}}_{i0}, \tilde{\mathbf{X}}_{i1}$, and $\tilde{\mathbf{Z}}_{i1}$, and the new parameters $\tilde{\beta}_0, \tilde{\mathbf{b}}_0, \tilde{\beta}_i$, and $\tilde{\mathbf{b}}_i$ are properly defined. We then reparameterize the constrained PLS criterion (6.96) as

$$\begin{aligned} \textstyle\sum_{i=1}^n \Big\{ &\Big[\mathbf{y}_i - (\tilde{\mathbf{X}}_{i0}\tilde{\beta}_0 + \tilde{\mathbf{Z}}_{i0}\tilde{\mathbf{b}}_0) - (\tilde{\mathbf{X}}_{i1}\tilde{\beta}_i + \tilde{\mathbf{Z}}_{i1}\tilde{\mathbf{b}}_i)\Big]^T \mathbf{R}_i^{-1} \\ &\times \Big[\mathbf{y}_i - (\tilde{\mathbf{X}}_{i0}\tilde{\beta}_0 + \tilde{\mathbf{Z}}_{i0}\tilde{\mathbf{b}}_0) - (\tilde{\mathbf{X}}_{i1}\tilde{\beta}_i + \tilde{\mathbf{Z}}_{i1}\tilde{\mathbf{b}}_i)\Big] \\ &\quad + \lambda_v\tilde{\mathbf{b}}_i^T\tilde{\mathbf{b}}_i\Big\} + \lambda\tilde{\mathbf{b}}_0^T\tilde{\mathbf{b}}_0, \\ &\text{Subject to } \textstyle\sum_{i=1}^n \tilde{\beta}_i = 0, \ \sum_{i=1}^n \tilde{\mathbf{b}}_i = 0. \end{aligned} \tag{6.99}$$

The identifiability condition is derived from the original identifiability condition in (6.96). To combine the parameter vectors, let

$$\tilde{\beta} = [\tilde{\beta}_0^T, \tilde{\beta}_1^T, \cdots, \tilde{\beta}_n^T]^T, \quad \tilde{\mathbf{b}} = [\tilde{\mathbf{b}}_0^T, \tilde{\mathbf{b}}_1^T, \cdots, \tilde{\mathbf{b}}_n^T]^T,$$

and

$$\mathbf{D} = \text{diag}[\mathbf{I}_{p-k}/\lambda, \mathbf{I}_{q-k_v}/\lambda_v, \cdots, \mathbf{I}_{q-k_v}/\lambda_v],$$

where k and k_v are the dimensions of $\tilde{\beta}_0$ and $\tilde{\beta}_1$, respectively. Then the constrained PLS criterion (6.99) can be further written as

$$\begin{aligned} (\mathbf{y} - \tilde{\mathbf{X}}\tilde{\beta} - \tilde{\mathbf{Z}}\tilde{\mathbf{b}})^T\mathbf{R}^{-1}(\mathbf{y} - \tilde{\mathbf{X}}\tilde{\beta} - \tilde{\mathbf{Z}}\tilde{\mathbf{b}}) + \tilde{\mathbf{b}}^T\mathbf{D}^{-1}\tilde{\mathbf{b}}, \\ \text{Subject to } \mathbf{H}_0\tilde{\beta} = 0, \ \mathbf{H}_1\tilde{\mathbf{b}} = 0, \end{aligned} \tag{6.100}$$

where $\mathbf{H}_0 = [\mathbf{0}_{k\times k}, \mathbf{I}_{k_v}, \cdots, \mathbf{I}_{k_v}]$ and $\mathbf{H}_1 = [\mathbf{0}_{(p-k)\times(p-k)}, \mathbf{I}_{(q-k_v)}, \cdots, \mathbf{I}_{(q-k_v)}]$. In the above expression, the design matrix $\tilde{\mathbf{X}}$ is defined as

$$\tilde{\mathbf{X}} = \begin{pmatrix} \tilde{\mathbf{X}}_{10}, & \tilde{\mathbf{X}}_{11}, & \mathbf{0}, & \cdots, & \mathbf{0}, & \mathbf{0} \\ \tilde{\mathbf{X}}_{20}, & \mathbf{0}, & \tilde{\mathbf{X}}_{21}, & \cdots, & \mathbf{0}, & \mathbf{0} \\ \vdots, & \vdots, & \vdots, & \vdots, & \vdots, & \vdots \\ \tilde{\mathbf{X}}_{(n-1)0}, & \mathbf{0}, & \mathbf{0}, & \cdots, & \tilde{\mathbf{X}}_{(n-1)1}, & \mathbf{0} \\ \tilde{\mathbf{X}}_{n0}, & \mathbf{0}, & \mathbf{0}, & \cdots, & \mathbf{0}, & \tilde{\mathbf{X}}_{n1} \end{pmatrix},$$

and the design matrix $\tilde{\mathbf{Z}}$ is defined similarly but with all the $\tilde{\mathbf{X}}_{i0}$ and $\tilde{\mathbf{X}}_{i1}$ replaced by $\tilde{\mathbf{Z}}_{i0}$ and $\tilde{\mathbf{Z}}_{i1}$ in the above definition.

Then the BLUPs of the following constrained LME model:

$$\begin{aligned} \mathbf{y} = \tilde{\mathbf{X}}\tilde{\beta} + \tilde{\mathbf{Z}}\tilde{\mathbf{b}} + \epsilon, \\ \tilde{\mathbf{b}} \sim N(0, \mathbf{D}), \ \ \epsilon \sim N(0, \mathbf{R}), \\ \text{Subject to } \mathbf{H}_0\tilde{\beta} = 0, \end{aligned} \tag{6.101}$$

are the minimizers of the constrained PLS criterion (6.99). The identifiability condition $\mathbf{H}_1\tilde{\mathbf{b}} = \sum_{i=1}^n \tilde{\mathbf{b}}_i = 0$ is dropped from the constrained LME model (6.101) since this condition is automatically satisfied at least asymtotically due to the fact that $E\tilde{\mathbf{b}}_i = 0,\ i = 1, 2, \cdots, n$ is assumed. Notice that there are 3 free variance components in the constrained LME model (6.101): σ^2, $1/\lambda$ and $1/\lambda_v$ for the measurement errors, the population mean function $\eta(t)$ and the subject-specific deviations $v_i(t)$, respectively. Thus, the λ and λ_v can be selected using the EM-algorithm for the constrained LME model (6.101).

6.6.4 General Degree MESS

Similar to the general degree ESS model, the general degree MESS model needs two bases $\mathbf{\Phi}_p(t)$ and $\mathbf{\Psi}_q(t)$, respectively, for $\eta(t)$ and $v_i(t),\ i = 1, 2, \cdots, n$. Based on the two bases, we can evaluate $\eta(t)$ and $v_i(t)$ at any time point t using (6.89) and (6.94), respectively, and compute the roughness of $\eta(t)$ and $v_i(t)$, and their associated roughness matrices $\mathbf{G}$ and $\mathbf{G}_v$ using (6.90), (6.95) and (6.91), respectively. Then the PGLL criterion (6.57) can be re-expressed as

$$\begin{aligned}\sum_{i=1}^n \{(\mathbf{y}_i - \mathbf{X}_i\boldsymbol{\beta} - \mathbf{Z}_i\mathbf{b}_i)^T\mathbf{R}_i^{-1}(\mathbf{y}_i - \mathbf{X}_i\boldsymbol{\beta} - \mathbf{Z}_i\mathbf{b}_i) \\ + \log|\mathbf{D}| + \mathbf{b}_i^T\mathbf{D}^{-1}\mathbf{b}_i + \log|\mathbf{R}_i| \\ + \lambda_v\mathbf{b}_i^T\mathbf{G}_v\mathbf{b}_i\} + \lambda\boldsymbol{\beta}^T\mathbf{G}\boldsymbol{\beta},\end{aligned} \tag{6.102}$$

where $\mathbf{X}_i = [\mathbf{x}_{i1}, \mathbf{x}_{i2}, \cdots, \mathbf{x}_{in_i}]^T$ and $\mathbf{Z}_i = [\mathbf{z}_{i1}, \mathbf{z}_{i2}, \cdots, \mathbf{z}_{in_i}]^T$ with $\mathbf{x}_{ij} = \mathbf{\Phi}_p(t_{ij})$, and $\mathbf{z}_{ij} = \mathbf{\Psi}_q(t_{ij})$ as defined before.

Comparing the PGLL criterion (6.102) to the PGLL criterion (6.58), we found that the general degree MESS model differs from the cubic MESS model in the following four aspects:

(a). The parameters are different. For the cubic MESS model, the fixed-effects and random-effects parameters are $\boldsymbol{\eta}$ and $\mathbf{v}_i$ which were obtained via evaluating $\eta(t)$ and $v_i(t)$ at all the distinct design time points (6.4) with some smoothness imposed. But for the general degree MESS model, the fixed-effects and random-effects parameters are $\boldsymbol{\beta}$ and $\mathbf{b}_i$ which are the coefficients of the bases with no smoothness property required.

(b). The covariate matrices are different. For the cubic MESS model, the fixed-effects and random-effects covariate matrices are $\mathbf{X} = [\mathbf{X}_1^T, \cdots, \mathbf{X}_n^T]^T$ and $\mathbf{Z} = \text{diag}(\mathbf{X}_1, \cdots, \mathbf{X}_n)$. But for the general degree MESS model, $\mathbf{X} = [\mathbf{X}_1^T, \cdots, \mathbf{X}_n^T]^T$ and $\mathbf{Z} = \text{diag}(\mathbf{Z}_1, \cdots, \mathbf{Z}_n)$.

(c). The roughness matrices are different. For the cubic MESS model, both the roughness matrices of $\eta(t)$ and $v_i(t),\ i = 1, 2, \cdots, n$ are $\mathbf{G}$. But for the general degree MESS model, the roughness matrices of $\eta(t)$ and $v_i(t),\ i = 1, 2, \cdots, n$ are $\mathbf{G}$ and $\mathbf{G}_v$, respectively.

(d). The connection between the covariance function $\gamma(s,t)$ and the random-effects covariance matrix $\mathbf{D}$ is different. For the cubic MESS model, $\gamma(\tau_r,\tau_s) = \mathbf{D}(r,s)$. But for the general degree MESS model, approximately, we have $\gamma(s,t) = \mathbf{\Psi}_q(s)^T\mathbf{D}\mathbf{\Psi}_q(t)$.

Nevertheless, the minimizers of (6.102) are quite similar in form to the minimizers of (6.58). Let

$$\begin{aligned}\mathbf{D}_v &= (\mathbf{D}^{-1} + \lambda_v\mathbf{G}_v)^{-1},\\ \tilde{\mathbf{V}}_i &= \mathbf{Z}_i\mathbf{D}_v\mathbf{Z}_i^T + \mathbf{R}_i,\ i = 1,2,\cdots,n.\end{aligned} \tag{6.103}$$

Then for given $\mathbf{D},\mathbf{R}_i,\ i = 1,2,\cdots,n$ and λ, λ_v, the minimizers of (6.102) are

$$\begin{aligned}\hat{\beta} &= \left(\sum_{i=1}^n \mathbf{X}_i^T\tilde{\mathbf{V}}_i^{-1}\mathbf{X}_i + \lambda\mathbf{G}\right)^{-1}\left(\sum_{i=1}^n \mathbf{X}_i^T\tilde{\mathbf{V}}_i^{-1}\mathbf{y}_i\right),\\ \hat{\mathbf{b}}_i &= \mathbf{D}_v\mathbf{Z}_i^T\tilde{\mathbf{V}}_i^{-1}\left(\mathbf{y}_i - \mathbf{X}_i\hat{\beta}\right),\ i = 1,2,\cdots,n,\end{aligned} \tag{6.104}$$

which are similar to the expressions in (6.61), but differ in their notation definitions.

The EM-algorithm for estimating the variance components of the cubic MESS model can be easily extended to the general degree MESS model. The resulting EM-algorithm can also be used for mixed-effects penalized splines that we shall explore in the next chapter. We shall defer the description of the EM-algorithm details until the next chapter. The AIC and BIC rules for selecting the smoothing parameters for the cubic MESS model can also be generalized to the current setup. Such a generalization can be employed for mixed-effects penalized splines, which will also be deferred until the next chapter.

6.6.5 Choice of the Bases

Theoretically, any basis that spans the Sobolev space $\mathcal{W}_2^k([a,b])$ (3.32) can be used as $\mathbf{\Phi}_p(t)$ or $\mathbf{\Psi}_q(t)$. A possible example for $\mathbf{\Phi}_p(t)$ or $\mathbf{\Psi}_q(t)$ is the $(2k-1)$-degree truncated power basis (3.18) with knots at all the distinct design time points. Another example is the reproducing kernel Hilbert space basis developed in Wahba (1990). The third example is the B-spline basis developed by de Boor (1978).

6.7 SUMMARY AND BIBLIOGRAPHICAL NOTES

In this chapter, we described the NSS, GSS, ESS and MESS models for longitudinal data. The first two models can be used to estimate the population mean function of the longitudinal data, while the latter two models can be employed to estimate both the population mean and the subject-specific deviation functions. For each of the models, smoothing parameter selection methods were considered. For the NSS and GSS models, the GCV rules were defined; for the ESS model, the smoothing parameter was proposed to be selected by an EM-algorithm based on a constrained LME model representation of the ESS model; and for the MESS model, the AIC and BIC rules were proposed.

Smoothing spline models for longitudinal data have been investigated for more than a decade in the literature. Rice and Silverman (1991) employed the NSS smoother to fit the population mean function of functional data, and their basic ideas are directly applicable to longitudinal data. Hoover et al. (1998) employed the NSS smoother mainly to fit a time-varying coefficient model which includes the nonparametric population mean model (6.1) as a special case. Although the GSS model proposed in Wang (1998b) was not originally designed for longitudinal data analysis, it is easy to adapt for longitudinal data as we discussed in Section 6.3. The ESS model has been actively investigated in the last decade, mainly due to the simple connection between a smoothing spline and a LME model suggested by Speed (1991), and popularized by Donnelly, Laird, and Ware (1995), Brumback and Rice (1998), Wang (1998a, b), Zhang et al. (1998), Verbyla et al. (1999), and Guo (2002a, b), among others. Ruppert, Wand and Carroll (2003) adapted this technique for semiparametric mixed-effects models using penalized splines. The MESS model is still in its infancy and more work needs to be done in this area.

For independent data, Silverman (1984) showed that smoothing splines are asymptotically equivalent to kernel regression with a specific higher-order kernel. However, for correlated longitudinal or clustered data, this is not true. Welsh, Lin and Carroll (2002) showed, via theoretical and numerical studies, that smoothing splines and P-splines are not local in their behavior, and it is necessary to account for data correlations when using spline methods. Lin et al. (2004) further showed that, for longitudinal data, the smoothing spline estimator is asymptotically equivalent to the kernel estimator of Wang (2003) in which the within-subject correlation is accounted for. Their theoretical analyses and numerical studies suggested that both Wang's kernel estimator and the smoothing spline estimator are nonlocal and have asymptotically negligible bias, and smoothing splines are asymptotically consistent for any arbitrary working covariance and have the smallest variance when the true covariance is specified. All these results provide evidences for that the correlation of longitudinal data needs to be accounted for in order to obtain an efficient estimator no matter for spline methods or kernel methods. The mixed-effects modeling approach for both kernel methods and spline methods is a natural way to incorporate the correlation of longitudinal data, and also has advantage of easy implementation using existing software packages.

6.8 APPENDIX: PROOFS

Proof of Theorem 6.1 Take the derivatives of (6.58) with respect to η and $\mathbf{v}_i$, $i = 1, 2, \cdots, n$ and make them 0. After some simplification, we have

$$-\sum_{i=1}^{n} \mathbf{X}_i^T \mathbf{R}_i^{-1}(\mathbf{y}_i - \mathbf{X}_i\eta - \mathbf{X}_i\mathbf{v}_i) + \lambda \mathbf{G}\eta = 0, \quad (6.105)$$

$$-\mathbf{X}_i^T \mathbf{R}_i^{-1}(\mathbf{y}_i - \mathbf{X}_i\eta - \mathbf{X}_i\mathbf{v}_i) + \mathbf{D}^{-1}\mathbf{v}_i + \lambda_v \mathbf{G}\mathbf{v}_i = 0. \quad (6.106)$$

By (6.106) and using the definition (6.59) of $\mathbf{D}_v$, we get

$$\mathbf{v}_i = (\mathbf{X}_i^T\mathbf{R}_i^{-1}\mathbf{X}_i + \mathbf{D}_v^{-1})^{-1}\mathbf{X}_i^T\mathbf{R}_i^{-1}(\mathbf{y}_i - \mathbf{X}_i\boldsymbol{\eta}). \qquad (6.107)$$

Directly applying Lemma 2.1, we get

$$\begin{aligned}(\mathbf{X}_i^T\mathbf{R}_i^{-1}\mathbf{X}_i + \mathbf{D}_v^{-1})^{-1} &= \mathbf{D}_v - \mathbf{D}_v\mathbf{X}_i^T(\mathbf{R}_i + \mathbf{X}_i\mathbf{D}_v\mathbf{X}_i)^{-1}\mathbf{X}_i\mathbf{D}_v \\ &= \mathbf{D}_v - \mathbf{D}_v\mathbf{X}_i^T\tilde{\mathbf{V}}_i^{-1}\mathbf{X}_i\mathbf{D}_v,\end{aligned}$$

using the definition (6.59) of $\tilde{\mathbf{V}}_i$. Thus

$$\begin{aligned}(\mathbf{X}_i^T\mathbf{R}_i^{-1}\mathbf{X}_i + \mathbf{D}_v^{-1})^{-1}\mathbf{X}_i^T\mathbf{R}_i^{-1} &= \mathbf{D}_v\mathbf{X}_i^T\mathbf{R}_i^{-1} - \mathbf{D}_v\mathbf{X}_i^T\tilde{\mathbf{V}}_i^{-1}\mathbf{X}_i\mathbf{D}_v\mathbf{X}_i^T\mathbf{R}_i^{-1} \\ &= \mathbf{D}_v\mathbf{X}_i^T\tilde{\mathbf{V}}_i^{-1}.\end{aligned}$$

Therefore, by (6.107), we get

$$\mathbf{v}_i = \mathbf{D}_v\mathbf{X}_i^T\tilde{\mathbf{V}}_i^{-1}(\mathbf{y}_i - \mathbf{X}_i\boldsymbol{\eta}), \; i = 1, 2, \cdots, n. \qquad (6.108)$$

Plugging (6.108) into (6.105), we get

$$-\sum_{i=1}^{n}\mathbf{X}_i^T\mathbf{R}_i^{-1}\left(\mathbf{I}_{n_i} - \mathbf{X}_i\mathbf{D}_v\mathbf{X}_i^T\tilde{\mathbf{V}}_i^{-1}\right)(\mathbf{y}_i - \mathbf{X}_i\boldsymbol{\eta}) + \lambda\mathbf{G}\boldsymbol{\eta} = 0.$$

Noticing that $\mathbf{I}_{n_i} - \mathbf{X}_i\mathbf{D}_v\mathbf{X}_i^T\tilde{\mathbf{V}}_i^{-1} = \mathbf{R}_i\tilde{\mathbf{V}}_i^{-1}$, we then have

$$-\sum_{i=1}^{n}\mathbf{X}_i^T\tilde{\mathbf{V}}_i^{-1}(\mathbf{y}_i - \mathbf{X}_i\boldsymbol{\eta}) + \lambda\mathbf{G}\boldsymbol{\eta} = 0.$$

Solving the above equation with respect to $\boldsymbol{\eta}$, we get the first equation of (6.62), and replacing $\boldsymbol{\eta}$ in (6.108) by $\hat{\boldsymbol{\eta}}$, we get the second equation. The theorem is proved.

Proof of Theorem 6.2 For simplicity, set $\mathbf{G}_\lambda = \lambda\mathbf{G}$ so that $\mathbf{G}_\lambda^- = \lambda^{-1}\mathbf{G}^-$. By (6.64) and (6.65), we have

$$\begin{pmatrix}\boldsymbol{\eta}\\ \mathbf{v}\\ \boldsymbol{\epsilon}\\ \mathbf{y}\end{pmatrix} \sim N\left[\mathbf{0}, \begin{pmatrix}\mathbf{G}_\lambda^- & \mathbf{0} & \mathbf{0} & \mathbf{G}_\lambda^-\mathbf{X}^T\\ \mathbf{0} & \tilde{\mathbf{D}}_v & \mathbf{0} & \tilde{\mathbf{D}}_v\mathbf{Z}^T\\ \mathbf{0} & \mathbf{0} & \mathbf{R} & \mathbf{R}\\ \mathbf{X}\mathbf{G}_\lambda^- & \mathbf{Z}\tilde{\mathbf{D}}_v & \mathbf{R} & \mathbf{W}^{-1}\end{pmatrix}\right], \qquad (6.109)$$

where $\mathbf{W}^{-1} = \tilde{\mathbf{V}} + \mathbf{X}\mathbf{G}_\lambda^-\mathbf{X}^T$ with $\tilde{\mathbf{V}}$ defined in (6.63). Then applying Lemma 2.2, we have

$$\boldsymbol{\eta}|\mathbf{y} \sim N\left[\mathbf{G}_\lambda^-\mathbf{X}^T\mathbf{W}\mathbf{y}, \mathbf{G}_\lambda^- - \mathbf{G}_\lambda^-\mathbf{X}^T\mathbf{W}\mathbf{X}\mathbf{G}_\lambda^-\right].$$

Applying Lemma 2.1, we have

$$\begin{aligned}\mathbf{W} &= \left(\tilde{\mathbf{V}} + \mathbf{X}\mathbf{G}_\lambda^-\mathbf{X}^T\right)^{-1} \\ &= \tilde{\mathbf{V}}^{-1} - \tilde{\mathbf{V}}^{-1}\mathbf{X}\mathbf{G}_\lambda^-(\mathbf{G}_\lambda^- + \mathbf{G}_\lambda^-\mathbf{X}^T\tilde{\mathbf{V}}^{-1}\mathbf{X}\mathbf{G}_\lambda^-)^{-1}\mathbf{G}_\lambda^-\mathbf{X}^T\tilde{\mathbf{V}}^{-1} \\ &= \tilde{\mathbf{V}}^{-1} - \tilde{\mathbf{V}}^{-1}\mathbf{X}(\mathbf{G}_\lambda + \mathbf{X}^T\tilde{\mathbf{V}}^{-1}\mathbf{X})^{-1}\mathbf{X}^T\tilde{\mathbf{V}}^{-1}.\end{aligned}$$

It follows that

$$\begin{aligned}
\mathbf{G}_\lambda^-\mathbf{X}^T\mathbf{W} &= \mathbf{G}_\lambda^-\mathbf{X}^T\tilde{\mathbf{V}}^{-1} - \mathbf{G}_\lambda^-\mathbf{X}^T\tilde{\mathbf{V}}^{-1}\mathbf{X}(\mathbf{G}_\lambda + \mathbf{X}^T\tilde{\mathbf{V}}^{-1}\mathbf{X})^{-1}\mathbf{X}^T\tilde{\mathbf{V}}^{-1} \\
&= \mathbf{G}_\lambda^-\mathbf{G}_\lambda(\mathbf{G}_\lambda + \mathbf{X}^T\tilde{\mathbf{V}}^{-1}\mathbf{X})^{-1}\mathbf{X}^T\tilde{\mathbf{V}}^{-1} \\
&= (\mathbf{G}_\lambda + \mathbf{X}^T\tilde{\mathbf{V}}^{-1}\mathbf{X})^{-1}\mathbf{X}^T\tilde{\mathbf{V}}^{-1} \\
&= (\lambda\mathbf{G} + \mathbf{X}^T\tilde{\mathbf{V}}^{-1}\mathbf{X})^{-1}\mathbf{X}^T\tilde{\mathbf{V}}^{-1}.
\end{aligned}$$

Moreover,

$$\begin{aligned}
\mathbf{G}_\lambda^- - \mathbf{G}_\lambda^-\mathbf{X}^T\mathbf{W}\mathbf{X}\mathbf{G}_\lambda^- &= \mathbf{G}_\lambda^- - (\mathbf{G}_\lambda + \mathbf{X}^T\tilde{\mathbf{V}}^{-1}\mathbf{X})^{-1}\mathbf{X}^T\tilde{\mathbf{V}}^{-1}\mathbf{X}\mathbf{G}_\lambda^- \\
&= (\mathbf{G}_\lambda + \mathbf{X}^T\tilde{\mathbf{V}}^{-1}\mathbf{X})^{-1} \\
&= (\lambda\mathbf{G} + \mathbf{X}^T\tilde{\mathbf{V}}^{-1}\mathbf{X})^{-1}.
\end{aligned}$$

Because $(\lambda\mathbf{G} + \mathbf{X}^T\tilde{\mathbf{V}}^{-1}\mathbf{X})^{-1}\mathbf{X}^T\tilde{\mathbf{V}}^{-1}\mathbf{y} = \hat{\boldsymbol{\eta}}$, we have (6.67) and the first equation of (6.66).

Similarly, by (6.109) and applying Lemma 2.2, we have

$$\begin{aligned}
\mathbf{v}|\mathbf{y} &\sim N\left[\tilde{\mathbf{D}}_v\mathbf{Z}^T\mathbf{W}\mathbf{y}, \tilde{\mathbf{D}}_v - \tilde{\mathbf{D}}_v\mathbf{Z}^T\mathbf{W}\mathbf{Z}\tilde{\mathbf{D}}_v\right], \\
\boldsymbol{\epsilon}|\mathbf{y} &\sim N\left[\mathbf{R}\mathbf{W}\mathbf{y}, \mathbf{R} - \mathbf{R}\mathbf{W}\mathbf{R}\right].
\end{aligned}$$

It follows that to show (6.68) and (6.69), it is sufficient to show

$$\tilde{\mathbf{D}}_v\mathbf{Z}^T\mathbf{W}\mathbf{y} = \hat{\mathbf{v}}, \quad \mathbf{R}\mathbf{W}\mathbf{y} = \hat{\boldsymbol{\epsilon}}. \tag{6.110}$$

Now using the formulas of $\mathbf{W}$ and $\hat{\boldsymbol{\eta}}$, we have

$$\begin{aligned}
\mathbf{W}\mathbf{y} &= \left(\tilde{\mathbf{V}}^{-1} - \tilde{\mathbf{V}}^{-1}\mathbf{X}(\mathbf{G}_\lambda + \mathbf{X}^T\tilde{\mathbf{V}}^{-1}\mathbf{X})^{-1}\mathbf{X}^T\tilde{\mathbf{V}}^{-1}\right)\mathbf{y} \\
&= \tilde{\mathbf{V}}^{-1}\mathbf{y} - \tilde{\mathbf{V}}^{-1}\mathbf{X}\hat{\boldsymbol{\eta}} \\
&= \tilde{\mathbf{V}}^{-1}(\mathbf{y} - \mathbf{X}\hat{\boldsymbol{\eta}}).
\end{aligned}$$

The first equation in (6.110) is valid since $\tilde{\mathbf{D}}_v\mathbf{Z}^T\tilde{\mathbf{V}}^{-1}(\mathbf{y} - \mathbf{X}\hat{\boldsymbol{\eta}}) = \hat{\mathbf{v}}$. The second equation in (6.110) is then verified as follows:

$$\begin{aligned}
\mathbf{R}\mathbf{W}\mathbf{y} &= \mathbf{R}\tilde{\mathbf{V}}^{-1}(\mathbf{y} - \mathbf{X}\hat{\boldsymbol{\eta}}) \\
&= \mathbf{R}(\mathbf{Z}\tilde{\mathbf{D}}_v\mathbf{Z}^T + \mathbf{R})^{-1}(\mathbf{y} - \mathbf{X}\hat{\boldsymbol{\eta}}) \\
&= (\mathbf{I}_N - \mathbf{Z}\tilde{\mathbf{D}}_v\mathbf{Z}^T\mathbf{V}^{-1})(\mathbf{y} - \mathbf{X}\hat{\boldsymbol{\eta}}) \\
&= \mathbf{y} - \mathbf{X}\hat{\boldsymbol{\eta}} - \mathbf{Z}\hat{\mathbf{v}} \\
&= \hat{\boldsymbol{\epsilon}}.
\end{aligned}$$

The theorem is then proved.

Proof of Theorem 6.3 Set $\tilde{\mathbf{y}} = \mathbf{y} - \mathbf{X}\eta$. Under the Bayesian framework (6.64) and (6.65), we have

$$\begin{pmatrix} \mathbf{v} \\ \epsilon \\ \tilde{\mathbf{y}} \end{pmatrix} \sim N\left[0, \begin{pmatrix} \tilde{\mathbf{D}}_v & \mathbf{0} & \tilde{\mathbf{D}}_v\mathbf{Z}^T \\ \mathbf{0} & \mathbf{R} & \mathbf{R} \\ \mathbf{Z}\tilde{\mathbf{D}}_v & \mathbf{R} & \tilde{\mathbf{V}} \end{pmatrix}\right].$$

Applying Lemma 2.2 directly, we have

$$\begin{aligned} \mathbf{v}|\tilde{\mathbf{y}} &\sim N\left[\tilde{\mathbf{D}}_v\mathbf{Z}^T\tilde{\mathbf{V}}^{-1}\tilde{\mathbf{y}}, \tilde{\mathbf{D}}_v - \tilde{\mathbf{D}}_v\mathbf{Z}^T\tilde{\mathbf{V}}^{-1}\mathbf{Z}\tilde{\mathbf{D}}_v\right], \\ \epsilon|\tilde{\mathbf{y}} &\sim N\left[\mathbf{R}\tilde{\mathbf{V}}^{-1}\tilde{\mathbf{y}}, \mathbf{R} - \mathbf{R}\tilde{\mathbf{V}}^{-1}\mathbf{R}\right]. \end{aligned}$$

Then (6.71) follows by replacing $\tilde{\mathbf{y}}$ by $\mathbf{y} - \mathbf{X}\eta$ in the above expressions. The theorem is proved.

Proof of Theorem 6.4 Because $\mathbf{R}, \tilde{\mathbf{V}}, \mathbf{Z}$ and $\tilde{\mathbf{D}}_v$ are all block diagonal, by Theorem 6.3, we have for $i = 1, 2, \cdots, n$ that

$$\epsilon_i|\mathbf{y}, \eta \sim N\left[\mathbf{R}_i\tilde{\mathbf{V}}^{-1}(\mathbf{y}_i - \mathbf{X}_i\eta), \mathbf{R}_i - \mathbf{R}_i\tilde{\mathbf{V}}_i^{-1}\mathbf{R}_i\right], \tag{6.111}$$

$$\mathbf{v}_i|\mathbf{y}, \eta \sim N\left[\mathbf{D}_v\mathbf{X}_i^T\tilde{\mathbf{V}}^{-1}(\mathbf{y}_i - \mathbf{X}_i\eta), \mathbf{D}_v - \mathbf{D}_v\mathbf{X}_i^T\tilde{\mathbf{V}}_i^{-1}\mathbf{X}_i\mathbf{D}_v\right] \tag{6.112}$$

From the proof of Theorem 6.2 and (6.111),

$$\mathrm{E}(\epsilon_i|\mathbf{y}, \eta = \hat{\eta}) = \mathbf{R}_i\tilde{\mathbf{V}}^{-1}(\mathbf{y}_i - \mathbf{X}_i\hat{\eta}) = \hat{\epsilon}_i,$$

and

$$\mathrm{Cov}(\epsilon_i|\mathbf{y}, \eta = \hat{\eta}) = \mathbf{R}_i - \mathbf{R}_i\tilde{\mathbf{V}}_i^{-1}\mathbf{R}_i = \sigma^2(\mathbf{I}_{n_i} - \sigma^2\tilde{\mathbf{V}}_i^{-1}).$$

Therefore,

$$\begin{aligned} &\mathrm{E}(\epsilon_i^T\epsilon_i|\mathbf{y}, \eta = \hat{\eta}) = \mathrm{tr}\left[\mathrm{E}(\epsilon_i\epsilon_i^T|\mathbf{y}, \eta = \hat{\eta})\right] \\ = &\ \mathrm{tr}\left[\mathrm{E}(\epsilon_i|\mathbf{y}, \eta = \hat{\eta})\mathrm{E}(\epsilon_i|\mathbf{y}, \eta = \hat{\eta})^T + \mathrm{Cov}(\epsilon_i|\mathbf{y}, \eta = \hat{\eta})\right] \\ = &\ \mathrm{tr}\left[\hat{\epsilon}_i\hat{\epsilon}_i^T + \sigma^2(\mathbf{I}_{n_i} - \sigma^2\tilde{\mathbf{V}}_i^{-1})\right] \\ = &\ \hat{\epsilon}_i^T\hat{\epsilon}_i + \sigma^2[n_i - \sigma^2\mathrm{tr}(\tilde{\mathbf{V}}_i^{-1})]. \end{aligned}$$

The claim (6.75) then follows directly from the definition of $\hat{\sigma}^2$ in (6.74). Similarly, from the proof of Theorem 6.2 and (6.112), we have

$$\mathrm{E}(\mathbf{v}_i|\mathbf{y}, \eta = \hat{\eta}) = \mathbf{D}_v\mathbf{X}_i^T\tilde{\mathbf{V}}^{-1}(\mathbf{y}_i - \mathbf{X}_i\hat{\eta}) = \hat{\mathbf{v}}_i,$$

and $\mathrm{Cov}(\mathbf{v}_i|\mathbf{y}, \eta = \hat{\eta}) = \mathbf{D}_v - \mathbf{D}_v\mathbf{X}_i^T\tilde{\mathbf{V}}_i^{-1}\mathbf{X}_i\mathbf{D}_v$. Therefore,

$$\begin{aligned} &\mathrm{E}(\mathbf{v}_i\mathbf{v}_i^T|\mathbf{y}, \eta = \hat{\eta}) \\ = &\ \mathrm{E}(\mathbf{v}_i|\mathbf{y}, \eta = \hat{\eta})\mathrm{E}(\mathbf{v}_i|\mathbf{y}, \eta = \hat{\eta})^T + \mathrm{Cov}(\mathbf{v}_i|\mathbf{y}, \eta = \hat{\eta}) \\ = &\ \hat{\mathbf{v}}_i\hat{\mathbf{v}}_i^T + [\mathbf{D}_v - \mathbf{D}_v\mathbf{X}_i^T\tilde{\mathbf{V}}_i^{-1}\mathbf{X}_i\mathbf{D}_v]. \end{aligned}$$

The claim (6.76) then follows. The theorem is proved.

Proof of Lemma 6.1 To show Lemma 6.1, let us first show that $\mathbf{X}$ is of full rank. Recall that $\mathbf{X}$ is the indicator matrix of size $N \times M$ for all the subjects. It has the same rank as the matrix $\mathbf{X}^T\mathbf{X}$. Recall that the entries of the r-column of $\mathbf{X}$ are $\{1_{\{t_{ij}=\tau_r\}}, j = 1, 2, \cdots, n_i; i = 1, 2, \cdots, n\}$. Therefore, the (r, s)-th entry of $\mathbf{X}^T\mathbf{X}$ is

$$\sum_{i=1}^{n}\sum_{j=1}^{n_i} 1_{\{t_{ij}=\tau_r\}} 1_{\{t_{ij}=\tau_s\}} = \begin{cases} m_r, & \text{if } r = s, \\ 0, & \text{if } r \neq s, \end{cases}$$

where $m_r \geq 1, r = 1, 2, \cdots, M$ since $\tau_r, r = 1, 2, \cdots, M$ are the distinct design time points (6.4) out of all the design time points (6.2). It follows that $\mathbf{X}^T\mathbf{X}$ is a diagonal matrix with rank M. Therefore, $\mathbf{X}$ has rank M, and is of full rank.

Since $\mathbf{X}$ is of full rank, the matrix $\mathbf{\Omega} = \mathbf{X}^T\tilde{\mathbf{V}}^{-1}\mathbf{X}$ is invertible. It follows that

$$\begin{aligned} \text{df} &= \text{tr}\left[(\mathbf{\Omega} + \lambda\mathbf{G})^{-1}\mathbf{\Omega}\right] \\ &= \text{tr}\left[(\mathbf{I}_M + \lambda\mathbf{\Omega}^{-1/2}\mathbf{G}\mathbf{\Omega}^{-1/2})^{-1}\right]. \end{aligned}$$

Let $\mathbf{H} = \mathbf{\Omega}^{-1/2}\mathbf{G}\mathbf{\Omega}^{-1/2}$ have a singular value decomposition $\mathbf{H} = \mathbf{U}\mathbf{W}\mathbf{U}^T$ where $\mathbf{U}$ is orthonormal and $\mathbf{W} = \text{diag}(w_1, \cdots, w_M)$. Because $\mathbf{H}$ and $\mathbf{G}$ (as the roughness matrix of a cubic smoothing spline, defined in (3.33)) have the same rank, $\mathbf{W}$ has two diagonal entries, say the first two, are 0. Therefore,

$$\begin{aligned} \text{df} &= \text{tr}\left[(\mathbf{I}_M + \lambda\mathbf{H})^{-1}\right] \\ &= \text{tr}\left[(\mathbf{I}_M + \lambda\mathbf{U}\mathbf{W}\mathbf{U}^T)^{-1}\right] \\ &= \sum_{l=1}^{M}(1 + \lambda w_l)^{-1} \\ &= 2 + \sum_{l=3}^{M}(1 + \lambda w_l)^{-1}. \end{aligned} \tag{6.113}$$

From above, we can see that for a fixed λ_v, when we increase λ, df becomes smaller because $w_l, l = 3, \cdots, M$ are positive. That is, the model for estimating $\eta(t)$ becomes simpler. Lemma 6.1 is proved.

Proof of Theorem 6.5 The first expression can be verified by simply checking the equation (6.113) and noticing that $w_l, l = 3, \cdots, M$ are positive.

To show the second expression, by (6.63), notice that

$$\begin{aligned} &\text{df}_v = \text{tr}(\mathbf{A}_v) \\ = \ &\text{tr}\left[\mathbf{Z}\mathbf{D}_v\mathbf{Z}^T\mathbf{D}_v\tilde{\mathbf{V}}^{-1}(\mathbf{I}_N - \mathbf{A})\right] \\ = \ &\text{tr}\left[\mathbf{Z}\mathbf{D}_v\mathbf{Z}^T\tilde{\mathbf{V}}^{-1}\right] \\ &- \text{tr}\left[\mathbf{D}_v^{1/2}\mathbf{Z}^T\tilde{\mathbf{V}}^{-1}\mathbf{X}(\mathbf{X}^T\tilde{\mathbf{V}}^{-1}\mathbf{X} + \lambda\mathbf{G})^{-1})\mathbf{X}^T\tilde{\mathbf{V}}^{-1}\mathbf{Z}\mathbf{D}_v^{1/2}\right] \end{aligned}$$

$$
\begin{aligned}
&\leq \operatorname{tr}\left[\mathbf{Z}\mathbf{D}_v\mathbf{Z}^T\tilde{\mathbf{V}}^{-1}\right] \\
&= \operatorname{tr}\left[\mathbf{H}\mathbf{H}^T(\mathbf{H}\mathbf{H}^T + \mathbf{I}_N)^{-1}\right] \\
&= N - \operatorname{tr}\left[(\mathbf{H}\mathbf{H}^T + \mathbf{I}_N)^{-1}\right] \\
&< N,
\end{aligned}
$$

where $\mathbf{H} = \mathbf{R}^{-1/2}\mathbf{Z}\mathbf{D}_v^{1/2}$. Therefore $0 < \mathrm{df}_v < N$.

On the other hand, using Lemma 2.1, we have

$$
(\mathbf{H}\mathbf{H}^T + \mathbf{I}_N)^{-1} = \mathbf{I}_N - \mathbf{H}(\mathbf{I}_{nM} + \mathbf{H}^T\mathbf{H})^{-1}\mathbf{H}^T.
$$

Therefore,

$$
\begin{aligned}
&\operatorname{tr}[\mathbf{H}\mathbf{H}^T(\mathbf{H}\mathbf{H}^T + \mathbf{I}_N)^{-1}] = \operatorname{tr}[\mathbf{H}^T(\mathbf{H}\mathbf{H}^T + \mathbf{I}_N)^{-1}\mathbf{H}] \\
&= \operatorname{tr}[\mathbf{H}^T\mathbf{H} - \mathbf{H}^T\mathbf{H}(\mathbf{I}_{nM} + \mathbf{H}^T\mathbf{H})^{-1}\mathbf{H}^T\mathbf{H}] \\
&= \operatorname{tr}[\mathbf{H}^T\mathbf{H}(\mathbf{I}_{nM} + \mathbf{H}^T\mathbf{H})^{-1}] \\
&= nM - \operatorname{tr}[(\mathbf{I}_{nM} + \mathbf{H}^T\mathbf{H})^{-1}] \\
&< nM.
\end{aligned}
$$

It follows that $0 < \mathrm{df}_v < nM$. The theorem is proved.

7

Penalized Spline Methods

7.1 INTRODUCTION

In Chapters 4, 5 and 6, we investigated local polynomial, regression spline and smoothing spline methods for longitudinal data analysis. Penalized splines or P-splines are one of the powerful smoothing techniques for uncorrelated or independent data, as demonstrated by Ruppert, Wand and Carroll (2003) and references therein. In this chapter, we shall focus on P-spline methods, which was briefly reviewed in Section 3.5 of Chapter 3, for longitudinal data analysis. Similar to the regression spline and smoothing spline methods, we shall investigate naive P-splines, generalized P-splines, extended P-splines, and mixed-effects P-splines. The first two methods are mainly used to estimate the population mean function of longitudinal data, while the latter two methods can be used to estimate both the population mean and the subject-specified individual functions.

7.2 NAIVE P-SPLINES

As with the NRS and NSS smoothers, we can directly apply P-splines to model the population mean function of longitudinal data by ignoring the within-subject correlation of the longitudinal data. The resulting smoothers are simple to construct and they perform rather well in many longitudinal data applications. For convenience, we refer to this technique as an naive P-spline (NPS) method, and the resulting smoother as a NPS smoother.

7.2.1 The NPS Smoother

Given a longitudinal data set:

$$(y_{ij}, t_{ij}),\ j = 1, 2, \cdots, n_i;\ i = 1, 2, \cdots, n, \tag{7.1}$$

where y_{ij} is the response at the design time point t_{ij}, we consider the following nonparametric population mean (NPM) model:

$$y_{ij} = \eta(t_{ij}) + e_{ij},\ j = 1, 2, \cdots, n_i;\ i = 1, 2, \cdots, n, \tag{7.2}$$

where $\eta(t)$ is the population mean function of the longitudinal data and e_{ij} is the error at t_{ij}.

In the regression spline (NRS) method described in Section 5.2 of Chapter 5, $\eta(t)$ was approximately expressed as a linear combination of some basis, e.g., the truncated power basis, $\mathbf{\Phi}_p(t)$, which was defined in (3.21). For easy reference, we rewrite the definition of $\mathbf{\Phi}_p(t)$ here as

$$\mathbf{\Phi}_p(t) = [1, t, \cdots, t^k, (t - \tau_1)_+^k, \cdots, (t - \tau_K)_+^k]^T, \tag{7.3}$$

where $w_+^k = [\max(0, w)]^k$, $p = K + k + 1$ is the number of basis functions, k is the degree of the truncated power basis, and K is the number of the knots:

$$\tau_1, \tau_2, \cdots, \tau_K, \tag{7.4}$$

which are scattered in the range of interest. The first $k+1$ basis functions are the usual polynomial functions up to degree k and the last K basis functions $(t-\tau_1)_+^k, \cdots, (t-\tau_K)_+^k$ are often referred to as the truncated power functions.

Then we can approximately express $\eta(t)$ as

$$\eta(t) = \mathbf{\Phi}_p(t)^T \boldsymbol{\beta} = \sum_{r=0}^{k} \beta_r t^r + \sum_{l=1}^{K} \beta_{k+l}(t - \tau_l)_+^k, \tag{7.5}$$

where $\boldsymbol{\beta} = [\beta_0, \beta_1, \cdots, \beta_k, \beta_{k+1}, \cdots, \beta_{k+K}]^T$ is the coefficient vector. For a given truncated power basis $\mathbf{\Phi}_p(t)$, the coefficient vector $\boldsymbol{\beta}$ can be determined by the ordinary least squares (OLS) method. Let $\hat{\boldsymbol{\beta}}$ be the resulting OLS estimator of $\boldsymbol{\beta}$. Then the NRS estimator of $\eta(t)$ is $\hat{\eta}_{nrs}(t) = \mathbf{\Phi}_p(t)^T \hat{\boldsymbol{\beta}}$.

The performance of $\hat{\eta}_{nrs}(t)$ strongly depends on K, the number of the knots of the k-th degree truncated power basis $\mathbf{\Phi}_p(t)$. Figure 7.1 (a) shows an example of the NRS fit (solid curve) to the population mean function of the ACTG 388 data, superimposed with the 95% pointwise standard deviation (SD) band (dashed curves). The quadratic ($k = 2$) truncated power basis $\mathbf{\Phi}_p(t)$ was used. We initially took $K = [M/2] = 21$ for $\mathbf{\Phi}_p(t)$ using the "equally spaced sample quantitles as knots" method as described in Section 5.2.4 of Chapter 5. We actually had $K = 16$ after removing the tied knots. It is seen that the NRS fit is rather rough and so is the 95% pointwise SD band, which indicates that the roughness of the NRS fit can be better traded-off with the associated goodness of fit.

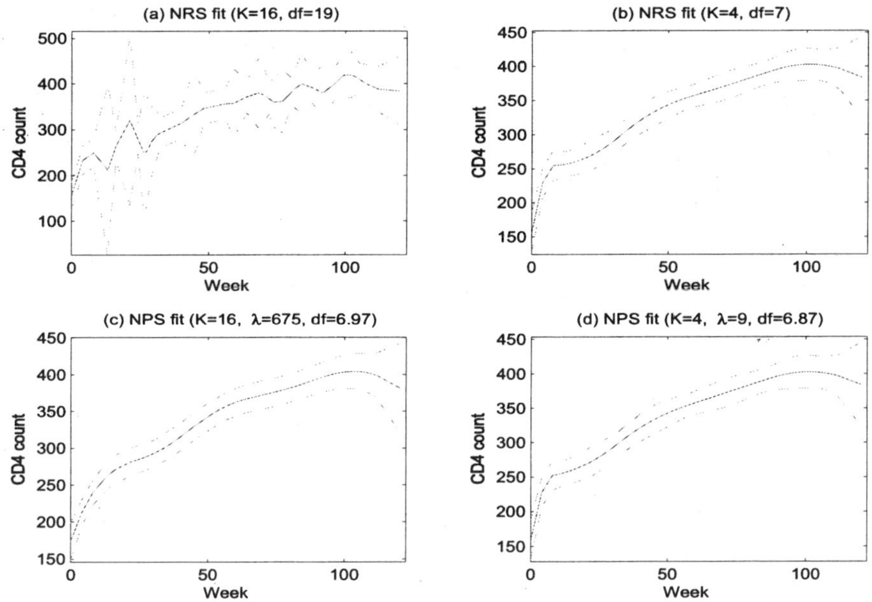

Fig. 7.1 Examples showing NRS and NPS fits (solid curves) to the population mean function of the ACTG 388 data, superimposed with the 95% pointwise SD bands (dashed curves).

In the NRS method described in Section 5.2 of Chapter 5, this issue was nicely handled via selecting K by a smoothing parameter selector, e.g., the GCV rule (5.22). The best K selected by the GCV rule is $K = 4$ (equivalently $p = 7$, see Figure 5.1 (b) for the GCV curve against p). Figure 7.1 (b) shows the NRS fit with $K = 4$, which provides a reasonably good fit to the ACTG 388 data.

Alternatively, the roughness of the NRS fit can be controlled via imposing some constraints on the coefficients $\beta_{k+1}, \beta_{k+2}, \cdots, \beta_{k+K}$ of the truncated power functions. Taking k-times derivative of $\eta(t)$ that has the expression (7.5), we have

$$\eta^{(k)}(\tau_l-) = k! \sum_{r=0}^{l-1} \beta_{k+r}, \text{ and } \eta^{(k)}(\tau_l+) = k! \sum_{r=0}^{l} \beta_{k+r},$$

implying that the coefficient β_{k+l} quantifies the jump

$$\eta^{(k)}(\tau_l+) - \eta^{(k)}(\tau_l-) = k!\beta_{k+l}$$

of the k-times derivative of $\eta(t)$ at τ_l up to a multiplying constant. Therefore, when the number of knots, K, is fixed, controlling the roughness of $\eta(t)$ is equivalent to controlling the size of $|\beta_{k+l}|, l = 1, 2, \cdots, K$. Ruppert et al. (2003) suggested the following three constraints on the coefficients:

(a) $\max_{l=1}^{K} |\beta_{k+l}| \leq C$,

(b) $\sum_{l=1}^{K} |\beta_{k+l}| \leq C$, and

(c) $\sum_{l=1}^{K} \beta_{k+l}^2 \leq C$.

For a good choice of $C > 0$, each of the above constraints will lead to a smoother fit of the data. However, the third constraint is much easier to implement than the first two. Let

$$\mathbf{G} = \begin{bmatrix} \mathbf{0}_{(k+1)\times(k+1)} & \mathbf{0}_{(k+1)\times K} \\ \mathbf{0}_{K\times(k+1)} & \mathbf{I}_K \end{bmatrix}. \tag{7.6}$$

Then the third constraint can be expressed as $\boldsymbol{\beta}^T \mathbf{G} \boldsymbol{\beta} \leq C$ so that the constrained least squares problem for determining $\boldsymbol{\beta}$ can be written as

$$\min_{\boldsymbol{\beta}} \sum_{i=1}^{n} \sum_{j=1}^{n_i} (y_{ij} - \mathbf{x}_{ij}^T \boldsymbol{\beta})^2, \text{ Subject to } \boldsymbol{\beta}^T \mathbf{G} \boldsymbol{\beta} \leq C, \tag{7.7}$$

where $\mathbf{x}_{ij} = \boldsymbol{\Phi}_p(t_{ij})$. It is not difficult to show, using a Lagrange multiplier argument, that the above minimization problem is equivalent to choosing $\boldsymbol{\beta}$ to minimize the following penalized least squares (PLS) criterion:

$$\sum_{i=1}^{n} \sum_{j=1}^{n_i} (y_{ij} - \mathbf{x}_{ij}^T \boldsymbol{\beta})^2 + \lambda \boldsymbol{\beta}^T \mathbf{G} \boldsymbol{\beta}, \tag{7.8}$$

where the first term is the sum of squared errors (SSE), quantifying the goodness of fit, and the second term is the roughness of the fit multiplied by a smoothing parameter λ. Use of λ aims to yield a smoother fit to the data via penalizing the roughness (the k-times derivative jumps at the knots) of the fit. The constrained least squares fit of $\eta(t)$ obtained in this way is often referred to as a P-spline (Ruppert et al. 2003). In our current setup, we call it a NPS for longitudinal data.

Figure 7.1 (c) displays the NPS fit (solid curve) to the ACTG 388 data with $K = 16$ and $\lambda = 675$. The smoothing parameter $\lambda = 675$ was selected by a GCV rule that we shall discuss briefly later. It is seen that the NPS fit is much smoother than the NRS fit in Figure 7.1 (a) and is comparable to the NRS fit in Figure 7.1 (b).

As a comparison, Figure 7.1 (d) displays the NPS fit to the ACTG 388 data with $K = 4$ and $\lambda = 9$. The smoothing parameter $\lambda = 9$ is much smaller than the smoothing parameter $\lambda = 675$ used for the NPS fit in Figure 7.1 (c).

7.2.2 NPS Fits and Smoother Matrix

Let

$$\begin{array}{rclrcl} \mathbf{X} & = & [\mathbf{X}_1^T, \cdots, \mathbf{X}_n^T]^T, & \mathbf{X}_i & = & [\mathbf{x}_{i1}, \cdots, \mathbf{x}_{in_i}]^T, \\ \mathbf{y} & = & [\mathbf{y}_1^T, \cdots, \mathbf{y}_n^T]^T, & \mathbf{y}_i & = & [y_{i1}, \cdots, y_{in_i}]^T. \end{array} \tag{7.9}$$

The PLS criterion (7.8) can be written as

$$\|\mathbf{y} - \mathbf{X}\boldsymbol{\beta}\|^2 + \lambda \boldsymbol{\beta}^T \mathbf{G} \boldsymbol{\beta}. \tag{7.10}$$

The minimizer of the above PLS criterion is

$$\hat{\beta} = (\mathbf{X}^T\mathbf{X} + \lambda\mathbf{G})^{-1}\mathbf{X}^T\mathbf{y}. \tag{7.11}$$

The NPS fit of $\eta(t)$ is then

$$\hat{\eta}_{nps}(t) = \mathbf{\Phi}_p(t)^T(\mathbf{X}^T\mathbf{X} + \lambda\mathbf{G})^{-1}\mathbf{X}^T\mathbf{y}. \tag{7.12}$$

The fits of $\eta(t)$ at all the design time points can be expressed as

$$\hat{\mathbf{y}}_{nps} = \mathbf{X}(\mathbf{X}^T\mathbf{X} + \lambda\mathbf{G})^{-1}\mathbf{X}^T\mathbf{y} \equiv \mathbf{A}_{nps}\mathbf{y}, \tag{7.13}$$

where $\mathbf{A}_{nps}$ is the associated NPS smoother matrix.

7.2.3 Variability Band Construction

The pointwise SD band (dashed curves) shown in Figure 7.1 (c) was constructed using a method similar to the method for constructing the pointwise SD band based on the NRS smoother described in Chapter 5. For simplicity, we generally impose a working independence assumption for the errors e_{ij} in the NPM model (7.2) so that $\mathbf{V} = \text{Cov}(\mathbf{y}) = \sigma^2\mathbf{I}_N$. Therefore,

$$\text{Var}(\hat{\eta}_{nps}(t)) = \sigma^2\mathbf{\Phi}_p(t)^T(\mathbf{X}^T\mathbf{X} + \lambda\mathbf{G})^{-1}\mathbf{X}^T\mathbf{X}(\mathbf{X}^T\mathbf{X} + \lambda\mathbf{G})^{-1}\mathbf{\Phi}_p(t), \tag{7.14}$$

where σ^2 may be estimated as $\hat{\sigma}^2 = \|\mathbf{y} - \hat{\mathbf{y}}_{nps}\|^2/[N - \text{tr}(\mathbf{A}_{nps})]$. Let $\widehat{\text{Var}}(\hat{\eta}_{nps}(t))$ denote the above expression (7.14) after replacing the σ^2 by its estimator $\hat{\sigma}^2$. Then the 95% pointwise SD band of $\eta(t)$ can be constructed as $\hat{\eta}_{nps}(t) \pm 1.96\sqrt{\widehat{\text{Var}}(\hat{\eta}_{nps}(t))}$.

7.2.4 Degrees of Freedom

The number of degrees of freedom (df) is usually defined as the trace of a smoother matrix. It quantifies approximately how many parameters are needed when a parametric model is fitted to the same data set to reach about the same amount of smoothing. It is also used to quantify the model complexity of a smoother, with smaller values indicating more complicated models. Ruppert et al. (2003) pointed out that the number of degrees of freedom is a more reasonable measure than the smoothing parameter λ to quantify the amount of smoothing imposed by a smoother, with smaller values indicating roughly more amount of smoothing.

Recall that $\mathbf{\Phi}_p(t)$ is a k-th degree truncated power basis with K knots as defined in (7.3). For a NRS smoother using $\mathbf{\Phi}_p(t)$, the associated number of degrees of freedom is

$$\text{df} = \text{tr}(\mathbf{A}_{nrs}) = p = K + k + 1.$$

For example, the numbers of df of the NRS fits presented in Figures 7.1 (a) and (b) are 19 and 7, respectively. For a NPS smoother using $\mathbf{\Phi}_p(t)$, the associated number of degrees of freedom is

$$\text{df}(\lambda) = \text{tr}(\mathbf{A}_{nps}) = \text{tr}\left[(\mathbf{X}^T\mathbf{X} + \lambda\mathbf{G})^{-1}\mathbf{X}^T\mathbf{X}\right], \tag{7.15}$$

which is often not an integer. The df of the NPS fit presented in Figure 7.1 (c) is 6.97. Therefore, the amount of smoothing imposed by the NRS fit in Figure 7.1 (a) is much less than that imposed by either the NRS or the NPS fits in Figures 7.1 (b) and (c), respectively, while the latter two fits have about the same amount of smoothing. This is consistent with our visual inspection of Figure 7.1: the NRS fit in Figure 7.1 (a) is much rougher than the NRS fit in Figure 7.1 (b) or the NPS fit in Figure 7.1 (c) while the latter two fits look quite similar.

Some properties of the df are given in the following lemma.

Lemma 7.1 *Assume that the design matrix* $\mathbf{X}$ *has full rank* $p = K + k + 1$. *Then given* $\mathbf{X}$, *we have*

(1) $df(\lambda)$ *is a monotonically decreasing continuous function of* λ.

(2) $\lim_{\lambda \to 0} df(\lambda) = K + k + 1$ *and* $\lim_{\lambda \to \infty} df(\lambda) = k + 1$.

(3) $k + 1 \leq df(\lambda) \leq K + k + 1$.

Proof: see Appendix.

Since $\mathbf{\Phi}_p(t)$ is a basis, to make $\mathbf{X}$ have full rank p, a sufficient condition is $M \geq p$ where M is the number of distinct design time points. This latter condition requires that the number of knots $K \leq M - k - 1$ when the degree k of $\mathbf{\Phi}_p(t)$ is fixed.

Lemma 7.1 indicates that when no smoothing ($\lambda = 0$) is imposed, the NPS model and the associated NRS model have about the same model complexity. However, as λ increases, the NPS model becomes simpler, and the simplest NPS model reduces to the k-th degree polynomial model when a huge amount of smoothing ($\lambda \to \infty$) is imposed.

7.2.5 Smoothing Parameter Selection

The smoothing parameter $\lambda = 675$ used in the NPS fit presented in Figure 7.1 (c) was selected by a GCV rule similar to the GCV rule (6.19) described in Section 6.2.5 of Chapter 6 for the NSS smoother. For the NPS smoother (7.12), the associated GCV can be written as

$$\mathrm{GCV}_{nps}(\lambda) = \frac{(\mathbf{y} - \hat{\mathbf{y}}_{nps})^T \mathbf{W} (\mathbf{y} - \hat{\mathbf{y}}_{nps})}{[1 - \mathrm{tr}(\mathbf{A}_{nps})/N]^2}, \tag{7.16}$$

where the weight matrix $\mathbf{W} = \mathrm{diag}(\mathbf{W}_1, \cdots, \mathbf{W}_n)$ with $\mathbf{W}_i = \mathrm{diag}(w_{i1}, \cdots, w_{in_i})$ giving weights to the subjects. For simplicity, we often take $\mathbf{W} = N^{-1}\mathbf{I}_N$ for the NPS model, i.e., using the same weight for different subjects.

Notice that the numerator of the right-hand side of the GCV score (7.16) is the weighted SSE, representing the goodness of fit of the NPS model and the denominator is a monotonically decreasing function of the number of degrees of freedom $\mathrm{tr}(\mathbf{A}_{nps}) = \mathrm{df}(\lambda)$, representing the model complexity of the NPS model. Therefore, choosing λ is equivalent to trading-off the goodness of fit with the model complexity of the NPS fit.

We use the GCV rule (7.16) for selecting λ when the number of knots K of the truncated power basis $\mathbf{\Phi}_p(t)$ is given in advance. When K needs to be determined, it can also be incorporated into (7.16) as an additional smoothing parameter.

The "leave-one-point-out" and "leave-one-subject-out" cross-validation (PCV and SCV) given in Section 6.2.5 of Chapter 6 can be similarly adapted here.

7.2.6 Choice of the Number of Knots

The number of knots, K, can be chosen in a rough manner. We need only to choose K so that it is not too small or too large. According to the discussion following Lemma 7.1, K can not be too large in order to guarantee the full-rank of the design matrix $\mathbf{X}$. That is, $K_{\max} = M - k - 1$ for a k-th degree truncated power basis where M is the number of distinct design time points. Let K^* be the optimal K for the NRS smoother (5.11). The number of degrees of freedom of the NRS smoother is $\text{df}^* = K^* + k + 1$. To get about the same amount of smoothing (measured by df) by the NPS smoother (7.12) using a k-th degree truncated power basis (7.3) with K knots, it is required that $K \geq K^*$. Therefore, $K_{\min} = K^*$. In summary, we should take K such that

$$K^* \leq K \leq M - k - 1. \tag{7.17}$$

In practice, K^* may not be known and it is often sufficient to take $K = [M/2]$ or $K = [M/3]$ initially. When the "equally spaced sample quantiles as knots" method is used to scatter the knots, some knots are tied together. After removing the tied knots, we get the actual number of knots, K. In the ACTG 388 data example presented in Figure 7.1, $k = 2, M = 43$ and $K^* = 4$, we initially took $K = [M/2] = 21$ and we got the actual $K = 16$ after removing the tied knots, which satisfies the condition (7.17). The NPS fit plotted in Figure 7.1 (c) had about the same amount of smoothing as the optimal NRS fit in Figure 7.1 (b). In fact, we can find a unique λ such that the associated NPS fit has exactly the same amount of smoothing as the optimal NRS fit.

Lemma 7.2 *For a fixed K satisfying the condition (7.17), there is a unique λ such that the associated NPS fit (7.12) has the same amount of smoothing as the optimal NRS fit (5.11).*

Lemma 7.2 can be easily proved. In fact, under the given conditions, $k + 1 \leq \text{df}^* = K^* + k + 1 \leq K + k + 1$ and by Lemma 7.1, $\text{df}(\lambda)$ is a monotonically decreasing continuous function of λ ($0 \leq \lambda < \infty$), and hence there is a unique λ^* such that $\text{df}(\lambda^*) = \text{df}^*$.

To illustrate Lemma 7.2 in more detail, we continue to use the ACTG 388 data as an example. Figure 7.2 displays the NRS fits (solid curves), superimposed with the associated 95% pointwise SD bands (dashed curves), to the ACTG 388 data for $K = 4, 5, 6, 7, 8, 10, 14, 16, 17$ knots. These K's are the actual numbers of knots although we initially used $K = [M/l], l = 9, 8, 7, 6, 5, 4, 3, 2, 4/3$ which satisfy the condition (7.17). It is seen that the NRS fit gets rougher as K increases. According to the GCV rule, the NRS fit with $K = 4$ is optimal, but visually, the NRS fits with $K = 5, 6, 7, 8$ show little difference. The NRS fits with $K = 10, 14, 16, 17$ show larger differences from the optimal NRS fit.

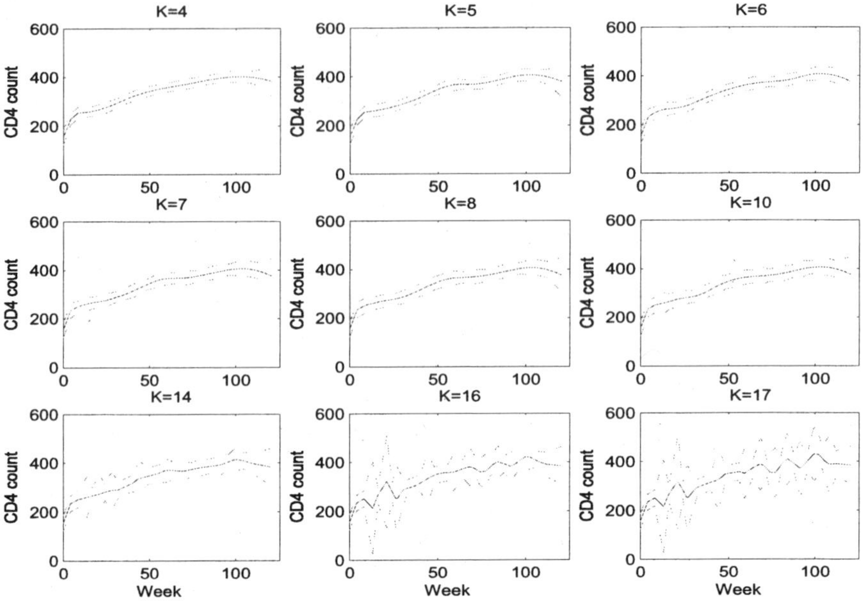

Fig. 7.2 NRS fits with $K = 4, 5, 6, 7, 8, 10, 14, 16, 17$ knots.

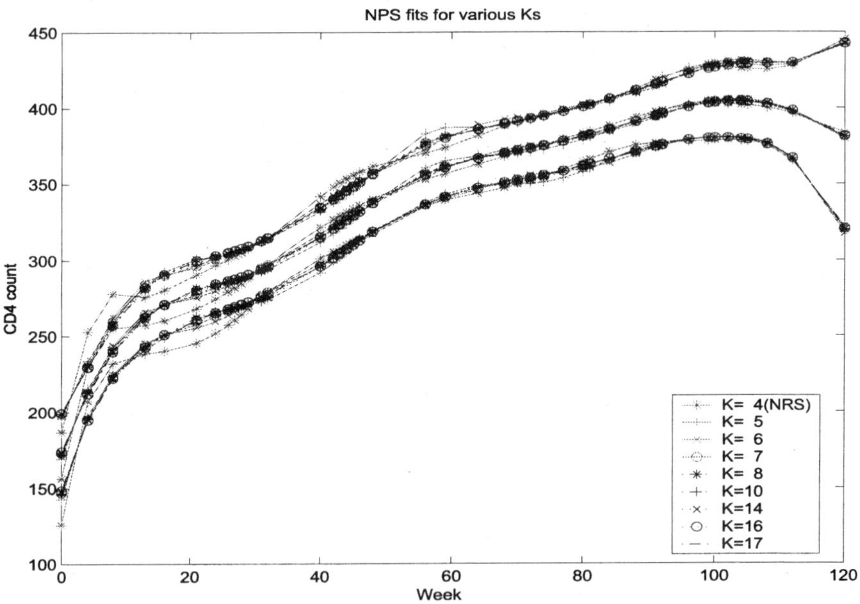

Fig. 7.3 NPS fits with $K = 4, 5, 6, 7, 8, 10, 14, 16, 17$ knots.

Figure 7.3 presents the NPS fits and the associated 95% pointwise SD bands for $K = 4, 5, 6, 7, 8, 10, 14, 16, 17$, superimposed with the optimal NRS fit and the associated 95% pointwise SD band. For each K, we chose a proper λ so that all these NPS fits had about the same amount of smoothing as the optimal NRS fit. That is, they all have about 7 degrees of freedom. It is seen that the NPS fits and the 95% pointwise SD bands are similar to each other, and to the optimal NRS fit.

Table 7.1 Values of λ, GCV, SSE and df for the NPS fits with $K = 4, 5, 6, 7, 8, 10, 14, 16, 17$ knots. The values of λ were chosen so that the associated dfs have about the same value 7. The quadratic ($k = 2$) truncated power basis was used in all these NPS fits.

K	λ	GCV/10^7	SSE/10^7	df
4	0	8.4092	8.3534	7.0000
5	99.492	8.4155	8.3597	6.9988
6	187.47	8.4182	8.3623	6.9986
7	290.45	8.4200	8.3642	7.0004
8	295.67	8.4207	8.3648	6.9977
10	395.31	8.4184	8.3625	7.0015
14	561.68	8.4204	8.3645	7.0000
16	659.86	8.4216	8.3658	6.9980
17	673.37	8.4218	8.3660	6.9974

Table 7.1 lists the values of λ, GCV, SSE and df for the NPS fits with $K = 4, 5, \cdots, 16, 17$ knots. The values of λ were chosen so that the associated dfs have about the same value 7. For the NPS fit with $K = 4$ (the optimal NRS fit), this λ-value must be 0 since by Lemma 7.1, for any $\lambda > 0$, the associated df < 7. The other λ-values are listed in the second column of the table and they increase as K increases. This is generally true since for a NRS fit, a larger K usually implies a rougher NRS fit (see Figure 7.2) and hence requires a larger λ to penalize the roughness so that the associated NPS fit becomes smoother. From the third, fourth and fifth columns, it is seen that when the df-values are about the same, the GCV-values are mainly determined by the associated SSE values, as expected.

We can also choose λ's so that the associated NPS fits with $K = 4, 5, 6, \cdots, 16, 17$ knots have the same amount of smoothing as the optimal NPS fit with $K = 16$ knots, which was presented in Figure 7.1 (c). Table 7.2 lists the associated values of λ, GCV, SSE and df for these NPS fits. It is seen that for the same K, the associated λ-value is slightly larger in Table 7.2 than that in Table 7.1. This is reasonable since in Table 7.2, the λ-values were chosen so that the associated df-values are about 6.97, while in Table 7.1, the chosen λ-values were smaller to guarantee that the associated df-values are about 7.

By the GCV rule (7.16), we can see that the NPS fit with $K = 4$ knots is the best among all the NPS fits listed in Tables 7.1 and 7.2.

We now consider a more practical problem. For a fixed K satisfying the condition (7.17), can we choose a proper λ to minimize the GCV score (7.16) such that the associated NPS fit to the ACTG 388 data is reasonably good? To answer this

Table 7.2 Values of λ, GCV, SSE and df for the NPS fits with $K = 4, 5, 6, \cdots, 16, 17$ knots. The values of λ were chosen so that the associated dfs have about the same value 6.97. The quadratic ($k = 2$) truncated power basis was used in all these NPS fits.

K	λ	GCV/10^7	SSE/10^7	df
4	1.9790	8.4090	8.3535	6.9700
5	105.91	8.4157	8.3601	6.9705
6	195.91	8.4182	8.3626	6.9714
7	301.75	8.4201	8.3645	6.9696
8	305.31	8.4207	8.3651	6.9717
10	411.02	8.4184	8.3628	6.9695
14	580.11	8.4204	8.3648	6.9717
16	680.95	8.4217	8.3660	6.9703
17	696.27	8.4218	8.3662	6.9676

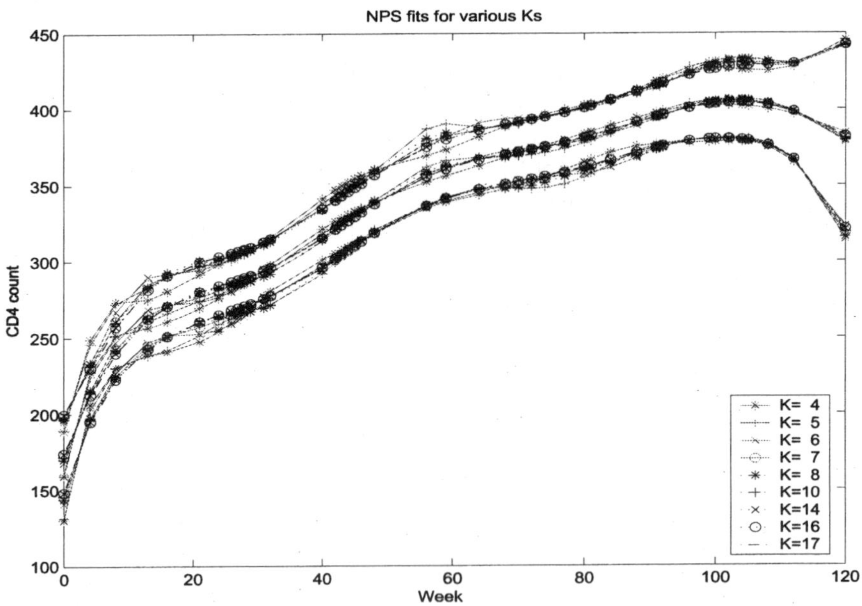

Fig. 7.4 NPS fits with $K = 4, 5, 6, \cdots, 16, 17$ knots.

Table 7.3 Values of λ, GCV, SSE and df for the NPS fits with $K = 4, 5, 6, \cdots, 16, 17$ knots. The values of λ were chosen for each fixed K so that the GCV score (7.16) was minimized. The quadratic ($k = 2$) truncated power basis was used in all these NPS fits.

K	λ	$\text{GCV}/10^7$	$\text{SSE}/10^7$	df
4	9.3314	8.4087	8.3540	6.8711
5	12.774	8.4129	8.3514	7.7181
6	97.269	8.4177	8.3590	7.3663
7	234.94	8.4199	8.3627	7.1716
8	146.56	8.4203	8.3599	7.5763
10	364.85	8.4183	8.3619	7.0697
14	561.68	8.4204	8.3645	7.0000
16	675.37	8.4217	8.3660	6.9775
17	690.03	8.4218	8.3662	6.9756

question, for each $K = 4, 5, 6, \cdots, 14, 16, 17$, we fit the ACTG 388 data using the NPS method with the smoothing parameter selected by the GCV rule (7.16). Figure 7.4 presents the NPS fits. All these NPS fits are comparable, and the NPS fits with $K = 5, 6, \cdots, 16, 17$ are just slightly different from the NPS fit with $K = 4$. Therefore, for different K's, the GCV rule will choose appropriate smoothing parameters so that the associated NPS fits are close to the optimal NPS fit. By the GCV rule (7.16) (see Table 7.3), the NPS fit with $K = 4$ is optimal.

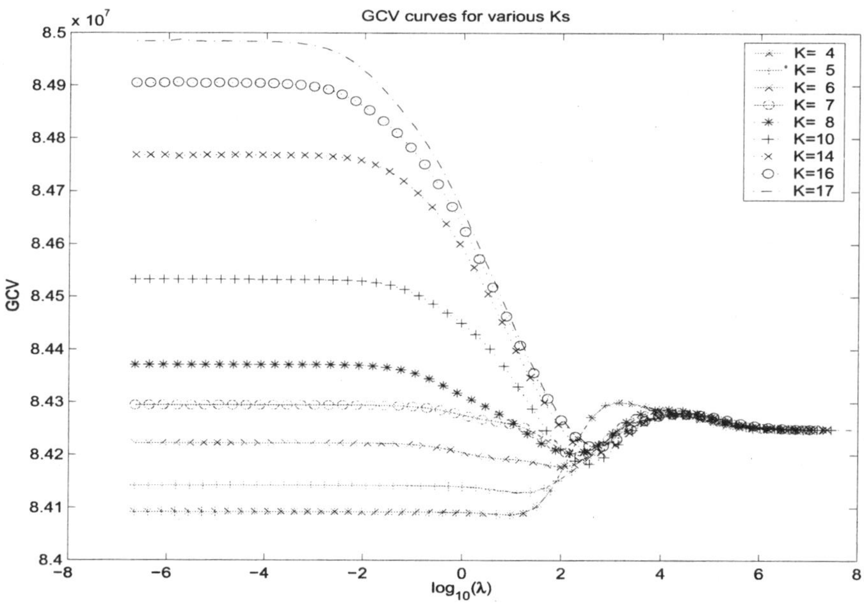

Fig. 7.5 GCV curves for the NPS fits with $K = 4, 5, 6, \cdots, 16, 17$ knots.

Figure 7.5 shows the GCV curves for the NPS fits with $K = 4, 5, 6, \cdots, 16, 17$ knots. From this figure, we can obtain the following interesting results:

(1) All the GCV curves are flat at the areas near the left and right ends. The GCV scores at the left end are different for different K but they are the same at the right end.

(2) The optimal GCV score generally increases as K increases when $K \geq 4$. This indicates that the NPS fit with $K = 4$ is optimal and it is better to choose K so that it is as close to K^* as possible when K^* is known.

Result (1) can be explained by the relationships between the k-th degree polynomial fit, the NPS fit and the NRS fit: when $\lambda \to 0$, the NPS fit tends to the NRS fit with the same truncated power basis $\mathbf{\Phi}_p(t)$, while when $\lambda \to \infty$, the NPS fit tends to the k-th degree polynomial fit. Therefore, when λ is small enough, the effect of the penalties is ignorable so that the GCV scores for the NPS fits are about the same as those for the associated NRS fits, which are different for different K. Similarly, when λ is large enough, the effect of the truncated power basis functions is ignorable so that the GCV scores for the NPS fits are about the same as those for the k-th degree polynomial fits. For all the GCV curves in Figure 7.5, $k = 2$.

In order to show that the number of knots K can not be too small, we plotted the GCV curves for the NPS fits with $K = 1, 2, 3, 4$ knots in Figure 7.6. It is seen that the optimal GCV score increases as K decreases when $K \leq 4$. Together with Figure 7.5, this figure shows that the NPS fit with $K = 4$ is the optimal fit among all the NPS fits we considered. Figure 7.7 shows that the NPS fits with $K = 1, 2, 3$ are different from the NPS fit with $K = 4$, and they oversmoothed the population mean function of the ACTG 388 data according to the GCV rule (7.16).

Table 7.4 Values of λ, GCV, SSE and df for the NPS fits with $K = 1, 2, 3, 4$ knots. The values of λ were chosen for each fixed K so that the GCV score (7.16) was minimized. The quadratic ($k = 2$) truncated power basis was used in all these NPS fits.

K	λ	GCV/10^7	SSE/10^7	df
1	717890	8.4250	8.4006	3.0500
2	2121.7	8.4237	8.3873	4.5529
3	158.55	8.4155	8.3694	5.7770
4	9.3314	8.4087	8.3540	6.8711

Table 7.4 lists the values of λ, GCV, SSE and df for the NPS fits with $K = 1, 2, 3, 4$. It is seen that the optimal GCV scores for $K = 1, 2, 3$ are larger than that for $K = 4$.

The lesson we learned from the above examples is that the number of knots K can be chosen in a rough manner; we can choose K such that it is neither too small nor too large. This is good enough to guarantee good performance of the NPS fit. When the optimal number of knots K^* for the NRS fit is known, we may take $K = K^*$ or $K^* + 1$; when K^* is unknown, it is often sufficient to take $K = [M/2]$, $[M/3]$, $[M/4]$ or $[M/5]$.

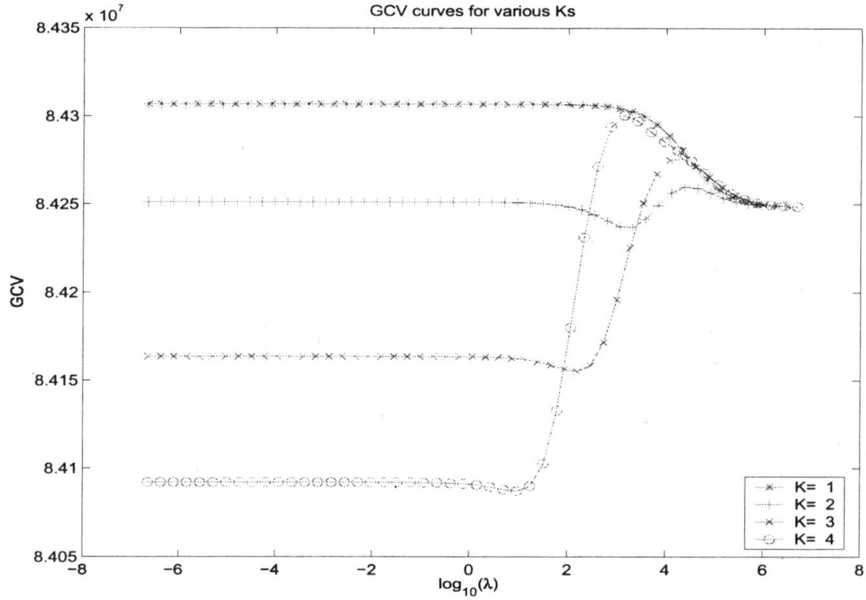

Fig. 7.6 GCV curves for the NPS fits with $K = 1, 2, 3, 4$ knots.

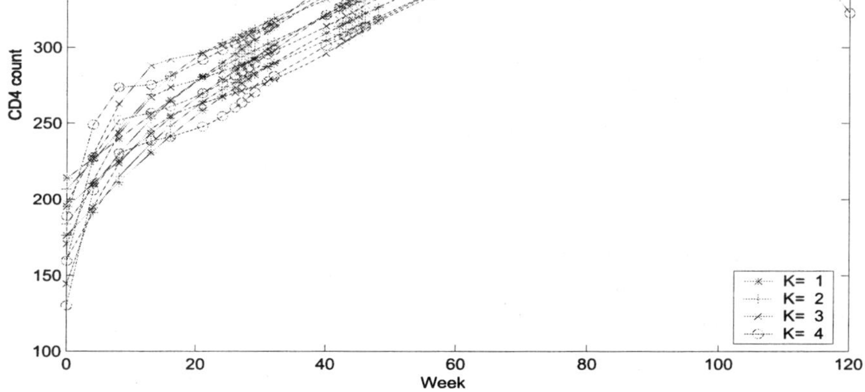

Fig. 7.7 NPS fits with $K = 1, 2, 3, 4$ knots.

7.2.7 NPS Fit as BLUP of a LME Model

We can reparameterize (7.10) so that its minimizer $\hat{\boldsymbol{\beta}}$ (7.11) can be connected to the BLUPs of a LME model. We then provide an alternative method to minimize the PLS criterion (7.10) and hence provide another way to construct the NPS smoother.

First we split the coefficient vector $\boldsymbol{\beta}$ into two parts:

$$\boldsymbol{\alpha} = [\beta_0, \beta_1, \cdots, \beta_k]^T, \ \mathbf{u} = [\beta_{k+1}, \cdots, \beta_{k+K}]^T,$$

so that the $\boldsymbol{\alpha}$ consists of all the coefficients of the $(k+1)$ polynomial basis functions, and the $\mathbf{u}$ consists of all the coefficients of the truncated power basis functions. We then denote $\tilde{\mathbf{X}}$ as the matrix consisting of the first $(k+1)$ columns of $\mathbf{X}$ and denote $\tilde{\mathbf{Z}}$ as the matrix consisting of the remaining columns. By the special structure of $\mathbf{G}$ (7.6), the PLS criterion (7.10) can be expressed as

$$\|\mathbf{y} - \tilde{\mathbf{X}}\boldsymbol{\alpha} - \tilde{\mathbf{Z}}\mathbf{u}\|^2 + \lambda \mathbf{u}^T\mathbf{u}. \tag{7.18}$$

Under the working independence assumption, i.e., the errors e_{ij} are independently and normally distributed, the minimizers of (7.18) can be connected to the BLUPs of the following LME model:

$$\begin{array}{rcl} \mathbf{y} & = & \tilde{\mathbf{X}}\boldsymbol{\alpha} + \tilde{\mathbf{Z}}\mathbf{u} + \mathbf{e}, \\ \mathbf{u} & \sim & N(0, \tau^2\mathbf{I}_K), \ \mathbf{e} \sim N(0, \sigma^2\mathbf{I}_N), \end{array} \tag{7.19}$$

where $\tau^2 = \sigma^2/\lambda$ and $\mathbf{e} = [\mathbf{e}_1^T, \cdots, \mathbf{e}_n^T]^T$ with $\mathbf{e}_i = [e_{i1}, \cdots, e_{in_i}]^T$. It is seen that the smoothing parameter λ can be interpreted as the ratio of the two variance components of the above LME model, i.e.,

$$\lambda = \sigma^2/\tau^2. \tag{7.20}$$

The LME model (7.19) can be solved using the S-PLUS function *lme* or the SAS procedure *PROC MIXED* to obtain the BLUPs $\hat{\boldsymbol{\alpha}}, \hat{\mathbf{u}}$ and the variance component estimates $\hat{\tau}^2, \hat{\sigma}^2$. Then the minimizers of (7.10) can be obtained as $\hat{\boldsymbol{\beta}} = [\hat{\boldsymbol{\alpha}}^T, \hat{\mathbf{u}}^T]^T$ so that the NPS fit (7.12) can be expressed as

$$\hat{\eta}_{nps}(t) = \boldsymbol{\Phi}_p(t)^T\hat{\boldsymbol{\beta}} = \sum_{r=0}^{k} \hat{\alpha}_r t^r + \sum_{l=1}^{K} \hat{u}_l (t - \tau_l)_+^k.$$

The fits of $\eta(t)$ at all the design time points can be summarized as $\hat{\mathbf{y}} = \tilde{\mathbf{X}}\hat{\boldsymbol{\alpha}} + \tilde{\mathbf{Z}}\hat{\mathbf{u}}$, the same expression as the one presented in (7.13). The standard deviations of $\hat{\mathbf{y}}$ can also be obtained by using the covariance matrices of $\hat{\boldsymbol{\alpha}}$ and $\hat{\mathbf{u}}$. By (7.20), the proper smoothing parameter obtained from solving the LME model is $\hat{\lambda} = \hat{\sigma}^2/\hat{\tau}^2$.

7.3 GENERALIZED P-SPLINES

The NPS smoother (7.12) was constructed under a working independence assumption. When a parametric model for the covariance function of the errors in the NPM model (7.2) can be specified, a generalized P-spline (GPS) smoother can be constructed. It is expected that the GPS smoother will outperform the NPS smoother since the GPS smoother accounts for the within-subject correlation while the NPS smoother does not.

7.3.1 Constructing the GPS Smoother

The GPS smoother of $\eta(t)$ is defined as follows. Using the notations, e.g., $\mathbf{y}, \mathbf{X}, \beta, \mathbf{G}$, specified in Section 7.2, we can write the penalized generalized least squares (PGLS) criterion as

$$(\mathbf{y} - \mathbf{X}\beta)^T\mathbf{V}^{-1}(\mathbf{y} - \mathbf{X}\beta) + \lambda\beta^T\mathbf{G}\beta, \tag{7.21}$$

where $\mathbf{V} = \text{Cov}(\mathbf{y}) = \text{diag}(\mathbf{V}_1, \cdots, \mathbf{V}_n)$ with $\mathbf{V}_i = \text{Cov}(\mathbf{y}_i)$. The first term in the above expression is the sum of squared errors, weighted by the covariance matrix of $\mathbf{y}$, representing the goodness of fit, and the second term is the roughness term, the same as the one in the PLS criterion (7.8) for the NPS smoother.

When $\mathbf{V}$ is known, minimizing (7.21) with respect to β leads to

$$\hat{\beta} = (\mathbf{X}^T\mathbf{V}^{-1}\mathbf{X} + \lambda\mathbf{G})^{-1}\mathbf{X}^T\mathbf{V}^{-1}\mathbf{y}. \tag{7.22}$$

The associated GPS smoother of $\eta(t)$ is then

$$\hat{\eta}_{gps}(t) = \mathbf{\Phi}_p(t)^T(\mathbf{X}^T\mathbf{V}^{-1}\mathbf{X} + \lambda\mathbf{G})^{-1}\mathbf{X}^T\mathbf{V}^{-1}\mathbf{y}, \tag{7.23}$$

where the basis $\mathbf{\Phi}_p(t)$ is given in (7.3). The $\hat{\eta}_{gps}(t)$ evaluated at all the design time points defined in (7.1) forms a fitted response vector:

$$\hat{\mathbf{y}}_{gps} = \mathbf{X}(\mathbf{X}^T\mathbf{V}^{-1}\mathbf{X} + \lambda\mathbf{G})^{-1}\mathbf{X}^T\mathbf{V}^{-1}\mathbf{y} \equiv \mathbf{A}_{gps}\mathbf{y}, \tag{7.24}$$

where $\mathbf{A}_{nps}$ is the associated GPS smoother matrix.

7.3.2 Degrees of Freedom

Similar to the NPS smoother (7.12), the number of degrees of freedom (df) for the GPS smoother (7.23) is usually defined as the trace of the GPS smoother matrix in (7.24), representing the model complexity of the GPS smoother. That is,

$$\text{df}(\lambda) = \text{tr}(\mathbf{A}_{gps}) = \text{tr}\left[(\mathbf{X}^T\mathbf{V}^{-1}\mathbf{X} + \lambda\mathbf{G})^{-1}\mathbf{X}^T\mathbf{V}^{-1}\mathbf{X}\right]. \tag{7.25}$$

We can show that for any fixed $\mathbf{V}$, the df(λ) defined above has the same properties as those listed in Lemma 7.1 for the df(λ) defined in (7.15). Therefore, as λ increases, the GPS model becomes simpler, with the simplest GPS model reducing to the k-th degree polynomial model when $\lambda \to \infty$.

7.3.3 Variability Band Construction

The pointwise SD band for $\eta(t)$ based on the GPS smoother (7.23) can be constructed as follows. Since Cov($\mathbf{y}$) $= \mathbf{V}$, by (7.23), we have

$$\text{Var}(\hat{\eta}_{gps}(t)) = \mathbf{\Phi}_p(t)^T(\mathbf{X}^T\mathbf{V}^{-1}\mathbf{X}+\lambda\mathbf{G})^{-1}\mathbf{X}^T\mathbf{V}^{-1}\mathbf{X}(\mathbf{X}^T\mathbf{V}^{-1}\mathbf{X}+\lambda\mathbf{G})^{-1}\mathbf{\Phi}_p(t).$$

Let $\widehat{\text{Var}}(\hat{\eta}_{gps}(t))$ denote the above variance after the $\mathbf{V}$ is replaced by its estimator. Then the 95% pointwise SD band for $\eta(t)$ can be approximately constructed as $\hat{\eta}_{gps}(t) \pm 1.96\widehat{\text{Var}}^{1/2}(\hat{\eta}_{gps}(t))$.

7.3.4 Smoothing Parameter Selection

The GCV rule (7.16) can be adapted for the GPS smoother (7.23):

$$\text{GCV}_{gps}(\lambda) = \frac{(\mathbf{y} - \hat{\mathbf{y}}_{gps})^T\mathbf{W}(\mathbf{y} - \hat{\mathbf{y}}_{gps})}{[1 - \text{tr}(\mathbf{A}_{gps})/N]^2}, \tag{7.26}$$

but with $\mathbf{W} = \mathbf{V}^{-1}$ if $\mathbf{V}$ is known. Similar to the GCV rule (7.16) for the NPS smoother, minimizing the above GCV rule (7.26) is equivalent to choosing a λ to trade-off the goodness of fit with the model complexity of the GPS smoother. In addition, we can incorporate K into (7.26) as an additional smoothing parameter if we want to select K as well.

7.3.5 Choice of the Number of Knots

Since Lemma 7.1 applies to the df (7.25) defined for the GPS smoother (7.23), choosing the number of knots K for the GPS smoother is similar to choosing K for the NPS smoother (7.12). That is K should be chosen such that it satisfies (7.17). It is often sufficient to initially take $K = [M/2]$, $[M/3]$, $[M/4]$ or $[M/5]$ where M is the number of distinct design time points.

7.3.6 GPS Fit as BLUP of a LME Model

It is easy to transform the parameters and covariates in the PGLS criterion (7.21) so that the associated minimizer $\hat{\boldsymbol{\beta}}$ (7.22) is connected to the BLUPs of a LME model. To this end, we just need to let $\tilde{\mathbf{y}} = \mathbf{V}^{-1/2}\mathbf{y}, \tilde{\mathbf{X}} = \mathbf{V}^{-1/2}\mathbf{X}^*$, and $\tilde{\mathbf{Z}} = \mathbf{V}^{-1/2}\mathbf{Z}^*$ where $\mathbf{X}^*$ consists of the first $(k+1)$ columns and $\mathbf{Z}^*$ the last K columns of $\mathbf{X}$, and to split the $\boldsymbol{\beta}$ into:

$$\boldsymbol{\alpha} = [\beta_0, \beta_1, \cdots, \beta_k]^T, \ \mathbf{u} = [\beta_{k+1}, \cdots, \beta_{k+K}]^T,$$

so that the PGLS criterion (7.21) becomes

$$\|\tilde{\mathbf{y}} - \tilde{\mathbf{X}}\boldsymbol{\alpha} - \tilde{\mathbf{Z}}\mathbf{u}\|^2 + \lambda\mathbf{u}^T\mathbf{u},$$

which is exactly the same as the PLS criterion (7.18). The remaining task runs exactly the same as that described in Section 7.2.7.

7.3.7 Estimating the Covariance Structure

In the above subsections, the covariance matrix $\mathbf{V}$ is assumed to be known. In practice, it has to be estimated. For the GPS smoother, it is often assumed that $\mathbf{V}$ follows a known parametric model with some unknown parameter vector $\boldsymbol{\theta}$. The parameter vector $\boldsymbol{\theta}$ can be estimated using the maximum likelihood method or restricted maximum likelihood method. See Section 5.3.4 of Chapter 5 for some details. See Wang and Taylor (1996) and Diggle et al. (2002) for some widely-used parametric models for $\mathbf{V}$.

7.4 EXTENDED P-SPLINES

The NPS and GPS smoothers discussed in the previous sections can be used to fit the population mean function only. In this section we consider a technique to estimate both the population mean and individual functions using P-splines as described in Ruppert et al. (2003), Chapter 9. We refer to the resulting smoother as an extended P-spline (EPS) smoother.

7.4.1 Subject-Specific Curve Fitting

For the longitudinal data (7.1), the subject-specific curve fitting (SSCF) model can be written as

$$y_{ij} = \eta(t_{ij}) + v_i(t_{ij}) + \epsilon_{ij}, \; j = 1, 2, \cdots, n_i; \; i = 1, 2, \cdots, n, \tag{7.27}$$

where $\eta(t)$ is the population mean function, and $v_i(t), \; i = 1, 2, \cdots, n$ are the subject-specific deviation functions. In the above SSCF model, we assume that $\eta(t)$ and $v_i(t)$ are smooth but not random. This means that only the smoothness of $\eta(t)$ and $v_i(t), \; i = 1, 2, \cdots, n$ will be taken into account when we fit the SSCF model (7.27). To reflect the fact that $\eta(t)$ is the population mean function, we also need to assume that all the $v_i(t), \; i = 1, 2, \cdots, n$ sum up to 0, i.e., the following identifiability condition:

$$\sum_{i=1}^{n} v_i(t) = 0. \tag{7.28}$$

In addition, we assume $\boldsymbol{\epsilon}_i = [\epsilon_{i1}, \cdots, \epsilon_{in_i}]^T \sim N(0, \mathbf{R}_i)$ and for simplicity, we assume

$$\mathbf{R}_i = \sigma^2 \mathbf{I}_{n_i}, \; i = 1, 2, \cdots, n. \tag{7.29}$$

The model (7.27) may be obtained via splitting the errors e_{ij} of the NPM model (7.2) into $v_i(t_{ij}) + \epsilon_{ij}$ so that we can fit not only $\eta(t)$ but also $v_i(t), \; i = 1, 2, \cdots, n$. This then allows obtaining the estimates of the individual functions $s_i(t) = \eta(t) + v_i(t), \; i = 1, 2, \cdots, n$.

Similar to the NPS and GPS smoothers of $\eta(t)$ introduced in Sections 7.2 and 7.3, we employ a k-th degree truncated power basis $\boldsymbol{\Phi}_p(t)$ (7.3) with K knots $\tau_1, \tau_2, \cdots, \tau_K$ to construct the EPS estimator for $\eta(t)$, and employ another, say, a k_v-th degree truncated power basis $\Psi_q(t)$ with K_v knots $\delta_1, \cdots, \delta_{K_v}$, to construct the EPS estimators for $v_i(t),\ i = 1, 2, \cdots, n$. Then approximately, we can write $\eta(t)$ and $v_i(t),\ i = 1, 2, \cdots, n$ as regression splines:

$$\begin{aligned} \eta(t) &= \Phi_p(t)^T\beta = \sum_{r=0}^k \beta_r t^r + \sum_{l=1}^K \beta_{k+l}(t - \tau_l)_+^k, \\ v_i(t) &= \Psi_q(t)^T\mathbf{b}_i = \sum_{r=0}^{k_v} b_{ir} t^r + \sum_{l=1}^{K_v} b_{i(k+l)}(t - \delta_l)_+^{k_v}, \end{aligned} \tag{7.30}$$

where

$$\begin{aligned} \beta &= [\beta_0, \beta_1, \cdots, \beta_k, \beta_{k+1}, \cdots, \beta_{k+K}]^T, \\ \mathbf{b}_i &= [b_{i0}, b_{i1}, \cdots, b_{ik_v}, b_{i(k_v+1)}, \cdots, b_{i(k_v+K_v)}]^T, \quad i = 1, 2, \cdots, n, \end{aligned}$$

are the associated coefficient vectors.

Notice that $\beta_{k+1}, \cdots, \beta_{k+K}$ quantify the k-times derivative jumps of the regression spline $\eta(t)$ at the knots $\tau_1, \cdots, \tau_K$ and $b_{i(k_v+1)}, \cdots, b_{i(k_v+K_v)}$ quantify the k_v-times derivative jumps of the regression spline $v_i(t)$ at the knots $\delta_1, \cdots, \delta_{K_v}$. When these jumps are large, the associated regression splines are rough. Following the previous sections, in order to obtain the P-spline estimators of $\eta(t)$ and $v_i(t),\ i = 1, 2, \cdots, n$, we can impose penalties to these jumps via using the following constrained PLS criterion:

$$\begin{aligned} &\sum_{i=1}^n (\mathbf{y}_i - \mathbf{X}_i\beta - \mathbf{Z}_i\mathbf{b}_i)^T\mathbf{R}_i^{-1}(\mathbf{y}_i - \mathbf{X}_i\beta - \mathbf{Z}_i\mathbf{b}_i) \\ &\quad + \lambda_v \sum_{i=1}^n \mathbf{b}_i^T\mathbf{G}_v\mathbf{b}_i + \lambda\beta^T\mathbf{G}\beta, \\ &\quad \text{Subject to } \sum_{i=1}^n \mathbf{b}_i = 0, \end{aligned} \tag{7.31}$$

where $\mathbf{X}_i$ and $\mathbf{Z}_i$ are the covariate matrices and $\mathbf{G}$ and $\mathbf{G}_v$ are the roughness matrices. The matrix $\mathbf{X}_i$ has been defined based on the truncated power basis $\Phi_p(t)$ as in the previous sections, and similarly we can define $\mathbf{Z}_i = [\mathbf{z}_{i1}, \cdots, \mathbf{z}_{in_i}]^T$ with $\mathbf{z}_{ij} = \Psi_q(t_{ij})$. The roughness matrix $\mathbf{G}$ is defined as in (7.6), and the roughness matrix $\mathbf{G}_v$ can be similarly defined as

$$\mathbf{G}_v = \text{diag}(0, \cdots, 0, 1, \cdots, 1), \tag{7.32}$$

a diagonal matrix with the first $k_v + 1$ diagonal entries as 0 and the remaining K_v diagonal entries as 1.

Notice that the first term in the PLS criterion (7.31) is the weighted sum of squared errors (SSE), representing the goodness of fit of the truncated power basis approximations of the individual curves $s_i(t),\ i = 1, 2, \cdots, n$; the second term excluding the smoothing parameter λ_v equals $\sum_{i=1}^n \sum_{l=1}^{K_v} b_{i(k_v+l)}^2$, representing the roughness of the subject-specific deviation functions $v_i(t),\ i = 1, 2, \cdots, n$; the third term excluding the smoothing parameter λ equals $\sum_{l=1}^K \beta_{k+l}^2$, representing the roughness of the population mean function $\eta(t)$; and the last term is the identifiability condition for $\mathbf{b}_i,\ i = 1, 2, \cdots, n$, which is derived from the identifiability condition (7.28) for $v_i(t),\ i = 1, 2, \cdots, n$. The smoothing parameters λ and λ_v are used to trade-off the goodness of fit and the roughness of $\eta(t)$ and $v_i(t),\ i = 1, 2, \cdots, n$.

7.4.2 Challenges for Computing the EPS Smoothers

In a matrix form, the constrained PLS criterion (7.31) can be rewritten as

$$\begin{gathered}(\mathbf{y}-\mathbf{X}\beta-\mathbf{Z}\mathbf{b})^T\mathbf{R}^{-1}(\mathbf{y}-\mathbf{X}\beta-\mathbf{Z}\mathbf{b})+\lambda_v\mathbf{b}^T\tilde{\mathbf{G}}_v\mathbf{b}+\lambda\beta^T\mathbf{G}\beta,\\ \text{Subject to }\mathbf{H}\mathbf{b}=0,\end{gathered} \tag{7.33}$$

where $\mathbf{H}=[\mathbf{I}_q,\mathbf{I}_q,\cdots,\mathbf{I}_q]$ so that $\mathbf{H}\mathbf{b}=0$ is equivalent to $\sum_{i=1}^n \mathbf{b}_i=0$; $\mathbf{y}$ and $\mathbf{X}$ are defined in the previous sections; and similarly we can define $\mathbf{Z},\mathbf{b},\ \tilde{\mathbf{G}}_v,\epsilon$ and $\mathbf{R}$ as below:

$$\begin{array}{rclrcl}\mathbf{Z} & = & \text{diag}(\mathbf{Z}_1,\cdots,\mathbf{Z}_n), & \mathbf{b} & = & [\mathbf{b}_1^T,\cdots,\mathbf{b}_n^T]^T,\\ \tilde{\mathbf{G}}_v & = & \text{diag}(\mathbf{G}_v,\cdots,\mathbf{G}_v), & \epsilon & = & [\epsilon_1^T,\cdots,\epsilon_n^T]^T,\\ \mathbf{R} & = & \text{diag}(\mathbf{R}_1,\cdots,\mathbf{R}_n). & & & \end{array} \tag{7.34}$$

Then the minimizers of the constrained PLS criterion (7.33) are the solutions to the following linear equation system:

$$\begin{pmatrix}\mathbf{X}^T\mathbf{R}^{-1}\mathbf{X}+\lambda\mathbf{G} & \mathbf{X}^T\mathbf{R}^{-1}\mathbf{Z}\\ \mathbf{Z}^T\mathbf{R}^{-1}\mathbf{X} & \mathbf{Z}^T\mathbf{R}^{-1}\mathbf{Z}+\lambda_v\tilde{\mathbf{G}}_v\\ \mathbf{0} & \mathbf{H}\end{pmatrix}\begin{pmatrix}\beta\\ \mathbf{b}\end{pmatrix}=\begin{pmatrix}\mathbf{X}^T\mathbf{R}^{-1}\mathbf{y}\\ \mathbf{Z}^T\mathbf{R}^{-1}\mathbf{y}\\ \mathbf{0}\end{pmatrix}. \tag{7.35}$$

It is not easy to solve the above linear equation system since it has $[p+(n+1)q]$ equations with $(p+nq)$ parameters. The numbers of equations and parameters increase as the number of subjects, n, increases. We can not take advantages of the block diagonal property of the matrix $\mathbf{Z},\mathbf{R}$ and $\tilde{\mathbf{G}}_v$ to simplify the above equation system. The matrix $\mathbf{Z}^T\mathbf{R}^{-1}\mathbf{Z}+\lambda_v\tilde{\mathbf{G}}_v$ is often not of full rank especially when there are some subjects which have only one or two measurements. Thus, we can not simplify the above linear equation system using $\mathbf{b}=(\mathbf{Z}^T\mathbf{R}^{-1}\mathbf{Z}+\lambda_v\tilde{\mathbf{G}}_v)^{-1}(\mathbf{y}-\mathbf{X}\beta)$ as we usually do.

When $\mathbf{X}$ is of full rank, the matrix $\mathbf{X}^T\mathbf{R}^{-1}\mathbf{X}+\lambda\mathbf{G}$ is of full rank. Then using the first p equations in (7.35), we can express β as $(\mathbf{X}^T\mathbf{R}^{-1}\mathbf{X}+\lambda\mathbf{G})^{-1}\mathbf{X}^T\mathbf{R}^{-1}(\mathbf{y}-\mathbf{Z}\mathbf{b})$. Let $\mathbf{A}=\mathbf{X}(\mathbf{X}^T\mathbf{R}^{-1}\mathbf{X}+\lambda\mathbf{G})^{-1}\mathbf{X}^T\mathbf{R}^{-1}$. Then we can solve the following linear equation system for $\mathbf{b}$:

$$\begin{array}{rcl}[\mathbf{Z}^T\mathbf{R}^{-1}(\mathbf{I}_N-\mathbf{A})\mathbf{Z}+\lambda_v\tilde{\mathbf{G}}_v]\mathbf{b} & = & \mathbf{Z}^T\mathbf{R}^{-1}(\mathbf{I}_N-\mathbf{A})\mathbf{y},\\ \mathbf{H}\mathbf{b} & = & \mathbf{0}.\end{array}$$

However, the above linear equation system has $[(n+1)q]$ equations with (nq) parameters, and the coefficient matrix, $\mathbf{Z}^T\mathbf{R}^{-1}(\mathbf{I}_N-\mathbf{A})\mathbf{Z}+\lambda_v\tilde{\mathbf{G}}_v$, is not of full rank and diagonal. Therefore, it is still a challenge to solve the above equation system.

7.4.3 EPS Fits as BLUPs of a LME Model

The linear equation system (7.35) may be solved via connecting the constrained PLS criterion (7.33) to a constrained LME model which may be solved using the S-PLUS function *lme* or the SAS procedure *PROC MIXED.*

We start from (7.33) by splitting β and $\mathbf{b}_i$, $i = 1, 2, \cdots, n$. Let $\boldsymbol{\alpha}_0$ and $\mathbf{u}_0$ denote the vectors consisting of the first $(k+1)$ and the remaining K entries of β, respectively. At the same time, let $\boldsymbol{\alpha}_i$ and $\mathbf{u}_i$ denote the vectors consisting of the first (k_v+1) and the remaining K_v entries of $\mathbf{b}_i$. Moreover, let $\tilde{\mathbf{X}}_{i0}$ and $\tilde{\mathbf{Z}}_{i0}$ denote the matrices consisting of the first $(k+1)$ and the remaining K columns of $\mathbf{X}_i$, and let $\tilde{\mathbf{X}}_{i1}$ and $\tilde{\mathbf{Z}}_{i1}$ denote the first (k_v+1) and the remaining K_v columns of $\mathbf{Z}_i$. Then by letting

$$\boldsymbol{\Delta}_i = \mathbf{y}_i - (\tilde{\mathbf{X}}_{i0}\boldsymbol{\alpha}_0 + \tilde{\mathbf{Z}}_{i0}\mathbf{u}_0) - (\tilde{\mathbf{X}}_{i1}\boldsymbol{\alpha}_i + \tilde{\mathbf{Z}}_{i1}\mathbf{u}_i),$$

we can write the constrained PLS criterion (7.33) as follows:

$$\begin{array}{c} \sum_{i=1}^n \boldsymbol{\Delta}_i^T \mathbf{R}_i^{-1} \boldsymbol{\Delta}_i + \lambda_v \sum_{i=1}^n \mathbf{u}_i^T \mathbf{u}_i + \lambda \mathbf{u}_0^T \mathbf{u}_0, \\ \text{Subject to } \sum_{i=1}^n \boldsymbol{\alpha}_i = 0, \ \ \sum_{i=1}^n \mathbf{u}_i = 0. \end{array} \tag{7.36}$$

Combining the parameter vectors, we get

$$\boldsymbol{\alpha} = [\boldsymbol{\alpha}_0^T, \boldsymbol{\alpha}_1^T, \cdots, \boldsymbol{\alpha}_n^T]^T, \quad \mathbf{u} = [\mathbf{u}_0^T, \mathbf{u}_1^T, \cdots, \mathbf{u}_n^T]^T,$$

and

$$\mathbf{D} = \text{diag}[\lambda^{-1}\mathbf{I}_K, \lambda_v^{-1}\mathbf{I}_{K_v}, \cdots, \lambda_v^{-1}\mathbf{I}_{K_v}].$$

Then the constrained PLS criterion (7.36) can be further written as

$$\begin{array}{c} (\mathbf{y} - \tilde{\mathbf{X}}\boldsymbol{\alpha} - \tilde{\mathbf{Z}}\mathbf{u})^T \mathbf{R}^{-1} (\mathbf{y} - \tilde{\mathbf{X}}\boldsymbol{\alpha} - \tilde{\mathbf{Z}}\mathbf{u}) + \mathbf{u}^T \mathbf{D}^{-1} \mathbf{u}, \\ \text{Subject to } \mathbf{H}_0\boldsymbol{\alpha} = 0, \ \ \mathbf{H}_1\mathbf{u} = 0, \end{array} \tag{7.37}$$

where $\mathbf{H}_0 = [\mathbf{0}_{(k+1)\times(k+1)}, \mathbf{I}_{k_v+1}, \cdots, \mathbf{I}_{k_v+1}]$, $\mathbf{H}_1 = [\mathbf{0}_{K\times K}, \mathbf{I}_{K_v}, \cdots, \mathbf{I}_{K_v}]$; $\tilde{\mathbf{X}}$ is defined as

$$\tilde{\mathbf{X}} = \begin{pmatrix} \tilde{\mathbf{X}}_{10}, & \tilde{\mathbf{X}}_{11}, & \mathbf{0}, & \cdots, & \mathbf{0}, & \mathbf{0} \\ \tilde{\mathbf{X}}_{20}, & \mathbf{0}, & \tilde{\mathbf{X}}_{21}, & \cdots, & \mathbf{0}, & \mathbf{0} \\ \vdots, & \vdots, & \vdots, & \vdots, & \vdots, & \vdots \\ \tilde{\mathbf{X}}_{(n-1)0}, & \mathbf{0}, & \mathbf{0}, & \cdots, & \tilde{\mathbf{X}}_{(n-1)1}, & \mathbf{0} \\ \tilde{\mathbf{X}}_{n0}, & \mathbf{0}, & \mathbf{0}, & \cdots, & \mathbf{0}, & \tilde{\mathbf{X}}_{n1} \end{pmatrix},$$

and $\tilde{\mathbf{Z}}$ is defined similarly but by replacing all the $\tilde{\mathbf{X}}_{i0}$ and $\tilde{\mathbf{X}}_{i1}$ with $\tilde{\mathbf{Z}}_{i0}$ and $\tilde{\mathbf{Z}}_{i1}$, respectively in the above definition. We then connect the constrained PLS criterion (7.37) to the following constrained LME model:

$$\begin{array}{c} \mathbf{y} = \tilde{\mathbf{X}}\boldsymbol{\alpha} + \tilde{\mathbf{Z}}\mathbf{u} + \boldsymbol{\epsilon}, \\ \mathbf{u} \sim N(0, \mathbf{D}), \ \ \boldsymbol{\epsilon} \sim N(0, \mathbf{R}), \\ \text{Subject to } \mathbf{H}_0\boldsymbol{\alpha} = 0, \ \ \mathbf{H}_1\mathbf{u} = 0, \end{array} \tag{7.38}$$

in the sense that the minimizers $\hat{\boldsymbol{\alpha}}$ and $\hat{\mathbf{u}}$ of the constrained PLS criterion (7.37) are the BLUPs of the above constrained LME model. Notice that in the constrained LME model (7.38), the identifiability condition $\mathbf{H}_1\mathbf{u} = \sum_{i=1}^n \mathbf{u}_i = 0$ may be dropped since the model assumes that $\mathbf{u}_i$, $i = 1, 2, \cdots, n$ are independent and $\text{E}\mathbf{u}_i = 0$, $i =$

$1, 2, \cdots, n$ so that $\sum_{i=1}^n \mathbf{u}_i = 0$ will be automatically and approximately satisfied when n is large.

It is still a challenge to fit the constrained LME model (7.38) since it is a constrained LME model with $[(k+1)+n(k_v+1)]$ fixed-effects parameters. Under the structure (7.29), there are three free variance components σ^2, $\sigma_\eta^2 = \lambda^{-1}$, and $\sigma_v^2 = \lambda_v^{-1}$. Although $\mathbf{D}$ is diagonal, the covariance matrix $\mathbf{V} = \tilde{\mathbf{Z}}\mathbf{D}\tilde{\mathbf{Z}}^T + \mathbf{R}$ may not be block diagonal since $\tilde{\mathbf{Z}}$ is not block diagonal. The matrix $\mathbf{V}$ is of size $N \times N$ where $N = \sum_{i=1}^n n_i$. It is generally difficult to solve (7.38) when N is large.

We may fit the constrained LME model (7.38) using the S-PLUS function *lme* or the SAS procedure *PROC MIXED*. Denote the resulting BLUPs and variance components as $\hat{\alpha}_i, \hat{\mathbf{u}}_i, i = 0, 1, \cdots, n$, and $\hat{\sigma}^2, \hat{\sigma}_\eta^2$ and $\hat{\sigma}_v^2$. Then the minimizers of the constrained PLS criterion (7.37) can be expressed as

$$\hat{\beta} = [\hat{\alpha}_0^T, \hat{\mathbf{u}}_0^T]^T, \quad \hat{\mathbf{b}}_i = [\hat{\alpha}_i^T, \hat{\mathbf{u}}_i^T]^T, \; i = 1, 2, \cdots, n.$$

The fits of $\eta(t)$ and $v_i(t), \; i = 1, 2, \cdots, n$ can be expressed as $\hat{\eta}(t) = \mathbf{\Phi}_p(t)^T \hat{\beta}$ and $\hat{v}_i(t) = \mathbf{\Psi}_q(t)^T \hat{\mathbf{b}}_i, \; i = 1, 2, \cdots, n.$ Furthermore, the associated smoothing parameters can be determined as $\lambda = 1/\hat{\sigma}_\eta^2$ and $\lambda_v = 1/\hat{\sigma}_v^2$.

7.5 MIXED-EFFECTS P-SPLINES

In the subject-specific curve fitting (SSCF) model (7.27), the subject-specific deviation functions $v_i(t), \; i = 1, 2, \cdots, n$ were assumed to be nonrandom. Since in longitudinal studies, these deviation functions are different from one subject to another, and thus it is more natural to assume that they are random. As in Chapters 4, 5 and 6, we can assume that $v_i(t), \; i = 1, 2, \cdots, n$ are i.i.d realizations of an underlying stochastic process. In particular, we may assume $v_i(t), \; i = 1, 2, \cdots, n$ follow a Gaussian stochastic process with mean 0 and covariance function $\gamma(s, t)$. That is, $v_i(t) \sim \text{GP}(0, \gamma)$. The resulting model is known as a nonparametric mixed-effects (NPME) model:

$$\begin{gathered} y_{ij} = \eta(t_{ij}) + v_i(t_{ij}) + \epsilon_{ij}, \\ v_i(t) \sim \text{GP}(0, \gamma), \quad \boldsymbol{\epsilon}_i = [\epsilon_{i1}, \cdots, \epsilon_{in_i}]^T \sim N(0, \mathbf{R}_i), \\ j = 1, 2, \cdots, n_i; \; i = 1, 2, \cdots, n. \end{gathered} \tag{7.39}$$

In the SSCF model (7.27), $\eta(t)$ is called a smooth population mean function, and $v_i(t), \; i = 1, 2, \cdots, n$ are called smooth subject-specific deviation functions, with an identifiability condition $\sum_{i=1}^n v_i(t) = 0$. In the NPME model (7.39), we call $\eta(t)$ and $v_i(t), \; i = 1, 2, \cdots, n$ as smooth fixed-effect and random-effect functions, respectively. Since we assume $\text{E}[v_i(t)] = 0, \; i = 1, 2, \cdots, n$, we no longer need the identifiability condition.

In Chapter 6, we employed a smoothing spline based-technique to fit the NPME model (7.39). In this section, we shall develop a P-spline based technique to fit this model. For simplicity, we refer to this technique as a mixed-effects P-spline (MEPS) method, and the resulting smoothers as MEPS smoothers.

7.5.1 The MEPS Smoothers

The only difference between the NPME model (7.39) and the SSCF model (7.27) is that in the NPME model, $v_i(t),\ i = 1, 2, \cdots, n$ are random. Therefore, the MEPS smoothers of $\eta(t)$ and $v_i(t),\ i = 1, 2, \cdots, n$ can be constructed via additionally accounting for the randomness of $v_i(t),\ i = 1, 2, \cdots, n$. That is, we can use the two truncated power bases $\mathbf{\Phi}_p(t)$ and $\mathbf{\Psi}_q(t)$ in the EPS smoothers so that we will have the approximations (7.30) for $\eta(t)$ and $v_i(t),\ i = 1, 2, \cdots, n$. Then we can re-write (7.39) as

$$\begin{gathered}\mathbf{y}_i = \mathbf{X}_i\boldsymbol{\beta} + \mathbf{Z}_i\mathbf{b}_i + \boldsymbol{\epsilon}_i,\\ \mathbf{b}_i \sim N(0, \mathbf{D}),\quad \boldsymbol{\epsilon}_i \sim N(0, \mathbf{R}_i),\\ i = 1, 2, \cdots, n,\end{gathered} \tag{7.40}$$

where $\mathbf{y}_i, \mathbf{X}_i, \mathbf{Z}_i, \boldsymbol{\beta}$ and $\mathbf{b}_i$ are defined as in the previous section for the EPS model, and $\mathbf{D}$ denotes the common covariance matrix of $\mathbf{b}_i,\ i = 1, 2, \cdots, n$, i.e., $\mathbf{D} = \text{Cov}(\mathbf{b}_i) = \text{E}(\mathbf{b}_i\mathbf{b}_i^T)$. We assume that $\mathbf{D}$ is invertible and hence positive definite. Based on (7.30), the relationship between $\mathbf{D}$ and $\gamma(s, t)$ is

$$\gamma(s, t) \approx \mathbf{\Psi}_q(s)^T\mathbf{D}\mathbf{\Psi}_q(t). \tag{7.41}$$

To account for the randomness of the random-effect functions, we define the following penalized generalized log-likelihood (PGLL) criterion:

$$\begin{gathered}\textstyle\sum_{i=1}^n \left\{(\mathbf{y}_i - \mathbf{X}_i\boldsymbol{\beta} - \mathbf{Z}_i\mathbf{b}_i)^T\mathbf{R}_i^{-1}(\mathbf{y}_i - \mathbf{X}_i\boldsymbol{\beta} - \mathbf{Z}_i\mathbf{b}_i) + \mathbf{b}_i^T\mathbf{D}^{-1}\mathbf{b}_i\right\}\\ + \lambda_v \textstyle\sum_{i=1}^n \mathbf{b}_i^T\mathbf{G}_v\mathbf{b}_i + \lambda\boldsymbol{\beta}^T\mathbf{G}\boldsymbol{\beta}.\end{gathered} \tag{7.42}$$

In the above expression, the first summation term is the twice negative logarithm of the generalized likelihood of $(\boldsymbol{\beta}, \mathbf{b}_i,\ i = 1, 2, \cdots, n)$(the joint probability density function of the responses $\mathbf{y}_i,\ i = 1, 2, \cdots, n$ and the random-effect vectors $\mathbf{b}_i,\ i = 1, 2, \cdots, n$) (up to a constant); the second term is the sum of the roughness of the random-effect functions $v_i(t),\ i = 1, 2, \cdots, n$ multiplied by a common smoothing parameter λ_v; and the last term is the roughness of the fixed-effect function $\eta(t)$ multiplied by another smoothing parameter λ, where the roughness matrices $\mathbf{G}$ and $\mathbf{G}_v$ are defined in (7.6) and (7.32), respectively. As in (7.31), λ is used to trade-off the goodness of fit with the roughness of $\eta(t)$, and λ_v is used to trade-off the goodness of fit with the roughness of $v_i(t),\ i = 1, 2, \cdots, n$. Notice that the main difference between (7.31) and (7.42) is that the latter takes the within-subject correlation into account via the term $\sum_{i=1}^n \mathbf{b}_i^T\mathbf{D}^{-1}\mathbf{b}_i$ within the first summation term in (7.42).

Let

$$\mathbf{D}_v = \left(\mathbf{D}^{-1} + \lambda_v\mathbf{G}_v\right)^{-1},\quad \tilde{\mathbf{V}}_i = \mathbf{Z}_i\mathbf{D}_v\mathbf{Z}_i^T + \mathbf{R}_i,\ i = 1, 2, \cdots, n. \tag{7.43}$$

Since $\mathbf{D}$ is positive definite, it is easy to show that $(\mathbf{D}^{-1} + \lambda_v\mathbf{G}_v)$ is also positive definite, and so is $\mathbf{D}_v$. By assuming $\mathbf{R}_i$ are positive definite, we have $\tilde{\mathbf{V}}_i$ are also positive definite and hence invertible.

Assuming that $\mathbf{D}, \mathbf{R}_i, \lambda$ and λ_v are known, we can show, via some simple linear algebra, that the minimizers of the PGLL criterion (7.42) are

$$\begin{aligned} \hat{\beta} &= \left(\sum_{i=1}^n \mathbf{X}_i^T \tilde{\mathbf{V}}_i^{-1} \mathbf{X}_i + \lambda \mathbf{G}\right)^{-1} \left(\sum_{i=1}^n \mathbf{X}_i^T \tilde{\mathbf{V}}_i^{-1} \mathbf{y}_i\right), \\ \hat{\mathbf{b}}_i &= \mathbf{D}_v \mathbf{Z}_i^T \tilde{\mathbf{V}}_i^{-1} (\mathbf{y}_i - \mathbf{X}_i \hat{\beta}), \ i = 1, 2, \cdots, n. \end{aligned} \tag{7.44}$$

In a matrix form, the minimizers (7.44) can be further written as

$$\begin{aligned} \hat{\beta} &= \left(\mathbf{X}^T \tilde{\mathbf{V}}^{-1} \mathbf{X} + \lambda \mathbf{G}\right)^{-1} \mathbf{X}^T \tilde{\mathbf{V}}^{-1} \mathbf{y}, \\ \hat{\mathbf{b}} &= \tilde{\mathbf{D}}_v \mathbf{Z}^T \tilde{\mathbf{V}}^{-1} (\mathbf{y} - \mathbf{X}\hat{\beta}), \end{aligned} \tag{7.45}$$

where $\mathbf{y}, \mathbf{X}$ and $\mathbf{Z}$ have been defined in the previous section, $\tilde{\mathbf{V}} = \mathrm{diag}(\tilde{\mathbf{V}}_1, \cdots, \tilde{\mathbf{V}}_n)$ and $\tilde{\mathbf{D}}_v = \mathrm{diag}(\mathbf{D}_v, \cdots, \mathbf{D}_v)$. Then the MEPS smoothers of $\eta(t)$ and $v_i(t),\ i = 1, 2, \cdots, n$ are

$$\hat{\eta}(t) = \mathbf{\Phi}_p(t)^T \hat{\beta}, \ \ \hat{v}_i(t) = \mathbf{\Psi}_q(t)^T \hat{\mathbf{b}}_i, \ i = 1, 2, \cdots, n. \tag{7.46}$$

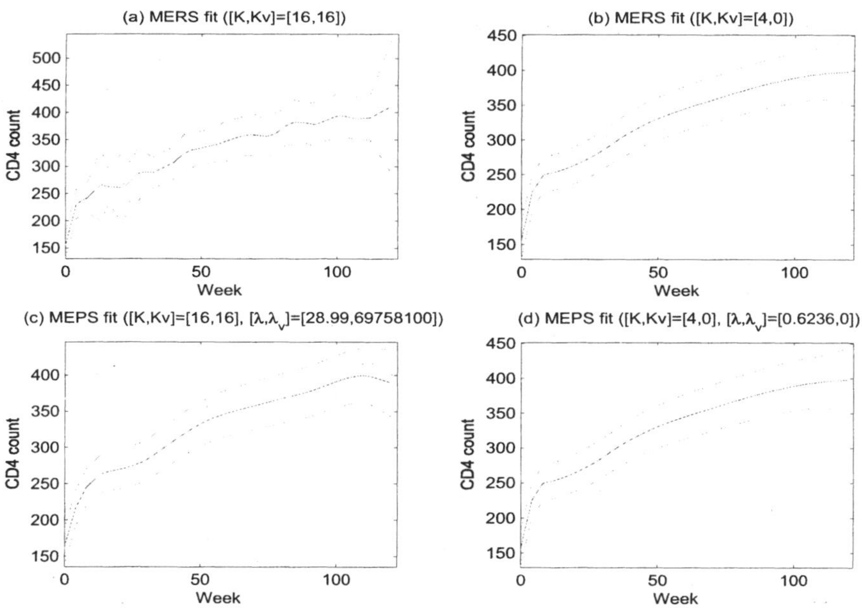

Fig. 7.8 Examples showing MERS and MEPS fits (solid curves) to the population mean function of the ACTG 388 data, superimposed with the associated 95% pointwise SD bands (dashed curves).

Notice that when both the smoothing parameters λ and λ_v equal 0, the above MEPS smoothers $\hat{\eta}(t)$ and $\hat{v}_i(t),\ i = 1, 2, \cdots, n$ reduce to the associated MERS (mixed-effects regression spline) smoothers as described in Chapter 5. In other words, a

MERS smoother is equivalent to a special MEPS smoother having the same bases $\mathbf{\Phi}_p(t)$ and $\mathbf{\Psi}_q(t)$, and with $\lambda = \lambda_v = 0$. Figure 7.8 (a) displays such a MERS fit (solid curve) of $\eta(t)$, together with the associated 95% pointwise SD band (dashed curves), where $\mathbf{\Phi}_p(t)$ and $\mathbf{\Psi}_q(t)$ are the quadratic ($k = 2$) truncated power bases with $K = K_v = 16$ knots. It is seen that this MERS fit (i.e., the MEPS fit with $\lambda = \lambda_v = 0$) and the pointwise SD band are rather rough. Figure 7.8 (c) displays a MEPS fit with properly selected λ and λ_v, and the associated 95% pointwise SD band. This MEPS fit improved the MERS fit in Figure 7.8 (a) since we selected the smoothing parameters as $[\lambda, \lambda_v] = [28.99, 69758100]$ by the BIC rule developed later in the chapter. It is seen that this MEPS fit in Figure 7.8 (c) is much smoother than the MERS fit in Figure 7.8 (a).

The above MEPS method is different from the MERS method described in Chapter 5 in the sense that the MERS method improves the MERS fit in Figure 7.8 (a) via selecting the proper number of knots for $\mathbf{\Phi}_p(t)$ and $\mathbf{\Psi}_q(t)$. Figure 7.8 (b) presents such a MERS fit where $[K, K_v] = [4, 0]$, selected by the BIC rule developed in Section 5.4.5 of Chapter 5. We can see that this MERS fit is much smoother than the MERS fit with $[K, K_v] = [16, 16]$ in Figure 7.8 (a), and is comparable with the MEPS fit in Figure 7.8 (c). This MERS fit with properly selected $[K, K_v]$ is already very good but it can be further improved by the MEPS method. Figure 7.8 (d) displays the associated MEPS fit with $[K, K_v] = [4, 0]$ and $[\lambda, \lambda_v] = [.6236, 0]$. The smoothing parameters for $[K, K_v] = [4, 0]$ are much smaller than those for $[K, K_v] = [16, 16]$ since the MERS fit in Figure 7.8 (b) is already quite smooth.

Table 7.5 Values of λ, λ_v, AIC and BIC for the MERS and MEPS fits presented in Figure 7.8.

$[K, K_v]$	$[\lambda, \lambda_v]$	AIC/10^4	BIC/10^4
[16, 16]	[0, 0]	2.7545	3.2180
[4, 0]	[0, 0]	2.5481	2.6582
[16, 16]	[28.99,69758100]	2.1618	2.2724
[4, 0]	[.6236,0]	2.1609	2.2709

Table 7.5 lists the values of λ, λ_v, AIC and BIC for the MERS and MEPS fits presented in Figure 7.8. The AIC and BIC rules will be defined later. They can be used to assess the performance of the MERS and MEPS fits. According to these two rules, the MERS fit with $[K, K_v] = [4, 0]$ is much better than the MERS fit with $[K, K_v] = [16, 16]$. However, their associated MEPS fits are comparable although the MEPS fit with $[K, K_v] = [4, 0]$ is slightly better.

7.5.2 Bayesian Interpretation

Theoretically, we can interpret the minimizers (7.45) of the PGLL criterion (7.42) as the posterior expectations of the parameters of a Bayesian problem. The associated posterior distributions will be used in the EM-algorithm for computing the variance components of the MEPS model.

The Bayesian problem can be defined as follows:

$$\mathbf{y}|\beta, \mathbf{b} \sim N(\mathbf{X}\beta + \mathbf{Z}\mathbf{b}, \mathbf{R}), \tag{7.47}$$

where the parameters β and $\mathbf{b}$ are independent, and they have the following prior distributions:

$$\beta \sim N(0, \lambda^{-1}\mathbf{G}^-), \quad \mathbf{b} \sim N(0, \tilde{\mathbf{D}}_v), \tag{7.48}$$

where $\mathbf{G}^-$ denotes one of the generalized inverses of $\mathbf{G}$. The prior distribution of β is improper but the prior distribution of $\mathbf{b}$ is proper since $\tilde{\mathbf{D}}_v$ is positive definite.

We can change the prior of β slightly so that the prior is proper. The new prior of β is $\beta \sim N\left[0, (\lambda_0\mathbf{I}_p + \lambda\mathbf{G})^{-1}\right]$. Under the new prior, each of the first $(k+1)$ components of β has a prior $N(0, 1/\lambda_0)$ and each of the last K components of β has a prior $N\left[0, 1/(\lambda_0 + \lambda)\right]$. This is because the roughness matrix $\mathbf{G} = \text{diag}(0, \cdots, 0, 1, \cdots, 1)$ with the last K entries being 1. Using limit theory arguments, it is easy to show that the limit posterior distributions (as $\lambda_0 \to 0$) are the same as the posterior distributions under the Bayesian framework (7.47) and (7.48).

Theorem 7.1 *Under the Bayesian framework defined in (7.47) and (7.48), we can write the minimizers (7.45) of the PGLL criterion (7.42) as*

$$\hat{\beta} = E(\beta|\mathbf{y}), \quad \hat{\mathbf{b}} = E(\mathbf{b}|\mathbf{y}). \tag{7.49}$$

In addition, we have

$$\begin{aligned} \beta|\mathbf{y} &\sim N\left[\hat{\beta}, (\mathbf{X}^T\tilde{\mathbf{V}}^{-1}\mathbf{X} + \lambda\mathbf{G})^{-1}\right], & (7.50) \\ \mathbf{b}|\mathbf{y} &\sim N\left[\hat{\mathbf{b}}, \tilde{\mathbf{D}}_v - \tilde{\mathbf{D}}_v\mathbf{Z}^T\mathbf{W}\mathbf{Z}\tilde{\mathbf{D}}_v\right], & (7.51) \\ \epsilon|\mathbf{y} &\sim N\left[\hat{\epsilon}, \mathbf{R} - \mathbf{R}\mathbf{W}\mathbf{R}\right], & (7.52) \end{aligned}$$

where $\epsilon = \mathbf{y} - \mathbf{X}\beta - \mathbf{Z}\mathbf{b}$, $\hat{\epsilon} = \mathbf{y} - \mathbf{X}\hat{\beta} - \mathbf{Z}\hat{\mathbf{b}}$, *and*

$$\mathbf{W} = \tilde{\mathbf{V}}^{-1} - \tilde{\mathbf{V}}^{-1}\mathbf{X}(\mathbf{X}^T\tilde{\mathbf{V}}^{-1}\mathbf{X} + \lambda\mathbf{G})^{-1}\mathbf{X}^T\tilde{\mathbf{V}}^{-1}. \tag{7.53}$$

Proof: see Appendix.

Let $\mathbf{W}_i = \tilde{\mathbf{V}}_i^{-1} - \tilde{\mathbf{V}}_i^{-1}\mathbf{X}_i(\mathbf{X}^T\tilde{\mathbf{V}}^{-1}\mathbf{X} + \lambda\mathbf{G})^{-1}\mathbf{X}_i^T\tilde{\mathbf{V}}_i^{-1}$, $i = 1, 2, \cdots, n$. Then it is easy to show that

$$\begin{array}{ll} \text{E}(\mathbf{b}_i|\mathbf{y}) = \hat{\mathbf{b}}_i, & \text{Cov}(\mathbf{b}_i|\mathbf{y}) = \mathbf{D}_v - \mathbf{D}_v\mathbf{Z}_i^T\mathbf{W}_i\mathbf{Z}_i\mathbf{D}_v, \\ \text{E}(\epsilon_i|\mathbf{y}) = \hat{\epsilon}_i, & \text{Cov}(\epsilon_i|\mathbf{y}) = \mathbf{R}_i - \mathbf{R}_i\mathbf{W}_i\mathbf{R}_i. \end{array} \tag{7.54}$$

When β is given, the conditional posterior distributions of $\mathbf{b}_i$ and ϵ_i are different from those when β is not given.

Theorem 7.2 *Under the Bayesian framework defined in (7.47) and (7.48), we have*

$$\begin{aligned} \mathbf{b}|\mathbf{y},\beta &\sim N\left[\tilde{\mathbf{D}}_v\mathbf{Z}^T\tilde{\mathbf{V}}^{-1}(\mathbf{y}-\mathbf{X}\beta), \tilde{\mathbf{D}}_v - \tilde{\mathbf{D}}_v\mathbf{Z}^T\tilde{\mathbf{V}}^{-1}\mathbf{Z}\tilde{\mathbf{D}}_v\right], \\ \epsilon|\mathbf{y},\beta &\sim N\left[\mathbf{R}\tilde{\mathbf{V}}^{-1}(\mathbf{y}-\mathbf{X}\beta), \mathbf{R}-\mathbf{R}\tilde{\mathbf{V}}^{-1}\mathbf{R}\right]. \end{aligned} \tag{7.55}$$

Proof: see Appendix.

Let $\hat{\beta}$ have the expression in (7.45). Then by the above theorem, we have

$$\begin{aligned} \mathrm{E}(\mathbf{b}|\mathbf{y},\beta=\hat{\beta}) &= \tilde{\mathbf{D}}_v\mathbf{Z}^T\tilde{\mathbf{V}}^{-1}(\mathbf{y}-\mathbf{X}\hat{\beta}) = \hat{\mathbf{b}} = \mathrm{E}(\mathbf{b}|\mathbf{y}), \\ \mathrm{E}(\epsilon|\mathbf{y},\beta=\hat{\beta}) &= \mathbf{R}\tilde{\mathbf{V}}^{-1}(\mathbf{y}-\mathbf{X}\hat{\beta}) = \hat{\epsilon} = \mathrm{E}(\epsilon|\mathbf{y}). \end{aligned}$$

Based on this and by Theorem 7.2, we have

$$\begin{aligned} &\mathrm{E}(\mathbf{b}_i|\mathbf{y},\beta=\hat{\beta}) = \hat{\mathbf{b}}_i, \qquad \mathrm{Cov}(\mathbf{b}_i|\mathbf{y},\beta=\hat{\beta}) = \mathbf{D}_v - \mathbf{D}_v\mathbf{Z}_i^T\tilde{\mathbf{V}}_i^{-1}\mathbf{Z}_i\mathbf{D}_v, \\ &\mathrm{E}(\epsilon_i|\mathbf{y},\beta=\hat{\beta}) = \hat{\epsilon}_i, \qquad \mathrm{Cov}(\epsilon_i|\mathbf{y},\beta=\hat{\beta}) = \tilde{\mathbf{R}}_i - \mathbf{R}_i\tilde{\mathbf{V}}_i^{-1}\mathbf{R}_i. \end{aligned}$$

Comparing this to (7.54), we see that $\mathrm{Cov}(\mathbf{b}_i|\mathbf{y},\beta=\hat{\beta})$ and $\mathrm{Cov}(\epsilon_i|\mathbf{y},\beta=\hat{\beta})$ can be obtained, respectively, from $\mathrm{Cov}(\mathbf{b}_i|\mathbf{y})$ and $\mathrm{Cov}(\epsilon_i|\mathbf{y})$ by replacing $\mathbf{W}_i$ with $\tilde{\mathbf{V}}_i^{-1}$.

Theorems 7.1 and 7.2 are similar to Theorems 6.2 and 6.3 of Chapter 6 in spirit and so are their proofs, although the former are more general in the sense that the matrices $\mathbf{X}, \mathbf{Z}, \mathbf{G}$ and $\mathbf{G}_v$, among others are defined differently from those in Theorems 6.2 and 6.3.

7.5.3 Variance Components Estimation

In the expression (7.44), we assumed that the variance components $\mathbf{D}$ and $\mathbf{R}_i = \sigma^2\mathbf{I}_{n_i}$, $i = 1, 2, \cdots, n$ [see (7.29)] are known. In practice, we need to estimate both $\mathbf{D}$ and σ^2. In Section 6.5.3 of Chapter 6, we have developed an EM-algorithm for estimating the variance components of the MESS model, which extends the EM algorithm developed by Laird, Lange and Stram (1987) for a general LME model. In this section, we adapt the EM-algorithm for the MEPS model.

We use the fact that under the model (7.39), both $\mathbf{b}_i$ and ϵ_i are normal. Thus, when $\mathbf{b}_i$ and ϵ_i were known, the MLEs of σ^2 and $\mathbf{D}$ would be

$$\hat{\sigma}^2 = N^{-1}\sum_{i=1}^n \epsilon_i^T\epsilon_i, \quad \hat{\mathbf{D}} = n^{-1}\sum_{i=1}^n \mathbf{b}_i\mathbf{b}_i^T. \tag{7.56}$$

Since ϵ_i and $\mathbf{b}_i$ are usually unknown, we replace the above MLEs either by their posterior expectations $\mathrm{E}(\hat{\sigma}^2|\mathbf{y})$ and $\mathrm{E}(\hat{\mathbf{D}}|\mathbf{y})$ or by their conditional posterior expectations $\mathrm{E}(\hat{\sigma}^2|\mathbf{y},\beta=\hat{\beta})$ and $\mathrm{E}(\hat{\mathbf{D}}|\mathbf{y},\beta=\hat{\beta})$.

Like any other EM-algorithms, the EM-algorithm that we shall adapt for the MEPS method consists of the maximum likelihood step (M-step) and the expectation step (E-step). The M-step is based on (7.56) and the E-step is based on either of the replacement methods for $\hat{\sigma}^2$ and $\hat{\mathbf{D}}$. When we use $\mathrm{E}(\hat{\sigma}^2|\mathbf{y})$ and $\mathrm{E}(\hat{\mathbf{D}}|\mathbf{y})$, the resulting EM-algorithm is traditionally called the restricted maximum likelihood (REML) based EM-algorithm; otherwise, it is called the maximum likelihood (ML) based EM-algorithm.

Theorem 7.3 *Assume that* $\mathbf{R}_i = \sigma^2\mathbf{I}_{n_i},\ i = 1, 2, \cdots, n$. *Then under the Bayesian framework defined in (7.47) and (7.48), we have*

$$E(\hat{\sigma}^2|\mathbf{y}) = N^{-1}\sum_{i=1}^{n}\left\{\hat{\epsilon}_i^T\hat{\epsilon}_i + \sigma^2\left[n_i - \sigma^2 tr(\mathbf{W}_i)\right]\right\}, \tag{7.57}$$

$$E(\hat{\mathbf{D}}|\mathbf{y}) = n^{-1}\sum_{i=1}^{n}\left\{\hat{\mathbf{b}}_i\hat{\mathbf{b}}_i^T + [\mathbf{D}_v - \mathbf{D}_v\mathbf{Z}_i^T\mathbf{W}_i\mathbf{Z}_i\mathbf{D}_v]\right\}. \tag{7.58}$$

In addition, $E[\hat{\sigma}^2|\mathbf{y}, \beta = \hat{\beta}]$ *and* $E[\hat{\mathbf{D}}|\mathbf{y}, \beta = \hat{\beta}]$ *are the same as (7.57) and (7.58), respectively, after replacing* $\mathbf{W}_i$ *with* $\tilde{\mathbf{V}}_i^{-1}$.

Proof: see Appendix.

Theorem 7.3 extends Theorem 6.4 of Chapter 6. When $\mathbf{Z}_i = \mathbf{X}_i,\ i = 1, 2, \cdots, n$ and $\mathbf{G}_v = \mathbf{G}$, the former will reduce to the latter. Based on the previous discussions, a practical REML-based EM-algorithm can be described as follows, which is more general than the one described in Section 6.5.3 of Chapter 6.

REML-Based EM-Algorithm for MEPS Modeling

Step 0. Initialize $\hat{\sigma}^2_{(r)} = 1$, and $\hat{\mathbf{D}}_{(r)} = \hat{\sigma}^2_{(r)}\mathbf{I}_q$ for $r = 0$.

Step 1. Set $r = r + 1$. Compute

$$\begin{aligned}\hat{\mathbf{D}}_{v(r-1)} &= (\hat{\mathbf{D}}_{(r-1)}^{-1} + \lambda_v\mathbf{G}_v)^{-1},\\ \tilde{\mathbf{V}}_{i(r-1)} &= \mathbf{Z}_i\hat{\mathbf{D}}_{v(r-1)}\mathbf{Z}_i^T + \hat{\sigma}^2_{(r-1)}\mathbf{I}_{n_i},\ i = 1, 2, \cdots, n.\end{aligned}$$

Denote $\tilde{\mathbf{V}}_{(r-1)} = \mathrm{diag}(\tilde{\mathbf{V}}_{1(r-1)}, \cdots, \tilde{\mathbf{V}}_{n(r-1)})$. Then update $\hat{\beta}_{(r)}$ and $\hat{\mathbf{b}}_{i(r)}$ using

$$\begin{aligned}\hat{\beta}_{(r)} &= (\mathbf{X}^T\tilde{\mathbf{V}}_{(r-1)}^{-1}\mathbf{X} + \lambda\mathbf{G})^{-1}(\mathbf{X}^T\tilde{\mathbf{V}}_{(r-1)}^{-1}\mathbf{y}),\\ \hat{\mathbf{b}}_{i(r)} &= \hat{\mathbf{D}}_{v(r-1)}\mathbf{Z}_i^T\tilde{\mathbf{V}}_{i(r-1)}^{-1}(\mathbf{y}_i - \mathbf{X}_i\hat{\beta}_{(r)}),\ i = 1, 2, \cdots, n.\end{aligned}$$

Step 2. Compute the residuals $\hat{\epsilon}_{i(r)} = \mathbf{y}_i - \mathbf{X}_i\hat{\beta}_{(r)} - \mathbf{Z}_i\hat{\mathbf{b}}_{i(r)}$ and the matrices $\mathbf{W}_{i(r-1)} = \tilde{\mathbf{V}}_{i(r-1)}^{-1} - \tilde{\mathbf{V}}_{i(r-1)}^{-1}\mathbf{X}_i(\mathbf{X}^T\tilde{\mathbf{V}}_{(r-1)}^{-1}\mathbf{X} + \lambda\mathbf{G})^{-1}\mathbf{X}_i^T\tilde{\mathbf{V}}_{i(r-1)}^{-1}$. Then update $\hat{\sigma}^2_{(r)}$ and $\hat{\mathbf{D}}_{(r)}$ using

$$\hat{\sigma}^2_{(r)} = N^{-1}\sum_{i=1}^{n}\left\{\hat{\epsilon}_{i(r)}^T\hat{\epsilon}_{i(r)} + \hat{\sigma}^2_{(r-1)}[n_i - \hat{\sigma}^2_{(r-1)}\mathrm{tr}(\mathbf{W}_{i(r-1)})]\right\},$$

$$
\begin{aligned}
\hat{\mathbf{D}}_{(r)} &= n^{-1}\sum_{i=1}^{n}\Big\{\hat{\mathbf{b}}_{i(r)}\hat{\mathbf{b}}_{i(r)}^T \\
&\quad + [\hat{\mathbf{D}}_{v(r-1)} - \hat{\mathbf{D}}_{v(r-1)}\mathbf{Z}_i^T\mathbf{W}_{i(r-1)}\mathbf{Z}_i\hat{\mathbf{D}}_{v(r-1)}]\Big\}.
\end{aligned}
$$

Step 3. When some convergence condition is satisfied, stop; otherwise, go back to Step 1.

Notice that the smoothing parameters λ and λ_v, and the roughness matrices $\mathbf{G}$ and $\mathbf{G}_v$ do not change during the whole iteration process. The ML-based EM-algorithm can be obtained from the above REML-based EM-algorithm simply by replacing the $\mathbf{W}_{i(\cdot)}$ with the $\tilde{\mathbf{V}}_{i(\cdot)}^{-1}$.

7.5.4 Fits and Smoother Matrices

The MEPS smoothers of $\eta(t)$ and $v_i(t),\ i = 1, 2, \cdots, n$ are given in (7.46). The smoothers of the individual functions $s_i(t),\ i = 1, 2, \cdots, n$ are then

$$
\hat{s}_i(t) = \hat{\eta}(t) + \hat{v}_i(t),\ i = 1, 2, \cdots, n.
$$

Using (7.45), the fits of $\eta(t), v_i(t),\ i = 1, 2, \cdots, n$ and $s_i(t),\ i = 1, 2, \cdots, n$ at all the design time points can be written as

$$
\begin{aligned}
\hat{\eta}^* &= \mathbf{X}\hat{\beta} = \mathbf{X}(\mathbf{X}^T\tilde{\mathbf{V}}^{-1}\mathbf{X} + \lambda\mathbf{G})^{-1}\mathbf{X}^T\tilde{\mathbf{V}}^{-1}\mathbf{y} \equiv \mathbf{A}\mathbf{y}, \\
\hat{\mathbf{v}}^* &= \mathbf{Z}\hat{\mathbf{b}} = \mathbf{Z}\hat{\mathbf{D}}_v\mathbf{Z}^T\tilde{\mathbf{V}}^{-1}(\mathbf{I}_N - \mathbf{A})\mathbf{y} \equiv \mathbf{A}_v\mathbf{y}, \\
\hat{\mathbf{s}}^* &= \mathbf{X}\hat{\beta} + \mathbf{Z}\hat{\mathbf{b}} = (\mathbf{A} + \mathbf{A}_v)\mathbf{y},
\end{aligned}
\tag{7.59}
$$

where $\mathbf{A}, \mathbf{A}_v$ and $\mathbf{A} + \mathbf{A}_v$ are the associated smoother matrices.

7.5.5 Variability Band Construction

Using the relationship (7.41), the estimator of $\gamma(s, t)$ can be written as

$$
\hat{\gamma}(s, t) = \mathbf{\Psi}_q(s)^T\hat{\mathbf{D}}_v\mathbf{\Psi}_q(t), \tag{7.60}
$$

where we estimate $\mathbf{D}$ using $\hat{\mathbf{D}}_v$ to account for the roughness of $v_i(t),\ i = 1, 2, \cdots, n$. Based on this and $\hat{\mathbf{D}}_v, \hat{\sigma}^2$, we can easily construct the estimators of $\mathbf{V}$ and $\tilde{\mathbf{V}}$.

The pointwise SD band of $\eta(t)$ can be constructed based on the covariance of $\hat{\beta}$. Based on (7.45), it is easy to see that

$$
\text{Cov}(\hat{\beta}) = (\mathbf{X}^T\tilde{\mathbf{V}}^{-1}\mathbf{X} + \lambda\mathbf{G})^{-1}\mathbf{X}^T\tilde{\mathbf{V}}^{-1}\mathbf{V}\tilde{\mathbf{V}}^{-1}\mathbf{X}(\mathbf{X}^T\tilde{\mathbf{V}}^{-1}\mathbf{X} + \lambda\mathbf{G})^{-1}, \tag{7.61}
$$

where $\mathbf{V} = \text{Cov}(\mathbf{y})$ which can be replaced by $\tilde{\mathbf{V}}$ to account for the roughness of $v_i(t),\ i = 1, 2, \cdots, n$. Let $\widehat{\text{Cov}}(\hat{\beta})$ denote the estimator of the above covariance matrix. Then the 95% pointwise SD band of $\eta(t)$ can be constructed approximately as $\hat{\eta}(t) \pm 1.96\widehat{\text{Var}}^{1/2}[\hat{\eta}(t)]$ where $\widehat{\text{Var}}[\hat{\eta}(t)] = \mathbf{\Phi}_p(t)^T\widehat{\text{Cov}}(\hat{\beta})\mathbf{\Phi}_p(t)$. Two examples of such pointwise SD bands are showed in Figures 7.8 (c) and (d).

To construct the Bayesian pointwise confidence band (Wahba 1983) for $\eta(t)$, we simply replace $\text{Cov}(\hat{\beta})$ by

$$\text{Cov}(\hat{\beta}|\mathbf{y}) = (\mathbf{X}^T\tilde{\mathbf{V}}^{-1}\mathbf{X} + \lambda\mathbf{G})^{-1}, \tag{7.62}$$

in the above pointwise SD band construction.

7.5.6 Choice of the Smoothing Parameters

As mentioned previously in Chapters 5 and 6, we extended the classic AIC (Akaike 1973) and BIC (Schwarz 1978) rules for the MERS and MESS models, respectively. We now adapt these rules for the MEPS model. The resulting AIC and BIC rules provide criteria for trading-off the goodness of fit with the model complexity of the MEPS model. As before, we quantify the goodness of fit by the log-likelihood of the MEPS model and quantify the model complexity by the sum of the traces of the associated smoother matrices.

Let $\boldsymbol{\eta}_i = [\eta(t_{i1}), \cdots, \eta(t_{in_i})]^T$ and $\mathbf{V}_i = \text{Cov}(\mathbf{y}_i)$. Then $\mathbf{y}_i \sim N(\boldsymbol{\eta}_i, \mathbf{V}_i),\ i = 1, 2, \cdots, n$. It follows that the log-likelihood of the MEPS model can be expressed as follows:

$$\text{Loglik} = -\frac{1}{2}\sum_{i=1}^{n}\left\{(\mathbf{y}_i - \boldsymbol{\eta}_i)^T\mathbf{V}_i^{-1}(\mathbf{y}_i - \boldsymbol{\eta}_i) + \log(|\mathbf{V}_i|)\right\} - \frac{N}{2}\log(2\pi), \tag{7.63}$$

with the unknown $\boldsymbol{\eta}_i$ and $\mathbf{V}_i$ replaced by their MEPS estimates:

$$\hat{\boldsymbol{\eta}}_i = \mathbf{X}_i\hat{\beta}, \quad \hat{\mathbf{V}}_i = \mathbf{Z}_i\hat{\mathbf{D}}_v\mathbf{Z}_i^T + \hat{\sigma}^2\mathbf{I}_{n_i},\ i = 1, 2, \cdots, n, \tag{7.64}$$

with the given smoothing parameters λ and λ_v.

In Figure 7.9 (a), five profile curves of Loglik against $\log_{10}(\lambda)$ for five selected values of λ_v are displayed. It is seen that for each fixed λ_v, the Loglik decreases as λ increases although the last two profile curves are overlapping. In Figure 7.9 (c), five profile curves of Loglik against $\log_{10}(\lambda_v)$ for five selected values of λ are displayed. It is seen that for each fixed λ, the Loglik decreases as λ_v increases. These observations are not surprising. In fact, the Loglik is expected to decrease as λ (or λ_v) increases. This is because when the smoothing parameter λ (or λ_v) increases, a heavier penalty is imposed for the roughness of $\eta(t)$ (or $v_i(t),\ i = 1, 2, \cdots, n$). As a result of the trade-off, the goodness of fit of the MEPS fit becomes poorer.

The model complexity for $\hat{\eta}(t)$ is defined as the trace of the smoother matrix $\mathbf{A}$ (7.59):

$$\text{df} = \text{tr}(\mathbf{A}) = \text{tr}\left[(\mathbf{X}^T\tilde{\mathbf{V}}^{-1}\mathbf{X} + \lambda\mathbf{G})^{-1}\mathbf{X}^T\tilde{\mathbf{V}}^{-1}\mathbf{X}\right]. \tag{7.65}$$

Note that the df is now a function of λ and λ_v. However, when λ_v is fixed, we have the following lemma which is almost the same as Lemma 7.1 for the NPS smoother described in Section 7.2.

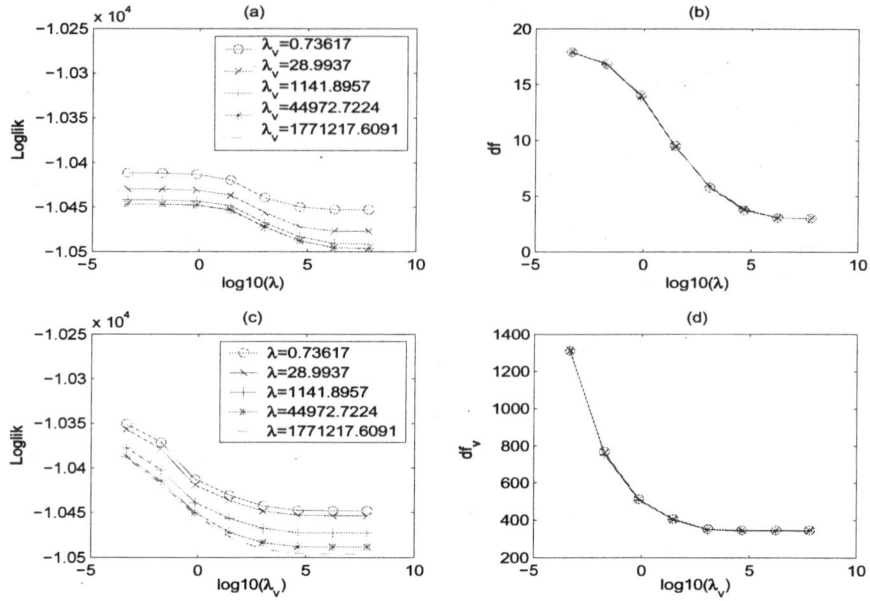

Fig. 7.9 Profile curves of Loglik, df and df_v against $\log_{10}(\lambda)$ and $\log_{10}(\lambda_v)$. The quadratic truncated power bases were used with $K = K_v = 16$.

Lemma 7.3 *Assume that the design matrix* $\mathbf{X}$ *has full rank* $p = K + k + 1$. *Then for fixed* $\mathbf{X}$ *and* λ_v, *we have*

(1) df is a monotonically decreasing continuous function in λ.

(2) $\lim_{\lambda\to 0} df(\lambda, \lambda_v) = K + k + 1$ *and* $\lim_{\lambda\to\infty} df(\lambda, \lambda_v) = k + 1$.

(3) $k + 1 \le df(\lambda, \lambda_v) \le K + k + 1$.

Proof: see Appendix.

Therefore, as λ increases, i.e., as more smoothing is imposed for $\hat{\eta}(t)$, the MEPS model for $\hat{\eta}(t)$ becomes simpler for any fixed λ_v. As we have known, a sufficient condition to guarantee the full-rank of $\mathbf{X}$ is $M \ge p$ where M is the number of distinct design time points. By the proof of Lemma 7.3 given in the Appendix of this chapter, it is known that

$$\mathrm{df} = k + 1 + \sum_{l=1}^{K} (1 + \lambda w_{k+1+l})^{-1},$$

where $w_{k+1+l}, l = 1, 2, \cdots, K$ are positive numbers. From the above expression, it is seen that the model complexity, df, is mainly dominated by λ. This can also be seen from Figure 7.9 (b) in which five profile curves of df against $\log_{10}(\lambda)$ for five different values of λ_v coincide. It is seen that the df indeed decreases as λ increases but it almost does not vary as λ_v changes.

The model complexity for $\hat{v}_i(t),\ i = 1, 2, \cdots, n$ is defined as the trace of the smoother matrix $\mathbf{A}_v$ (7.59):

$$\text{df}_v = \text{tr}(\mathbf{A}_v) = \text{tr}\left[\mathbf{Z}\tilde{\mathbf{D}}_v\mathbf{Z}^T\tilde{\mathbf{V}}^{-1}(\mathbf{I}_N - \mathbf{A})\right]. \tag{7.66}$$

As expected, while the df decreases with increasing λ and is mainly dominated by λ, the df_v decreases with increasing λ_v and is mainly dominated by λ_v. This relationship can be observed from Figure 7.9 (d) in which five profile curves of df_v against $\log_{10}(\lambda_v)$ for five different values of λ coincide. It is seen that the df_v decreases as λ_v increases but it changes little as λ increases. The following lemma indicates that the df_v must be less than the total number of measurements $N = \sum_{i=1}^n n_i$.

Lemma 7.4 *For the MEPS model,*

$$0 < df_v < n \min(q, N/n), \tag{7.67}$$

where $q = K_v + k_v + 1$.

Proof: see Appendix.

Since in the MEPS model, both $\eta(t)$ and $v_i(t),\ i = 1, 2, \cdots, n$ are fitted, the overall MEPS model complexity is then the sum of df and df_v, which is also the number of degrees of freedom of the MEPS fits of the individual curves $s_i(t),\ i = 1, 2, \cdots, n$. Like in Chapters 5 and 6, we can extend the classical AIC (Akaike 1973) and BIC (Schwarz 1978) rules for the MEPS model as

$$\begin{aligned} \text{AIC}(\lambda, \lambda_v) &= -2\text{Loglik} + 2(\text{df} + \text{df}_v), \\ \text{BIC}(\lambda, \lambda_v) &= -2\text{Loglik} + \log(n)(\text{df} + \text{df}_v), \end{aligned} \tag{7.68}$$

where n is the number of subjects of the longitudinal data set.

Figures 7.10 (a) and (b) show the profile curves (associated with different values of λ_v) of AIC and BIC against $\log_{10}(\lambda)$, and Figures 7.10 (c) and (d) show the profile curves (associated with different values of λ) of AIC and BIC against $\log_{10}(\lambda_v)$. Based on the numerical outputs and by the BIC rule, the best pair of $[\lambda, \lambda_v]$ is $[28.99, 69758100]$ where 69758100 is the the largest value in the candidates for λ_v. These candidates were selected so that they were scatterd uniformly on $\log_{10}$-scale in the range of interest for λ_v. Figures 7.10 (c) and (d) further indicate that both AIC and BIC favor large values for λ_v. Notice that when $\lambda_v \to \infty$, the MEPS model for $v_i(t),\ i = 1, 2, \cdots, n$ will tend to the k-th degree polynomial model. Therefore, the above result suggests that the quadratic model is adequate to the random-effect curves of the ACTG 388 data, a result we obtained in Chapter 5 using the MERS model.

For the MEPS model, we usually fix the truncated power bases $\mathbf{\Phi}_p(t)$ and $\mathbf{\Psi}_q(t)$ by fixing the degrees k and k_v and the numbers of knots K and K_v. We often use $k = k_v = 2$, i.e., the quadratic truncated power bases and initially take $K = K_v = [M/2]$ or $[M/3]$ where M is the number of distinct design time points (the actual K and K_v will reduce after removing the tied knots). For example, we applied the MEPS model to the ACTG 388 data using the quadratic truncated power basis via

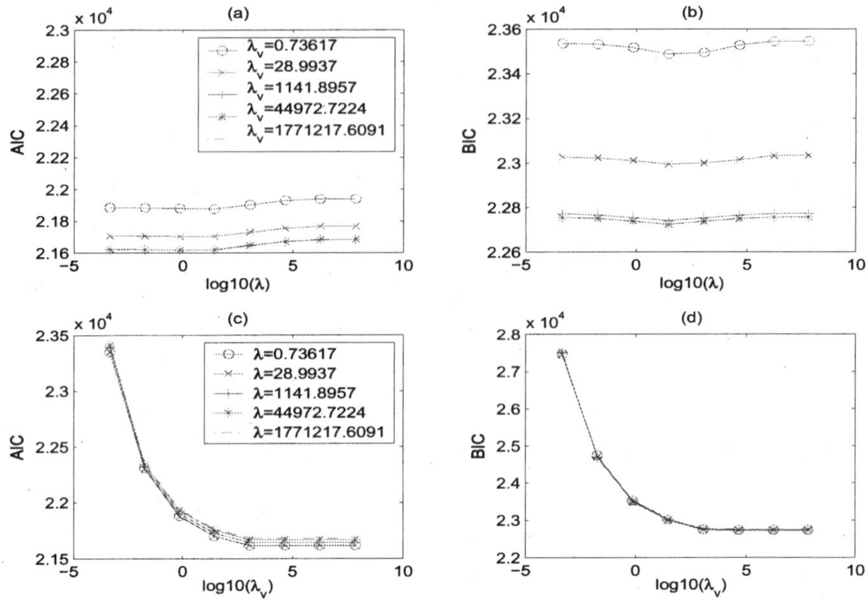

Fig. 7.10 Profile curves of AIC and BIC against $\log_{10}(\lambda)$ or $\log_{10}(\lambda_v)$. The quadratic truncated power bases were used with $K = K_v = 16$.

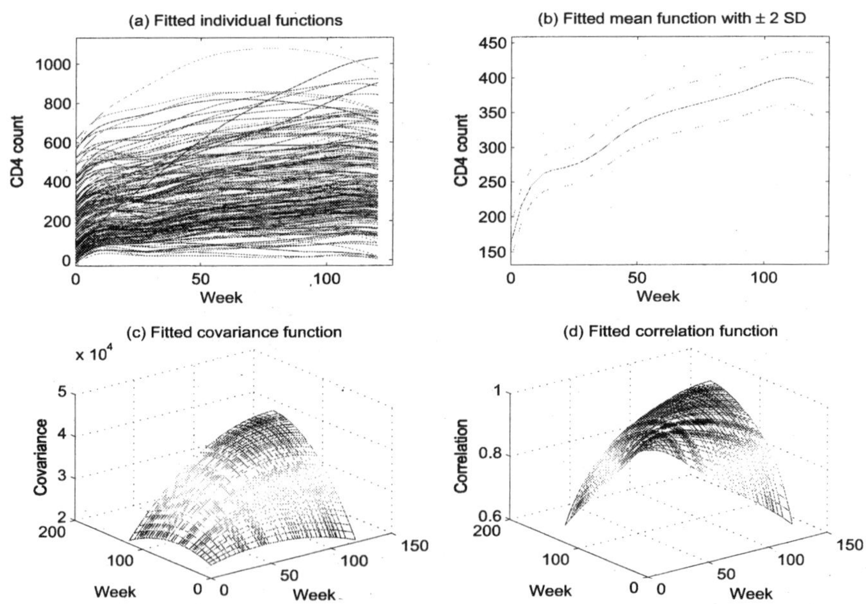

Fig. 7.11 Overall MEPS fits to the ACTG 388 data. Quadratic truncated power bases were used with $[K, K_v] = [16, 16]$ and $[\lambda, \lambda_v] = [28.99, 69758100]$.

initially taking $K = K_v = [M/2] = 21$ (the actual $K = K_v = 16$ after removing the tied knots). The smoothing parameters $[\lambda, \lambda_v] = [28.99, 69758100]$ were selected by the BIC rule (7.68). Figures 7.11 and 7.12 display, respectively, the associated overall fits and six selected individual fits. It is seen that they are nearly the same as those presented in Figures 5.5 and 5.6 in Chapter 5 using the MERS model with $[K, K_v] = [4, 0]$. Therefore, the MEPS method performs well for fitting the ACTG 388 data.

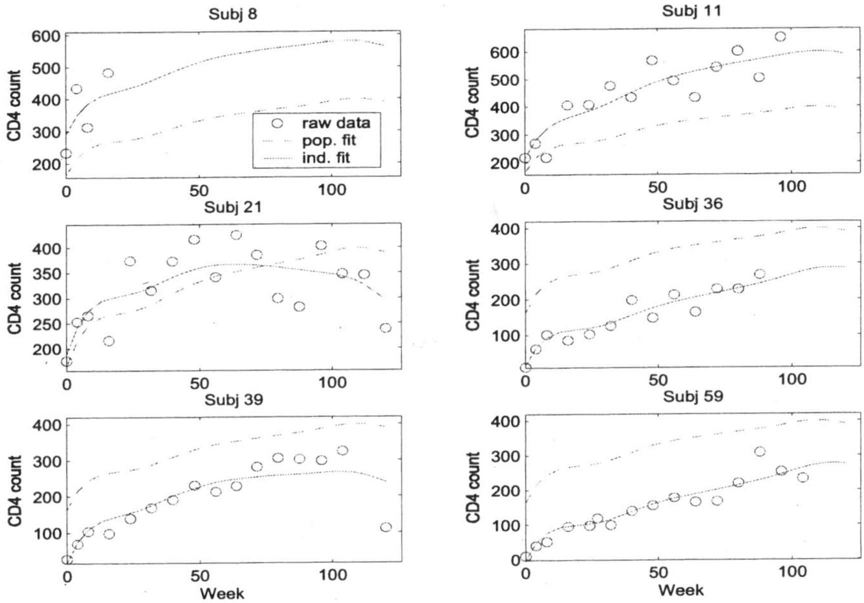

Fig. 7.12 Individual fits for six randomly selected subjects from the ACTG 388 data. Quadratic truncated power bases were used with $[K, K_v] = [16, 16]$ and $[\lambda, \lambda_v] = [28.99, 69758100]$.

7.5.7 Choosing the Numbers of Knots

In Section 7.2.6, we found that the number of knots, K, for $\mathbf{\Phi}_p(t)$ can be chosen in a rough manner provided that it is neither too small nor too large. The proper range was given in (7.17). That is, $K^* \leq K \leq M - k - 1$ where K^* is the optimal K for the NRS fit to the same data set using $\mathbf{\Phi}_p(t)$. Only when $K \geq K^*$, the associated NRS fit will not oversmooth the data, so that the associated NPS model can well fit the data via selecting a proper smoothing parameter λ; but when $K > M - k - 1$, the associated design matrix $\mathbf{X}$ is not of full rank so that $\mathbf{X}^T\mathbf{X} + \lambda\mathbf{G}$ may not be of full rank. This may cause computation difficulties or yield an unstable NPS fit to the data.

Similarly, for the MEPS model, the K and K_v can be chosen in a rough manner provided that they are neither too small nor too large. Let $[K^*, K_v^*]$ be the optimal pair of $[K, K_v]$ for the MERS fit to the data using the bases $\mathbf{\Phi}_p(t)$ and $\mathbf{\Psi}_q(t)$. On the one hand, when $K < K^*$, the population mean function may be oversmoothed, and on the other hand, when $K > M - k - 1$, the associated design matrix $\mathbf{X}$ must be degenerated so that $\mathbf{X}^T\tilde{\mathbf{V}}^{-1}\mathbf{X} + \lambda\mathbf{G}$ may not be of full rank. This may cause computation difficulties or yield an unstable MEPS fit to the data. Therefore, the range (7.17) is still appropriate for the MEPS model. Similarly, we require that $K_v \geq K_v^*$ to avoid the situation where the subject-specific random effects functions are already oversmoothed by the MERS model. However, the upper limit of K_v can not be determined based on $\mathbf{Z}$. This is because it does not matter if $\mathbf{Z}$ is of full rank or not, $\tilde{\mathbf{V}} = \mathbf{Z}\tilde{\mathbf{D}}_v\mathbf{Z}^T + \mathbf{R}$ is always of full rank and invertible since $\mathbf{R}$ is assumed to be diagonal and invertible. To overcome this difficulty, we simply take the upper limit of K_v as $M - k_v - 1$, mimicking the upper limit of K. We believe that such an upper limit for K_v is often sufficient. In summary, we use the following ranges for K and K_v:

$$\begin{aligned} K^* \leq K \leq M - k - 1, \\ K_v^* \leq K_v \leq M - k_v - 1. \end{aligned} \tag{7.69}$$

In practice, when $[K^*, K_v^*]$ is known, we can take

$$[K, K_v] = [K^*, K_v^*], [K^* + 1, K_v^*], [K^*, K_v^* + 1], \text{ or } [K^* + 1, K_v^* + 1];$$

otherwise, we can take $K, K_v = [M/2], [M/3], [M/4]$ or $[M/5]$ initially as long as the actual K, K_v are smaller than their upper limits, respectively.

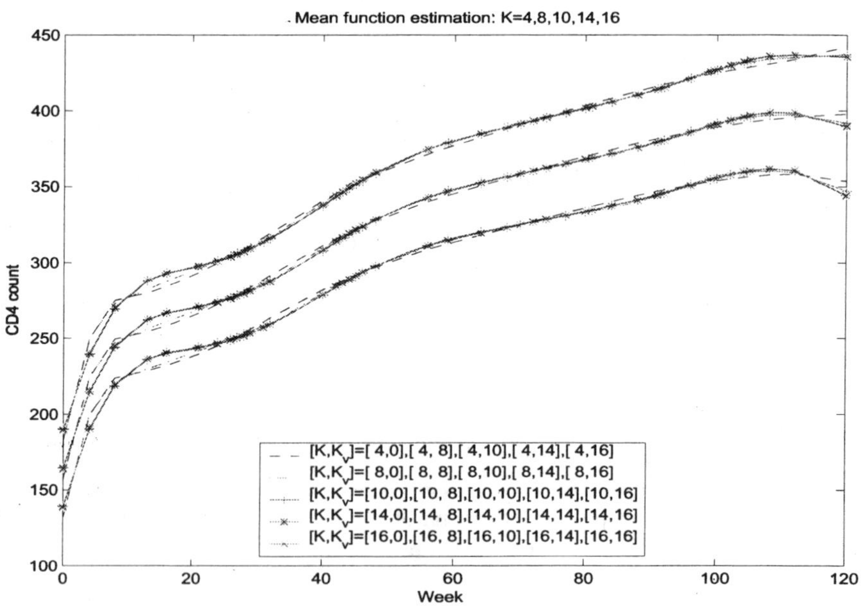

Fig. 7.13 Population mean function fits (with 95% pointwise SD bands) to the ACTG 388 data.

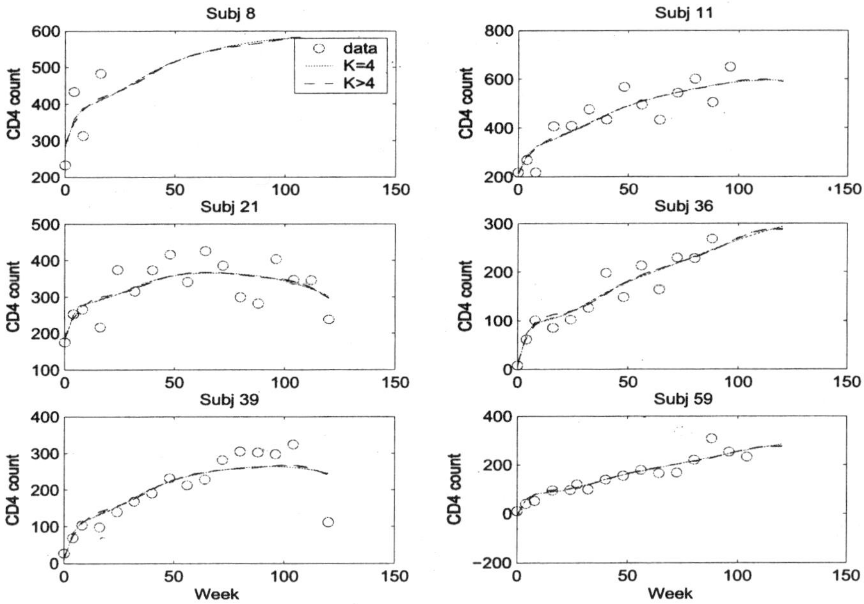

Fig. 7.14 Individual fits for six selected subjects.

As an illustration, we fit the MEPS model to the ACTG 388 data with different choices of K and K_v. Since $[K^*, K_v^*] = [4, 0]$, we initially took $K = 4, 8, 10, 14, 21$ and $K_v = 0, 8, 10, 14, 21$ that satisfy the ranges in (7.69), where $8 = [M/5], 10 = [M/4], 14 = [M/3], 21 = [M/2]$ and $M = 43$. We actually got $K = 4, 8, 10, 14, 16$ and $K_v = 0, 8, 10, 14, 16$ after removing the tied knots.

Table 7.6 lists the values of λ, λ_v, AIC and BIC for $K = 4, 8, 10, 14, 16$ and $K_v = 0, 8, 10, 14, 16$. The values of λ and λ_v were selected so that they minimized the BIC rule. It is seen that all the AIC and BIC values are around 2.16 and 2.27, respectively, and slightly increase as K increases but change little as K_v increases. Although these AIC and BIC values are slightly different, the MEPS fits to the ACTG 388 data are nearly the same, as seen from Figures 7.13 and 7.14. Figure 7.13 displays all the associated MEPS fits, together with the associated 95% pointwise SD bands, of the population mean function of the ACTG 388 data. It is seen that although $[K, K_v]$ took different values, these MEPS fits are overlaping, and so are their 95% pointwise SD bands. Figure 7.14 displays the individual fits for six selected subjects. It is seen that for each subject, the individual fits with different $[K, K_v]$'s are nearly overlapping. Therefore, the choice of $[K, K_v]$ is not crucial provided that both K and K_v satisfy the range limits (7.69).

As a comparison, we also fitted the MEPS model to the ACTG 388 data with $K = 0, 1, 2, 3, 4$ and $K_v = 0$. These $[K, K_v]$ pairs except $[K, K_v] = [4, 0]$ do not fall into the required ranges given in (7.69). As a result, the associated MEPS fits (except with $[K, K_v] = [4, 0]$) oversmoothed the ACTG 388 data. Figure 7.15

Table 7.6 Values of λ, λ_v, AIC and BIC for the MEPS fits with $K = 4, 8, 10, 14, 16$ and $K_v = 0, 8, 10, 14, 16$. The values of λ and λ_v were chosen for each fixed pair $[K, K_v]$ so that the BIC score was minimized. Quadratic truncated power bases were used in all these MEPS fits.

$[K, K_v]$	$[\lambda, \lambda_v/10^8]$	AIC/10^4	BIC/10^4
[4, 0]	[.6236, 0]	2.1609	2.2709
[4, 8]	[.6236, 1.1395]	2.1609	2.2709
[4, 10]	[.6236, 1.5471]	2.1609	2.2709
[4, 14]	[.6236, .00071]	2.1609	2.2709
[4, 16]	[.6236, .01771]	2.1609	2.2709
[8, 0]	[.8470, 0]	2.1611	2.2720
[8, 8]	[.8470, 1.1395]	2.1611	2.2720
[8, 10]	[.8470, 1.5471]	2.1611	2.2720
[8, 14]	[.8470, 1.3302]	2.1611	2.2720
[8, 16]	[.8470, .69758]	2.1611	2.2720
[10, 0]	[40.790, 0]	2.1618	2.2721
[10, 8]	[40.790, 1.1395]	2.1618	2.2720
[10, 10]	[40.790, 1.5471]	2.1618	2.2720
[10, 14]	[40.790, 1.3302]	2.1618	2.2720
[10, 16]	[40.790, .01771]	2.1618	2.2721
[14, 0]	[38.234, 0]	2.1618	2.2723
[14, 8]	[38.234, 1.1395]	2.1618	2.2723
[14, 10]	[38.234, 1.5471]	2.1618	2.2723
[14, 14]	[38.234, 1.3302]	2.1618	2.2723
[14, 16]	[38.234, .01771]	2.1618	2.2723
[16, 0]	[28.994, 0]	2.1618	2.2724
[16, 8]	[28.994, .00064]	2.1618	2.2724
[16, 10]	[28.994, 1.5471]	2.1618	2.2724
[16, 14]	[28.994, 1.3302]	2.1618	2.2724
[16, 16]	[28.994, .69758]	2.1618	2.2724

displays all the associated MEPS fits of the population mean function. Compared with the optimal MEPS fit (dashed curves) with $[K, K_v] = [4, 0]$, most of the MEPS fits with $K = 0, 1, 2, 3$ oversmoothed the data, especially within the first 20 weeks of the study. Figure 7.16 displays the individual fits for six selected subjects. It is seen that these MEPS models fitted the data well although some of them oversmoothed the data within the first 20 weeks.

Table 7.7 lists the values of λ, λ_v, AIC and BIC for the MEPS fits with $K = 0, 1, 2, 3, 4$ and $K_v = 0$. It is seen that the AIC and BIC values for $K = 0, 1, 2, 3$ are generally much larger than those for $K = 8, 10, 14, 16$ listed in Table 7.6. Therefore, K and K_v for MEPS modeling should not be too small.

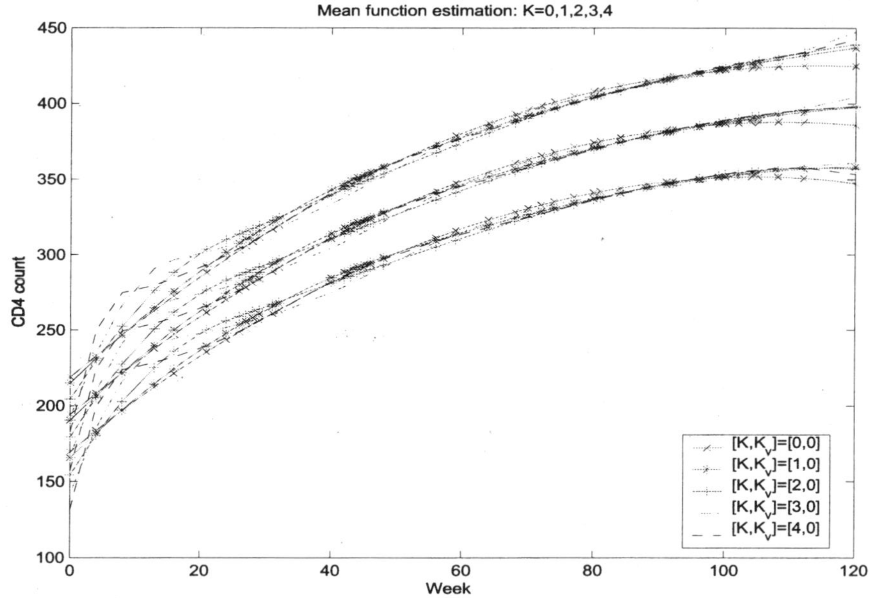

Fig. 7.15 Population mean function fits (with 95% pointwise SD bands) to the ACTG 388 data.

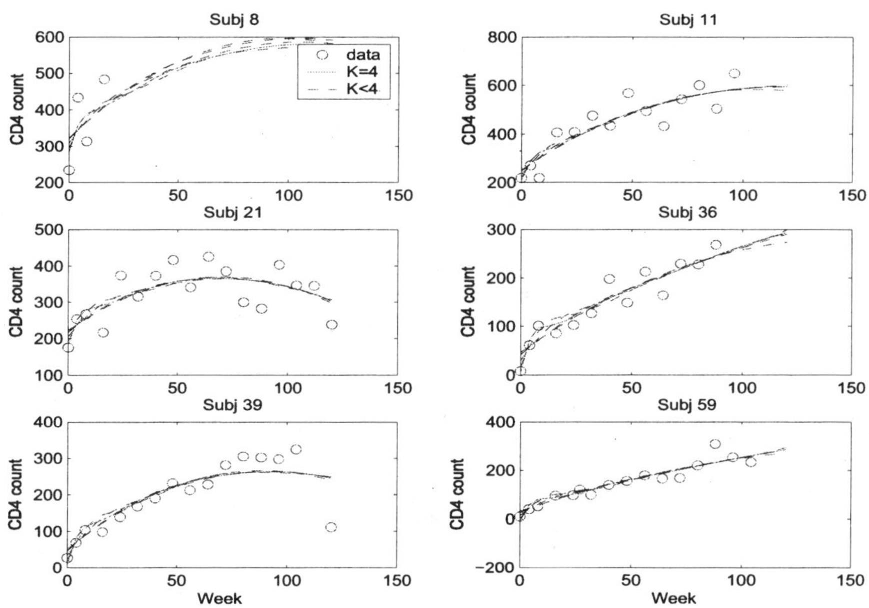

Fig. 7.16 Individual fits for six selected subjects.

Table 7.7 Values of λ, λ_v, AIC and BIC for the MEPS fits with $K = 0, 1, 2, 3, 4$ and $K_v = 0$. The values of λ and λ_v were chosen for each fixed pair $[K, K_v]$ so that the BIC score was minimized. Quadratic truncated power bases were used in all these MEPS fits.

$[K, K_v]$	$[\lambda, \lambda_v]$	AIC/10^4	BIC/10^4
$[0, 0]$	$[0, 0]$	2.1683	2.2756
$[1, 0]$	$[10494, 0]$	2.1674	2.2751
$[2, 0]$	$[509.34, 0]$	2.1653	2.2739
$[3, 0]$	$[19.514, 0]$	2.1629	2.2722
$[4, 0]$	$[.6236, 0]$	2.1609	2.2709

7.6 SUMMARY AND BIBLIOGRAPHICAL NOTES

In this chapter, we described the NPS, GPS, EPS and MEPS models for longitudinal data. These models are, respectively, similar to the NSS, GSS, ESS and MESS models described in Chapter 6. Therefore, we essentially adapted the associated methodologies in Chapter 6 to the P-spline based models.

By using the ACTG 388 data, we mainly discussed the effect of the choice of the numbers of knots for the NPS and MEPS models, respectively. For the NPS model, we concluded that the number of knots K can be chosen in a rough manner provided it is not too small nor too large. When K is too small, the associated NRS fit already oversmooths the data, so the resulting NPS fit will also oversmooth the data. However, when K is too large, the associated design matrix will be degenerated. This may cause an unstable NPS fit to the data. For a proper K, the NRS fit will undersmooth the data so that it can be improved by a NPS fit via choosing a proper smoothing parameter λ. The choice of such a K is often not crucial. For the MEPS model, the lessons are similar: we can choose K and K_v in a rough manner provided they are neither too small nor too large. When both K and K_v are too small, the associated MERS fit oversmooths the data, so that the associated MEPS fit will oversmooth the data too. When K is too large, the design matrix will be degenerated. This may cause the associated MEPS fit to be unstable. For proper K and K_v, the associated MERS fit undersmooths the data. Good MEPS fitting can always be obtained via choosing proper smoothing parameters λ and λ_v.

As a smoothing technique for uncorrelated or independent data, P-spline was proposed at least a decade ago (e.g., Eilers and Marx 1996). It can be regarded as a natural extension from ridge regression (e.g., Drap and Smith 1998). Eilers and Marx (1996) used B-splines with equally spaced knots and penalized higher order difference in the B-spline coefficients. Ruppert and Carroll (2000) used truncated power bases and penalized the coefficients associated with the truncated power functions. They proposed a myopic algorithm for selecting the number of knots. This myopic algorithm, together with a full-search algorithm, was studied in details in Ruppert (2002). A comprehensive survey on theory and applications of P-splines was given by Ruppert, Wand and Carroll (2003).

Applications of P-splines to longitudinal data are still in their infancy. To our knowledge, most of the materials presented in this chapter are rather new. Some of the materials in Section 7.4 are similar to those presented in Chapter 9 of Ruppert, Wand and Carroll (2003).

7.7 APPENDIX: PROOFS

Proof of Lemma 7.1 Since $\mathbf{X}$ has full rank $p = K + k + 1$, the matrix $\mathbf{H} = \mathbf{X}^T\mathbf{X}$ is invertible. It follows that

$$\begin{aligned} \text{df} &= \text{tr}\left[(\mathbf{H} + \lambda\mathbf{G})^{-1}\mathbf{H}\right] \\ &= \text{tr}\left[(\mathbf{I}_p + \lambda\mathbf{H}^{-1/2}\mathbf{G}\mathbf{H}^{-1/2})^{-1}\right]. \end{aligned}$$

Let $\mathbf{H}^{-1/2}\mathbf{G}\mathbf{H}^{-1/2}$ have a singular value decomposition $\mathbf{U}\mathbf{W}\mathbf{U}^T$ where $\mathbf{U}$ is orthonormal and $\mathbf{W} = \text{diag}(w_1, \cdots, w_p)$. Since $\text{rank}(\mathbf{H}^{-1/2}\mathbf{G}\mathbf{H}^{-1/2}) = \text{rank}(\mathbf{G}) = K$, the $\mathbf{W}$ has $k + 1$ diagonal entries, say the first $k + 1$, are 0 and the remaining K diagonal entries are positive. It follows that

$$\begin{aligned} \text{df}(\lambda) &= \text{tr}\left[(\mathbf{I}_p + \lambda\mathbf{H}^{-1/2}\mathbf{G}\mathbf{H}^{-1/2})^{-1}\right] \\ &= \text{tr}\left[(\mathbf{I}_p + \lambda\mathbf{U}\mathbf{W}\mathbf{U}^T)^{-1}\right] \\ &= \sum_{l=1}^{p}(1 + \lambda w_l)^{-1} \\ &= k + 1 + \sum_{r=1}^{K}(1 + \lambda w_{k+1+r})^{-1}. \end{aligned} \tag{7.70}$$

From the above expression, it is easy to see that (1) $\text{df}(\lambda)$ is a monotonically decreasing function of λ; (2) $\text{df}(\lambda) \to K + k + 1$ as $\lambda \to 0$ and $\text{df}(\lambda) \to k + 1$ as $\lambda \to \infty$; and (3) $k + 1 \leq \text{df}(\lambda) \leq K + k + 1$. Lemma 7.1 is proved.

Proof of Theorem 7.1 The proof of Theorem 7.1 is along the same lines as those of Theorem 6.2 of Chapter 6 except now $\mathbf{D}_v = (\mathbf{D}^{-1} + \lambda_v\mathbf{G}_v)^{-1}$.

Proof of Theorem 7.2 The proof of Theorem 7.2 is along the same lines as those of Theorem 6.3 of Chapter 6 except now $\mathbf{D}_v = (\mathbf{D}^{-1} + \lambda_v\mathbf{G}_v)^{-1}$.

Proof of Theorem 7.3 The proof of Theorem 7.3 is along the same lines as those of Theorem 6.4 of Chapter 6 except now $\mathbf{D}_v = (\mathbf{D}^{-1} + \lambda_v\mathbf{G}_v)^{-1}$ and $\mathbf{Z} = \text{diag}(\mathbf{Z}_1, \cdots, \mathbf{Z}_n)$.

Proof of Lemma 7.3 The proof of Lemma 7.3 is along the same lines as those of Lemma 7.1 as long as we replace $\mathbf{H} = \mathbf{X}^T\mathbf{X}$ by $\mathbf{H} = \mathbf{X}^T\tilde{\mathbf{V}}^{-1}\mathbf{X}$.

Proof of Lemma 7.4 On the one hand, we have

$$\text{df}_v = \text{tr}(\mathbf{A}_v)$$

$$
\begin{aligned}
&= \operatorname{tr}\left[\mathbf{Z}\mathbf{D}_v\mathbf{Z}^T\mathbf{D}_v\tilde{\mathbf{V}}^{-1}(\mathbf{I}_N - \mathbf{A})\right] \\
&= \operatorname{tr}\left[\mathbf{Z}\mathbf{D}_v\mathbf{Z}^T\tilde{\mathbf{V}}^{-1}\right] \\
&\quad - \operatorname{tr}\left[\mathbf{D}_v^{1/2}\mathbf{Z}^T\tilde{\mathbf{V}}^{-1}\mathbf{X}(\mathbf{X}^T\tilde{\mathbf{V}}^{-1}\mathbf{X} + \lambda\mathbf{G})^{-1})\mathbf{X}^T\tilde{\mathbf{V}}^{-1}\mathbf{Z}\mathbf{D}_v^{1/2}\right] \\
&\leq \operatorname{tr}\left[\mathbf{Z}\mathbf{D}_v\mathbf{Z}^T\tilde{\mathbf{V}}^{-1}\right] \\
&= \operatorname{tr}\left[\mathbf{H}\mathbf{H}^T(\mathbf{H}\mathbf{H}^T + \mathbf{I}_N)^{-1}\right] \\
&= N - \operatorname{tr}\left[(\mathbf{H}\mathbf{H}^T + \mathbf{I}_N)^{-1}\right] \\
&< N,
\end{aligned}
$$

where $\mathbf{H} = \mathbf{R}^{-1/2}\mathbf{Z}\mathbf{D}_v^{1/2}$. Therefore $0 < \mathrm{df}_v < N$.

On the other hand, using Lemma 2.1, we have

$$
(\mathbf{H}\mathbf{H}^T + \mathbf{I}_N)^{-1} = \mathbf{I}_N - \mathbf{H}(\mathbf{I}_{nq} + \mathbf{H}^T\mathbf{H})^{-1}\mathbf{H}^T,
$$

where $q = K_v + k_v + 1$. Therefore,

$$
\begin{aligned}
\operatorname{tr}\left[\mathbf{H}\mathbf{H}^T(\mathbf{H}\mathbf{H}^T + \mathbf{I}_N)^{-1}\right] &= \operatorname{tr}\left[\mathbf{H}^T(\mathbf{H}\mathbf{H}^T + \mathbf{I}_N)^{-1}\mathbf{H}\right] \\
&= \operatorname{tr}[\mathbf{H}^T\mathbf{H} - \mathbf{H}^T\mathbf{H}(\mathbf{I}_{nq} + \mathbf{H}^T\mathbf{H})^{-1}\mathbf{H}^T\mathbf{H}] \\
&= \operatorname{tr}[\mathbf{H}^T\mathbf{H}(\mathbf{I}_{nq} + \mathbf{H}^T\mathbf{H})^{-1}] \\
&= nq - \operatorname{tr}[(\mathbf{I}_{nq} + \mathbf{H}^T\mathbf{H})^{-1}] \\
&< nq.
\end{aligned}
$$

It follows that $0 < \mathrm{df}_v < nq$. The lemma is proved.

8

Semiparametric Models

8.1 INTRODUCTION

In Chapters 4~7, we investigated methodologies for fitting a simple nonparametric population mean (NPM) model or a simple nonparametric mixed-effects (NPME) model for longitudinal data. We say that the NPM model or the NPME model is simple since the NPM model (7.2) or the NPME model (7.39) includes only a single covariate, time, to explain the mean response or the random-effects via smooth nonparametric functions. For many practical applications, the mean response or the random-effects may also depend on covariates other than time.

To incorporate the effects of other covariates, various parametric models such as linear/nonlinear regression models and linear/nonlinear mixed-effects models have been proposed for longitudinal data; see for example Lindsey (1993), Diggle, Liang and Zeger (1994), Davidian and Giltinan (1995), Vonesh and Chinchilli (1996), Pinheiro and Bates (2000), Verbeke and Molcnbcrghs (2000) and Diggle, Heagerty, Liang and Zeger (2002), among others. A brief review of linear/nonlinear mixed-effects models was given in Chapter 2. Parametric models are very useful for longitudinal data analysis since they provide a parsimonious description of the relationship between the response variable and its covariates. However, when the parametric model does not fit the data well, the resulting estimates may be biased. To relax the assumptions on parametric forms, various nonparametric models such as nonparametric population mean models and nonparametric mixed-effects models have been proposed for longitudinal data; see for example Müller (1988), Hoover, Rice, Wu and Yang (1998), Wang (1998a, b), Wu, Chiang and Hoover (1998), Fan and Zhang (2000), Chiang, Rice and Wu (2001), Rice and Wu (2001), Guo (2002a,b), Huang, Wu and Zhou (2002), and Wu and Zhang (2002a), among others. Basic nonparametric regression techniques for independent data were briefly reviewed in Chapter 3.

Parametric models are usually restrictive and less robust against model assumptions, but they are parsimonious and efficient when the models are correctly specified. In contrast, nonparametric models are more robust against the model assumptions but they are usually more complex and less efficient. Semiparametric models are a good compromise and retain nice features of both parametric and nonparametric models. In semiparametric models, the parametric components are often used to model important factors that affect the responses parametrically and the nonparametric components are often used for nuisance factors which are usually less important. In this chapter, we shall discuss three classes of semiparametric models, including semiparametric population mean models (Martinussen and Scheike 1999, Cheng and Wei 2000, Lin and Carroll 2001a,b, Lin and Ying 2001, He, Zhu and Fung 2002, Fan and Li 2004, Hu, Wang and Carroll 2004, Lin and Carroll 2005), semiparametric mixed-effects models (Zeger and Diggle 1994, Zhang et al. 1998, Tao et al. 1999, Jacqmin-Gadda et al. 2002, Ruppert, Wand and Carroll 2003, Durban et al. 2005) and semiparametric nonlinear mixed-effects models (Ke and Wang 2001, Wu and Zhang 2002b) for longitudinal data.

8.2 SEMIPARAMETRIC POPULATION MEAN MODEL

8.2.1 Model Specification

Let y_{ij} denote the response for the i-th subject at time t_{ij}, and let $\mathbf{c}_{ij} = [c_{1ij}, \cdots, c_{p_0 ij}]^T$ denote p_0 covariates observed at time t_{ij}. If the mean response is known to depend on $\mathbf{c}_{ij}$ linearly and on t_{ij} nonlinearly but the particular form of the nonlinear function is unknown, then we can model the relationship between the response and the covariates using the following semiparametric population mean (SPM) model:

$$\begin{aligned} y_{ij} &= \mathbf{c}_{ij}^T\boldsymbol{\alpha} + \eta(t_{ij}) + e_{ij}, \\ j &= 1, 2, \cdots, n_i;\ i = 1, 2, \cdots, n, \end{aligned} \tag{8.1}$$

where $\boldsymbol{\alpha} = [\alpha_1, \cdots, \alpha_{p_0}]^T$ consists of the coefficients of the covariates, $\eta(\cdot)$ is a smooth function of time, and e_{ij} is the error that is not explained by either the covariates $\mathbf{c}_{ij}$ or the smooth function $\eta(\cdot)$. When the covariates $\mathbf{c}_{ij}$ are centered, the smooth function $\eta(t)$ may be called an intercept function. The SPM model (8.1) is so-called since the mean response

$$\mathrm{E}y_{ij} = \mathbf{c}_{ij}^T\boldsymbol{\alpha} + \eta(t_{ij}) \tag{8.2}$$

consists of two parts:

$$\begin{array}{ll} \text{parametric component} & \mathbf{c}_{ij}^T\boldsymbol{\alpha}, \\ \text{nonparametric component} & \eta(t_{ij}). \end{array}$$

In the expression (8.2) and throughout this section, the expectation, variance and covariance are understood as the expectation, variance and covariance conditional to the given covariates $\mathbf{c}_{ij}, t_{ij}, \ j = 1, 2, \cdots, n_i; \ i = 1, 2, \cdots, n$, and the coefficient vector $\boldsymbol{\alpha}$ is also known as the effect of the covariates.

8.2.2 Local Polynomial Method

In this subsection, we briefly describe two local polynomial kernel (LPK) methods for fitting the SPM model (8.1). The first one is based on the backfitting algorithm of Hastie and Tishirani (1990), which generalizes the first kernel method of Speckman (1988), and the second one is based on profile smoothing which extends the second kernel method of Speckman (1988). Following Fan and Li (2004) and Hu, Wang and Carroll (2004), the second method may be called a profile LPK method. Similarly, we can call the first method a backfitting LPK method.

Speckman (1988) proposed two kernel methods for fitting a semiparametric regression model for independent data. Let $\mathbf{S}$ be a kernel smoother for fitting the standard nonparametric regression model $z_i = f(t_i) + \epsilon_i, \ i = 1, 2, \cdots, n$ so that $\hat{\mathbf{f}} = \mathbf{S}\mathbf{z}$ where $\mathbf{z} = [z_1, \cdots, z_n]^T$, $\hat{\mathbf{f}} = [\hat{f}(t_1), \cdots, \hat{f}(t_n)]^T$ and $\hat{f}(t)$ is the kernel estimator, e.g., the Nadaraya-Watson kernel estimator (3.9) described in Section 3.2.2 of Chapter 3. Motivated by Green et al. (1985), Speckman's first kernel method for the semiparametric model $z_i = \mathbf{u}_i^T\boldsymbol{\delta} + f(t_i) + \epsilon_i, \ i = 1, 2, \cdots, n$ results in the following estimators:

$$\begin{aligned} \hat{\boldsymbol{\delta}} &= \left[\mathbf{U}^T(\mathbf{I}_n - \mathbf{S})\mathbf{U}\right]^{-1}\mathbf{U}^T(\mathbf{I}_n - \mathbf{S})\mathbf{z}, \\ \hat{\mathbf{f}} &= \mathbf{S}(\mathbf{z} - \mathbf{U}\hat{\boldsymbol{\delta}}), \end{aligned} \tag{8.3}$$

where $\mathbf{U} = [\mathbf{u}_1, \cdots, \mathbf{u}_n]^T$. Speckman's second kernel method is based on residual analysis and results in the following estimators:

$$\begin{aligned} \hat{\boldsymbol{\delta}} &= \left[\mathbf{U}^T(\mathbf{I}_n - \mathbf{S})^T(\mathbf{I}_n - \mathbf{S})\mathbf{U}\right]^{-1}\mathbf{U}^T(\mathbf{I}_n - \mathbf{S})^T(\mathbf{I}_n - \mathbf{S})\mathbf{z}, \\ \hat{\mathbf{f}} &= \mathbf{S}(\mathbf{z} - \mathbf{U}\hat{\boldsymbol{\delta}}). \end{aligned} \tag{8.4}$$

Speckman (1988) showed that under appropriate conditions, both methods achieve comparable fits to $f(t)$, and produce $\sqrt{n}$ - consistent estimators for $\boldsymbol{\delta}$; however, the second method is more appropriate since the first method may produce estimates which are biased as noticed by Rice (1986), but this bias problem for the second method is asymptotically negligible with no need for undersmoothing $f(t)$.

It is straightforward to generalize Speckman's two kernel methods to fit the SPM model (8.1). Hu, Wang and Carroll (2004) called the resulting methods as the backfitting kernel method and the profile kernel method, respectively. Moreover, one can replace the kernel smoother with the LPK smoother (3.6) [see Section 3.2 of Chapter 3], resulting in the backfitting LPK method and the profile LPK method, respectively. For convenience, we refer to Speckman's first kernel estimators (8.3) as backfitting kernel estimators and Speckman's second kernel estimators (8.4) as profile kernel estimators.

Let $\mathbf{A}$ be the LPK smoother matrix associated with the LPK smoother $\hat{\eta}(t)$ (4.7) as described in Section 4.2.1 of Chapter 4 for fitting the following NPM model:

$$y_{ij} = \eta(t_{ij}) + \epsilon_{ij}, \ j = 1, 2, \cdots, n_i; \ i = 1, 2, \cdots, n.$$

Then $\hat{\eta} = \mathbf{A}\mathbf{y}$ where

$$\begin{array}{rclrcl} \mathbf{y} &=& [\mathbf{y}_1^T, \cdots, \mathbf{y}_n^T]^T, & \mathbf{y}_i &=& [y_{i1}, \cdots, y_{in_i}]^T, \\ \hat{\eta} &=& [\hat{\eta}_1^T, \cdots, \hat{\eta}_n^T]^T, & \hat{\eta}_i &=& [\hat{\eta}(t_{i1}), \cdots, \hat{\eta}(t_{in_i})]^T. \end{array} \tag{8.5}$$

Thus, for the SPM model (8.1), the backfitting LPK estimators are

$$\begin{array}{rcl} \hat{\alpha} &=& (\mathbf{C}^T(\mathbf{I}_N - \mathbf{A})\mathbf{C})^{-1}\mathbf{C}^T(\mathbf{I}_N - \mathbf{A})\mathbf{y}, \\ \hat{\eta} &=& \mathbf{A}(\mathbf{y} - \mathbf{C}\hat{\alpha}), \end{array} \tag{8.6}$$

and the profile LPK estimators are

$$\begin{array}{rcl} \hat{\alpha} &=& (\mathbf{C}^T(\mathbf{I}_N - \mathbf{A})^T(\mathbf{I}_N - \mathbf{A})\mathbf{C})^{-1}\mathbf{C}^T(\mathbf{I}_N - \mathbf{A})^T(\mathbf{I}_N - \mathbf{A})\mathbf{y}, \\ \hat{\eta} &=& \mathbf{A}(\mathbf{y} - \mathbf{C}\hat{\alpha}), \end{array} \tag{8.7}$$

where $\mathbf{C} = [\mathbf{C}_1^T, \cdots, \mathbf{C}_n^T]^T$ with $\mathbf{C}_i = [\mathbf{c}_{i1}, \cdots, \mathbf{c}_{in_i}]^T$, $i = 1, 2, \cdots, n$.

Following Hu, Wang and Carroll (2004), we can show that the backfitting LPK estimators (8.6) can be obtained via applying the backfitting algorithm of Hastie and Tishirani (1990) to the SPM model (8.1). Observe that when α is known, the SPM model (8.1) reduces to the population mean model (4.2); and when $\eta(t)$ is known, the SPM model reduces to a standard linear model for longitudinal data. A standard linear model can be fitted easily by using a least squares (LS) or generalized least squares (GLS) method while the NPM model (4.2) can be fitted using a LPK method as described in Section 4.2.1 of Chapter 4. In the spirit of Zeger and Diggle (1994), we can then fit the SPM model (8.1) iteratively by using the backfitting algorithm of Hastie and Tishirani (1990). The associated backfitting procedure can be written as

(a) Given the current estimate $\hat{\alpha}$, calculate the residuals $r_{ij} = y_{ij} - \mathbf{c}_{ij}^T\hat{\alpha}$ and use these residuals as responses to calculate the LPK estimator $\hat{\eta}(t)$ using the methodologies described in Section 4.2.1 of Chapter 4.

(b) Given the current LPK estimator $\hat{\eta}(t)$, calculate the residuals $s_{ij} = y_{ij} - \hat{\eta}(t_{ij})$ and use these residuals as responses to calculate a LS or GLS estimator $\hat{\alpha}$.

The above procedure is easy to implement provided a good initial value for $\hat{\alpha}$ is at hand. A good initial value for $\hat{\alpha}$ may be obtained by using the difference-based method of Fan and Li (2004). A simple version of Fan and Li (2004)'s difference-based method will be described in Section 8.2.6. A drawback of the above backfitting procedure is that the estimators $\hat{\alpha}$ and $\hat{\eta}(t)$ are defined via backfitting iterations. Following Hu, Wang and Carroll (2004), we can show that at convergence, the above backfitting algorithm will result in the backfitting LPK estimators (8.6).

The profile LPK estimators (8.7) can be derived via using the profile smoothing method of Fan and Li (2004), Hu, Wang and Carroll (2004) and Wang, Carroll and Lin (2005); see also Lin and Carroll (2001a, 2005) for profile smoothing in a GEE setting. In vector notation, the SPM model (8.1) can be written as

$$\mathbf{y} = \mathbf{C}\alpha + \eta + \mathbf{e}, \tag{8.8}$$

where η and $\mathbf{e}$ are defined in the same manner as that for defining $\mathbf{y}$ as in (8.5). When α is known, we can fit the NPM model $y_{ij} - \mathbf{c}_{ij}^T\alpha = \eta(t_{ij}) + e_{ij}, \ j = 1, 2, \cdots, n_i; \ i = 1, 2, \cdots, n$, or in vector notion, the model

$$\mathbf{y} - \mathbf{C}\alpha = \eta + \mathbf{e}, \tag{8.9}$$

to obtain $\hat{\eta}$ using LPK smoothing as described in Section 4.2.1 of Chapter 4. By the definition of the LPK smoother $\mathbf{A}$ defined eariler, we can express $\hat{\eta}$ as a linear function of $\mathbf{y} - \mathbf{C}\alpha$, i.e.,

$$\hat{\eta} = \mathbf{A}(\mathbf{y} - \mathbf{C}\alpha), \tag{8.10}$$

which is called the profile LPK smoother of η.

Replacing η in (8.8) with its profile LPK smoother $\hat{\eta}$, and after a simplification, we have

$$(\mathbf{I}_N - \mathbf{A})\mathbf{y} = (\mathbf{I}_N - \mathbf{A})\mathbf{C}\alpha + \mathbf{e}.$$

The LS estimator $\hat{\alpha}$ based on the above linear model is then given by the first expression in (8.7) while the second expression is obtained by replacing α in (8.10) with $\hat{\alpha}$.

When the local kernel smoother is used and under some regular conditions, Hu, Wang and Carroll (2004) showed that the profile kernel estimator $\hat{\alpha}$ is $\sqrt{n}$-consistent and asymptotically normal. Moreover, they showed that the profile kernel estimator (8.7) is more efficient than its backfitting kernel estimator (8.6) in the sense that the latter has larger asymptotic variance. As a result, the common practice of using the backfitting method to compute the profile kernel estimator is no longer advised. When the local linear smoother is used and under some regular conditions, Fan and Li (2004) showed that the profile estimator $\hat{\alpha}$ is $\sqrt{n}$ - consistent and asymptotically normally distributed. To select a proper bandwidth for $\hat{\eta}$, Fan and Li (2004) suggested the following simple method. They first replaced $\mathbf{y} - \mathbf{C}\alpha$ in (8.9) with $\tilde{\mathbf{y}} = \mathbf{y} - \mathbf{C}\hat{\alpha}_0$ where $\hat{\alpha}_0$ is the difference-based estimator of α proposed in Fan and Li (2004). They then proceed to select a proper bandwidth h for the standard nonparametric regression model $\tilde{\mathbf{y}} = \eta + \mathbf{e}$ from (8.9) using a standard bandwidth selector such as the "leave-one-point-out" cross-validation or the bandwidth selector proposed by Ruppert, Sheather and Wand (1995). To avoid using $\hat{\alpha}_0$, we can adapt the GCV rule (7.16) here. By (8.7), the smoother matrices for the parametric and nonparametric components of the SPM model (8.1) are

$$\begin{aligned} \mathbf{A}_{par} &= \mathbf{C}\left\{\mathbf{C}^T(\mathbf{I}_N - \mathbf{A})^T(\mathbf{I}_N - \mathbf{A})\mathbf{C}\right\}^{-1}\mathbf{C}^T(\mathbf{I}_N - \mathbf{A})^T(\mathbf{I}_N - \mathbf{A}), \\ \mathbf{A}_{npar} &= \mathbf{A}(\mathbf{I}_N - \mathbf{A}_{par}). \end{aligned}$$

It follows that $\mathbf{A}^* = \mathbf{A}_{par} + \mathbf{A}(\mathbf{I}_N - \mathbf{A}_{par})$ is the smoother matrix for the responses at all the design time points. Therefore, we can adapt the GCV rule (7.16) here as

$$\text{GCV}(h) = \frac{(\mathbf{y} - \hat{\mathbf{y}})^T\mathbf{W}(\mathbf{y} - \hat{\mathbf{y}})}{[1 - \text{df}(h)/N]^2}, \tag{8.11}$$

where $\hat{\mathbf{y}} = \mathbf{A}^*\mathbf{y}$, $\text{df}(h) = \text{tr}(\mathbf{A}^*)$, the weight matrix $\mathbf{W} = \text{diag}(\mathbf{W}_1, \cdots, \mathbf{W}_n)$ with $\mathbf{W}_i = \text{diag}(w_{i1}, \cdots, w_{in_i})$ and $N = \sum_{i=1}^n n_i$ is the total number of the observations for all the subjects. For simplicity, we often take $\mathbf{W} = N^{-1}\mathbf{I}_N$, i.e., using the same weight for all the subjects.

8.2.3 Regression Spline Method

In Section 5.2 of Chapter 5, we discussed a regression spline method for the NPM model (5.1). The advantage of the regression spline method is that the method is simple. The key idea of the method is to express $\eta(t)$ as a regression spline. It is straightforward to generalize the method for the SPM model (8.1). Let $\mathbf{\Phi}_p(t)$ be a basis, e.g., a k-th degree truncated power basis with knots $\tau_1, \cdots, \tau_K$:

$$\mathbf{\Phi}_p(t) = [1, t, \cdots, t^k, (t - \tau_1)_+^k, \cdots, (t - \tau_K)_+^k]^T, \tag{8.12}$$

where $w_+^k = [\max(0, w)]^k$ and $p = K + k + 1$ is the number of basis functions in $\mathbf{\Phi}_p(t)$. Then we can express $\eta(t)$ as a regression spline:

$$\eta(t) = \mathbf{\Phi}_p(t)^T \boldsymbol{\beta}, \tag{8.13}$$

where $\boldsymbol{\beta} = [\beta_1, \cdots, \beta_p]^T$ is the associated coefficient vector. The SPM model (8.1) can then be approximately expressed as a standard linear regression model:

$$y_{ij} = \mathbf{c}_{ij}^T \boldsymbol{\alpha} + \mathbf{x}_{ij}^T \boldsymbol{\beta} + e_{ij},$$
$$j = 1, 2, \cdots, n_i; \; i = 1, 2, \cdots, n,$$

where $\mathbf{x}_{ij} = \mathbf{\Phi}_p(t_{ij})$. For fixed $\mathbf{\Phi}_p(t)$, the above linear regression model can be fitted by LS or GLS methods as those in Section 5.2 of Chapter 5. Moreover, the knot-spacing methods described in Chapter 5 can be used to specify the knots for $\mathbf{\Phi}_p(t)$ when $\mathbf{\Phi}_p(t)$ is a k-th degree truncated power basis (8.12), and the GCV rule (5.22) can be adapted for selecting p. For more details, readers are referred to Section 5.2 of Chapter 5.

8.2.4 Penalized Spline Method

Like the regression spline method for fitting the SPM model (8.1) described in the previous subsection, the penalized spline (or P-spline) method uses a basis $\mathbf{\Phi}_p(t)$ to express $\eta(t)$ as a regression spline $\eta(t) = \mathbf{\Phi}_p(t)^T \boldsymbol{\beta}$ where $\boldsymbol{\beta}$ is the associated coefficient vector. Unlike the regression spline method where a good regression spline smoother $\hat{\eta}(t)$ is obtained via properly selecting the basis $\mathbf{\Phi}_p(t)$, a good P-spline smoother $\hat{\eta}(t)$ is obtained via properly controlling the roughness of $\hat{\eta}(t)$. Specifically, when $\mathbf{\Phi}_p(t)$ is a k-th degree truncated power basis (8.12) with knots $\tau_1, \cdots, \tau_K$, a P-spline smoother $\hat{\eta}(t)$ can be obtained as $\mathbf{\Phi}_p(t)^T \hat{\boldsymbol{\beta}}$ where $\hat{\boldsymbol{\beta}}$, together with $\hat{\boldsymbol{\alpha}}$, minimizes the following penalized least squares (PLS) criterion:

$$\sum_{i=1}^{n} \sum_{j=1}^{n_i} \left(y_{ij} - \mathbf{c}_{ij}^T \boldsymbol{\alpha} - \mathbf{x}_{ij}^T \boldsymbol{\beta} \right)^2 + \lambda \boldsymbol{\beta}^T \mathbf{G} \boldsymbol{\beta}, \tag{8.14}$$

where $\mathbf{x}_{ij} = \mathbf{\Phi}_p(t_{ij})$, λ is the smoothing parameter and $\mathbf{G}$ is the roughness matrix for the P-spline based on $\mathbf{\Phi}_p(t)$. As in (7.6), $\mathbf{G}$ is defined as

$$\mathbf{G} = \begin{bmatrix} \mathbf{0}_{(k+1)\times(k+1)} & \mathbf{0}_{(k+1)\times K} \\ \mathbf{0}_{K\times(k+1)} & \mathbf{I}_K \end{bmatrix}. \tag{8.15}$$

Notice again that the roughness matrix $\mathbf{G}$ is chosen so that we can penalize the k-times derivative jumps of the regression spline $\eta(t) = \mathbf{\Phi}_p(t)^T\boldsymbol{\beta}$; see Chapter 5 for details.

The minimizers of the PLS criterion (8.14) may not be unique. A sufficient condition for the uniqueness is that the matrix $[\mathbf{C}, \mathbf{X}]$ is of full rank where $\mathbf{C}$ is defined as in the last subsection, while

$$\mathbf{X} = [\mathbf{X}_1^T, \cdots, \mathbf{X}_n^T]^T, \ \mathbf{X}_i = [\mathbf{x}_{i1}, \cdots, \mathbf{x}_{in_i}]^T. \tag{8.16}$$

A necessary condition for $[\mathbf{C}, \mathbf{X}]$ to be of full rank is that both $\mathbf{C}$ and $\mathbf{X}$ are of full rank, and $\mathbf{X}$ is linearly independent of $\mathbf{C}$. Without loss of generality, we always assume that $\mathbf{C}$ is of full rank since when the columns of $\mathbf{C}$ are linearly dependent, and we always can remove some columns of $\mathbf{C}$ so that the remaining columns of $\mathbf{C}$ are linearly independent. We can also assume that $\mathbf{X}$ is linearly independent of $\mathbf{C}$ since when there are some columns of $\mathbf{X}$ that are linearly dependent with $\mathbf{C}$, we always can remove these columns. For example, when the first column of $\mathbf{C}$ is $\mathbf{1}_N$, we can remove the first column of $\mathbf{X}$, equivalently the first basis function of $\mathbf{\Phi}_p(t)$. Since $\mathbf{\Phi}_p(t)$ is a basis, a sufficient and necessary condition for $\mathbf{X}$ to be of full rank is that $M \geq p$ where M is the number of distinct design time points. This gives the upper limit of K as $M - k - 1$.

In a compact way, the PLS criterion (8.14) can be further written as

$$\|\mathbf{y} - \mathbf{C}\boldsymbol{\alpha} - \mathbf{X}\boldsymbol{\beta}\|^2 + \lambda\boldsymbol{\beta}^T\mathbf{G}\boldsymbol{\beta}, \tag{8.17}$$

where as before, $\mathbf{y} = [\mathbf{y}_1^T, \cdots, \mathbf{y}_n^T]^T$ with $\mathbf{y}_i = [y_{i1}, \cdots, y_{in_i}]^T$. By some simple algebra, we can show that the minimizers of (8.17) can be expressed as

$$\begin{aligned} \hat{\boldsymbol{\alpha}} &= (\mathbf{C}^T\mathbf{P}\mathbf{C})^{-1}\mathbf{C}^T\mathbf{P}\mathbf{y}, \\ \hat{\boldsymbol{\beta}} &= (\mathbf{X}^T\mathbf{Q}\mathbf{X} + \lambda\mathbf{G})^{-1}\mathbf{X}^T\mathbf{Q}\mathbf{y}, \end{aligned} \tag{8.18}$$

where $\mathbf{P} = \mathbf{I}_N - \mathbf{X}(\mathbf{X}^T\mathbf{X}+\lambda\mathbf{G})^{-1}\mathbf{X}^T$ and $\mathbf{Q} = \mathbf{I}_N - \mathbf{C}(\mathbf{C}^T\mathbf{C})^{-1}\mathbf{C}$. It follows that the smoother matrix for the parametric component is $\mathbf{A}_{par} = \mathbf{C}(\mathbf{C}^T\mathbf{P}\mathbf{C})^{-1}\mathbf{C}^T\mathbf{P}$ with $\text{df}_{par} = \text{tr}(\mathbf{A}_{par}) = p_0$; moreover,

$$\hat{\eta}(t) = \mathbf{\Phi}_p(t)^T(\mathbf{X}^T\mathbf{Q}\mathbf{X} + \lambda\mathbf{G})^{-1}\mathbf{X}^T\mathbf{Q}\mathbf{y}, \tag{8.19}$$

with the smoother matrix for the nonparametric component $\mathbf{A}_{npar} = \mathbf{X}(\mathbf{X}^T\mathbf{Q}\mathbf{X} + \lambda\mathbf{G})^{-1}\mathbf{X}^T\mathbf{Q}$.

To construct the approximate confidence intervals (CIs) for each of the components of $\boldsymbol{\alpha}$, we need to compute $\text{Cov}(\hat{\boldsymbol{\alpha}})$. As mentioned earlier, here and throughout, the terms "expectation, variance and covariance" are understood to be conditional to the given covariate values:

$$(\mathbf{c}_{ij}, t_{ij}),\ j = 1, 2, \cdots, n_i;\ i = 1, 2, \cdots, n.$$

For this purpose, we may make a working independence assumption for the errors e_{ij} in the SPM model (8.1). That is, e_{ij} are independently and identically distributed with mean 0 and a common working variance σ^2. Under this working independence assumption, we have $\text{Cov}(\mathbf{y}) = \sigma^2 \mathbf{I}_N$. By (8.18), we then have

$$\text{Cov}(\hat{\boldsymbol{\alpha}}) = \sigma^2 (\mathbf{C}^T \mathbf{P} \mathbf{C})^{-1} \mathbf{C}^T \mathbf{P}^2 \mathbf{C} (\mathbf{C}^T \mathbf{P} \mathbf{C})^{-1}.$$

Based on the above expression, when σ^2 can be estimated by its estimator $\hat{\sigma}^2$, e.g., (8.23) given later, for some given λ, it is easy to construct the approximate CIs for each of the components of $\boldsymbol{\alpha}$.

To construct the pointwise standard deviation (SD) band of $\eta(t)$, by (8.19) and under the working independence assumption, we have

$$\text{Var}(\hat{\eta}(t)) = \sigma^2 \boldsymbol{\Phi}_p(t)^T (\mathbf{X}^T \mathbf{Q} \mathbf{X} + \lambda \mathbf{G})^{-1} \mathbf{X}^T \mathbf{Q}^2 \mathbf{X} (\mathbf{X}^T \mathbf{Q} \mathbf{X} + \lambda \mathbf{G})^{-1} \boldsymbol{\Phi}_p(t). \tag{8.20}$$

Let $\widehat{\text{Var}}(\hat{\eta}(t))$ denote the above expression after replacing the σ^2 by its estimator $\hat{\sigma}^2$, say (8.23). Then the 95% pointwise SD band of $\eta(t)$ can be constructed as

$$\hat{\eta}(t) \pm 1.96 \sqrt{\widehat{\text{Var}}(\hat{\eta}(t))}. \tag{8.21}$$

To select the smoothing parameter λ, we can adapt the GCV rule (8.11). To this end, we need to define the fitted response vector $\hat{\mathbf{y}}$ and the number of degrees of freedom df based on the above P-spline method. It is easy to define

$$\hat{\mathbf{y}} = \mathbf{C}\hat{\boldsymbol{\alpha}} + \mathbf{X}\hat{\boldsymbol{\beta}} = \mathbf{A}_{par}\mathbf{y} + \mathbf{A}_{npar}\mathbf{y} = \mathbf{A}\mathbf{y}, \tag{8.22}$$

where $\mathbf{A} = \mathbf{A}_{par} + \mathbf{A}_{npar}$ is the smoother matrix for the fitted responses at all the design time points while we define $\text{df} = \text{df}(\lambda) = \text{tr}(\mathbf{A}) = p_0 + \text{tr}(\mathbf{A}_{npar})$. We then adapt the GCV rule (8.11) as

$$\text{GCV}(\lambda) = \frac{(\mathbf{y} - \hat{\mathbf{y}})^T \mathbf{W} (\mathbf{y} - \hat{\mathbf{y}})}{[1 - \text{df}(\lambda)/N]^2}.$$

In the above GCV rule, the numerator measures the goodness of fit of the P-spline estimator of the responses, and the denominator is a monotone function of the model complexity.

By using $\hat{\mathbf{y}}$ and $\text{df}(\lambda)$, an estimator of the working noise variance σ^2 can be constructed as

$$\hat{\sigma}^2 = \|\mathbf{y} - \hat{\mathbf{y}}\|^2 / [N - \text{df}(\lambda)]. \tag{8.23}$$

The P-spline estimators $\hat{\boldsymbol{\alpha}}$ and $\hat{\eta}(t)$ can be connected to the BLUPs of a LME model when we reparameterize the PLS criterion (8.17). This provides an alternative method for minimizing the PLS criterion (8.17) and conducting inferences about the parametric covariate effect $\boldsymbol{\alpha}$ and the nonparametric effect $\eta(t)$.

Let $\boldsymbol{\delta}$ and $\mathbf{u}$ contain the first $(k+1)$ and the last K components of $\boldsymbol{\beta}$, respectively. Correspondingly, let $\tilde{\mathbf{X}}$ and $\tilde{\mathbf{Z}}$ be the matrices consisting of the first $(k+1)$ and the last K columns of $\mathbf{X}$. Due to the special structure of $\mathbf{G}$ (8.15), the PLS criterion (8.17) can be re-expressed as

$$\|\mathbf{y} - \mathbf{C}\boldsymbol{\alpha} - \tilde{\mathbf{X}}\boldsymbol{\delta} - \tilde{\mathbf{Z}}\mathbf{u}\|^2 + \lambda \mathbf{u}^T\mathbf{u}. \tag{8.24}$$

Under the working independence assumption, i.e., the errors e_{ij} are independently and normally distributed, the minimizers of (8.24) are the BLUPs of the following LME model:

$$\begin{array}{rcl} \mathbf{y} & = & \mathbf{C}\boldsymbol{\alpha} + \tilde{\mathbf{X}}\boldsymbol{\delta} + \tilde{\mathbf{Z}}\mathbf{u} + \mathbf{e}, \\ \mathbf{u} & \sim & N(0, \tau^2\mathbf{I}_K), \ \mathbf{e} \sim N(0, \sigma^2\mathbf{I}_N), \end{array} \tag{8.25}$$

where $\tau^2 = \sigma^2/\lambda$ and $\mathbf{e} = [\mathbf{e}_1^T, \cdots, \mathbf{e}_n^T]^T$ with $\mathbf{e}_i = [e_{i1}, \cdots, e_{in_i}]^T$. In the above LME model, the fixed-effects and random-effects vectors are $[\boldsymbol{\alpha}^T, \boldsymbol{\delta}^T]^T$ and $\mathbf{u}$, respectively. Correspondingly, the fixed-effects and random-effects covariate matrices are $[\mathbf{C}, \tilde{\mathbf{X}}]$ and $\tilde{\mathbf{Z}}$, respectively, and the smoothing parameter λ can be interpreted as the ratio of the two variance components, i.e., $\lambda = \sigma^2/\tau^2$.

The LME model (8.25) can be solved using the S-PLUS function *lme* or the SAS procedure *PROC MIXED* to obtain the BLUPs $\hat{\boldsymbol{\alpha}}, \hat{\boldsymbol{\delta}}, \hat{\mathbf{u}}$ and the variance component estimates $\hat{\tau}^2, \hat{\sigma}^2$. Then $\hat{\boldsymbol{\beta}} = [\hat{\boldsymbol{\delta}}^T, \mathbf{u}^T]^T$ so that

$$\hat{\eta}(t) = \boldsymbol{\Phi}_p(t)^T\hat{\boldsymbol{\beta}} = \sum_{r=0}^{k} \hat{\delta}_r t^r + \sum_{l=1}^{K} \hat{u}_l (t - \tau_l)_+^k.$$

The fitted responses at all the design time points can be summarized as $\hat{\mathbf{y}} = \mathbf{C}\hat{\boldsymbol{\alpha}} + \tilde{\mathbf{X}}\hat{\boldsymbol{\delta}} + \tilde{\mathbf{Z}}\hat{\mathbf{u}}$ where $\mathbf{C}\hat{\boldsymbol{\alpha}}$ is the estimated parametric component and $\tilde{\mathbf{X}}\hat{\boldsymbol{\delta}} + \tilde{\mathbf{Z}}\hat{\mathbf{u}}$ is the estimated nonparametric component. The standard deviations of $\hat{\mathbf{y}}$ can also be obtained by using the covariance matrices of $\hat{\boldsymbol{\alpha}}, \hat{\boldsymbol{\delta}}$ and $\hat{\mathbf{u}}$. The proper smoothing parameter obtained by solving the LME model (8.25) is $\hat{\sigma}^2/\hat{\tau}^2$.

8.2.5 Smoothing Spline Method

For independent data, Green et al. (1985), Engle et al. (1986) and Heckman (1986) considered fitting a semiparametric model $z_i = \mathbf{u}_i^T\boldsymbol{\delta} + f(t_i) + \epsilon_i, \ i = 1, 2, \cdots, n$ using smoothing splines. Let $\mathbf{S}$ be the smoothing spline smoother matrix associated with the smoothing spline smoother $\hat{f}(t)$ for the standard nonparametric regression model $z_i = f(t_i) + \epsilon_i, \ i = 1, 2, \cdots, n$ such that $\hat{\mathbf{f}} = \mathbf{S}\mathbf{z}$ where $\mathbf{z} = [z_1, \cdots, z_n]^T$ is the response vector and $\hat{\mathbf{f}} = [\hat{f}(t_1), \cdots, \hat{f}(t_n)]^T$ is the estimator of $\mathbf{f} = [f(t_1), \cdots, f(t_n)]^T$. For detailed construction of such a smoothing spline smoother, see Section 3.4 of Chapter 3 or Wahba (1990), Green and Silverman (1994), and Eubank (1988, 1999) among others. Then Heckman (1986)'s estimators of $\boldsymbol{\delta}$ and

$\mathbf{f}$ can be written as

$$\begin{aligned}\hat{\delta} &= \left[\mathbf{U}^T(\mathbf{I}_n - \mathbf{S})\mathbf{U}\right]^{-1}\mathbf{U}^T(\mathbf{I}_n - \mathbf{S})\mathbf{z}, \\ \hat{\mathbf{f}} &= \mathbf{S}(\mathbf{z} - \mathbf{U}\hat{\delta}),\end{aligned} \tag{8.26}$$

where $\mathbf{U} = [\mathbf{u}_1, \cdots, \mathbf{u}_n]^T$. The above estimators (8.26) are similar to Speckman's backfitting kernel estimators (8.3). Heckman (1986) showed that under mild conditions, $\hat{\delta}$ is a $\sqrt{n}$ - consistent estimator of δ with no need of undersmoothing $f(t)$ and it is asymptotically normal. Since $\mathbf{S}$ is symmetric, the estimators, similar to Speckman's profile kernel estimators (8.4), can be expressed as

$$\begin{aligned}\hat{\delta} &= \left[\mathbf{U}^T(\mathbf{I}_n - \mathbf{S})^2\mathbf{U}\right]^{-1}\mathbf{U}^T(\mathbf{I}_n - \mathbf{S})^2\mathbf{z}, \\ \hat{\mathbf{f}} &= \mathbf{S}(\mathbf{z} - \mathbf{U}\hat{\delta}).\end{aligned} \tag{8.27}$$

To extend the above backfitting and profile smoothing spline estimators for the SPM model (8.1), let $\mathbf{A}$ be the smoothing spline smoother matrix associated with the smoothing spline smoother $\hat{\eta}(t)$ for the NPM model $y_{ij} = \eta(t_{ij}) + e_{ij},\ j = 1, 2, \cdots, n_i;\ i = 1, 2, \cdots, n$. Such a smoothing spline smoother has been described in Section 6.2 of Chapter 6. Then the backfitting smoothing spline estimators of $\boldsymbol{\alpha}$ and $\boldsymbol{\eta} = [\boldsymbol{\eta}_1^T, \cdots, \boldsymbol{\eta}_n^T]^T, \boldsymbol{\eta}_i = [\eta(t_{i1}), \cdots, \eta(t_{in_i})]^T$ for the SPM model (8.1) are

$$\begin{aligned}\hat{\boldsymbol{\alpha}} &= \left(\mathbf{C}^T(\mathbf{I}_N - \mathbf{A})\mathbf{C}\right)^{-1}\mathbf{C}^T(\mathbf{I}_N - \mathbf{A})\mathbf{y}, \\ \hat{\boldsymbol{\eta}} &= \mathbf{A}(\mathbf{y} - \mathbf{C}\hat{\boldsymbol{\alpha}}),\end{aligned} \tag{8.28}$$

and the profile smoothing spline estimators are

$$\begin{aligned}\hat{\boldsymbol{\alpha}} &= (\mathbf{C}^T(\mathbf{I}_N - \mathbf{A})^2\mathbf{C})^{-1}\mathbf{C}^T(\mathbf{I}_N - \mathbf{A})^2\mathbf{y}, \\ \hat{\boldsymbol{\eta}} &= \mathbf{A}(\mathbf{y} - \mathbf{C}\hat{\boldsymbol{\alpha}}),\end{aligned} \tag{8.29}$$

where $\mathbf{C} = [\mathbf{C}_1^T, \cdots, \mathbf{C}_n^T]^T$ with $\mathbf{C}_i = [\mathbf{c}_{i1}, \cdots, \mathbf{c}_{in_i}]^T,\ i = 1, 2, \cdots, n$, and $\mathbf{y} = [\mathbf{y}_1^T, \cdots, \mathbf{y}_n^T]^T$ with $\mathbf{y}_i = [y_{i1}, \cdots, y_{in_i}]^T$.

It is straightforward to show that, when $\hat{\eta}(t)$ is the $(2k-1)$-st degree smoothing spline smoother for the NPM model $y_{ij} = \eta(t_{ij}) + e_{ij},\ j = 1, 2, \cdots, n_i;\ i = 1, 2, \cdots, n$, the backfitting smoothing spline estimators $\hat{\boldsymbol{\alpha}}$ and $\hat{\eta}(t)$ (or equivalently $\hat{\boldsymbol{\eta}}$) defined in (8.28) are the minimizers of the following PLS criterion:

$$\sum_{i=1}^{n}\sum_{j=1}^{n_i}[y_{ij} - \mathbf{c}_{ij}^T\boldsymbol{\alpha} - \eta(t_{ij})]^2 + \lambda\int_a^b [\eta^{(k)}(t)]^2 dt,$$

with respect to $\boldsymbol{\alpha}$ and $\eta(t)$ by assuming that $e_{ij} \sim_{i.i.d} N(0, \sigma^2),\ j = 1, 2, \cdots, n_i;\ i = 1, 2, \cdots, n$ and $\eta(t)$ is a function of the k-th order Sobolev space $\mathcal{W}_2^k[a, b]$ defined in (3.32) of Chapter 3.

Similarly, the profile smoothing spline estimators defined in (8.29) can be obtained by using profile smoothing as described in Section 8.2.2 except now the original LPK smoother matrix $\mathbf{A}$ should be replaced by the currently defined smoothing spline smoother $\mathbf{A}$.

The smoothing parameter λ can be selected using the GCV rule (8.11) if we re-define the smoother matrix $\mathbf{A}^*$ and the number of degrees of freedom df based on the estimators (8.28) or (8.29). We believe that under mild conditions, the estimators of $\boldsymbol{\alpha}$ defined in (8.28) and (8.29) are $\sqrt{n}$ - consistent and semiparametrically efficient (Lin and Carroll 2001a).

8.2.6 Methods Involving No Smoothing

In this section, we describe two methods which involve no smoothing in the estimation scheme. The first one, due to Lin and Ying (2001) [see also Martinussen and Scheike (1999), Cheng and Wei (2000), He, Zhu and Fung (2002) and Sun and Wu (2005), among others], uses a counting process and is called the counting process method. The second one, due to Fan and Li (2004), is a difference-based method.

To describe the counting process method, we write the continuous version of the SPM model (8.1) as

$$y_i(t) = \mathbf{c}_i(t)^T\boldsymbol{\alpha} + \eta(t) + e_i(t),\ i = 1, 2, \cdots, n. \tag{8.30}$$

Alternatively, we call the SPM model (8.1) as a discrete version of the SPM model (8.30) where $y_{ij}, \mathbf{c}_{ij}$ and e_{ij} are the values of $y_i(t), \mathbf{c}_i(t)$ and $e_i(t)$ at the design time point t_{ij}. The counting process method assumes that the design time points $t_{ij}, j = 1, \cdots, n_i$ for subject i are the realizations of the following counting process:

$$N_i(t) = \sum_{j=1}^{n_i} 1_{\{t_{ij}\leq t\}},$$

where 1_B is the indicator function of a set B. Both $y_i(t)$ and $\mathbf{c}_i(t)$ are observed at the jump points of $N_i(t)$. The observation times are considered as realizations from an arbitrary counting process that is censored at the end of follow-up. Specifically, $N_i(t) = N_i^*(\min(t, \tau_i))$ where $N_i^*(t)$ is a counting process in discrete or continuous time, and τ_i is the follow-up or censoring time. The censoring time τ_i may depend on the covariate vector $\mathbf{c}_i(t)$ in an arbitary manner. The censoring mechanism is assumed to be noninformative in the sense that

$$\mathrm{E}[y_i(t)|\mathbf{c}_i(t), \tau_i \geq t] = \mathrm{E}[y_i(t)|\mathbf{c}_i(t)].$$

Using the counting process notation, one can fit the SPM model (8.30) using the following weighted least squares (WLS) criterion:

$$\sum_{i=1}^{n} \int w_i(t)\left[y_i(t) - \mathbf{c}_i(t)^T\boldsymbol{\alpha} - \eta(t)\right]^2 dN_i(t), \tag{8.31}$$

where $w_i(t)$ are weight functions.

Lin and Ying (2001) first estimated $\eta(t)$ under two situations, based on whether the potential observation times depend on the covariates $\mathbf{c}_i(t)$. When the potential observation times depend on the covariates, they assumed that

$$\mathrm{E}\left[dN_i^*(t)|\mathbf{c}_i(t), y_i(t), \tau_i \geq t\right] = \exp\left(\mathbf{c}_i(t)^T\boldsymbol{\gamma}\right)dF(t),\ i = 1, 2, \cdots, n,$$

where γ is an unknown parameter vector and $F(t)$ is an arbitary nondecreasing function. When γ is 0, the observation times are independent of the covariates. Let

$$\bar{\mathbf{c}}(t;\gamma) = \frac{\sum_{i=1}^{n} 1_{\{\tau_i \geq t\}} \exp(\mathbf{c}_i(t)^T\gamma)\mathbf{c}_i(t)}{\sum_{i=1}^{n} 1_{\{\tau_i \geq t\}} \exp(\mathbf{c}_i(t)^T\gamma)},$$

$$\bar{y}(t;\gamma) = \frac{\sum_{i=1}^{n} 1_{\{\tau_i \geq t\}} \exp(\mathbf{c}_i(t)^T\gamma)y_i(t)}{\sum_{i=1}^{n} 1_{\{\tau_i \geq t\}} \exp(\mathbf{c}_i(t)^T\gamma)}.$$

Then Lin and Ying (2001) estimated $\eta(t)$ by

$$\hat{\eta}(t;\boldsymbol{\alpha},\gamma) = \bar{y}(t;\gamma) - \bar{\mathbf{c}}(t;\gamma)^T\boldsymbol{\alpha}.$$

In practice, $\bar{\mathbf{c}}(t;\gamma)$ and $\bar{y}(t;\gamma)$ may not be evaluable because $\mathbf{c}_i(t)$ and $y_i(t)$ are not fully observed until censoring time τ_i. Lin and Ying (2001) replaced $\mathbf{c}_i(t)$ and $y_i(t)$ with the associated observed values at the nearest time points. Therefore, replacing $\eta(t)$ with $\hat{\eta}(t;\boldsymbol{\alpha},\gamma)$ in (8.31) yields

$$\sum_{i=1}^{n} \int w_i(t) \Big[\{y_i(t) - \bar{y}(t;\gamma)\} - \{\mathbf{c}_i(t) - \bar{\mathbf{c}}(t;\gamma)\}^T\boldsymbol{\alpha}\Big]^2 dN_i(t), \qquad (8.32)$$

where the parameter γ can be estimated, using the method of moments, as the solution to

$$\sum_{i=1}^{n} \int [\mathbf{c}_i(t) - \bar{\mathbf{c}}(t;\gamma)]dN_i(t) = 0.$$

An explicit form for $\hat{\boldsymbol{\alpha}}$ can then be obtained by replacing $\hat{\gamma}$ for γ in (8.32). Lin and Ying (2001) showed that $\hat{\boldsymbol{\alpha}}$ is $\sqrt{n}$ - consistent and asymptotically normally distributed. The standard error of $\hat{\boldsymbol{\alpha}}$ can be constructed using the corresponding sandwich formula given by Lin and Ying (2001).

One drawback of Lin and Ying's method is the unrealistic assumption that the responses $y_i(t)$ and the covariates $\mathbf{c}_i(t)$ are fully observable until the censoring time τ_i. To overcome this difficulty, as mentioned previously, Lin and Ying (2001) replaced $y_i(t)$ and $\mathbf{c}_i(t)$ with the associated values at the nearest time points as their observed values. Although this treatment helps with the practical implementation of their method, it introduces biases due to the nearest-neighbor approximations. Moreover, because the spaces among the observation times for each subject do not tend to 0 even when the number of subjects n tends to infinity, the approximation biases will not always be negligible in practice. To overcome these problems, Fan and Li (2004) proposed a difference-based approach for estimating $\boldsymbol{\alpha}$ for the SPM model (8.30), which is briefly described as follows.

Sort $(y_{ij}, t_{ij}, \mathbf{c}_{ij}),\ j = 1, 2, \cdots, n_i;\ i = 1, 2, \cdots, n$ according to the values of $t_{ij},\ i = 1, 2, \cdots, n$ and denote the resultant data set as $(y_{(l)}, t_{(l)}, \mathbf{c}_{(l)}), l = 1, 2, \cdots, N$ where $t_{(1)} \leq t_{(2)} \leq \cdots \leq t_{(N)}$ are the order statistics of the design time points $t_{ij},\ j = 1, 2, \cdots, n_i;\ i = 1, 2, \cdots, n$. Then we can write the SPM model (8.1) as

$$y_{(l)} = \mathbf{c}_{(l)}^T\boldsymbol{\alpha} + \eta(t_{(l)}) + e_{(l)}, l = 1, \cdots, N,$$

where $e_{(l)}, l = 1, \cdots, N$ may be regarded as the sorted measurement errors. Then for $l = 1, \cdots, N-1$,

$$y_{(l+1)} - y_{(l)} = [\mathbf{c}_{(l+1)} - \mathbf{c}_{(l)}]^T \boldsymbol{\alpha} + [\eta(t_{(l+1)}) - \eta(t_{(l)})] + [e_{(l+1)} - e_{(l)}]. \quad (8.33)$$

When the differences $t_{(l+1)} - t_{(l)}$ are small, the differences $\eta(t_{(l+1)}) - \eta(t_{(l)})$ in the above expression are also small by the continuity assumption of $\eta(t)$. In this case, the differences $[\eta(t_{(l+1)}) - \eta(t_{(l)})]$ in the expression (8.33) can be ignored, and the LS or GLS method can be used to estimate $\boldsymbol{\alpha}$. To further improve the above difference-based estimator $\hat{\boldsymbol{\alpha}}$, Fan and Li (2004) suggested approximating the difference $\eta(t_{(l+1)}) - \eta(t_{(l)})$ by a linear function $\delta_0 + \delta_1[t_{(l+1)} - t_{(l)}]$ to correct the finite-sample bias when the spacing gap is wide. That is, the improved estimator $\hat{\boldsymbol{\alpha}}$ can be obtained via fitting the following linear model for $\hat{\boldsymbol{\alpha}}$ and $\boldsymbol{\delta} = [\delta_0, \delta_1]^T$:

$$y_{(l+1)} - y_{(l)} = [\mathbf{c}_{(l+1)} - \mathbf{c}_{(l)}]^T \boldsymbol{\alpha} + [\delta_0 + \delta_1(t_{(l+1)} - t_{(l)})] + [e_{(l+1)} - e_{(l)}].$$

The difference-based method is simple to implement. However it requires that the adjacent $t_{(l+1)}$ and $t_{(l)}$ are close to each other for all $l = 1, 2, \cdots, N$, which is not a realistic assumption for many longitudinal data sets. Therefore, we should use the difference-based method with care.

8.2.7 MACS Data

The MACS data were introduced in Section 1.1.3 of Chapter 1. The data are from the Multi-center AIDS Cohort Study (MACS) and contain the HIV status of 283 homosexual men who were infected with HIV during the follow-up period between 1984 and 1991. Three covariates were assessed in this study, namely "Smoking" or X_1, "Age" or X_2, and "PreCD4" or X_3. We are first interested in applying the SPM model (8.1) to the CD4 data via fitting the following SPM model:

$$y_{ij} = X_{1i}\beta_1 + X_{2i}\beta_2 + X_{3i}\beta_3 + \eta(t_{ij}) + e_{ij}, \quad (8.34)$$

where y_{ij} is the percentage of CD4 cells for subject i at time t_{ij}, the coefficients β_1, β_2 and β_3 are used to model the constant effects of the three covariates, the smooth function $\eta(t)$ is used to model the time effect, and the e_{ij} are the errors.

The aim of this study was to assess the effects of cigarette smoking, age at HIV infection and pre-HIV infection CD4 cell percentage on the CD4 cell percentage depletion among the HIV infected homosexual men. For a clear biological interpretation of the covariate effects, a centralization of the covariates Age and PreCD4 is needed before the above SPM model can be fitted. The new covariates Age and PreCD4 are obtained via subtracting the old ones by their sample means, respectively. As a result, the nonparametric component function $\eta(t)$ in the above model can be interpretated as the intercept CD4 percentage, i.e., the mean CD4 percentage curve for a nonsmoker with an average pre-infection CD4 percentage and an average age at HIV infection. See Wu and Chiang (2000) for a detailed explanation for advantages of such a normalization.

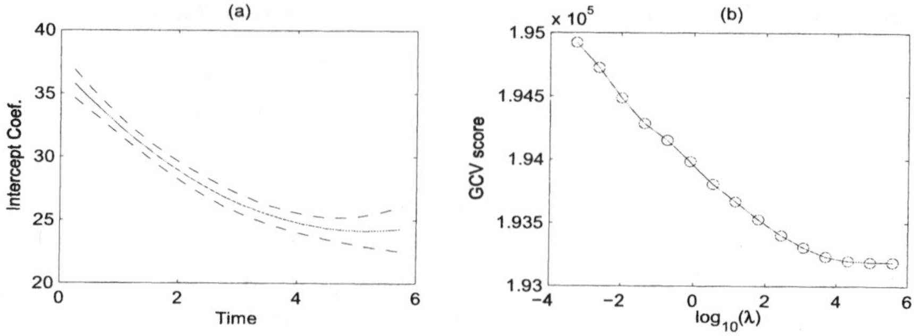

Fig. 8.1 SPM model (8.34) fit to the MACS data: (a) estimated intercept CD4 percentage curve with 95% pointwise SD band , and (b) GCV curve.

We fit the SPM model (8.34) to the MACS data using the P-spline method described previously. The number of knots for the quadratic truncated power basis was taken as $K = 16$, and the knots were scattered using the "equally spaced sample quantiles as knots" method. The smoothing parameter $\lambda = 3.8703 \times 10^5$ was selected by the GCV rule (8.11). The GCV curve presented in Figure 8.1 (b) indicates that the GCV rule favors large smoothing parameters. This implies that the quadratic polynomial model may be adequate to fit the intercept CD4 cell percentage curve. In fact, the fitted intercept CD4 cell percentage curve in Figure 8.1 (a) looks like a quadratic polynomial. Moreover, the estimated number of degrees of freedom for the estimated mean responses is about 6, with 3 dfs for the three covariates and another 3 dfs for the intercept CD4 percentage curve.

Table 8.1 SPM model (8.34) fit to the MACS data: estimated covariate effects, standard deviations and approximate z-test values for the parametric component. Quadratic truncated power basis with $K = 16$ knots was used. The smoothing parameter selected by GCV is $\lambda = 3.8703 \times 10^5$.

Covariate	Effect	SD	z-test value
Smoking	.6202	.5297	1.1709
Age	−.0722	.0322	−2.2438
PreCD4	.3756	.0300	12.5030

The estimated effects, standard deviations (SD) and approximate z-test values of the Smoking, Age and PreCD4 covaiates are presented in Table 8.1. It is seen that, although the Smoking effect is positive, it is not significant; the Age effect is significantly negative and the PreCD4 effect is significantly positive. These results indicate that patients' smoking habits may not affect the CD4 cell percentage levels after HIV infection, but the age and the CD4 percentage level before HIV infection may have a negative or positive impact, respectively. This is reasonable since it is expected that CD4 percentage levels among HIV-infected patients with older ages and lower pre-CD4 percentage levels will decline faster than those among patients with younger ages and higher pre-CD4 percentage levels.

Fan and Zhang (2000) fit a time-varying coefficient model to the above MACS data to illustrate their two-step method. That is, they assumed that the Smoking, Age, PreCD4 effects, and the intercept CD4 percentage level are smooth functions of time. In order to illustrate a more complex SPM model than the SPM model (8.34) for the MACS data, we introduce the following SPM model:

$$y_{ij} = X_{1i}\beta_1(t_{ij}) + X_{2i}\beta_2(t_{ij}) + X_{3i}\beta_3(t_{ij}) + \eta(t_{ij}) + e_{ij}, \tag{8.35}$$

where

$$\begin{aligned}\beta_1(t) &= \beta_{10} + \beta_{11}t,\\ \beta_2(t) &= \beta_{20} + \beta_{21}t,\\ \beta_3(t) &= \beta_{30} + \beta_{31}t + \beta_{32}t^2.\end{aligned}$$

That is, the effects of the Smoking and Age covariates are linear functions of time, while the effect of the PreCD4 covariate is a quadratic polynomial function of time. For illustration purpose, we continue to model the intercept CD4 percentage level as a nonparametric function of time, namely $\eta(t)$ in the SPM model (8.35) although based on the previous analysis, a quadratic polynomial model may be more appropriate to fit the intercept CD4 percentage level.

We used the same method to fit the SPM model (8.34) as to fit the SPM model (8.35). The smoothing parameter is now $\lambda = 3.8397 \times 10^5$, selected by the GCV rule (8.11). Again, the GCV rule favors large smoothing parameters, indicating that a quadratic polynomial model may be appropriate for fitting the intercept CD4 percentage levels. The estimated number of degrees of freedom for the estimated mean responses is now about 10, with $2 + 2 + 3 = 7$ dfs for the three covariates and another 3 dfs for the intercept CD4 percentage curve.

Table 8.2 SPM model (8.35) fit to the MACS data: estimated coefficients, standard deviations and approximate z-test values for the parametric component. Quadratic truncated power basis with $K = 16$ knots was used. The smoothing parameter selected by GCV is $\lambda = 3.8397 \times 10^5$.

Covariate	Coef.	Estimated Coef.	SD	z-test value
Smoking	β_{10}	$-.7286$	.9434	$-.7723$
	β_{11}	.6672	.3438	1.9406
Age	β_{20}	.0741	.0569	1.3025
	β_{21}	$-.0679$	.0211	-3.2187
PreCD4	β_{30}	.5244	.0780	6.7230
	β_{31}	$-.1067$	.0665	-1.6050
	β_{32}	.0132	.0116	1.1348

The estimated coefficients of the new SPM model (8.35), together with their SDs and approximate z-test values are presented in Table 8.2, and the fitted intercept

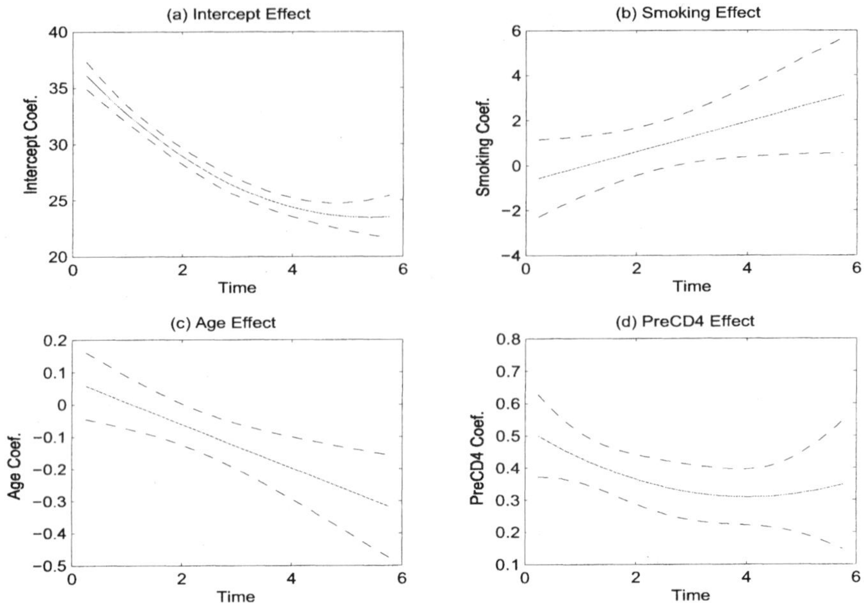

Fig. 8.2 SPM model (8.35) fit to the MACS data: (a) intercept CD4 percentage curve; (b) Smoking effect; (c) Age effect; and (d) PreCD4 effect. The dashed curves are 95% pointwise SD bands.

CD4 percentage curve and the covariate effects (solid curves), together with their approximate 95% pointwise SD bands (dashed-curves), are displayed in Figure 8.2. It is seen that the smoking effect increases over time but it is not significant since the approximate 95% pointwise SD band contains 0 at all time; the age effect decreases over time and the approximate 95% pointwise SD band does not contain 0 most of the time; and the pre-infection CD4 percentage levels are always positive and the approximate 95% pointwise SD band never contains 0. These results are consistent with those obtained by fitting the simpler SPM model (8.34) to the MACS data. In particular, both SPM model fits suggest that a quadratic polynomial model is adequate to fit the intercept CD4 percentage level.

8.3 SEMIPARAMETRIC MIXED-EFFECTS MODEL

8.3.1 Model Specification

In the SPM model (8.1), the mean response function at time t_{ij} depends on time t_{ij} nonparametrically via a smooth function $\eta(t)$, and linearly on some other observable covariates $\mathbf{c}_{ij} = [c_{1ij}, \cdots, c_{p_0ij}]^T$. To incorporate the within-subject correlation, we can extend the SPM model (8.1) by adding a random-effect component. The random-effect component at time t_{ij} may depend on time t_{ij} nonparametrically via a smooth process $v_i(\cdot)$ and linearly on some other covariates, namely $\mathbf{h}_{ij} = [h_{1ij}, \cdots, h_{q_0ij}]^T$.

The resulting model extends the SPM model (8.1) in one way, and extends the NPME model (7.39) in another. We call it a semiparametric mixed-effects (SPME) model, and write it as

$$\begin{aligned} y_{ij} &= \mathbf{c}_{ij}^T\boldsymbol{\alpha} + \eta(t_{ij}) + \mathbf{h}_{ij}^T\mathbf{a}_i + v_i(t_{ij}) + \epsilon_{ij}, \\ & \quad j = 1, 2, \cdots, n_i;\ i = 1, 2, \cdots, n, \end{aligned} \tag{8.36}$$

where $\boldsymbol{\alpha}$ and $\eta(t)$ are defined as in (8.1), $\mathbf{a}_i = [a_{1i}, \cdots, a_{q_0 i}]^T$ consists of the coefficients of the covariate vectors $\mathbf{h}_{ij}$, $v_i(t)$ is a smooth process of time, and ϵ_{ij} is the error at time t_{ij} that is not explained by either the fixed-effects component $\mathbf{c}_{ij}^T\boldsymbol{\alpha} + \eta(t_{ij})$ or the random-effects component $\mathbf{h}_{ij}^T\mathbf{a}_i + v_i(t_{ij})$. In the above SPME model, for simplicity, we assume that

$$\begin{aligned} \mathbf{a}_i \sim N(0, \mathbf{D}_a), \ \ v_i(t) \sim \mathrm{GP}(0, \gamma), \ \ \mathrm{E}[\mathbf{a}_i v_i(t)] = \boldsymbol{\gamma}_a(t), \\ \boldsymbol{\epsilon}_i = [\epsilon_{i1}, \cdots, \epsilon_{in_i}]^T \sim N(0, \mathbf{R}_i), \ \ i = 1, 2, \cdots, n, \end{aligned} \tag{8.37}$$

where as before $\mathrm{GP}(\mu, \gamma)$ denotes a Gaussian process with mean function $\mu(t)$ and covariance function $\gamma(s,t)$. That is, $\mathbf{a}_i,\ i = 1, 2, \cdots, n$ are independently normally distributed with mean 0 and covariance matrix $\mathbf{D}_a$, and $v_i(t),\ i = 1, 2, \cdots, n$ are independent and identical copies of a Gaussian process with mean function 0 and covariance function $\gamma(s,t)$, and $\boldsymbol{\epsilon}_i$ are independently normally distributed.

When the covariance function $\gamma(s,t)$ is known except an unknown parameter vector, the above SPME model has been discussed by Zhang et al. (1998) using a smoothing spline method. They used a connection of a smoothing spline to a LME model for estimating $\eta(t)$ and the variance components. Such a connection can be found in Section 3.4 of Chapter 3 and was discussed in Chapters 6 and 7. Jacqmin-Gadda et al. (2002) fit the above SPME model (8.36), also using smoothing splines but with smoothing parameters selected by a "leave-one-subject-out" likelihood based cross-validation method.

The SPME model (8.36) is very general. It consists of the following four major components:

parametric fixed-effects	$\mathbf{c}_{ij}^T\boldsymbol{\alpha}$,
nonparametric fixed-effects	$\eta(t_{ij})$,
parametric random-effects	$\mathbf{h}_{ij}^T\mathbf{a}_i$,
nonparametric random-effects	$v_i(t_{ij})$.

When some components are dropped, the SPME model will reduce to some special SPME models. When both the nonparametric fixed-effects and random-effects components are dropped from the model, it reduces to the usual LME model that we reviewed in Chapter 2. When both the parametric fixed-effects and random-effects components are dropped, it reduces to the usual NPME model that we investigated in Chapters 4~7. When both the parametric and nonparametric random-effects components are dropped from the model, it reduces to the SPM model (8.1).

For further discussion, we can write the SPME model (8.36) in vector or matrix notation. In vector notation, the SPME model (8.36) can be written as

$$\begin{array}{rcl} \mathbf{y}_i &=& \mathbf{C}_i\boldsymbol{\alpha} + \boldsymbol{\eta}_i + \mathbf{H}_i\mathbf{a}_i + \mathbf{v}_i + \boldsymbol{\epsilon}_i, \\ \mathbf{a}_i &\sim& N(0, \mathbf{D}_a),\ \mathbf{v}_i \sim N(0, \boldsymbol{\Gamma}_i), \mathrm{E}(\mathbf{a}_i\mathbf{v}_i^T) = \boldsymbol{\Gamma}_{ia}, \\ \boldsymbol{\epsilon}_i &\sim& N(0, \mathbf{R}_i),\ \ i = 1, 2, \cdots, n, \end{array} \tag{8.38}$$

where $\mathbf{y}_i = [y_{i1}, \cdots, y_{in_i}]^T$, $\boldsymbol{\eta}_i = [\eta(t_{i1}), \cdots, \eta(t_{in_i})]^T$, and $\mathbf{v}_i$ and $\boldsymbol{\epsilon}_i$ are similarly defined; $\mathbf{C}_i = [\mathbf{c}_{i1}, \cdots, \mathbf{c}_{in_i}]^T$, $\mathbf{H}_i = [\mathbf{h}_{i1}, \cdots, \mathbf{h}_{in_i}]^T$; the (k, l)-th entry of $\boldsymbol{\Gamma}_i$ is $\gamma(t_{ik}, t_{il})$ and the l-th column of $\boldsymbol{\Gamma}_{ia}$ is $\boldsymbol{\gamma}_a(t_{il})$. In matrix notation, the SPME model (8.38) can be further written as

$$\begin{array}{l} \mathbf{y} = \mathbf{C}\boldsymbol{\alpha} + \boldsymbol{\eta} + \mathbf{H}\mathbf{a} + \mathbf{v} + \boldsymbol{\epsilon}, \\ \mathbf{a} \sim N(0, \tilde{\mathbf{D}}_a),\ \mathbf{v} \sim N(0, \boldsymbol{\Gamma}), \\ \mathrm{E}(\mathbf{a}\mathbf{v}^T) = \boldsymbol{\Gamma}_a,\ \boldsymbol{\epsilon} \sim N(0, \mathbf{R}), \end{array} \tag{8.39}$$

where $\mathbf{y} = [\mathbf{y}_1^T, \cdots, \mathbf{y}_n^T]^T$, $\boldsymbol{\eta} = [\boldsymbol{\eta}_1^T, \cdots, \boldsymbol{\eta}_n^T]^T$, and $\mathbf{v}$ and $\boldsymbol{\epsilon}$ are similarly defined; $\mathbf{C} = [\mathbf{C}_1^T, \cdots, \mathbf{C}_n^T]^T$, $\mathbf{H} = \mathrm{diag}(\mathbf{H}_1, \cdots, \mathbf{H}_n)$, $\boldsymbol{\Gamma} = \mathrm{diag}(\boldsymbol{\Gamma}_1, \cdots, \boldsymbol{\Gamma}_n)$, $\boldsymbol{\Gamma}_a = \mathrm{diag}(\boldsymbol{\Gamma}_{1a}, \cdots, \boldsymbol{\Gamma}_{na})$, and $\mathbf{R} = \mathrm{diag}(\mathbf{R}_1, \cdots, \mathbf{R}_n)$.

Other special SPME models are obtained as follows when one or two SPME components are dropped from the general SPME model (8.36).

Type 1 SPME model: When the parametric fixed-effects component and the nonparametric random-effects component are dropped, the SPME model (8.36) reduces to the following Type 1 SPME model:

$$\begin{array}{c} y_{ij} = \eta(t_{ij}) + \mathbf{h}_{ij}^T\mathbf{a}_i + \epsilon_{ij}, \\ j = 1, 2, \cdots, n_i;\ i = 1, 2, \cdots, n. \end{array} \tag{8.40}$$

The random-effects component is purely parametric so that when $\mathbf{R}_i = \sigma^2\mathbf{I}_{n_i}$, $i = 1, 2, \cdots, n$, the covariance matrices $\mathbf{V}_i = \mathrm{Cov}(\mathbf{y}_i) = \mathbf{H}_i\mathbf{D}_a\mathbf{H}_i^T + \mathbf{R}_i$ just depend on a parameter vector. This allows incorporating the within-subject correlation to estimate $\eta(t)$ using a generalized smoothing method. See, for example, Section 6.3 of Chapter 6 for a generalized smoothing spline smoothing method. In fact, Wang (1998a) estimated $\eta(t)$ in the above model using a generalized smoothing spline method. Following Wang (1998a), Guo (2002b) considered likelihood ratio tests for the above Type 1 SPME model. Recently, this model was also considered by Durban et al. (2005) using P-splines.

Type 2 SPME model: When only the parametric fixed-effects component is dropped, the SPME model (8.36) reduces to the following Type 2 SPME model:

$$\begin{array}{c} y_{ij} = \eta(t_{ij}) + \mathbf{h}_{ij}^T\mathbf{a}_i + v_i(t_{ij}) + \epsilon_{ij}, \\ j = 1, 2, \cdots, n_i;\ i = 1, 2, \cdots, n. \end{array} \tag{8.41}$$

Type 3 SPME model: When the nonparametric fixed-effects and the parametric random-effects components are dropped, the SPME model (8.36) reduces to the following Type 3 SPME model:

$$\begin{aligned} y_{ij} &= \mathbf{c}_{ij}^T \boldsymbol{\alpha} + v_i(t_{ij}) + \epsilon_{ij}, \\ j &= 1, 2, \cdots, n_i;\ i = 1, 2, \cdots, n. \end{aligned} \tag{8.42}$$

This model will be used when a known parametric model is adequate for the fixed-effects component but a parametric model for the random-effects component is not available.

Type 4 SPME model: When only the nonparametric fixed-effects component is dropped, the SPME model (8.36) reduces to the following Type 4 SPME model:

$$\begin{aligned} y_{ij} &= \mathbf{c}_{ij}^T \boldsymbol{\alpha} + \mathbf{h}_{ij}^T \mathbf{a}_i + v_i(t_{ij}) + \epsilon_{ij}, \\ j &= 1, 2, \cdots, n_i;\ i = 1, 2, \cdots, n. \end{aligned} \tag{8.43}$$

Diggle (1988) considered a special case of the above model when the number of random-effects covariates $q_0 = 1$, and when $v_i(t)$ has a parametric covariance structure such as an auto-regression AR(1) model. This special case was also discussed by Donnelly, Laird and Ware (1995).

Type 5 SPME model: When only the nonparametric random-effects component is dropped, the SPME model (8.36) reduces to the following Type 5 SPME model:

$$\begin{aligned} y_{ij} &= \mathbf{c}_{ij}^T \boldsymbol{\alpha} + \eta(t_{ij}) + \mathbf{h}_{ij}^T \mathbf{a}_i + \epsilon_{ij}, \\ j &= 1, 2, \cdots, n_i;\ i = 1, 2, \cdots, n. \end{aligned} \tag{8.44}$$

The Type 5 SPME model is slightly more complicated than the Type 1 SPME model since it involves an extra term, i.e., the parametric fixed-effects component. Ruppert, Wand and Carroll (2003) dealt with a simple version (with $q_0 = 1$) of this type of SPME model using P-splines.

Type 6 SPME model: When the parametric random-effects component is dropped, the SPME model (8.36) reduces to the following Type 6 SPME model:

$$\begin{aligned} y_{ij} &= \mathbf{c}_{ij}^T \boldsymbol{\alpha} + \eta(t_{ij}) + v_i(t_{ij}) + \epsilon_{ij}, \\ j &= 1, 2, \cdots, n_i;\ i = 1, 2, \cdots, n. \end{aligned} \tag{8.45}$$

In Zeger and Diggle (1994), backfitting is used to estimate $\boldsymbol{\alpha}, \eta(t)$ and $v_i(t)$ separately. In particular, the kernel method is used for estimating the nonparametric fixed-effects component $\eta(t)$ with the bandwidth selected by "leave-one-subject-out" cross-validation. However, they did not take the smoothness of $v_i(t)$ into account. Guo (2002b) considered the above SPME model for his likelihood ratio test.

8.3.2 Local Polynomial Method

In this subsection, we consider methods for fitting the SPME model (8.36) using local polynomial kernel (LPK) smoothing (Fan and Gijbels 1996). We first briefly review a backfitting algorithm proposed by Zeger and Diggle (1994) for dealing with a special case of the SPME model (8.36), and then outline the ideas to generalize the algorithm for the general SPME model. In addition, we outline a profile LPK method for fitting the general SPME model.

Zeger and Diggle (1994) considered a special case of the SPME model (8.36) when the parametric random-effects $\mathbf{h}_{ij}^T\mathbf{a}_i$ are not present and when the nonparametric random-effect functions $v_i(t),\ i = 1, 2, \cdots, n$ are assumed to be stationary with a parametric covariance function $\gamma(s,t) = \sigma_v^2\rho(s-t;\boldsymbol{\theta})$ where σ_v^2 and $\boldsymbol{\theta}$ are parameters, and $\rho(\cdot)$ is a known function. It is clear that the Zeger and Diggle (ZD) model is a special case of the Type 6 SPME model (8.45). Notice that $\mathbf{V}_i = \text{Cov}(\mathbf{y}_i) = \mathbf{\Gamma}_i + \mathbf{R}_i$ and $\mathbf{V} = \text{Cov}(\mathbf{y}) = \text{diag}(\mathbf{V}_1, \cdots, \mathbf{V}_n)$ depend on parameters $\sigma_v^2, \boldsymbol{\theta}$ and σ^2 when $\mathbf{R}_i = \sigma^2\mathbf{I}_{n_i},\ i = 1, 2, \cdots, n$.

Zeger and Diggle proposed the following procedure to iteratively fit the ZD model:

(a) Given the current estimate $\hat{\boldsymbol{\alpha}}$, calculate the residuals $r_{ij} = y_{ij} - \mathbf{c}_{ij}^T\hat{\boldsymbol{\alpha}}$ and use these residuals as responses to calculate the kernel estimator (Nadaraya 1964, Watson 1964, Hart and Wehrly 1986, Fan 1992, 1993):

$$\hat{\eta}(t) = \frac{\sum_{i=1}^{n}\sum_{j=1}^{n_i} K_h(t_{ij}-t)r_{ij}}{\sum_{i=1}^{n}\sum_{j=1}^{n_i} K_h(t_{ij}-t)},$$

for a fixed design time point t in the range of interest where $K_h(\cdot) = K(\cdot/h)/h$ with $K(\cdot)$ a kernel function and h a bandwidth. The bandwith h is chosen using the "leave-one-subject-out" cross-validation proposed by Rice and Silverman (1991).

(b) Given the current kernel estimator $\hat{\eta}(t)$, calculate the residuals $s_{ij} = y_{ij} - \hat{\eta}(t_{ij})$ and obtain $\hat{\boldsymbol{\alpha}}$ by the GLS method as follows: $\hat{\boldsymbol{\alpha}} = (\mathbf{C}^T\mathbf{V}^{-1}\mathbf{C})^{-1}\mathbf{C}^T\mathbf{V}^{-1}\mathbf{s}$ where $\mathbf{s} = [\mathbf{s}_1^T, \cdots, \mathbf{s}_n^T]^T$ with $\mathbf{s}_i = [s_{i1}, \cdots, s_{in_i}]^T$.

Recall that $\mathbf{v}_i$ denotes the vector of the values of $v_i(t)$ for the i-th subject. Zeger and Diggle (1994) estimated $\mathbf{v}_i$ using the conditional expectation of $\mathbf{v}_i$ given $\mathbf{y}_i, \hat{\boldsymbol{\alpha}}$ and $\hat{\boldsymbol{\eta}}_i$, i.e.,

$$\hat{\mathbf{v}}_i = \text{Cov}(\mathbf{v}_i|\mathbf{y}_i, \hat{\boldsymbol{\alpha}}, \hat{\boldsymbol{\eta}}_i) = \mathbf{\Gamma}_i\mathbf{V}_i^{-1}(\mathbf{y}_i - \mathbf{C}_i\hat{\boldsymbol{\alpha}} - \hat{\boldsymbol{\eta}}_i),\ i = 1, 2, \cdots, n,$$

where $\hat{\boldsymbol{\eta}}_i$ denotes the values of $\hat{\eta}(t)$ for subject i as defined previously.

It is straightfoward to generalize the above Zeger and Diggle (1994)'s approach to the general SPME model (8.36). Moreover, the employed simple kernel method can be replaced by the general LPK method (Fan and Gijbels 1996).

Notice that when the parametric fixed-effects component $\mathbf{c}_{ij}^T\boldsymbol{\alpha}$ and the random-effects component $\mathbf{h}_{ij}^T\mathbf{a}_i$ are given, the SPME model (8.36) essentially reduces to the NPME model that we discussed in Chapter 4 by using LPK smoothing. Also notice that when the nonparametric fixed-effects/random-effects components $\eta(t_{ij})$ and $v_i(t_{ij})$ are given, the SPME model (8.36) essentially reduces to the standard LME model that we briefly reviewed in Chapter 2. Therefore, we can employ backfitting strategies (Hastie and Tibshirani 1990) to fit the SPME model (8.36) via iteratively fitting a standard LME model and an NPME model. The LPK backfitting algorithm for fitting the SPME model (8.36) can be written as

(a) Given the current estimates of $\boldsymbol{\alpha}$ and $\mathbf{a}_i,\ i = 1, 2, \cdots, n$, fit (8.36) using the LPK method described in Chapter 4, and obtain the estimates of $\eta(t)$ and $v_i(t),\ i = 1, 2, \cdots, n$.

(b) Given the current LPK estimates of $\eta(t)$ and $v_i(t),\ i = 1, 2, \cdots, n$, fit (8.36) using a standard LME fitting routine such as the S-PLUS function *lme* or the SAS procedure *PROC MIXED*, and obtain the estimates of $\boldsymbol{\alpha}$ and $\mathbf{a}_i,\ i = 1, 2, \cdots, n$.

Intuitively, the above LPK backfitting algorithm is simple. However, like all other backfitting based approaches, inferences about the SPME model (8.36) are often challenging. More research in this direction is definitely warranted.

We now describe a profile LPK method for fitting the SPME model (8.36). Let $\mathbf{A}$ and $\mathbf{A}_v$ be the LPK smoother matrices for fitting the NPME model $y_{ij} = \eta(t_{ij}) + v_i(t_{ij}) + \epsilon_{ij},\ j = 1, 2, \cdots, n_i;\ i = 1, 2, \cdots, n$ or in matrix notation $\mathbf{y} = \boldsymbol{\eta} + \mathbf{v} + \boldsymbol{\epsilon}$. See Chapter 4 for details for constructing $\mathbf{A}$ and $\mathbf{A}_v$. Then $\hat{\boldsymbol{\eta}} = \mathbf{A}\mathbf{y}$ and $\hat{\mathbf{v}} = \mathbf{A}_v\mathbf{y}$ where $\hat{\boldsymbol{\eta}}$ and $\hat{\mathbf{v}}$ are the LPK estimators of $\boldsymbol{\eta}$ and $\mathbf{v}$. It follows that when $\boldsymbol{\alpha}$ and $\mathbf{a}$ are known, by fitting (8.39) using LPK smoothing, we have the following profile smoothers of $\boldsymbol{\eta}$ and $\mathbf{v}$:

$$\begin{aligned} \hat{\boldsymbol{\eta}} &= \mathbf{A}(\mathbf{y} - \mathbf{C}\boldsymbol{\alpha} - \mathbf{H}\mathbf{a}), \\ \hat{\mathbf{v}} &= \mathbf{A}_v(\mathbf{y} - \mathbf{C}\boldsymbol{\alpha} - \mathbf{H}\mathbf{a}). \end{aligned} \tag{8.46}$$

Notice that both smoother matrices $\mathbf{A}$ and $\mathbf{A}_v$ depend on the covariance matrix $\mathbf{V}_v = \text{Cov}(\mathbf{v} + \boldsymbol{\epsilon}) = \boldsymbol{\Gamma} + \mathbf{R}$.

Replacing $\boldsymbol{\eta}$ and $\mathbf{v}$ in (8.39) with their profile smoothers, letting $\mathbf{P} = \mathbf{I}_N - \mathbf{A}^*$ where $\mathbf{A}^* = \mathbf{A} + \mathbf{A}_v$, and after simplification, we have

$$\mathbf{P}\mathbf{y} = \mathbf{P}\mathbf{C}\boldsymbol{\alpha} + \mathbf{P}\mathbf{H}\mathbf{a} + \boldsymbol{\epsilon}, \tag{8.47}$$

where again $\mathbf{a} \sim N(0, \tilde{\mathbf{D}}_a)$ and $\boldsymbol{\epsilon} \sim N(0, \mathbf{R})$. For given $\mathbf{A}^*$, (8.47) is a standard LME model. Let $\mathbf{V}_a = \mathbf{P}\mathbf{H}\tilde{\mathbf{D}}_a\mathbf{H}^T\mathbf{P}^T + \mathbf{R}$. Solving the above LME model (8.46) leads to the following estimators of $\boldsymbol{\alpha}$ and $\mathbf{a}$:

$$\begin{aligned} \hat{\boldsymbol{\alpha}} &= (\mathbf{C}^T\mathbf{P}^T\mathbf{V}_a^{-1}\mathbf{P}\mathbf{C})^{-1}\mathbf{C}^T\mathbf{P}^T\mathbf{V}_a^{-1}\mathbf{P}\mathbf{y}, \\ \hat{\mathbf{a}} &= \tilde{\mathbf{D}}_a\mathbf{H}^T\mathbf{P}^T\mathbf{V}_a^{-1}\mathbf{P}(\mathbf{y} - \mathbf{C}\hat{\boldsymbol{\alpha}}). \end{aligned} \tag{8.48}$$

Plugging the above estimators into the profile LPK smoothers in (8.46), we have

$$\begin{aligned} \hat{\boldsymbol{\eta}} &= \mathbf{A}(\mathbf{y} - \mathbf{C}\hat{\boldsymbol{\alpha}} - \mathbf{H}\hat{\mathbf{a}}), \\ \hat{\mathbf{v}} &= \mathbf{A}_v(\mathbf{y} - \mathbf{C}\hat{\boldsymbol{\alpha}} - \mathbf{H}\hat{\mathbf{a}}). \end{aligned} \tag{8.49}$$

Note that the LPK method used in the LPK backfitting algorithm and in the profile LPK method is not crucial and it can be replaced by any smoothing technique such as smoothing splines (Wahba 1990, Green and Silverman 1994), regression splines (Eubank 1988, 1999) and penalized splines (Ruppert, Wand and Carroll 2003), among others. See Chapters 5~7 for details on these techniques.

8.3.3 Regression Spline Method

In the previous subsection, the nonparametric fixed-effect function $\eta(t)$ and the random-effect functions $v_i(t),\ i = 1, 2, \cdots, n$ were estimated using LPK smoothers. Backfitting strategies of Hastie and Tishrani (1990) were used in the LPK backfitting algorithm. As previously mentioned, the LPK method in the LPK backfitting algorithm and in the profile LPK method can be replaced by regression splines, smoothing splines and penalized splines. Alternatively, these latter smoothing techniques can be used to fit the SPME model (8.36) directly without backfitting, as we shall show here and in the next several subsections.

As in Chapter 5, to fit $\eta(t)$ and $v_i(t),\ i = 1, 2, \cdots, n$, we need two bases $\mathbf{\Phi}_p(t)$ and $\mathbf{\Psi}_q(t)$ where for example, $\mathbf{\Phi}_p(t)$ and $\mathbf{\Psi}_q(t)$ can be taken as a k-th degree truncated power basis with K knots $\tau_1, \tau_2, \cdots, \tau_K$ and a k_v-th degree truncated power basis with K_v knots $\delta_1, \cdots, \delta_{K_v}$. See (8.12) for a general definition of a truncated power basis. Then we can approximately express $\eta(t)$ and $v_i(t),\ i = 1, 2, \cdots, n$ as regression splines:

$$\eta(t) = \mathbf{\Phi}_p(t)^T\boldsymbol{\beta},\ v_i(t) = \mathbf{\Psi}_q(t)^T\mathbf{b}_i,\ \ i = 1, 2, \cdots, n, \tag{8.50}$$

where

$$\boldsymbol{\beta} = [\beta_1, \cdots, \beta_p]^T,\ \mathbf{b}_i = [b_{i1}, \cdots, b_{iq}]^T,\ \ i = 1, 2, \cdots, n,$$

are the associated coefficient vectors. Since $v_i(t),\ i = 1, 2, \cdots, n$ are assumed to be independent and identical copies of a Gaussian process $v(t)$ with mean function 0 and covariance function $\gamma(s,t)$, the coefficients $\mathbf{b}_i$ are independently and identically distributed with mean 0 and covariance matrix, say, $\mathbf{D}_b$. Let $\mathrm{E}[\mathbf{a}_i\mathbf{b}_i^T] = \mathbf{D}_{ab}$. Then we have

$$\begin{bmatrix} \mathbf{a}_i \\ \mathbf{b}_i \end{bmatrix} \sim N(\mathbf{0}, \mathbf{D}), \quad \text{where } \mathbf{D} = \begin{bmatrix} \mathbf{D}_a & \mathbf{D}_{ab} \\ \mathbf{D}_{ab}^T & \mathbf{D}_b \end{bmatrix}. \tag{8.51}$$

By using (8.37), we approximately have the following relationships:

$$\gamma(s,t) = \mathbf{\Psi}_q(s)^T\mathbf{D}_b\mathbf{\Psi}_q(t), \ \text{ and } \ \boldsymbol{\gamma}_a(t) = \mathbf{D}_{ab}\mathbf{\Psi}_q(t). \tag{8.52}$$

Then we can approximately express the SPME model (8.36) as a standard LME model:

$$\begin{gathered} \mathbf{y}_{ij} = \mathbf{c}_{ij}^T\boldsymbol{\alpha} + \mathbf{x}_{ij}^T\boldsymbol{\beta} + \mathbf{h}_{ij}^T\mathbf{a}_i + \mathbf{z}_{ij}^T\mathbf{b}_i + \epsilon_{ij}, \\ \begin{bmatrix} \mathbf{a}_i \\ \mathbf{b}_i \end{bmatrix} \sim N(\mathbf{0}, \mathbf{D}),\ \boldsymbol{\epsilon}_i \sim N(0, \mathbf{R}_i), \\ j = 1, 2, \cdots, n_i;\ i = 1, 2, \cdots, n, \end{gathered} \tag{8.53}$$

where $\mathbf{x}_{ij} = \mathbf{\Phi}_p(t_{ij})$ and $\mathbf{z}_{ij} = \mathbf{\Psi}_q(t_{ij})$. Methods for fitting the above standard LME model (8.53) can be found in several books including Davidian and Giltinan (1995), Vonesh and Chinchilli (1996), Pinheiro and Bates (2000), and Verbeke and Molenberghs (2000), among others. See Chapter 2 for details on LME and NLME models. For implementation, the S-PLUS function *lme* and the SAS procedure *PROC MIXED* can be used.

For given $\mathbf{\Phi}_p(t)$ and $\mathbf{\Psi}_q(t)$, statistical inferences for the fixed-effects components $\boldsymbol{\alpha}, \eta(t)$ and the variance components $\mathbf{D}_a, \gamma(s,t)$ and σ^2 are straightfoward. The numbers of basis functions p and q can be chosen using the AIC and BIC rules proposed in Chapter 5 with Loglik, df and df_v defined under the approximate LME model (8.53).

8.3.4 Penalized Spline Method

Similar to the regression spline method for fitting the SPME model (8.36), the penalized spline (or P-spline) method uses two bases $\mathbf{\Phi}_p(t)$ and $\mathbf{\Psi}_q(t)$ to express $\eta(t)$ and $v_i(t),\ i = 1, 2, \cdots, n$ as regression splines (8.50). Notice that in the regression spline method, good regression spline smoothers of $\eta(t)$ and $v_i(t),\ i = 1, 2, \cdots, n$ are obtained via properly selecting bases $\mathbf{\Phi}_p(t)$ and $\mathbf{\Psi}_q(t)$ and the numbers of basis functions. In contrast, in the P-spline method, good P-spline smoothers are obtained via properly trading-off the goodness of fit and the roughness of the smoothers of $\eta(t)$ and $v_i(t),\ i = 1, 2, \cdots, n$ while the two bases $\mathbf{\Phi}_p(t)$ and $\mathbf{\Psi}_q(t)$ are taken as two truncated power bases of form (8.12), one being degree k with K knots, and the other being degree k_v with K_v knots. The methodologies are similar to those for the NPME model (7.39) presented in Section 7.5 of Chapter 7.

By expressing $\eta(t)$ and $v_i(t),\ i = 1, 2, \cdots, n$ as regression splines (8.50), the SPME model (8.36) can be expressed as the approximate LME model (8.53). In the spirit of P-splines, we fit the approximate LME model (8.53) via minimizing the following penalized generalized log-likelihood (PGLL) criterion:

$$
\begin{array}{c}
\sum_{i=1}^n \left\{ \mathbf{\Delta}_i^T \mathbf{R}_i^{-1} \mathbf{\Delta}_i + \mathbf{r}_i^T \mathbf{D}^{-1} \mathbf{r}_i \right\} + \lambda_v \sum_{i=1}^n \mathbf{b}_i^T \mathbf{G}_v \mathbf{b}_i + \lambda \boldsymbol{\beta}^T \mathbf{G} \boldsymbol{\beta}, \\
\mathbf{r}_i = [\mathbf{a}_i^T, \mathbf{b}_i^T]^T, \ \mathbf{\Delta}_i = \mathbf{y}_i - \mathbf{C}_i \boldsymbol{\alpha} - \mathbf{X}_i \boldsymbol{\beta} - \mathbf{H}_i \mathbf{a}_i - \mathbf{Z}_i \mathbf{b}_i,
\end{array}
\tag{8.54}
$$

where $\mathbf{X}_i = [\mathbf{x}_{i1}, \cdots, \mathbf{x}_{in_i}]^T$ and $\mathbf{Z}_i = [\mathbf{z}_{i1}, \cdots, \mathbf{z}_{in_i}]^T$; the roughness matrix $\mathbf{G}$ for $\eta(t)$ has been defined in (8.15), and similarly the roughness matrix $\mathbf{G}_v$ for $v_i(t),\ i = 1, 2, \cdots, n$ can be defined as

$$
\mathbf{G}_v = \begin{bmatrix} \mathbf{0}_{(k_v+1)\times(k_v+1)} & \mathbf{0}_{(k_v+1)\times K_v} \\ \mathbf{0}_{K_v\times(k_v+1)} & \mathbf{I}_{K_v} \end{bmatrix}. \tag{8.55}
$$

In the expression (8.54), by assuming that the variance components $\mathbf{D}$ and $\mathbf{R}$ are known, the first term is proportional to the twice negative logarithm of the generalized likelihood of $(\boldsymbol{\alpha}, \boldsymbol{\beta}, \mathbf{a}_i, \mathbf{b}_i,\ i = 1, 2, \cdots, n)$ (the joint probability density of the responses $\mathbf{y}_i,\ i = 1, 2, \cdots, n$ and the random-effects vectors $\mathbf{a}_i, \mathbf{b}_i,\ i = 1, 2, \cdots, n$); the second term is the sum of the roughness of the random-effect functions $v_i(t),\ i = 1, 2, \cdots, n$ multiplied by a common smoothing parameter λ_v; and the last term is the roughness of the fixed-effect function $\eta(t)$ multiplied by another smoothing parameter λ. Clearly, the roughness matrices $\mathbf{G}$ and $\mathbf{G}_v$ are well designed to allow the second term and the last term to penalize, respectively the k-times and k_v-times derivative jumps of $\eta(t)$ and $v_i(t),\ i = 1, 2, \cdots, n$ when they are expressed as regression splines (8.50). The smoothing parameter λ is used to trade-off between the goodness of fit and the roughness of $\eta(t)$, and the smoothing parameter λ_v is used to trade-off between the goodness of fit and the roughness of $v_i(t),\ i = 1, 2, \cdots, n$.

Denote

$$\begin{array}{rclrcl}\boldsymbol{\delta} &=& [\boldsymbol{\alpha}^T,\boldsymbol{\beta}^T]^T, & \mathbf{r} &=& [\mathbf{r}_1^T,\cdots,\mathbf{r}_n^T]^T,\\ \mathbf{P}_i &=& [\mathbf{C}_i,\mathbf{X}_i], & \mathbf{P} &=& [\mathbf{P}_1^T,\cdots,\mathbf{P}_n^T]^T,\\ \mathbf{Q}_i &=& [\mathbf{H}_i,\mathbf{Z}_i], & \mathbf{Q} &=& \text{diag}(\mathbf{Q}_1,\cdots,\mathbf{Q}_n).\end{array}$$

Moreover, denote

$$\mathbf{G}^* = \begin{bmatrix} \mathbf{0}_{p_0\times p_0} & \mathbf{0}_{p_0\times p} \\ \mathbf{0}_{p\times p_0} & \mathbf{G}\end{bmatrix},\ \mathbf{G}_v^* = \begin{bmatrix} \mathbf{0}_{q_0\times q_0} & \mathbf{0}_{q_0\times q} \\ \mathbf{0}_{q\times q_0} & \mathbf{G}_v\end{bmatrix}. \tag{8.56}$$

Then the PGLL criterion (8.54) can be re-written as

$$\begin{array}{c}\sum_{i=1}^n \left\{(\mathbf{y}_i - \mathbf{P}_i\boldsymbol{\delta} - \mathbf{Q}_i\mathbf{r}_i)^T\mathbf{R}_i^{-1}(\mathbf{y}_i - \mathbf{P}_i\boldsymbol{\delta} - \mathbf{Q}_i\mathbf{r}_i) + \mathbf{r}_i^T\mathbf{D}^{-1}\mathbf{r}_i\right\}\\ \quad + \lambda_v\sum_{i=1}^n \mathbf{r}_i^T\mathbf{G}_v^*\mathbf{r}_i + \lambda\boldsymbol{\delta}^T\mathbf{G}^*\boldsymbol{\delta}.\end{array} \tag{8.57}$$

This PGLL criterion with $\mathbf{P}_i, \boldsymbol{\delta}, \mathbf{G}^*, \mathbf{Q}_i, \mathbf{r}_i, \mathbf{D}$, and $\mathbf{G}_v^*$ is essentially equivalent to the PGLL criterion (7.42) with $\mathbf{X}_i, \boldsymbol{\beta}, \mathbf{G}, \mathbf{Z}_i, \mathbf{b}_i, \mathbf{D}$, and $\mathbf{G}_v$. Therefore, essentially our problem reduces to the MEPS fitting problem discussed in Section 7.5 of Chapter 7. Therefore, statistical inferences for the SPME model (8.36) will be based on the results for the MEPS model described in Section 7.5 in Chapter 7. We now describe methods for extracting the estimators for the parametric components $\boldsymbol{\alpha}, \mathbf{a}_i$ and the nonparametric components $\eta(t), v_i(t),\ i = 1, 2, \cdots, n$ from the MEPS estimators $\hat{\boldsymbol{\delta}}$ and $\hat{\mathbf{r}}_i,\ i = 1, 2, \cdots, n$.

Define

$$\begin{array}{rclrcl}\mathbf{D}_v &=& (\mathbf{D}^{-1} + \lambda_v\mathbf{G}_v^*)^{-1}, & \tilde{\mathbf{D}}_v &=& \text{diag}(\mathbf{D}_v,\mathbf{D}_v,\cdots,\mathbf{D}_v),\\ \tilde{\mathbf{V}}_i &=& \mathbf{Q}_i\mathbf{D}_v\mathbf{Q}_i^T + \mathbf{R}_i, & \tilde{\mathbf{V}} &=& \mathbf{Q}\tilde{\mathbf{D}}_v\mathbf{Q}^T + \mathbf{R}.\end{array} \tag{8.58}$$

Then by (7.45), we can express the minimizers of (8.57) as

$$\begin{array}{rcl}\hat{\boldsymbol{\delta}} &=& \left(\mathbf{P}^T\tilde{\mathbf{V}}^{-1}\mathbf{P} + \lambda\mathbf{G}^*\right)^{-1}\mathbf{P}^T\tilde{\mathbf{V}}^{-1}\mathbf{y},\\ \hat{\mathbf{r}} &=& \tilde{\mathbf{D}}_v\mathbf{Q}^T\tilde{\mathbf{V}}^{-1}(\mathbf{y} - \mathbf{P}\hat{\boldsymbol{\delta}}),\end{array} \tag{8.59}$$

so that $\hat{\mathbf{r}}_i = \mathbf{D}_v\mathbf{Q}_i^T\tilde{\mathbf{V}}_i^{-1}(\mathbf{y}_i - \mathbf{P}_i\hat{\boldsymbol{\delta}}),\ i = 1, 2, \cdots, n$.

Without loss of generality, from now on, we assume that $\mathbf{P}$ and $\mathbf{Q}$ are fixed; otherwise, we can regard that all the results are obtained when $\mathbf{P}$ and $\mathbf{Q}$ are given. By (8.59), we have

$$\text{Cov}(\hat{\boldsymbol{\delta}}) = (\mathbf{P}^T\tilde{\mathbf{V}}^{-1}\mathbf{P} + \lambda\mathbf{G}^*)^{-1}\mathbf{P}^T\tilde{\mathbf{V}}^{-1}\mathbf{V}\tilde{\mathbf{V}}^{-1}\mathbf{P}(\mathbf{P}^T\tilde{\mathbf{V}}^{-1}\mathbf{P} + \lambda\mathbf{G}^*)^{-1},$$

where $\mathbf{V} = \text{Cov}(\mathbf{y})$. In computation, it is natural to replace $\mathbf{V}$ with $\tilde{\mathbf{V}}$ in the above expression to take the roughness of $v_i(t),\ i = 1, 2, \cdots, n$ into account so that approximately we have

$$\begin{array}{rcl}\text{Cov}(\hat{\boldsymbol{\delta}}) &\approx& (\mathbf{P}^T\tilde{\mathbf{V}}^{-1}\mathbf{P} + \lambda\mathbf{G}^*)^{-1}\mathbf{P}^T\tilde{\mathbf{V}}^{-1}\mathbf{P}(\mathbf{P}^T\tilde{\mathbf{V}}^{-1}\mathbf{P} + \lambda\mathbf{G}^*)^{-1}\\ &\equiv& \boldsymbol{\Omega}.\end{array} \tag{8.60}$$

Let $\mathbf{R} = \sigma^2 \mathbf{I}_N$. In general, the unknown variance components $\mathbf{D}$ and σ^2 can be estimated using the methods described in Section 7.5.3 of Chapter 7; and the smoothing parameters λ and λ_v can be selected using the AIC or BIC rules developed in Section 7.5.6 of Chapter 7.

To extract the estimators of $\boldsymbol{\alpha}, \eta(t), \mathbf{a}_i, v_i(t),\ i = 1, 2, \cdots, n$, we use the matrices defined as follows:

$$\begin{array}{rclrcl} \mathbf{E}_1 & = & [\mathbf{I}_{p_0}, \mathbf{0}_{p_0 \times p}], & \mathbf{E}_2 & = & [\mathbf{0}_{p \times p_0}, \mathbf{I}_p], \\ \mathbf{F}_1 & = & [\mathbf{I}_{q_0}, \mathbf{0}_{q_0 \times q}], & \mathbf{F}_2 & = & [\mathbf{0}_{q \times q_0}, \mathbf{I}_q], \end{array} \tag{8.61}$$

so that $\boldsymbol{\alpha} = \mathbf{E}_1 \boldsymbol{\delta}, \boldsymbol{\beta} = \mathbf{E}_2 \boldsymbol{\delta}, \mathbf{a}_i = \mathbf{F}_1 \mathbf{r}_i$ and $\mathbf{b}_i = \mathbf{F}_2 \mathbf{r}_i,\ i = 1, 2, \cdots, n$. We then have

$$\begin{array}{rcl} \hat{\boldsymbol{\alpha}} & = & \mathbf{E}_1 \hat{\boldsymbol{\delta}} = \mathbf{E}_1 (\mathbf{P}^T \tilde{\mathbf{V}}^{-1} \mathbf{P} + \lambda \mathbf{G}^*)^{-1} \mathbf{P}^T \tilde{\mathbf{V}}^{-1} \mathbf{y}, \\ \hat{\boldsymbol{\beta}} & = & \mathbf{E}_2 \hat{\boldsymbol{\delta}} = \mathbf{E}_2 (\mathbf{P}^T \tilde{\mathbf{V}}^{-1} \mathbf{P} + \lambda \mathbf{G}^*)^{-1} \mathbf{P}^T \tilde{\mathbf{V}}^{-1} \mathbf{y}, \\ \hat{\mathbf{a}}_i & = & \mathbf{F}_1 \hat{\mathbf{r}}_i = \mathbf{F}_1 \mathbf{D}_v \mathbf{Q}_i^T \tilde{\mathbf{V}}_i^{-1} (\mathbf{y}_i - \mathbf{C}_i \hat{\boldsymbol{\alpha}} - \mathbf{X}_i \hat{\boldsymbol{\beta}}), \\ \hat{\mathbf{b}}_i & = & \mathbf{F}_2 \hat{\mathbf{r}}_i = \mathbf{F}_2 \mathbf{D}_v \mathbf{Q}_i^T \tilde{\mathbf{V}}_i^{-1} (\mathbf{y}_i - \mathbf{C}_i \hat{\boldsymbol{\alpha}} - \mathbf{X}_i \hat{\boldsymbol{\beta}}). \end{array}$$

It follows that the nonparametric components can be expressed as

$$\begin{array}{rcl} \hat{\eta}(t) & = & \boldsymbol{\Phi}_p(t)^T \mathbf{E}_2 (\mathbf{P}^T \tilde{\mathbf{V}}^{-1} \mathbf{P} + \lambda \mathbf{G}^*)^{-1} \mathbf{P}^T \tilde{\mathbf{V}}^{-1} \mathbf{y}, \\ \hat{v}_i(t) & = & \boldsymbol{\Psi}_q(t)^T \mathbf{F}_2 \mathbf{D}_v \mathbf{Q}_i^T \tilde{\mathbf{V}}_i^{-1} (\mathbf{y}_i - \mathbf{C}_i \hat{\boldsymbol{\alpha}} - \mathbf{X}_i \hat{\boldsymbol{\beta}}). \end{array}$$

To construct the approximate confidence region of $\boldsymbol{\alpha}$ or the approximate confidence intervals (CIs) of the components of $\boldsymbol{\alpha}$, we compute

$$\text{Cov}(\hat{\boldsymbol{\alpha}}) \approx \mathbf{E}_1 \boldsymbol{\Omega} \mathbf{E}_1^T.$$

To construct the pointwise SD band of $\eta(t)$, we can use

$$\text{Var}(\hat{\eta}(t)) \approx \boldsymbol{\Phi}_p(t)^T \mathbf{E}_2 \boldsymbol{\Omega} \mathbf{E}_2^T \boldsymbol{\Phi}_p(t). \tag{8.62}$$

For example, the 95% pointwise SD band of $\eta(t)$ can be constructed as $\hat{\eta}(t) \pm 1.96 \widehat{\text{Var}}(\hat{\eta}(t))^{1/2}$ where $\widehat{\text{Var}}(\hat{\eta}(t))$ denotes the estimate of the variance (8.62).

8.3.5 Smoothing Spline Method

In the previous subsection, we studied the P-spline method for fitting the general SPME model (8.36). The same ideas apply to fitting the SPME model (8.36) using smoothing splines. We choose not to repeat them to save space. However, in this section, we shall outline the basic ideas of Zhang et al. (1998) who proposed methods to fit the SPME model (8.36) in a different manner. Jacqmin-Gadda et al. (2002) discussed the same SPME model and used the same smoothing spline based approach as Zhang et al. (1998) except that the former used the "leave-one-subject-out" cross-validation method for choosing the associated smoothing parameter.

Unlike the general SPME model (8.36), Zhang et al. (1998) assumed that $v_i(t),\ i = 1, 2, \cdots, n$ are stationary or nonstationary Gaussian processes whose common covariance structure can be specified by a parametric model with a parameter vector $\boldsymbol{\xi}$. For example, they proposed to use the nonhomogeneous Ornstein-Uhlenbeck process which has a correlation function decaying exponentially over time, i.e., $\text{Cor}(v_i(s), v_i(t)) = \exp(-\rho|s-t|)$ with $\boldsymbol{\xi} = \rho > 0$. Moreover, they assumed that $\mathbf{D}_a$ depends on a parameter vector $\boldsymbol{\phi}$, $\mathbf{R} = \sigma^2 \mathbf{I}_N$, and the parametric random-effects $\mathbf{a}_i$ and the nonparametric random-effects $v_i(t),\ i = 1, 2, \cdots, n$ are uncorrelated, which often restricts the use of the SPME model (8.36). In this case, the covariance matrices

$$\mathbf{V}_i = \text{Cov}(\mathbf{y}_i) = \mathbf{H}_i \mathbf{D}_a \mathbf{H}_i^T + \boldsymbol{\Gamma}_i + \mathbf{R}_i,\ i = 1, 2, \cdots, n, \tag{8.63}$$

depend only on the parameter vector $\boldsymbol{\theta} = [\boldsymbol{\phi}^T, \boldsymbol{\xi}^T, \sigma^2]^T$.

We shall use the notations as defined previously unless otherwise specified. Let

$$\tau_1, \tau_2, \cdots, \tau_M \tag{8.64}$$

be all the distinct design time points among $\{t_{ij},\ j = 1, 2, \cdots, n_i;\ i = 1, 2, \cdots, n\}$ where $a < \tau_1 < \tau_2 < \cdots < \tau_M < b$ with a and b specifying the data range of interest. Let $x_{ijr} = 1$ if $t_{ij} = \tau_r$ and 0 otherwise. Then $\mathbf{x}_{ij} = [x_{ij1}, \cdots, x_{ijM}]^T$ is the incidence vector for t_{ij} with respect to all the distinct design time points, $\mathbf{X}_i = [\mathbf{x}_{i1}, \cdots, \mathbf{x}_{in_i}]^T$ is the associated incidence matrix for the i-th subject, and $\mathbf{X} = [\mathbf{X}_1^T, \cdots, \mathbf{X}_n^T]^T$ is the associated incidence matrix for the whole data set. Let $\boldsymbol{\eta} = [\eta(\tau_1), \cdots, \eta(\tau_M)]^T$ denote the values of $\eta(t)$ evaluated at all the distinct design time points, we have $\eta(t_{ij}) = \mathbf{x}_{ij}^T \boldsymbol{\eta},\ j = 1, 2, \cdots, n_i;\ i = 1, 2, \cdots, n$. Then the SPME model (8.38) can be written as

$$\begin{aligned} \mathbf{y}_i &= \mathbf{C}_i \boldsymbol{\alpha} + \mathbf{X}_i \boldsymbol{\eta} + \mathbf{H}_i \mathbf{a}_i + \mathbf{v}_i + \boldsymbol{\epsilon}_i, \\ \mathbf{a}_i &\sim N(0, \mathbf{D}_a),\ \mathbf{v}_i \sim N(0, \boldsymbol{\Gamma}_i), \boldsymbol{\epsilon}_i \sim N(0, \mathbf{R}_i), \\ & \quad i = 1, 2, \cdots, n. \end{aligned} \tag{8.65}$$

In a more compact form, we can further write

$$\begin{aligned} \mathbf{y} &= \mathbf{C}\boldsymbol{\alpha} + \mathbf{X}\boldsymbol{\eta} + \mathbf{H}\mathbf{a} + \mathbf{v} + \boldsymbol{\epsilon}, \\ \mathbf{a} &\sim N(0, \tilde{\mathbf{D}}_a),\ \mathbf{v} \sim N(0, \boldsymbol{\Gamma}), \boldsymbol{\epsilon} \sim N(0, \mathbf{R}). \end{aligned} \tag{8.66}$$

Notice that $\boldsymbol{\eta}$ in the above models is different from the one in the SPME model (8.39).

To estimate $\boldsymbol{\alpha}$ and $\eta(t)$, for given variance components $\tilde{\mathbf{D}}_a, \boldsymbol{\Gamma}$ and $\mathbf{R}$, Zhang et al. (1998) proposed to minimize the following penalized log-likelihood (PLL) criterion:

$$\log|\mathbf{V}| + (\mathbf{y} - \mathbf{C}\boldsymbol{\alpha} - \mathbf{X}\boldsymbol{\eta})^T \mathbf{V}^{-1} (\mathbf{y} - \mathbf{C}\boldsymbol{\alpha} - \mathbf{X}\boldsymbol{\eta}) + \lambda \boldsymbol{\eta}^T \mathbf{G} \boldsymbol{\eta}, \tag{8.67}$$

where $\mathbf{G}$ is the cubic smoothing spline roughness matrix with knots at (8.64) which can be computed using (3.33) of Chapter 3 so that $\boldsymbol{\eta}^T \mathbf{G} \boldsymbol{\eta} = \int_a^b [\eta''(t)]^2 dt$ is the roughness of the underlying nonparametric fixed-effect function $\eta(t)$. In the PLL criterion (8.67), the first two terms, up to a constant, are the log-likelihood function of

$\boldsymbol{\alpha}$ and $\boldsymbol{\eta}$, and the last term is the cubic smoothing spline roughness penalty multiplied by the smoothing parameter λ. Zhang et al. (1998) called the minimizers of (8.67) the maximum penalized likelihood (MPL) estimators of $\boldsymbol{\alpha}$ and $\eta(t)$.

For given variance components and smoothing parameter, the MPL estimators of $\boldsymbol{\alpha}$ and $\boldsymbol{\eta}$ can be obtained via solving the following normal equation system of (8.67):

$$\begin{bmatrix} \mathbf{C}^T\mathbf{V}^{-1}\mathbf{C} & \mathbf{C}^T\mathbf{V}^{-1}\mathbf{X} \\ \mathbf{X}^T\mathbf{V}^{-1}\mathbf{C} & \mathbf{X}^T\mathbf{V}^{-1}\mathbf{X}+\lambda\mathbf{G} \end{bmatrix} \begin{bmatrix} \boldsymbol{\alpha} \\ \boldsymbol{\eta} \end{bmatrix} = \begin{bmatrix} \mathbf{C}^T\mathbf{V}^{-1}\mathbf{y} \\ \mathbf{X}^T\mathbf{V}^{-1}\mathbf{y} \end{bmatrix}. \tag{8.68}$$

The above normal equation system has a unique solution if and only if the matrix $[\mathbf{C}, \mathbf{XT}]$ is of full rank where

$$\mathbf{T} = \begin{bmatrix} 1 & \tau_1 \\ 1 & \tau_2 \\ \vdots & \vdots \\ 1 & \tau_M \end{bmatrix}. \tag{8.69}$$

Some simple algebra gives the MPL estimators of $\boldsymbol{\alpha}$ and $\boldsymbol{\eta}$ as

$$\begin{aligned} \hat{\alpha} &= (\mathbf{C}^T\mathbf{P}\mathbf{C})^{-1}\mathbf{C}^T\mathbf{P}\mathbf{y}, \\ \hat{\eta} &= (\mathbf{X}^T\mathbf{Q}\mathbf{X}+\lambda\mathbf{G})^{-1}\mathbf{X}^T\mathbf{Q}\mathbf{y}, \end{aligned} \tag{8.70}$$

where

$$\begin{aligned} \mathbf{P} &= \mathbf{V}^{-1} - \mathbf{V}^{-1}\mathbf{X}(\mathbf{X}^T\mathbf{V}^{-1}\mathbf{X}+\lambda\mathbf{G})^{-1}\mathbf{X}^T\mathbf{V}^{-1}, \\ \mathbf{Q} &= \mathbf{V}^{-1} - \mathbf{V}^{-1}\mathbf{C}(\mathbf{C}^T\mathbf{V}^{-1}\mathbf{C})^{-1}\mathbf{C}^T\mathbf{V}^{-1}. \end{aligned} \tag{8.71}$$

The value of $\eta(t)$ at any given time point t can be obtained via simple interpolation.

Based on the SPME model (8.65), the estimates of the random-effects $\mathbf{a}_i$ and $\mathbf{v}_i$ can be obtained via their conditional expectations given the response vector $\mathbf{y}$ and the MPL estimators $\hat{\alpha}$ and $\hat{\eta}$:

$$\begin{aligned} \hat{\mathbf{a}}_i &= \mathbf{D}_a\mathbf{H}_i^T\mathbf{V}_i^{-1}(\mathbf{y}_i - \mathbf{C}_i\hat{\alpha} - \mathbf{X}_i\hat{\eta}), \\ \hat{\mathbf{v}}_i &= \boldsymbol{\Gamma}_i\mathbf{V}_i^{-1}(\mathbf{y}_i - \mathbf{C}_i\hat{\alpha} - \mathbf{X}_i\hat{\eta}), \; i = 1, 2, \cdots, n. \end{aligned} \tag{8.72}$$

One advantage of Zhang et al. (1998)'s approach is that the variance components and the smoothing parameter λ can be estimated simultaneously via using a modified LME model. They employed the LME model representation of a cubic smoothing spline obtained by Green (1987), which is slightly different from but equivalent to the LME model representation obtained in Section 3.4.3. That is, under the PLL criterion (8.67) where the underlying function $\eta(t)$ of the unknown vector $\boldsymbol{\eta}$ is a cubic smoothing spline, we can write $\boldsymbol{\eta}$ as

$$\boldsymbol{\eta} = \mathbf{T}\boldsymbol{\beta} + \mathbf{B}\mathbf{u},$$

where $\mathbf{T}$ is defined in (8.69), $\mathbf{B} = \mathbf{L}(\mathbf{L}^T\mathbf{L})^{-1}$ with $\mathbf{L}$ a full-rank matrix satisfying $\mathbf{G} = \mathbf{L}\mathbf{L}^T$ and $\mathbf{L}^T\mathbf{T} = 0$, and $\mathbf{u}$ is a $(M-2)$-dimensional vector. It follows that $\boldsymbol{\eta}^T\mathbf{G}\boldsymbol{\eta} = \mathbf{u}^T\mathbf{u}$. Therefore, the PLL criterion (8.67) can now be written as

$$\log|\mathbf{V}| + (\mathbf{y} - \mathbf{C}\boldsymbol{\alpha} - \mathbf{X}\mathbf{T}\boldsymbol{\beta} - \mathbf{X}\mathbf{B}\mathbf{u})^T\mathbf{V}^{-1}(\mathbf{y} - \mathbf{C}\boldsymbol{\alpha} - \mathbf{X}\mathbf{T}\boldsymbol{\beta} - \mathbf{X}\mathbf{B}\mathbf{u}) + \lambda\mathbf{u}^T\mathbf{u}.$$

This indicates that the SPME model (8.66) can now be written as a modified LME model:

$$\begin{aligned} \mathbf{y} &= \mathbf{C}^*\boldsymbol{\alpha}^* + \mathbf{Z}^*\mathbf{a}^* + \boldsymbol{\epsilon}, \\ \mathbf{a}^* &\sim N(0, \mathbf{D}^*),\ \boldsymbol{\epsilon} \sim N(0, \mathbf{R}), \end{aligned} \tag{8.73}$$

where $\mathbf{D}^* = \text{diag}(\tilde{\mathbf{D}}_a, \lambda^{-1}\mathbf{I}_{M-2}, \boldsymbol{\Gamma})$, $\boldsymbol{\alpha}^* = [\boldsymbol{\alpha}^T, \boldsymbol{\beta}^T]^T$, $\mathbf{a}^* = [\mathbf{a}^T, \mathbf{u}^T, \mathbf{v}^T]^T$ and $\mathbf{C}^* = [\mathbf{C}, \mathbf{XT}]$ and $\mathbf{Z}^* = [\mathbf{H}, \mathbf{XB}, \mathbf{I}_N]$.

It is easy to show that the BLUPs $\hat{\boldsymbol{\alpha}}$ and $\hat{\boldsymbol{\eta}} = \mathbf{T}\hat{\boldsymbol{\beta}} + \mathbf{B}\hat{\mathbf{u}}$, from the LME model (8.73) are identical to the MPL estimators $\hat{\boldsymbol{\alpha}}$ and $\hat{\boldsymbol{\eta}}$ given in (8.70). The same is true for the estimators of the random-effects $\hat{\mathbf{a}}_i$ and $\hat{\mathbf{v}}_i$ given in (8.72).

It is often challenging to solve $\hat{\boldsymbol{\alpha}}^*$ and $\hat{\mathbf{a}}^*$ from (8.73) because the random-effects vector $\mathbf{a}^*$ is of high dimensionality [size $nq + (M-2) + N$], and the covariance matrix $\mathbf{V}^* = \lambda^{-1}\mathbf{X}^*\mathbf{X}^{*T} + \mathbf{V}$ is not block diagonal where $\mathbf{X}^* = \mathbf{XB}$. However, it is easy to show that when the variance components $\boldsymbol{\theta}$ and the smoothing parameter λ are given, the BLUPs $\hat{\boldsymbol{\alpha}}^*$ and $\hat{\mathbf{u}}$ solve the following normal equation system of (8.73):

$$\begin{bmatrix} \mathbf{C}^{*T}\mathbf{V}^{-1}\mathbf{C}^* & \mathbf{C}^{*T}\mathbf{V}^{-1}\mathbf{X}^* \\ \mathbf{X}^{*T}\mathbf{V}^{-1}\mathbf{C}^* & \mathbf{X}^{*T}\mathbf{V}^{-1}\mathbf{X}^* + \lambda\mathbf{I}_{M-2} \end{bmatrix} \begin{bmatrix} \boldsymbol{\alpha}^* \\ \mathbf{u} \end{bmatrix} = \begin{bmatrix} \mathbf{C}^{*T}\mathbf{V}^{-1}\mathbf{y} \\ \mathbf{X}^{*T}\mathbf{V}^{-1}\mathbf{y} \end{bmatrix}. \tag{8.74}$$

Under the LME model representation (8.73) of the SPME model (8.65), Zhang et al. (1998) proposed to estimate the variance components $\boldsymbol{\theta}$ and the smoothing parameter λ simultaneously via maximizing the following restricted log-likelihood function of $\boldsymbol{\theta}$ and λ:

$$-\frac{1}{2}\log|\mathbf{V}^*| - \frac{1}{2}\log|\mathbf{C}^{*T}\mathbf{V}^{*-1}\mathbf{C}^*| - \frac{1}{2}(\mathbf{y} - \mathbf{C}^*\hat{\boldsymbol{\alpha}}^*)^T\mathbf{V}^{*-1}(\mathbf{y} - \mathbf{C}^*\hat{\boldsymbol{\alpha}}^*).$$

They also showed how to estimate the covariance of $\hat{\boldsymbol{\theta}}$ using the Fisher information matrix obtained from the normal equation system of the above restricted log-likelihood function.

By (8.70) and the SPME model (8.65), it is straightforward to show that, given $\mathbf{C}, \mathbf{H}$ and $\mathbf{X}$, the frequentist covariances of $\hat{\alpha}$ and $\hat{\eta}$ are

$$\begin{aligned} \text{Cov}(\hat{\boldsymbol{\alpha}}) &= (\mathbf{C}^T\mathbf{PC})^{-1}\mathbf{C}^T\mathbf{PVPC}(\mathbf{C}^T\mathbf{PC})^{-1}, \\ \text{Cov}(\hat{\boldsymbol{\eta}}) &= (\mathbf{X}^T\mathbf{QX} + \lambda\mathbf{G})^{-1}\mathbf{X}^T\mathbf{QVQX}(\mathbf{X}^T\mathbf{QX} + \lambda\mathbf{G})^{-1}, \end{aligned} \tag{8.75}$$

where $\mathbf{P}$ and $\mathbf{Q}$ are defined in (8.71). The estimates of the above covariance matrices can be obtained via replacing $\mathbf{V}$ by $\hat{\mathbf{V}} = \mathbf{H}\hat{\tilde{\mathbf{D}}}_a\mathbf{H}^T + \hat{\boldsymbol{\Gamma}} + \hat{\mathbf{R}}$. They can be used to construct approximate confidence intervals of the components of $\boldsymbol{\alpha}$ and approximate pointwise SD bands of $\eta(t)$.

Alternatively, we can estimate the covariance of $\hat{\boldsymbol{\alpha}}$ and $\hat{\boldsymbol{\eta}}$ using their Bayesian covariances in the spirit of Wahba (1983). To this end, following Green and Silverman (1994), we first interpret the estimates (8.70) and (8.72) as the limit posterior expectations (as $\lambda_0 \to 0$) of the parameters in the following finite dimensional Bayesian problem:

$$\begin{aligned} \mathbf{y} &= \mathbf{C}\boldsymbol{\alpha} + \mathbf{X}\boldsymbol{\eta} + \mathbf{Ha} + \mathbf{v} + \boldsymbol{\epsilon}, \\ \boldsymbol{\epsilon} &\sim N(0, \mathbf{R}), \end{aligned} \tag{8.76}$$

where the parameters have the following prior distributions:

$$\begin{array}{rcl}\boldsymbol{\alpha} & \sim & N(0, \lambda_0^{-1}\mathbf{I}_{p_0}),\ \boldsymbol{\eta} \sim N(0, (\lambda_0 \mathbf{I}_M + \lambda \mathbf{G})^{-1}), \\ \mathbf{a} & \sim & N(0, \tilde{\mathbf{D}}_a),\ \mathbf{v} \sim N(0, \boldsymbol{\Gamma}),\end{array} \tag{8.77}$$

where p_0 is the dimension of $\boldsymbol{\alpha}$ as defined previously. With some algebra, it is straightforward to show that as $\lambda_0 \to 0$, for given $\mathbf{C}, \mathbf{H}$, and $\mathbf{X}$, we have

$$\begin{array}{rcl}\boldsymbol{\alpha}|\mathbf{y} & \to & N[\hat{\boldsymbol{\alpha}}, (\mathbf{C}^T\mathbf{P}\mathbf{C})^{-1}], \\ \boldsymbol{\eta}|\mathbf{y} & \to & N[\hat{\boldsymbol{\eta}}, (\mathbf{X}^T\mathbf{Q}\mathbf{X} + \lambda\mathbf{G})^{-1}], \\ \mathbf{a}|\mathbf{y} & \to & N[\hat{\mathbf{a}}, \tilde{\mathbf{D}}_a - \tilde{\mathbf{D}}_a\mathbf{H}^T\boldsymbol{\Omega}\mathbf{H}\tilde{\mathbf{D}}_a], \\ \mathbf{v}|\mathbf{y} & \to & N[\hat{\mathbf{v}}, \boldsymbol{\Gamma} - \boldsymbol{\Gamma}\boldsymbol{\Omega}\boldsymbol{\Gamma}], \\ \boldsymbol{\epsilon}|\mathbf{y} & \to & N[\hat{\boldsymbol{\epsilon}}, \mathbf{R} - \mathbf{R}\boldsymbol{\Omega}\mathbf{R}],\end{array} \tag{8.78}$$

where $\boldsymbol{\Omega} = \mathbf{P} - \mathbf{P}\mathbf{C}(\mathbf{C}^T\mathbf{P}\mathbf{C})^{-1}\mathbf{C}^T\mathbf{P} = \mathbf{Q} - \mathbf{Q}\mathbf{X}(\mathbf{X}^T\mathbf{Q}\mathbf{X} + \lambda\mathbf{G})^{-1}\mathbf{X}^T\mathbf{Q}$. Therefore, the Bayesian covariances of $\hat{\boldsymbol{\alpha}}$ and $\hat{\boldsymbol{\eta}}$ are $(\mathbf{C}^T\mathbf{P}\mathbf{C})^{-1}$ and $(\mathbf{X}^T\mathbf{Q}\mathbf{X} + \lambda\mathbf{G})^{-1}$, respectively. Their estimates can be used to construct approximate Bayesian confidence intervals for the components of $\boldsymbol{\alpha}$ and approximate Bayesian pointwise confidence bands for $\eta(t)$. Since the roughness of $\eta(t)$ is accounted for, the Bayesian confidence intervals for the components of $\boldsymbol{\alpha}$ and the pointwise confidence bands of $\eta(t)$ are usually wider than their associated frequentist counterparts. See Zhang et al. (1998) for more details and for a continuous version of the above Bayesian formulation constructed in the spirit of Wahba (1978, 1983).

8.3.6 ACTG 388 Data Revisited

In Section 7.5 of Chapter 7, we suspected that the quadratic polynomial model was adequate for the subject-specific deviations of the ACTG 388 data. In this section, we present the results of the Type 1 SPME model (8.40) fit to the ACTG 388 data using the P-spline method. For the ACTG 388 data, the Type 1 SPME model (8.40) can be written as

$$\begin{array}{c} y_{ij} = \eta(t_{ij}) + \mathbf{h}_{ij}^T\mathbf{a}_i + \epsilon_{ij}, \\ \mathbf{h}_{ij} = [1, t_{ij}, t_{ij}^2]^T,\ \ \mathbf{a}_i = [a_{i1}, a_{i2}, a_{i3}]^T \sim N(0, \mathbf{D}), \\ j = 1, 2, \cdots, n_i;\ i = 1, 2, \cdots, n,\end{array} \tag{8.79}$$

where $\eta(t)$ is the smooth fixed-effect function and $\mathbf{D}$ is an unstructured random-effects' covariance matrix. The parametric random-effects component $h_{ij}^T\mathbf{a}_i = a_{i1} + a_{i2}t_{ij} + a_{i3}t_{ij}^2$ denotes a quadratic polynomial model for the random-effect functions $v_i(t),\ i = 1, 2, \cdots, n$.

We used quadratic ($k = 2$) truncated power basis (8.12) with $K = 12$ knots, scattered by the "equally spaced sample quantiles as knots" method, for fitting the smooth fixed-effect function $\eta(t)$. To select the smoothing parameter λ for the $\eta(t)$, we computed the Loglik, df, AIC and BIC values for a grid of smoothing parameters selected properly so that they represent a wide range of possible smoothing parameters. Figure 8.3 shows the Loglik, df, AIC and BIC curves against λ in $\log_{10}$-scale. It

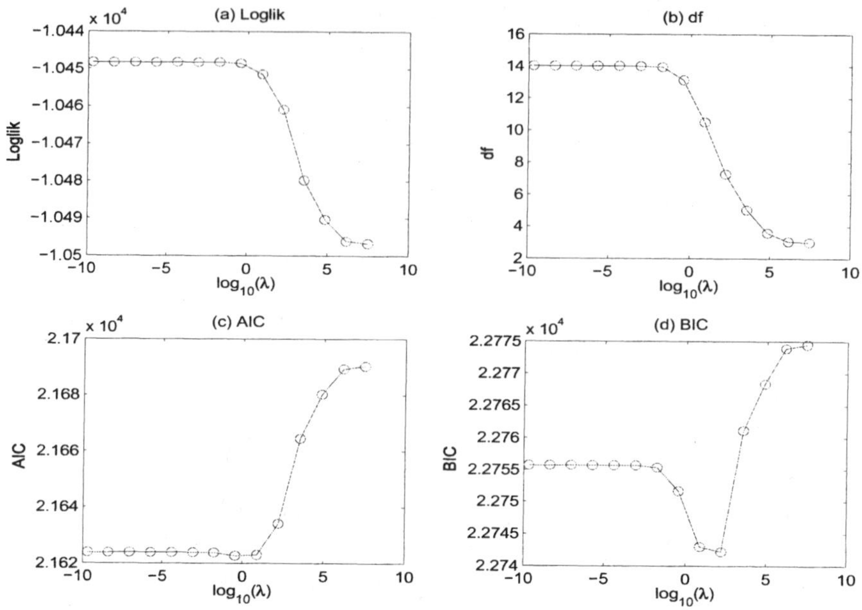

Fig. 8.3 Type 1 SPME model (8.79) fit to the ACTG 388 data: Loglik, df, AIC and BIC curves. Quadratic truncated power bases were used with $K = 12$ knots.

is seen that both Loglik and df decrease as λ increases, indicating that as λ increases, the goodness of fit gets worse while the SPME model becomes simpler. However, both AIC and BIC curves indicate that there exist proper smoothing parameters to trade-off well these two aspects. As expected the smoothing parameter $\lambda = .3631$ selected by AIC is smaller than the smoothing parameter $\lambda = 160.2$ selected by BIC. As an illustration, we present just the results using the smoothing parameter selected by BIC as follows.

The variance components estimated by the EM-algorithm are

$$\hat{\sigma}^2 = 4642.1, \quad \text{and} \quad \hat{\mathbf{D}} = \begin{bmatrix} 24350 & 41.567 & -.5939 \\ 41.567 & 8.2851 & -.0448 \\ -.5939 & -.0448 & .0003 \end{bmatrix}.$$

It is seen that the variation of the intercept term of the random-effects component is much larger than the noise variance, indicating that the between-subject variations dominate the within-subject variations.

Figure 8.4 shows the results of fitting the SPME model (8.79) to the ACTG 388 data. These plots are almost the same as those plots presented in Figure 7.11 which were obtained by fitting a MEPS model to the ACTG 388 data. Figure 8.5 shows the individual SPME fits for six subjects selected from the ACTG 388 study. These individual SPME model fits are nearly the same as those MEPS fits for the same six subjects, which were presented in Figure 7.12. It seems that the SPME model (8.42)

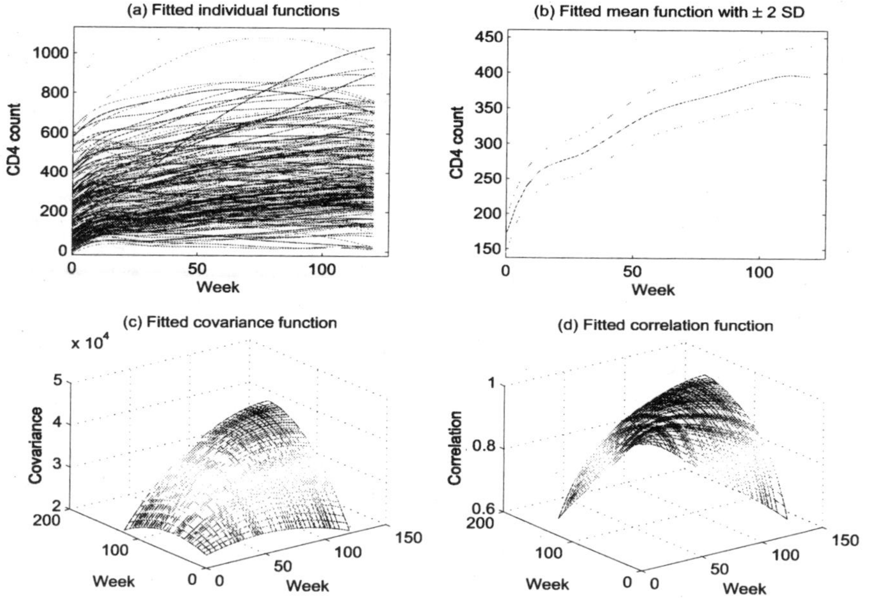

Fig. 8.4 Type 1 SPME model (8.79) fit to the ACTG 388 data: overall fits. Quadratic truncated power bases were used with $K = 12$ knots and $\lambda = 160.2$.

is indeed adequate to fit the ACTG 388 data. That is, a quadratic polynomial model provides an adequate structure to handle the subject-specific deviations of the ACTG 388 data.

The appropriateness of the SPME model (8.79) is also partially verified by the residual plots presented in Figure 8.6. The standardized residuals are plotted against the SPME fit, time, and response in Figures 8.6 (a), (b) and (c), respectively. Figure 8.6 (d) presents the histogram of the standard residuals. Here the SPME model fits are computed as

$$\hat{y}_{ij} = \hat{\eta}(t_{ij}) + \mathbf{h}_{ij}^T \hat{\mathbf{a}}_i, \ j = 1, 2, \cdots, n_i; \ i = 1, 2, \cdots, n,$$

and the standardized residuals are computed as $r_{ij} = (y_{ij} - \hat{y}_{ij})/\hat{\sigma}, j = 1, \cdots, n_i; i = 1, 2, \cdots, n$ where $\hat{\sigma}^2$ is the noise variance and is computed using the EM-algorithm described in Section 7.5.3. Figure 8.6 (d) indicates that the measurement errors are nearly normally distributed. These plots further show that the Type 1 SPME model (8.79) is indeed an adequate model for the ACTG 388 data.

8.3.7 MACS Data Revisted

In Section 8.2.7, we fit both SPM models (8.34) and (8.35) to the MACS data but we did not incorporate the within-subject correlation. Both SPM models suggested that a quadratic polynomial model is appropriate for the intercept CD4 percentage

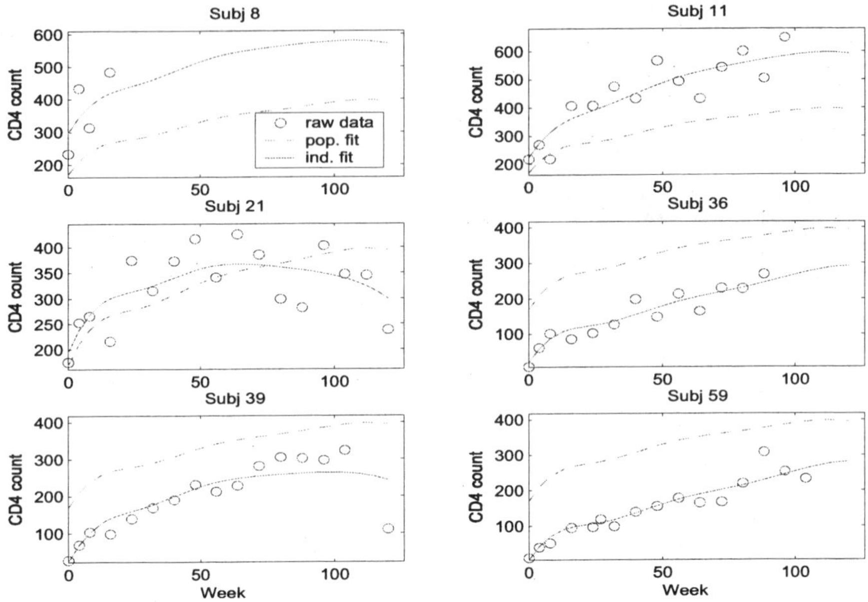

Fig. 8.5 Type 1 SPME model (8.79) fit to the ACTG 388 data: individual fits for six randomly selected subjects from the ACTG 388 data. Quadratic truncated power bases were used with $K = 12$ knots and $\lambda = 160.2$.

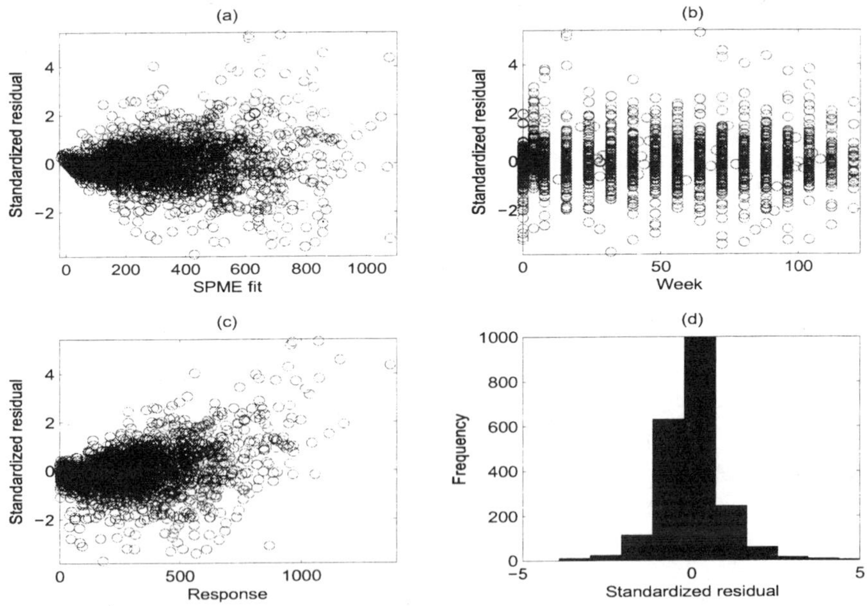

Fig. 8.6 Type 1 SPME model (8.79) fit to the ACTG 388 data: residual plots.

levels. Therefore, in order to incorporate the within-subject correlation with unknown covariance structure, we fit the Type 3 SPME model (8.42) to the MACS data with a quadratic polynomial model for the intercept CD4 percentage levels. Recall that the three covariates of interest are Smoking, Age and PreCD4 which are denoted as X_1, X_2 and X_3, respectively. Following the SPM model (8.35), we can now write the Type 3 SPME model (8.41) for the MACS data as

$$
\begin{gathered}
y_{ij} = \beta_0(t_{ij}) + X_{1i}\beta_1(t_{ij}) + X_{2i}\beta_2(t_{ij}) + X_{3i}\beta_3(t_{ij}) + v_i(t_{ij}) + \epsilon_{ij}, \\
j = 1, 2, \cdots, n_i;\ i = 1, 2, \cdots, n,
\end{gathered}
\tag{8.80}
$$

where $\beta_0(t) = \beta_{00} + \beta_{01}t + \beta_{02}t^2$ denotes a quadratic polynomial model for the intercept CD4 percentage levels; the coefficient functions $\beta_1(t), \beta_2(t), \beta_3(t)$ have been defined in (8.35); $v_i(t)$ denotes a Gaussian process with mean 0 and unknown covariance function $\gamma(s,t)$; and the errors ϵ_{ij} are assumed to be i.i.d with a variance σ^2.

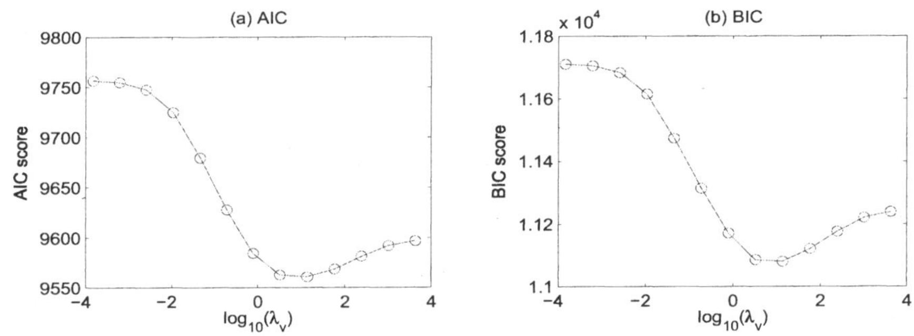

Fig. 8.7 Type 3 SPME model (8.80) fit to the MACS data: smoothing parameter selectors, (a) AIC, and (b) BIC.

The P-spline method was used to fit the Type 3 SPME model (8.80). As usual, the quadratic truncated power basis was used with $K = 16$ knots. There is only one smoothing parameter λ_v, which was used to control the roughness of the random-effect functions $v_i(t),\ i = 1, 2, \cdots, n$. The proper value of λ_v suggested by both AIC and BIC rules was 13.917. Figure 8.7 shows the AIC and BIC curves against λ_v in $\log_{10}$-scale.

The parametric component estimation results of the Type 3 SPME model (8.80) fit to the MACS data are presented in Table 8.3. The parameters for the intercept CD4 percentage level curve are highly significant with a negative slope, indicating that the intercept CD4 percentage level decreases over time after HIV infection. Compared with the parametric component fitting results of the SPM model (8.35) presented in Table 8.2, the parameters for the smoking effect coefficient function are still not significant although the estimated intercept parameter is now positive. The parameters for the age effect coefficient function are now insignificant although the estimated intercept and slope parameters have the same signs as those presented in Table 8.2. The results for the pre-infection CD4 percentage level are comparable as

Table 8.3 Type 3 SPME model (8.80) fit to the MACS data: Estimated coefficients, standard deviations and approximate z-test values for the parametric fixed-effects components. Quadratic truncated power basis with $K = 16$ knots was used. The smoothing parameter for the random-effects component, selected by GCV, is $\lambda_v = 13.92$.

Covariate	Coef.	Estimated Coef.	SD	z-test value
intercept	β_{00}	36.8520	0.6455	57.0940
	β_{01}	−4.8549	0.4046	−12.0000
	β_{02}	0.3733	0.0708	5.2730
Smoking	β_{10}	0.4117	1.0777	0.3821
	β_{11}	0.1427	0.4712	0.3030
Age	β_{20}	0.0529	0.0664	0.7959
	β_{21}	−0.0297	0.0291	−1.0205
PreCD4	β_{30}	0.5574	0.0657	8.4876
	β_{31}	−0.1264	0.0445	−2.8364
	β_{32}	0.0112	0.0082	1.3598

those presented in Table 8.2. Therefore, the SPME model fitting suggests that after accounting for the large within-subject correlation, both smoking habit and age at HIV-infection have little predictive power on the CD4 percentage levels after HIV-infection, while the pre-infection CD4 percentage levels have positive effect although the effect decreases over time.

The above conclusion can be seen more clearly by comparing Figure 8.8 against Figure 8.2. Figure 8.8 displays the fitted coefficient functions (solid curves), together with their 95% pointwise SD bands of the intercept CD4 percentage levels, Smoking, Age and PreCD4 covariates. The fitted coefficient functions are comparable with those presented in Figure 8.2 but the 95% pointwise SD bands are wider than those presented in Figure 8.2. This is not surprising since by fitting the SPME model (8.80), we accounted for the within-subject correlation in the MACS data while we did not when fitting the SPM model (8.35). As a result, both SD bands of the Smoking and Age coefficient functions contain 0 all time, indicating again that patient's smoking status and age have limited predictive power over the CD4 percentage levels after HIV-infection. The SD band of the PreCD4 coefficient function is above 0 most of time, indicating again that the PreCD4 levels have a positive effect on CD4 percentage levels after HIV-infection.

Figure 8.9 displays the individual fits (solid curves), superimposed by the raw individual data (dots) for six randomly selected subjects from the CD4 percentage study. These individual fits provide good predictions to the raw individual data. The individual plots show that the SPME model (8.80) fit well the raw individual data.

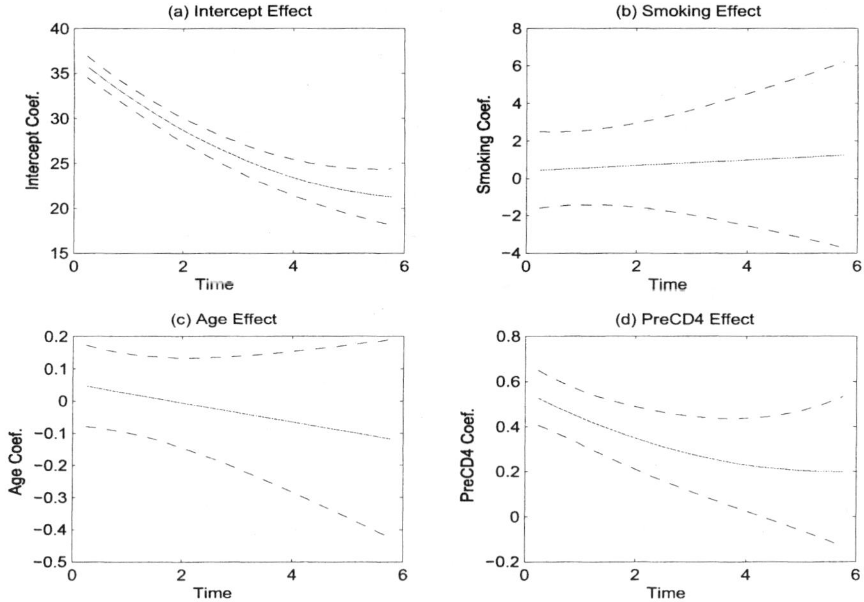

Fig. 8.8 Type 3 SPME model (8.80) fit to the MACS data: fitted coefficient functions (solid curves) and their 95% pointwise SD bands (dashed curves).

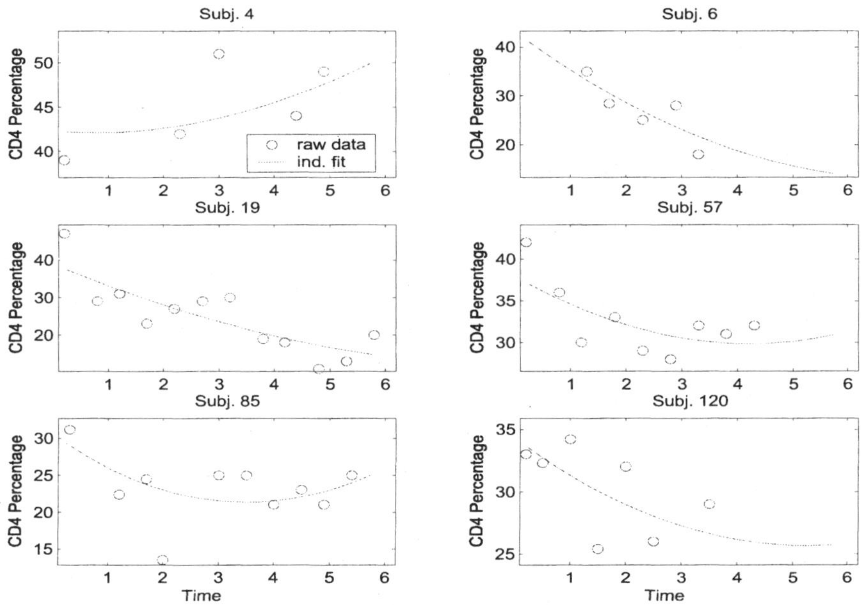

Fig. 8.9 Type 3 SPME model (8.80) fit to the MACS data: Individual fits (solid curves) and the raw data (dots) for six selected subjects.

8.4 SEMIPARAMETRIC NONLINEAR MIXED-EFFECTS MODEL

In the previous section, we investigated how to fit the SPME model (8.36) and make statistical inferences about the model. In the SPME model (8.36), the relationships between the response and the parametric/nonparametric components are linear. In some applications, the relationships between the response and the parametric/nonparametric components can be nonlinear as illustrated by Ke and Wang (2001) and Wu and Zhang (2002b), among others. In this section, we shall investigate how to fit a semiparametric nonlinear mixed-effects (SNLME) model and how to conduct statistical inferences under the SNLME models.

8.4.1 Model Specification

Wu and Zhang (2002b) proposed the following two-stage SNLME model:

Stage 1 (Intra-individual variation)

$$\begin{aligned} y_{ij} &= f[\boldsymbol{\alpha}_{ij}, \eta_i(t_{ij})] + \epsilon_{ij}, \\ \boldsymbol{\epsilon}_i &= [\epsilon_{i1}, \cdots, \epsilon_{in_i}]^T \sim N(0, \mathbf{R}_i), \\ & \quad j = 1, 2, \cdots, n_i; \ i = 1, 2, \cdots, n, \end{aligned} \tag{8.81}$$

where $\boldsymbol{\alpha}_{ij}$ and $\eta_i(t_{ij})$ denote the parametric and nonparametric mixed-effects components, and f is a known nonlinear function of $\boldsymbol{\alpha}_{ij}$ and $\eta_i(t_{ij})$.

Stage 2 (Inter-individual variation)

(1) Parametric mixed-effects component:

$$\boldsymbol{\alpha}_{ij} = \mathbf{d}(\boldsymbol{\alpha}, \mathbf{a}_i, \mathbf{c}_{ij}, \mathbf{h}_{ij}), \quad j = 1, 2, \cdots, n_i; \ i = 1, 2, \cdots, n, \tag{8.82}$$

where the (possibly time-dependent) mixed-effects parameters $\boldsymbol{\alpha}_{ij}$ are functions of fixed-effects $\boldsymbol{\alpha}$ and random-effects $\mathbf{a}_i$ with their corresponding covariates $\mathbf{c}_{ij}$ and $\mathbf{h}_{ij}$. The link function $\mathbf{d}$ may be nonlinear, but a simple linear example can be written as

$$\boldsymbol{\alpha}_{ij} = \mathbf{c}_{ij}^T \boldsymbol{\alpha} + \mathbf{h}_{ij}^T \mathbf{a}_i, \quad j = 1, 2, \cdots, n_i; \ i = 1, 2, \cdots, n. \tag{8.83}$$

In general, the random-effects $\mathbf{a}_i$, $i = 1, 2, \cdots, n$ are assumed to be iid copies of a zero-mean normal random vector, $\mathbf{a}$, say, with covariance matrix $\mathbf{D}_a$, which models the parametric part of the between-subject variation.

(2) Nonparametric mixed-effects component:

$$\eta_i(t) = g[\eta(t), v_i(t)], \quad i = 1, 2, \cdots, n, \tag{8.84}$$

where $\eta(t)$ and $v_i(t)$ are unknown smooth fixed-effects and random-effect functions, respectively. Although the known link function g can be complicated, a simple example is an additive model:

$$\eta_i(t) = \eta(t) + v_i(t), \quad i = 1, 2, \cdots, n. \tag{8.85}$$

In general, the random-effect functions $v_i(t),\ i = 1, 2, \cdots, n$ are assumed to be iid copies of a zero-mean Gaussian stochastic process $v(t)$ with covariance function $\gamma(s,t)$. We also assume that the parametric and nonparametric random-effects $\mathbf{a}$ and $v(t)$ are correlated but they are uncorrelated with the measurement errors ϵ_{ij}. For simplicity, we write $\mathrm{Cov}(\mathbf{a}, v(t)) = \boldsymbol{\gamma}_a(t)$.

The above two-stage SNLME model can be written in a compact way:

$$\begin{aligned} y_{ij} &= f[\mathbf{d}(\boldsymbol{\alpha}, \mathbf{a}_i, \mathbf{c}_{ij}, \mathbf{h}_{ij}), g(\eta(t_{ij}), v_i(t_{ij}))] + \epsilon_{ij}, \\ \mathbf{a}_i &\sim N(0, \mathbf{D}_a),\ v_i(t) \sim \mathrm{GP}(0, \gamma),\ \mathrm{E}[\mathbf{a}_i v_i(t)] = \boldsymbol{\gamma}_a(t), \\ \epsilon_i &\sim N(0, \mathbf{R}_i),\ j = 1, 2, \cdots, n_i;\ i = 1, 2, \cdots, n. \end{aligned} \tag{8.86}$$

The known link functions $f, \mathbf{d}$ and g reflect the associations between the responses and the mixed-effects. It is interesting to note that when we relax the link functions $f, \mathbf{d}$ and g to be linear or nonlinear functions, the above SNLME model (8.86) can be regarded as a unified mixed-effects model in the sense that it includes almost all mixed-effects models as its special cases.

For instance, when f is a simple linear function of $\boldsymbol{\alpha}_{ij}$ and $\eta_i(t_{ij})$ so that we have $f[\boldsymbol{\alpha}_{ij}, \eta_i(t_{ij})] = \boldsymbol{\alpha}_{ij} + \eta_i(t_{ij})$, and $\boldsymbol{\alpha}_{ij}$ and $\eta_i(t_{ij})$ take the forms of (8.83) and (8.85), respectively, then the above SNLME model (8.86) reduces to the general SPME model (8.36) discussed in the previous section. The latter itself include six types of special SPME models listed in Section 8.3, some of which are investigated by different authors including Zeger and Diggle (1994), Zhang et al. (1998), and Wang (1998a), among others.

Similarly, when f is a known nonlinear link function and there are no nonparametric random-effects $v_i(t)$, the SNLME model (8.86) reduces to the SNLME model of Ke and Wang (2001). In addition, if both nonparametric fixed-effects $\eta(t)$ and random-effects $v_i(t)$ are not in the SNLME model (8.86), the SNLME model reduces to a parametric NLME model studied by Davidian and Giltinan (1995) and Vonesh and Chinchilli (1996), among others.

Due to the presence of the nonparametric mixed-effects component, the SNLME models (8.86) are more flexible when fitting longitudinal data than standard NLME models. They are also more flexible than the SNLME models proposed by Ke and Wang (2001) and the semiparametric self-models by Ladd and Lindstrom (2000). In Ke and Wang (2001)'s models, the nonparametric mixed-effects component only consists of a fixed-effect function and no random-effect functions are included.

8.4.2 Wu and Zhang's Approach

It is challenging to fit the SNLME model (8.81) since we need to deal with the nonparametric smooth fixed-effect function $\eta(t)$ and the random-effect functions $v_i(t),\ i = 1, 2, \cdots, n$ under a nonlinear framework. Wu and Zhang (2002b) proposed a basis-based approach to reduce the SNLME model (8.81) to a standard NLME model that can be fitted using existing statistical software such as S-PLUS. The main ideas of the approach are similar to those for fitting the NPME model (5.36) using regression splines as described in Section 5.4 of Chapter 5.

Let $\mathbf{\Phi}_p(t)$ and $\mathbf{\Psi}_q(t)$ be two bases with p and q basis functions, respectively. For example, $\mathbf{\Phi}_p(t)$ can be a k-th degree truncated power basis (8.12) with K knots $\tau_1, \tau_2, \cdots, \tau_K$, and $\mathbf{\Psi}_q(t)$ can be a k_v-th degree truncated power basis (8.12) with K_v knots $\delta_1, \cdots, \delta_{K_v}$. Then approximately, we can express

$$\begin{aligned} \eta(t) &= \mathbf{\Phi}_p(t)^T \boldsymbol{\beta}, \\ v_i(t) &= \mathbf{\Psi}_q(t)^T \mathbf{b}_i, \ i = 1, 2, \cdots, n, \end{aligned} \tag{8.87}$$

where $\boldsymbol{\beta}$ is a p-dimensional coefficient vector of $\eta(t)$ and $\mathbf{b}_i$ are q-dimensional coefficient vectors of $v_i(t), \ i = 1, 2, \cdots, n$. Since $v_i(t), \ i = 1, 2, \cdots, n$ are iid copies of a Gaussian zero-mean process $v(t)$, they should have similar functional structure. Thus, it is reasonable to approximate all the random effects functions $v_i(t)$ using the same basis $\mathbf{\Psi}_q(t)$; moreover, it is reasonable to assume that the coefficients $\mathbf{b}_i, \ i = 1, 2, \cdots, n$ are iid copies of a normal zero-mean random vector $\mathbf{b}$ with covariance matrix $\mathbf{D}_b$, satisfying $v(t) = \mathbf{\Psi}_q(t)^T \mathbf{b}$ approximately. Let $\text{Cov}(\mathbf{a}_i, \mathbf{b}_i^T) = \mathbf{D}_{ab}$. Then the following relationships are approximately true:

$$\gamma(s, t) = \mathbf{\Psi}_q(s)^T \mathbf{D}_b \mathbf{\Psi}_q(t), \ \boldsymbol{\gamma}_a(t) = \mathbf{D}_{ab} \mathbf{\Psi}_q(t). \tag{8.88}$$

Moreover, by letting $\mathbf{r}_i = [\mathbf{a}_i^T, \mathbf{b}_i^T]^T, \ i = 1, 2, \cdots, n$, we have

$$\mathbf{r}_i \sim N(0, \mathbf{D}), \text{ where } \quad \mathbf{D} = \begin{bmatrix} \mathbf{D}_a & \mathbf{D}_{ab} \\ \mathbf{D}_{ab}^T & \mathbf{D}_b \end{bmatrix}. \tag{8.89}$$

Then the SNLME model (8.86) can be approximated by the following standard parametric NLME model:

$$\begin{aligned} y_{ij} &= f\left\{\mathbf{d}(\boldsymbol{\alpha}, \mathbf{a}_i, \mathbf{c}_{ij}, \mathbf{h}_{ij}), g\left[\Psi_p(t_{ij})^T \boldsymbol{\beta}, \mathbf{\Phi}_q(t_{ij})^T \mathbf{b}_i\right]\right\} + \epsilon_{ij}, \\ \mathbf{r}_i &= [\mathbf{a}_i^T, \mathbf{b}_i^T]^T \sim N(0, \mathbf{D}), \ \boldsymbol{\epsilon}_i \sim N(0, \mathbf{R}_i), \\ & \quad j = 1, 2, \cdots, n_i; \ i = 1, 2, \cdots, n. \end{aligned} \tag{8.90}$$

It can be further written as

$$\begin{aligned} y_{ij} &= f\left[\mathbf{d}^*(\boldsymbol{\alpha}^*, \mathbf{a}_i^*, \mathbf{c}_{ij}^*, \mathbf{h}_{ij}^*)\right] + \epsilon_{ij}, \\ \mathbf{a}_i^* &= \mathbf{r}_i \sim N(0, \mathbf{D}), \ \boldsymbol{\epsilon}_i \sim N(0, \mathbf{R}_i), \\ & \quad j = 1, 2, \cdots, n_i; \ i = 1, 2, \cdots, n, \end{aligned}$$

via combining the associated terms, for example, $\boldsymbol{\alpha}^* = [\boldsymbol{\alpha}^T, \boldsymbol{\beta}^T]^T$.

From the previous discussions, we can see that for two given bases $\mathbf{\Phi}_p(t)$ and $\mathbf{\Psi}_q(t)$, we can approximate the SNLME model (8.86) by the standard parametric NLME model (8.90). A brief review about NLME models is given in Section 2.3 of Chapter 2. The standard NLME model (8.90) can be fitted by any existing NLME algorithm such as Lindstrom and Bates (1990), ready available in S-PLUS, SAS and other statistical packages, resulting in the estimators $\hat{\boldsymbol{\alpha}}, \hat{\mathbf{a}}_i, \hat{\boldsymbol{\beta}}, \hat{\mathbf{b}}_i$, the variance components estimates $\hat{\mathbf{D}}_a, \hat{\mathbf{D}}_{ab}, \hat{\mathbf{D}}_b$ and $\hat{\sigma}^2$, the estimated covariance matrices of the estimators such as $\hat{\Sigma}_\alpha = \widehat{\text{Cov}}(\hat{\boldsymbol{\alpha}})$ and $\hat{\Sigma}_\beta = \widehat{\text{Cov}}(\hat{\boldsymbol{\beta}})$ and other useful quantities.

Thus, we can obtain the estimates of parametric mixed-effects, $\hat{\alpha}$ and $\hat{\mathbf{a}}_i$, as well as the estimators of nonparametric mixed-effects,

$$\hat{\eta}(t) = \mathbf{\Phi}_p(t)^T\hat{\boldsymbol{\beta}}, \ \hat{v}_i(t) = \mathbf{\Psi}_q(t)^T\hat{\mathbf{b}}_i, \ i = 1, 2, \cdots, n.$$

These results allow us to conduct some further statistical inferences. For example, using $\hat{\Sigma}_\alpha$, we can construct approximate CIs for the components of $\boldsymbol{\alpha}$. Moreover, by using $\hat{\Sigma}_\beta$, we can compute the pointwise variance of $\hat{\eta}(t)$ as $\widehat{\text{Var}}(\hat{\eta}(t)) = \mathbf{\Phi}_p(t)^T\hat{\Sigma}_\beta\mathbf{\Phi}_p(t)$. This allows for construction of approximate pointwise SD bands for $\eta(t)$.

Based on the variance components estimates $\hat{\mathbf{D}}_a, \hat{\mathbf{D}}_{ab}, \hat{\mathbf{D}}_b$ and $\hat{\sigma}^2$ of the approximate NLME model (8.90), we can compute the variance components estimates of the original SNLME model (8.81). For example, using the relationships specified in (8.88), we have

$$\hat{\gamma}(s,t) = \mathbf{\Psi}_q(s)^T\hat{\mathbf{D}}_b\mathbf{\Psi}_q(t), \ \hat{\gamma}_a(t) = \hat{\mathbf{D}}_{ab}\mathbf{\Psi}_q(t).$$

More statistical inferences are discussed in Wu and Zhang (2002b). For example, they proposed Wald, likelihood-ratio and F-type tests for testing a reduced SNLME model against a full SNLME model. For the F-type test, they proposed an innovative bootstrap procedure to approximate the null distribution of the F-type test statistic. Wu and Zhang (2002b) illustrated the methodologies via applying them to a real data set about long-term HIV dynamics.

How well the parametric NLME model (8.90) approximates the original SNLME model (8.86) is mainly determined by the choice of the bases $\mathbf{\Phi}_p(t)$ and $\mathbf{\Psi}_q(t)$. Wu and Zhang (2002b) proposed to use B-spline bases with knots specified by the "equally spaced sample quantiles as knots" rule (see Section 5.2.4 of Chapter 5) to construct $\mathbf{\Phi}_p(t)$ and $\mathbf{\Psi}_q(t)$. As mentioned previously, we can also use truncated power bases (8.12) for $\mathbf{\Phi}_p(t)$ and $\mathbf{\Psi}_q(t)$. Wu and Zhang (2002b) also proposed to use AIC and BIC (Davidian and Giltinan 1995, pages 156 and 207) to choose the numbers of basis functions p and q. The AIC and BIC values can be extracted from the output when fitting the approximate NLME model (8.90) using S-PLUS or SAS.

8.4.3 Ke and Wang's Approach

Ke and Wang (2001) considered a special case of the SNLME model (8.86) when (a) the function $\mathbf{d}$ takes the form (8.83), (b) the function g takes the form (8.85) and (c) all the nonparametric random-effect functions $v_i(t), \ i = 1, 2, \cdots, n$ are zero. That is, Ke and Wang's SNLME model can be written as

$$\begin{array}{rcl} y_{ij} & = & f[\mathbf{c}_{ij}^T\boldsymbol{\alpha} + \mathbf{h}_{ij}^T\mathbf{a}_i, \eta(t_{ij})] + \epsilon_{ij}, \\ \mathbf{a}_i & \sim & N(0, \mathbf{D}_a), \ \boldsymbol{\epsilon}_i = [\epsilon_{i1}, \cdots, \epsilon_{in_i}]^T \sim N(0, \mathbf{R}_i), \\ & & j = 1, 2, \cdots, n_i; \ i = 1, 2, \cdots, n. \end{array} \tag{8.91}$$

The above SNLME model can be written as

$$\begin{array}{rcl} \mathbf{y}_i & = & f[\mathbf{C}_i\boldsymbol{\alpha} + \mathbf{H}_i\mathbf{a}_i, \boldsymbol{\eta}_i] + \boldsymbol{\epsilon}_i, \\ \mathbf{a}_i & \sim & N(0, \mathbf{D}_a), \ \boldsymbol{\epsilon}_i \sim N(0, \mathbf{R}_i), \ \ i = 1, 2, \cdots, n, \end{array} \tag{8.92}$$

where $\mathbf{y}_i = [y_{i1}, \cdots, y_{in_i}]^T$, $\boldsymbol{\eta}_i = [\eta(t_{i1}), \cdots, \eta(t_{in_i})]^T$, and $f[\mathbf{C}_i\boldsymbol{\alpha} + \mathbf{H}_i\mathbf{a}_i, \boldsymbol{\eta}_i]$ denotes a column vector consisting of the components

$$f\left[\mathbf{c}_{i1}^T\boldsymbol{\alpha} + \mathbf{h}_{i1}^T\mathbf{a}_i, \eta(t_{i1})\right], \cdots, f\left[\mathbf{c}_{in_i}^T\boldsymbol{\alpha} + \mathbf{h}_{in_i}^T\mathbf{a}_i, \eta(t_{in_i})\right].$$

Further define $\mathbf{y} = [\mathbf{y}_1^T, \cdots, \mathbf{y}_n^T]^T$, and define $\mathbf{a}, \boldsymbol{\eta}, \boldsymbol{\epsilon}$ similarly. Define

$$\begin{aligned} \mathbf{C} &= [\mathbf{C}_1^T, \cdots, \mathbf{C}_n^T]^T, & \mathbf{H} &= \text{diag}(\mathbf{H}_1, \cdots, \mathbf{H}_n), \\ \tilde{\mathbf{D}}_a &= \text{diag}(\mathbf{D}_a, \cdots, \mathbf{D}_a), & \mathbf{R} &= \text{diag}(\mathbf{R}_1, \cdots, \mathbf{R}_n). \end{aligned}$$

Then the above SNLME model can be further written as

$$\begin{aligned} \mathbf{y} &= f[\mathbf{C}\boldsymbol{\alpha} + \mathbf{H}\mathbf{a}, \boldsymbol{\eta}] + \boldsymbol{\epsilon}, \\ \mathbf{a} &\sim N(0, \tilde{\mathbf{D}}_a), \ \boldsymbol{\epsilon} \sim N(0, \mathbf{R}). \end{aligned} \tag{8.93}$$

Given the variance components $\tilde{\mathbf{D}}_a$ and $\mathbf{R}$, Ke and Wang (2001) essentially define the estimates of $\boldsymbol{\alpha}$, $\mathbf{a}$ and $\eta(t)$ as the minimizers of the following penalized generalized log-likelihood (PGLL) criterion:

$$\begin{aligned} Q(\boldsymbol{\alpha}, \mathbf{a}, \eta) &= (\mathbf{y} - f[\mathbf{C}\boldsymbol{\alpha} + \mathbf{H}\mathbf{a}, \boldsymbol{\eta}])^T \mathbf{R}^{-1} (\mathbf{y} - f[\mathbf{C}\boldsymbol{\alpha} + \mathbf{H}\mathbf{a}, \boldsymbol{\eta}]) \\ &\quad + \mathbf{a}^T \tilde{\mathbf{D}}_a^{-1} \mathbf{a} + \lambda \mathcal{P}(\eta), \end{aligned} \tag{8.94}$$

with respect to $\boldsymbol{\alpha}$, $\mathbf{a}$ and $\eta(t)$ where $\eta(t)$ is a function in a reproducing kernel Hilbert space which is constructed based on a basis and a reproducing kernel (Wahba 1990). In the expression (8.94), the first two terms are proportional to the twice negative logarithm of the generalized likelihood of $(\boldsymbol{\alpha}, \mathbf{a}, \boldsymbol{\eta})$, and the last term is a penalty mutiplied by a smoothing parameter λ to account for the roughness of $\eta(t)$. The form of the penalty $\mathcal{P}(\eta)$ depends on specific applications and prior knowledge. For example, $\mathcal{P}(\eta) = \int [\eta^{(k)}(t)]^2 dt$ when fitting a $(2k-1)$-st degree smoothing spline (Wahba 1990, Green and Silverman 1994) and $\mathcal{P}(\eta) = \int [L\eta(t)]^2 dt$ when fitting a L-spline with a linear operator L (Wahba 1990, Heckman and Ramsay 2000). Lin and Zhang (1999) called (8.94) as a doubled penalized log-likelihood since one can also regard the term $\mathbf{a}^T\tilde{\mathbf{D}}^{-1}\mathbf{a}$ as a penalty on the parametric random-effects.

To estimate the variance components $\tilde{\mathbf{D}}_a$ and $\mathbf{R}$, Ke and Wang (2001) derived an approximate log-likelihood function of $\tilde{\mathbf{D}}_a$ and $\mathbf{R}$. Denote $\mathbf{a}^*$ as the solution to the following equation system:

$$\partial Q / \partial \mathbf{a} = -\mathbf{Z}^T\mathbf{R}^{-1}(\mathbf{y} - f[\mathbf{C}\boldsymbol{\alpha} + \mathbf{H}\mathbf{a}, \boldsymbol{\eta}]) + \tilde{\mathbf{D}}_a^{-1}\mathbf{a} = 0, \tag{8.95}$$

where $\mathbf{Z} = \partial f[\mathbf{C}\boldsymbol{\alpha} + \mathbf{H}\mathbf{a}, \boldsymbol{\eta}]/\partial \mathbf{a}$. Notice that $\mathbf{a}^*$ is the minimizer of $Q(\boldsymbol{\alpha}, \mathbf{a}, \eta)$ when $\boldsymbol{\alpha}$ and $\eta(t)$ are given. Define $\mathbf{Z}^*$ as $\mathbf{Z}$ when $\mathbf{a} = \mathbf{a}^*$. Further define $\mathbf{V}^* = \mathbf{Z}^*\tilde{\mathbf{D}}_a\mathbf{Z}^{*T} + \mathbf{R}$ and $\mu^* = f[\mathbf{C}\boldsymbol{\alpha} + \mathbf{H}\mathbf{a}^*, \boldsymbol{\eta}]$. Then Ke and Wang (2001) showed that

$$\begin{aligned} &(\mathbf{y} - \mu^* + \mathbf{Z}^*\mathbf{a}^*)^T\mathbf{V}^{*-1}(\mathbf{y} - \mu^* + \mathbf{Z}^*\mathbf{a}^*) \\ &= (\mathbf{y} - \mu^*)^T\mathbf{R}^{-1}(\mathbf{y} - \mu^*) + \mathbf{a}^{*T}\tilde{\mathbf{D}}^{-1}\mathbf{a}^*. \end{aligned} \tag{8.96}$$

Then one can obtain the following approximate log-likelihood function (up to a constant) of $\boldsymbol{\alpha}, \eta(t), \tilde{\mathbf{D}}_a$ and $\mathbf{R}$:

$$-\frac{1}{2}\log|\mathbf{V}^*| - \frac{1}{2}(\mathbf{y} - f[\mathbf{C}\boldsymbol{\alpha} + \mathbf{H}\mathbf{a}^*, \eta] + \mathbf{Z}^*\mathbf{a}^*)^T \times \mathbf{V}^{*-1}(\mathbf{y} - f[\mathbf{C}\boldsymbol{\alpha} + \mathbf{H}\mathbf{a}^*, \eta] + \mathbf{Z}^*\mathbf{a}^*). \tag{8.97}$$

Denote $\boldsymbol{\alpha}^*$ and η^* be the minimizers of $Q(\boldsymbol{\alpha}, \mathbf{a}^*, \eta)$ with respect to $\boldsymbol{\alpha}$ and $\eta(t)$. Then replacing $\boldsymbol{\alpha}$ and $\eta(t)$ by $\boldsymbol{\alpha}^*$ and $\eta^*(t)$, respectively, Ke and Wang obtained the following approximate log-likelihood function of $\tilde{\mathbf{D}}_a$ and $\mathbf{R}$:

$$-\frac{1}{2}\log|\mathbf{V}^*| - \frac{1}{2}(\mathbf{y} - f[\mathbf{C}\boldsymbol{\alpha}^* + \mathbf{H}\mathbf{a}^*, \eta^*] + \mathbf{Z}^*\mathbf{a}^*)^T \times \mathbf{V}^{*-1}(\mathbf{y} - f[\mathbf{C}\boldsymbol{\alpha}^* + \mathbf{H}\mathbf{a}^*, \eta^*] + \mathbf{Z}^*\mathbf{a}^*). \tag{8.98}$$

For easy implementation, Ke and Wang (2001) essentially assumed that $\mathbf{R} = \sigma^2\mathbf{I}_N$ and $\tilde{\mathbf{D}}_a$ depends on a parameter vector $\boldsymbol{\theta}$. The estimates of the variance components of $\tilde{\mathbf{D}}_a$ and $\mathbf{R}$ are then obtained by maximizing the approximate log-likelihood function (8.98) with respect to σ^2 and $\boldsymbol{\theta}$ when $\boldsymbol{\alpha}^*, \mathbf{a}^*$ and η^* are given.

To implement the above methodology, Ke and Wang (2001) proposed two useful procedures. The first procedure estimates $\boldsymbol{\alpha}, \mathbf{a}, \eta(t), \boldsymbol{\theta}$ and σ^2 iteratively using the following three steps:

(a) Given the current estimates of $\boldsymbol{\alpha}, \mathbf{a}, \boldsymbol{\theta}$ and σ^2, update $\eta(t)$ by minimizing (8.94) with respect to $\eta(t)$.

(b) Given the current estimates of $\eta(t), \boldsymbol{\theta}$ and σ^2, update $\boldsymbol{\alpha}$ and $\mathbf{a}$ by minimizing (8.94) with respect to $\boldsymbol{\alpha}$ and $\mathbf{a}$.

(c) Given the current estimates of $\eta(t), \boldsymbol{\alpha}$ and $\mathbf{a}$, update $\boldsymbol{\theta}$ and σ^2 by maximizing (8.98) with respect to $\boldsymbol{\theta}$ and σ^2.

Ke and Wang (2001) pointed out that Step (b) corresponds to the pseudo-data step and Step (c) corresponds to part of the LME step of Lindstrom and Bates (1990). Therefore, the procedure proposed by Lindstrom and Bates (1990) can be used. Ke and Wang (2001) also discussed how to implement Step (a) using the Fortran routine RKPACK by Gu (1989), in which they chose the smoothing parameter λ using a data-driven criterion such as generalized cross-validation (GCV), unbiased risk (UBR) or the restricted maximum likelihood (REML). See Wahba (1990), Gu (1989) and Wang (1998b) for more details.

The second procedure depends on minimizing the following approximate twice negative penalized log-likelihood (PLL) criterion:

$$\log|\mathbf{V}^*| + (\mathbf{y} - f[\mathbf{C}\boldsymbol{\alpha} + \mathbf{H}\mathbf{a}^*, \eta] + \mathbf{Z}^*\mathbf{a}^*)^T \times \mathbf{V}^{*-1}(\mathbf{y} - f[\mathbf{C}\boldsymbol{\alpha} + \mathbf{H}\mathbf{a}^*, \eta] + \mathbf{Z}^*\mathbf{a}^*) + \lambda\mathcal{P}(\eta), \tag{8.99}$$

where the first two terms are the expression (8.97) multiplied by -2 and the penalty term is the same as the penalty term in the PGLL criterion (8.94). As in the first procedure, the second procedure estimates $\boldsymbol{\alpha}, \mathbf{a}, \eta(t), \boldsymbol{\theta}$ and σ^2 using the following three steps:

(a) Given the current estimates of $\eta(t), \boldsymbol{\alpha}, \boldsymbol{\theta}$ and σ^2, update $\mathbf{a}$ by solving (8.95).

(b) Given the current estimates of $\boldsymbol{\alpha}, \mathbf{a}, \boldsymbol{\theta}$ and σ^2, update $\eta(t)$ by minimizing (8.99) with respect to $\eta(t)$.

(c) Given the current estimates of $\eta(t)$ and $\mathbf{a}$, update $\boldsymbol{\alpha}, \boldsymbol{\theta}$ and σ^2 by minimizing (8.99) with respect to $\boldsymbol{\alpha}, \boldsymbol{\theta}$ and σ^2.

Ke and Wang (2001) pointed out that Step (a) corresponds to part of the pseudo-data step and Step (c) corresponds to the LME step of Lindstrom and Bates (1990). Therefore, these two steps can be combined and solved by the *nlme* routine in S-PLUS. Ke and Wang (2001) also described how to implement Step (b) using the Fortran routine RKPACK by Gu (1989).

Ke and Wang (2001) noted that the two procedures are closely related. The first procedure is stable and performs well but the second procedure needs fewer iterations and is faster for small sample sizes. They generally recommanded the first procedure.

8.4.4 Generalizations of Ke and Wang's Approach

In the last two subsections, we described Wu and Zhang's approach and Ke and Wang's approach. Both approaches involve (1) fitting standard NLME models using the method proposed by Lindstrom and Bates (1990); and (2) fitting nonparametric regression models using one of the smoothing techniques. Wu and Zhang's approach uses regression splines. Due to its simplicity, Wu and Zhang's approach applies to the general SNLME model (8.86). Ke and Wang's approach uses smoothing splines, which is more complicated, involving backfitting strategies (Hastie and Tibshirani 1990). Ke and Wang's approach applies to the special SNLME model (8.91). For this special SNLME model, Ke and Wang's approach can be easily generalized via replacing the smoothing spline techniques with other smoothing techniques such as regression splines (Eubank 1988, 1999), P-splines (Ruppert, Wand and Carroll 2003), local polynomial smoothing (Fan and Gijbels 1996) and orthogonal series (Eubank 1988, 1999), among others.

Notice that when the nonparametric fixed-effect function $\eta(t)$ is known, Ke and Wang's SNLME model (8.93) essentially reduces to a standard NLME model, which can be solved by the standard NLME fitting routine proposed by Lindstrom and Bates (1990) and we can obtain the estimates of $\boldsymbol{\alpha}, \mathbf{a}$ and the estimates of the variance components $\tilde{\mathbf{D}}_a$ and $\mathbf{R}$. Also notice that when $\boldsymbol{\alpha}, \mathbf{a}, \tilde{\mathbf{D}}_a$ and $\mathbf{R}$ are known, the SNLME model (8.93) reduces to a nonparametric regression model for longitudinal data with an unknown function $\eta(t)$, which can be fitted by any one of the smoothing techniques mentioned earlier. Therefore, a general procedure for fitting the SNLME model (8.93) can be written as

(a) Given the current estimate of $\eta(t)$, fit (8.93) using a standard NLME model fitting routine such as the one proposed by Lindstrom and Bates (1990) and obtain the estimates of $\boldsymbol{\alpha}, \mathbf{a}, \tilde{\mathbf{D}}_a$ and $\mathbf{R}$.

(b) Given the current estimates of $\boldsymbol{\alpha}$, $\mathbf{a}$, $\tilde{\mathbf{D}}_a$ and $\mathbf{R}$, fit (8.93) using one of the smoothing techniques while accounting for the between and within-subject correlations, and obtain the estimate of $\eta(t)$.

It is not difficult to see that Step (a) here is equivalent to the combination of Steps (a) and (c) in Ke and Wang's second procedure described in the previous subsection. When the smoothing spline technique is used, Step (b) here is equivalent to Ke and Wang's second procedure, Step (b).

As with Ke and Wang's procedure, the above procedure is simple to understand but challenging to implement. First of all, due to the nonlinearity of the function $f(\cdot)$, both steps need good initial values. However, how to find good initial values is problem-specific. Secondly, as with any other nonlinear problems, nonconvergence issues within each step often occur. More research is definitely warranted in this area.

8.5 SUMMARY AND BIBLIOGRAPHICAL NOTES

In this chapter, we mainly reviewed semiparametric models for longitudinal data. We discussed three classes of semiparametric models and surveyed various methods for fitting them.

Semiparametric models for independent responses are also known as partial linear models, partially linear models, partly linear models and partial spline models. Theories and methodologies for fitting semiparametric models and their applications have been discussed in Engle, Granger, Rice and Weiss (1986), Heckman (1986), Rice (1986), Green (1987), Speckman (1988), Eubank (1988, 1999), Severini and Staniswalis (1994), Carroll et al. (1997), and Fan and Huang (2001), among others. Important surveys on semiparametric models for independent data can be found in Green and Silverman (1994), Härdle, Liang and Gao (2000), and Ruppert, Wand and Carroll (2003) and references therein.

Recently, much attention has been paid to semiparametric population mean models for clustered or longitudinal data. Martinussen and Scheike (1999) fit a semiparametric model and derived the asymptotic properties of the estimators using martingale techniques for the cumulative regression functions, which are much easier to estimate and study than the regression functions themselves. Cheng and Wei (2000) considerd a semiparametric model for panel data. Lin and Ying (2001) proposed a counting process technique. Fan and Li (2004) proposed a difference-based approach when the design time points for all subjects are not too sparse. Su and Wu (2005) studied a semiparametric time varying coefficient model. A common feature of these techniques is the obsence of smoothing for the nuisance intercept function estimation. Lin and Ying (2001), Lin and Carroll (2001a) and Fan and Li (2004) realized that efficiency can be gained by incorporating smoothing techniques into the nuisance intercept function estimation. Lin and Carroll (2001a, b) proposed a kernel GEE method. Hu, Wang and Carroll (2004) and Lin and Carroll (2005) considered a backfitting kernel method and a profile kernel method. Fan and Li (2004) proposed a profile local linear method. Moreover, Fan and Li (2004) considered variable selection in semiparametric population mean models using a technique developed in

Fan and Li (2001). He, Zhu and Fung (2002) proposed a robust M-estimator for a semiparametric model for longitudinal data with unspecified dependence structure. Rotnitzky and Jewell (1990) considered hypothesis testing of regression parameters in generalized semiparametric models.

In semiparametric population mean modeling mentioned above, the within-subject correlation is often ignored or modeled using a known parametric model for the correlation or covariance function. In the last decade, a number of methods have been developed for semiparametric mixed-effects (SPME) models for longitudinal data where the within-subject correlations are properly modeled using random-effects parametrically or nonparametrically. Most of the literature focuses on some special SPME models listed in Section 8.3 of this chapter. For example, Diggle (1988) considered a special case of the Type 4 SPME model (8.43) when the number of random-effects covariates $q_0 = 1$, and $v_i(t)$ has a parametric covariance structure such as AR(1) model. This special case was also discussed by Donnelly, Laird and Ware (1995). Zeger and Diggle (1994) considered fitting the Type 6 SPME model (8.45) using kernel smoothing and a backfitting algorithm. Moyeed and Diggle (1994) investigated the rate of convergence for such estimators. Wang (1998a) considered fitting the Type 1 SPME model (8.40) using reproducing kernel based smoothing splines of Wahba (1990) and a connection of a smoothing spline to a LME model. Following Wang (1998a), Guo (2002b) considered likelihood ratio tests about the Type 1 SPME model. Zhang et al. (1998) extended the Zeger and Diggle (1994)'s SPME model to the general SPME model (8.36) that we considered in this chapter, but they assumed a parametric covariance structure for the nonparametric random-effects component $v_i(t),\ i = 1, 2, \cdots, n$. They employed smoothing splines to fit the nonparametric fixed-effects component $\eta(t)$ and used a connection of a smoothing spline to a LME model. Jacqmin-Gadda et al. (2002) investigated the same SPME model but they computed the roughness term using a basis of cubic M-splines. Recently, Ruppert, Wand and Carroll (2003) dealt with a simple version (with $q_0 = 1$) of the Type 5 SPME model (8.44) using P-splines and a connection of a P-spline to a LME model, and Durban et al. (2005) considered fitting the general SPME model (8.36) using P-splines smoothing and a connection of a P-spline to a LME model. Tao et al. (1999) investigated a semiparametric mixed-effects model for analysis of clustered or longitudinal data with a continuous, ordinal or binary outcome. Lin and Zhang (1999) studied generalized semiparametric additive mixed-models using smoothing splines.

It is more challenging to fit and conduct statistical inferences about semiparametric nonlinear mixed-effects (SNLME) models. More work is needed on this topic. Key references on SNLME models include Ladd and Lindstrom (2000), Ke and Wang (2001), and Wu and Zhang (2002b), among others. Ladd and Lindstrom (2000) considered semiparametric self-modeling for two-dimensional curves. Ke and Wang (2001) employed smoothing splines to fit a special case of the SNLME model (8.86) discussed in this chapter. In Ke and Wang's SNLME model, the only nonparametric component is the nonparametric fixed-effect function. Wu and Zhang (2002b) fit the general SNLME model (8.81) using a basis-based approach. This approach essentially transforms the original SNLME model into a series of standard NLME models indexed by the bases, which can be solved by the existing NLME software. Wu and Zhang (2002b) proposed to choose the bases using the AIC and BIC rules for standard NLME models defined in Davidian and Giltinan (1995).

9

Time-Varying Coefficient Models

9.1 INTRODUCTION

In Chapter 8, we surveyed various methods for fitting semiparametric models. In this chapter, we shall introduce various methods for fitting time-varying coefficient models.

Hastie and Tibshirani (1993) proposed a general varying coefficient model:

$$\eta = \mathbf{x}(\mathbf{u})^T\boldsymbol{\beta}(\mathbf{r}) = x_0(u_0)\beta_0(r_0) + x_1(u_1)\beta_1(r_1) + \cdots + x_d(u_d)\beta_d(r_d),$$

where η is a parameter of the distribution of the response variable y, e.g., the expectation, and $\mathbf{x}(\mathbf{u}) = [x_0(u_0), x_1(u_1), \cdots, x_d(u_d)]^T$ is the covariate vector, and $\boldsymbol{\beta}(\mathbf{r}) = [\beta_0(r_0), \beta_1(r_1), \cdots, \beta_d(r_d)]^T$ is the associated coefficient function vector with $\mathbf{u} = [u_0, \cdots, u_d]^T$ and $\mathbf{r} = [r_0, \cdots, r_d]^T$. When the components of $\mathbf{u}$ and $\mathbf{r}$'s are the same variable such as time t, it reduces to the following special case:

$$\eta = \mathbf{x}(t)^T\boldsymbol{\beta}(t) = x_0(t)\beta_0(t) + x_1(t)\beta_1(t) + \cdots + x_d(t)\beta_d(t), \tag{9.1}$$

which is known as a time-varying coefficient (TVC) model or a "dynamic generalized linear model" (West et al. 1985). In the above model, the response variable y and the parameter η are also time-varying and can be expressed as $y(t)$ and $\eta(t)$, respectively. In particular, when $\eta(t) = \mathrm{E}[y(t)|\mathbf{x}(t)]$, the TVC model (9.1) can be further expressed as

$$y(t) = \mathbf{x}(t)^T\boldsymbol{\beta}(t) + e(t), \tag{9.2}$$

where $e(t)$ denotes the error at t. When $x_0(t) \equiv 1$, the coefficient function $\beta_0(t)$ is used to model the intercept or baseline function when the covariates $x_r(t), r = 1, \cdots, d$ are centered or normalized. To model cross-sectional data using the TVC model (9.2), Fan and Zhang (1999, 2000) proposed a two-step method. Cai, Fan and Yao (2000) applied the TVC model (9.2) to nonlinear time-series data. Hypothesis tests under the TVC model framework were considered in Fan, Zhang and Zhang (2001). Efficient estimation and inferences for generalized time-varying coefficient models for discrete cross-sectional data were considered in Cai, Fan and Li (2000).

For longitudinal data analysis, Hoover et al. (1998) and Wu, Chiang and Hoover (1998) revisited the TVC model (9.2). A smoothing spline method and a local polynomial kernel (LPK) method were proposed by Hoover et al. (1998) and Wu, Chiang and Hoover (1998). Notice that at any given time point t, the TVC model (9.2) is a simple linear model. Using this property, Fan and Zhang (2000) proposed a two-step method. When the covariates are time-independent, Wu and Chiang (2000), and Chiang, Rice and Wu (2001) proposed componentwise methods. Huang, Wu and Zhou (2002) proposed a regression spline method. For curve data, Brumback and Rice (1998) proposed a smoothing spline method.

However, none of these methods efficiently considered the important features of longitudinal data such as the within-subject correlation and the between-subject/within-subject variations. Lin and Carroll (2000), however, showed that using standard kernel methods to incorporate correlations is typically the wrong thing to do, as the methods inflate rather than deflate the variability. Welsh, Lin and Carroll (2002) showed that regression and smoothing splines do not suffer from this difficulty. Guo (2002a), Liang et al. (2003) and Wu and Liang (2004) extended the TVC models to time-varying coefficient mixed-effects models, thereby incorporating the advantages of mixed-effects models for longitudinal data analysis.

In this chapter we shall consider two classes of TVC models. The first class focuses on modeling the population mean function of the longitudinal data nonparametrically or semiparametrically, called time-varying coefficient nonparametric or semiparametric population mean models. The second class focuses on modeling the population mean and the subject-specific deviation functions nonparametrically or semiparametrically, called time-varying coefficient nonparametric or semiparametric mixed-effects models.

9.2 TIME-VARYING COEFFICIENT NPM MODEL

For longitudinal data, the TVC model (9.2) can be written as

$$y_i(t) = \mathbf{x}_i(t)^T \boldsymbol{\beta}(t) + e_i(t), \quad i = 1, 2, \cdots, n, \tag{9.3}$$

where n is the number of subjects. Since $\mathbf{x}_i(t)^T\boldsymbol{\beta}(t)$ is used to model the population mean of the data, we refer to the above model as a time-varying coefficient nonparametric population mean (TVC-NPM) model.

Let t_{ij} denote the time when the j-th measurement of the i-th subject was recorded where $j = 1, 2, \cdots, n_i$ with n_i denoting the number of measurements of the i-th subject. Let $y_{ij} = y_i(t_{ij})$ and $e_{ij} = e_i(t_{ij})$. Then the discrete version of the above TVC-NPM model (9.3) can be written as

$$y_{ij} = \mathbf{x}_i(t_{ij})^T \boldsymbol{\beta}(t_{ij}) + e_{ij}, \quad j = 1, 2, \cdots, n_i; \; i = 1, 2, \cdots, n. \tag{9.4}$$

For fitting the above TVC-NPM model, Hoover et al. (1998) proposed a LPK method and a smoothing spline method. Fan and Zhang (2000) proposed a two-step method. Huang et al. (2002) proposed a regression spline method. When the covariates are time-independent, Wu and Chiang (2000) and Chiang, Rice and Wu (2001) proposed, respectively, a componentwise method. For ANOVA-type analysis, Brumback and Rice (1998) proposed a smoothing spline method.

9.2.1 Local Polynomial Kernel Method

The LPK method for the TVC-NPM model (9.4) was first proposed and studied by Hoover et al. (1998). The key idea of the LPK method is to fit a polynomial of some degree to the coefficient functions $\beta_r(t), r = 0, 1, \cdots, d$ locally.

For an arbitrary fixed point t, assume that $\beta_r(t), r = 0, 1, \cdots, d$ all have up to $(p+1)$-st continuous derivatives for some nonnegative integer p. Then by Taylor expansion, we have

$$\beta_r(t_{ij}) \approx \mathbf{h}_{ij}^T \boldsymbol{\alpha}_r, \; j = 1, 2, \cdots, n_i; \; i = 1, 2, \cdots, n, \tag{9.5}$$

where $\mathbf{h}_{ij} = [1, t_{ij} - t, \cdots, (t_{ij} - t)^p]^T$ and $\boldsymbol{\alpha}_r = [\alpha_{r0}, \cdots, \alpha_{rp}]^T$ with $\alpha_{rl} = \beta_r^{(l)}(t)/l!$, $l = 0, \cdots, p$. Notice that $\mathbf{h}_{ij}$ is common for all the coefficient functions due to the use of the same degree of polynomial approximation. It follows that within a local neighborhood of t, the TVC-NPM model (9.4) can be approximately expressed as

$$y_{ij} = \mathbf{x}_{ij}^T \boldsymbol{\alpha} + e_{ij}, \quad j = 1, 2, \cdots, n_i; \; i = 1, 2, \cdots, n, \tag{9.6}$$

where

$$\boldsymbol{\alpha} = [\boldsymbol{\alpha}_0^T, \boldsymbol{\alpha}_1^T, \cdots, \boldsymbol{\alpha}_d^T]^T, \; \mathbf{x}_{ij} = \mathbf{x}_i(t_{ij}) \otimes \mathbf{h}_{ij}.$$

Above and throughout this book, $\mathbf{A} \otimes \mathbf{B}$ denotes the Kronecker product, $(a_{ij}\mathbf{B})$, of two matrices $\mathbf{A}$ and $\mathbf{B}$.

Let $\hat{\boldsymbol{\alpha}}$ be the estimator of $\boldsymbol{\alpha}$ obtained by minimizing the following weighted least squares (WLS) criterion:

$$\sum_{i=1}^{n} \sum_{j=1}^{n_i} w_i \left(y_{ij} - \mathbf{x}_{ij}^T \boldsymbol{\alpha} \right)^2 K_h(t_{ij} - t), \tag{9.7}$$

where w_i are weights, $K_h(\cdot) = K(\cdot/h)/h$ with a kernel function K.

The weights $w_i, \; i = 1, 2, \cdots, n$ can be specified by two schemes. Let $N = \sum_{i=1}^n n_i$ denote the total number of observations for all the subjects. Then the two weight schemes can be expressed as

$$\begin{aligned} &\text{Type 1 weight scheme: } w_i = 1/N, \; i = 1, 2, \cdots, n, \\ &\text{Type 2 weight scheme: } w_i = 1/(n n_i), \; i = 1, 2, \cdots, n. \end{aligned} \tag{9.8}$$

The Type 1 weight scheme uses the same weight for all the subjects while the Type 2 weight scheme uses different weights for different subjects. Hoover et al. (1998) used the Type 1 weight scheme, but Huang et al. (2002) considered both the weight schemes and they showed that the Type 1 weight scheme may lead to inconsistent estimators for some choices of $n_i,\ i = 1, 2, \cdots, n$ while this is not the case for the Type 2 weight scheme.

To give an explicit expression for $\hat{\boldsymbol{\alpha}}$, let

$$\mathbf{X}_i = [\mathbf{x}_{i1}, \mathbf{x}_{i2}, \cdots, \mathbf{x}_{in_i}]^T,\ \mathbf{K}_{ih} = w_i \text{diag}\Big(K_h(t_{i1} - t), \cdots, K_h(t_{in_i} - t)\Big),$$

be the design matrix and the weight matrix for the i-th subject, respectively. In addition, denote $\mathbf{X} = [\mathbf{X}_1^T, \cdots, \mathbf{X}_n^T]^T$ and $\mathbf{K}_h = \text{diag}(\mathbf{K}_{1h}, \cdots, \mathbf{K}_{nh})$. Then the WLS criterion (9.7) can be rewritten as

$$(\mathbf{y} - \mathbf{X}\boldsymbol{\alpha})^T \mathbf{K}_h (\mathbf{y} - \mathbf{X}\boldsymbol{\alpha}), \tag{9.9}$$

where $\mathbf{y} = [\mathbf{y}_1^T, \cdots, \mathbf{y}_n^T]^T$ with $\mathbf{y}_i = [y_{i1}, \cdots, y_{in_i}]^T$ being the i-th subject response vector. Minimizing (9.9) with respect to $\boldsymbol{\alpha}$ leads to

$$\hat{\boldsymbol{\alpha}} = (\mathbf{X}^T \mathbf{K}_h \mathbf{X})^{-1} \mathbf{X}^T \mathbf{K}_h \mathbf{y}, \tag{9.10}$$

where $\hat{\boldsymbol{\alpha}} = [\hat{\boldsymbol{\alpha}}_0^T, \hat{\boldsymbol{\alpha}}_1^T, \cdots, \hat{\boldsymbol{\alpha}}_d^T]^T$.

For each $r = 0, 1, \cdots, d$, to extract the estimators of $\beta_r(t)$ and its derivatives, we let $\mathbf{e}_l$ denote a $(p+1)$-dimensional unit vector whose l-th entry is 1 and others are 0. Then it is easy to see from the definitions of $\alpha_{rl}, l = 0, 1, \cdots, p$ that the estimators of the derivatives $\beta_r^{(l)}(t), l = 0, 1, \cdots, p$ are

$$\hat{\beta}_r^{(l)}(t) = l! \mathbf{e}_{l+1}^T \hat{\boldsymbol{\alpha}}_r, l = 0, 1, \cdots, p. \tag{9.11}$$

In particular, the LPK estimators of $\beta_r(t), r = 0, 1, \cdots, d$ are

$$\hat{\beta}_r(t) = \mathbf{e}_1^T \hat{\boldsymbol{\alpha}}_r, r = 0, 1, \cdots, d. \tag{9.12}$$

When $p = 0$, we approximate the coefficient functions $\beta_r(t), r = 0, 1, \cdots, d$ locally using $(d+1)$ constants. The resulting LPK estimator of $\boldsymbol{\beta}(t)$ is known as a kernel estimator. By (9.10) and (9.12), the kernel estimator of $\boldsymbol{\beta}(t)$ can be expressed as

$$\hat{\boldsymbol{\beta}}(t) = \left\{ \sum_{i=1}^n \sum_{j=1}^{n_i} K_h(t_{ij} - t) \mathbf{x}_i(t_{ij}) \mathbf{x}_i(t_{ij})^T \right\}^{-1} \left\{ \sum_{i=1}^n \sum_{j=1}^{n_i} K_h(t_{ij} - t) \mathbf{x}_i(t_{ij}) y_{ij} \right\}.$$

Hoover et al. (1998) showed that under some regularity conditions and when $w_i = 1/N,\ i = 1, 2, \cdots, n$, for the above kernel estimator, one has

$$\begin{aligned} \text{Var}[\hat{\beta}_r(t)] &= O_P[1/(Nh)] + O_P\left[\textstyle\sum_{i=1}^n n_i^2/N^2\right], \\ \text{Bias}[\hat{\beta}_r(t)] &= O_P(h^2),\ r = 0, 1, \cdots, d, \end{aligned} \tag{9.13}$$

where Var$(\cdot)$ and Bias$(\cdot)$ were defined slightly different from their regular definitions (see Hoover et al. (1998) for details). Notice that the first order term $O_P[1/(Nh)]$ in the expression of Var$[\hat{\beta}_r(t)]$ is related to the within-subject variation only, and the second order term $O_P\left[\sum_{i=1}^n n_i^2/N^2\right]$ is associated with the between-subject variation. Therefore, the asymptotic properties of $\hat{\beta}_r(t)$ are different when n_i are bounded, compared to the case when n_i are not bounded. Actually, when all the n_i are bounded, the Var$[\hat{\beta}_r(t)]$ in (9.13) is dominated by its first order term so that the mean squared error(MSE) of $\hat{\beta}_r(t)$ is

$$\text{MSE}[\hat{\beta}_r(t)] = \text{Bias}^2[\hat{\beta}_r(t)] + \text{Var}[\hat{\beta}_r(t)] = O_P\left(N^{-4/5}\right).$$

When all the n_i tend to infinity, the Var$[\hat{\beta}_r(t)]$ is dominated by the second order term $O_P\left(\sum_{i=1}^n n_i^2/N^2\right)$ so that MSE$(\hat{\beta}_r(t)) = O_P\left(\sum_{i=1}^n n_i^2/N^2\right)$. In particular, by assuming that $n_i = m, i = 1, 2, \cdots, n$, we have MSE$[\hat{\beta}_r(t)] = O_P(n^{-1})$ as $m \to \infty$. In this case, $\hat{\beta}_r(t)$ is $\sqrt{n}$-consistent.

By (9.13), the theoretical optimal bandwidth that minimizes the following weighted MSE (with weights $c_r, r = 0, 1, \cdots, d$)

$$\sum_{r=0}^{d} c_r \text{MSE}[\hat{\beta}_r(t)] = \sum_{r=0}^{d} c_r \text{Bias}^2[\hat{\beta}_r(t)] + \sum_{r=0}^{d} c_r \text{Var}[\hat{\beta}_r(t)],$$

is of order $O_P\left(N^{-1/5}\right)$ when n_i are bounded.

Compared with the smoothing spline method proposed in Hoover et al. (1998) which will be briefly described later, the above LPK method has some advantages. It is computationally less intensive and theoretically more tractable. However, since the LPK method involves only a single smoothing parameter h, it often cannot adequately fit some of the underlying coefficient functions when these coefficient functions admit different degrees of smoothness. See Hoover et al. (1998) for a typical example in which some coefficient functions are not adequately fitted. In addition, the LPK method requires intensive computations for large datasets or when the degree of local polynomials, p, is too large. This is especially the case when the "leave-one-subject-out" criterion (9.35) (see Section 9.2.5) is used to select the smoothing parameter h.

9.2.2 Regression Spline Method

The LPK method proposed by Hoover et al. (1998) uses only a single smoothing parameter for all the coefficient functions. Huang, Wu and Zhou (2002) proposed a regression spline method for fitting the TVC-NPM model (9.2) which allows different smoothing parameters for different coefficient functions.

The key concept of the regression spline method is to express each of the coefficient functions $\beta_r(t), r = 0, \cdots, d$ as a regression spline using a regression spline basis. Let $\mathbf{\Phi}_{rp_r}(t) = [\phi_{r1}(t), \cdots, \phi_{rp_r}(t)]^T$ be a basis for $\beta_r(t)$ where p_r is an nonnegative integer denoting the number of the basis functions. Then as a regression spline, we can approximately express $\beta_r(t)$ as

$$\beta_r(t) = \mathbf{\Phi}_{rp_r}(t)^T \boldsymbol{\alpha}_r, \; r = 0, 1, \cdots, d, \tag{9.14}$$

where $\boldsymbol{\alpha}_r = [\alpha_{r1}, \cdots, \alpha_{rp_r}]^T$ denotes the associated coefficient vector. Let $\mathbf{h}_{rij} = \boldsymbol{\Phi}_{rp_r}(t_{ij})$. Then we have $\beta_r(t_{ij}) = \mathbf{h}_{rij}^T \boldsymbol{\alpha}_r, r = 0, 1, \cdots, d$. When $\boldsymbol{\alpha}_r, r = 0, 1, \cdots, d$ are specified, all the coefficient functions are specified.

To estimate $\boldsymbol{\alpha}_r, r = 0, 1, \cdots, d$, let $\mathbf{x}_{ij} = [\mathbf{x}_{0ij}^T, \mathbf{x}_{1ij}^T, \cdots, \mathbf{x}_{dij}^T]^T$ where $\mathbf{x}_{rij} = x_{ri}(t_{ij})\mathbf{h}_{rij}$ and $x_{ri}(t)$ denotes the r-th entry of $\mathbf{x}_i(t)$. Denote $\boldsymbol{\alpha} = [\boldsymbol{\alpha}_0^T, \boldsymbol{\alpha}_1^T, \cdots, \boldsymbol{\alpha}_d^T]^T$. Then the TVC-NPM model (9.4) can be approximately written as

$$y_{ij} = \mathbf{x}_{ij}^T \boldsymbol{\alpha} + e_{ij}, \;\; j = 1, 2, \cdots, n_i; \; i = 1, 2, \cdots, n. \tag{9.15}$$

For fixed bases $\boldsymbol{\Phi}_{rp_r}(t), r = 0, 1, \cdots, d$, the above model is a standard linear regression model.

Huang et al. (2002) estimated $\boldsymbol{\alpha}$ via minimizing the following WLS criterion:

$$\sum_{i=1}^{n} \sum_{j=1}^{n_i} w_i \left(y_{ij} - \mathbf{x}_{ij}^T \boldsymbol{\alpha}\right)^2,$$

where $w_i, \; i = 1, 2, \cdots, n$ are weights that can be specified using one of the weight schemes in (9.8). The resulting WLS estimator of $\boldsymbol{\alpha}$ is

$$\hat{\alpha} = (\mathbf{X}^T \mathbf{W} \mathbf{X})^{-1} \mathbf{X}^T \mathbf{W} \mathbf{y},$$

where

$$\begin{array}{rclcrcl} \mathbf{y} & = & [\mathbf{y}_1^T, \cdots, \mathbf{y}_n^T]^T, & \text{with } \mathbf{y}_i & = & [y_{i1}, \cdots, y_{in_i}]^T, \\ \mathbf{W} & = & \text{diag}(\mathbf{W}_1, \cdots, \mathbf{W}_n), & \text{with } \mathbf{W}_i & = & w_i \mathbf{I}_{n_i}, \\ \mathbf{X} & = & [\mathbf{X}_1^T, \cdots, \mathbf{X}_n^T]^T, & \text{with } \mathbf{X}_i & = & [\mathbf{x}_{i1}, \cdots, \mathbf{x}_{in_i}]^T. \end{array} \tag{9.16}$$

To extract the estimators $\hat{\boldsymbol{\alpha}}_r, r = 0, 1, \cdots, d$, let

$$\mathbf{E}_r = [\mathbf{0}_{p_r \times p_0}, \cdots, \mathbf{0}_{p_r \times p_{(r-1)}}, \mathbf{I}_{p_r}, \mathbf{0}_{p_r \times p_{(r+1)}}, \cdots, \mathbf{0}_{p_r \times p_d}], \tag{9.17}$$

where $\mathbf{0}_{k \times l}$ denotes a $k \times l$ zero's matrix and $\mathbf{I}_k$ denotes a $k \times k$ identity matrix. Then we have

$$\hat{\boldsymbol{\alpha}}_r = \mathbf{E}_r (\mathbf{X}^T \mathbf{W} \mathbf{X})^{-1} \mathbf{X}^T \mathbf{W} \mathbf{y}, r = 0, 1, \cdots, d.$$

By (9.14), we further have

$$\hat{\beta}_r(t) = \boldsymbol{\Phi}_{rp_r}(t)^T \mathbf{E}_r (\mathbf{X}^T \mathbf{W} \mathbf{X})^{-1} \mathbf{X}^T \mathbf{W} \mathbf{y}, \; r = 0, 1, \cdots, d. \tag{9.18}$$

Based on the above expressions, for fixed bases $\boldsymbol{\Phi}_{rp_r}, r = 0, 1, \cdots, d$, it is easy to construct the smoother matrices and the standard deviation bands for the coefficient functions $\beta_r(t), r = 0, \cdots, d$ and the mean responses at all the design time points. The choice of the bases can be flexible. When the underlying coefficient functions exhibit periodicity, the Fourier basis may be used. When the underlying coefficient functions are smooth with a simple global structure, the polynomial basis can be used. However, these global bases may not be sensitive enough to exhibit certain local features without using a large number of basis functions. In these cases, local bases such as the truncated power bases [see (8.12)], B-spline bases (de Boor 1978) and wavelet bases are more desirable. See Ramsay and Silverman (1997, Chapter 3) for more discussions on the selection of bases.

Huang et al. (2002) derived the convergence rates of the regression spline estimators $\hat{\beta}_r(t), r = 0, 1, \cdots, d$ for general regression spline bases and in particular for truncated power bases. They showed that when the Type 1 weight scheme $w_i = 1/N,\ i = 1, 2, \cdots, n$ is used, the regression spline estimators may be inconsistent but when the Type 2 weight scheme $w_i = 1/(nn_i),\ i = 1, 2, \cdots, n$ is used, the regression spline estimators are consistent for different choices of $n_i,\ i = 1, 2, \cdots, n$. Thus, they demonstrated that the Type 2 weight scheme is more preferable.

9.2.3 Penalized Spline Method

The regression spline method proposed by Huang et al. (2002) is a simple smoothing technique for fitting the TVC-NPM model (9.4). It uses different bases for different coefficient functions and the smoothness of a coefficient function is controlled by the number of the associated basis fucntions. It is sometimes challenging to specify $(d + 1)$ bases simultaneously especially when d is large. Alternatively, we can use the same basis and the same number of basis functions for all the coefficient functions while controlling the smoothness of a coefficient function by some other means. In this subsection, we describe the penalized spline (P-spline) method.

The central idea of the P-spline method is to express all the coefficient functions $\beta_r(t), r = 0, 1, \cdots, d$ as regression splines using the same truncated power basis and penalize the highest order derivative jumps of the regression splines. Let $\mathbf{\Phi}_p(t)$ be a truncated power basis (8.12) of degree k with K knots $\tau_1, \cdots, \tau_K$ so that $p = K + k + 1$. Then we can approximately express $\beta_r(t), r = 0, 1, \cdots, d$ as regression splines:

$$\beta_r(t) = \mathbf{\Phi}_p(t)^T \boldsymbol{\alpha}_r,\ r = 0, 1, \cdots, d, \tag{9.19}$$

where $\boldsymbol{\alpha}_r = [\alpha_{r1}, \cdots, \alpha_{rp}]^T$. Recall that the last K entries of $\boldsymbol{\alpha}_r$ are proportional to the k-times derivative jumps of the regression spline $\beta_r(t)$.

To penalize these jumps, let $\mathbf{G}$ be a P-spline roughness matrix associated with $\mathbf{\Phi}_p(t)$. That is,

$$\mathbf{G} = \begin{bmatrix} \mathbf{0}_{(k+1)\times(k+1)} & \mathbf{0}_{(k+1)\times K} \\ \mathbf{0}_{K\times(k+1)} & \mathbf{I}_K \end{bmatrix}. \tag{9.20}$$

Then the estimators of $\boldsymbol{\alpha}_r, r = 0, 1, \cdots, d$ can be obtained via minimizing the following penalized weighted least squares (PWLS) criterion:

$$\sum_{i=1}^{n}\sum_{j=1}^{n_i} w_i \left(y_{ij} - \sum_{r=0}^{d} \mathbf{x}_{rij}^T \boldsymbol{\alpha}_r \right)^2 + \sum_{r=0}^{d} \lambda_r \boldsymbol{\alpha}_r^T \mathbf{G} \boldsymbol{\alpha}_r, \tag{9.21}$$

where again $w_i,\ i = 1, 2, \cdots, n$ are positive weights that can be specified using (9.8), $\mathbf{x}_{rij} = x_{ri}(t_{ij})\mathbf{\Phi}_p(t_{ij})$ with $x_{ri}(t)$ denoting the r-entry of $\mathbf{x}_i(t)$, and $\lambda_r, r = 0, 1, \cdots, d$ are the smoothing parameters. In the PWLS criterion (9.21), the first term is the weighted sum of squared residuals, representing the goodness of fit, and the second term equals to $\sum_{r=0}^{d} \lambda_r \sum_{l=k+2}^{k+K+1} \alpha_{rl}^2$, representing the weighted sum of the roughness of the regression splines.

Let $\boldsymbol{\alpha} = [\boldsymbol{\alpha}_0^T, \boldsymbol{\alpha}_1^T, \cdots, \boldsymbol{\alpha}_d^T]^T$, $\mathbf{x}_{ij} = \mathbf{x}_i(t_{ij}) \otimes \boldsymbol{\Phi}_p(t_{ij})$, and

$$\tilde{\mathbf{G}} = \text{diag}(\lambda_0 \mathbf{G}, \lambda_1 \mathbf{G}, \cdots, \lambda_d \mathbf{G}).$$

Then the WLS criterion (9.21) can be further written as

$$(\mathbf{y} - \mathbf{X}\boldsymbol{\alpha})^T \mathbf{W} (\mathbf{y} - \mathbf{X}\boldsymbol{\alpha}) + \boldsymbol{\alpha}^T \tilde{\mathbf{G}} \boldsymbol{\alpha}, \tag{9.22}$$

where $\mathbf{y}, \mathbf{W}$ and $\mathbf{X}$ are similarly defined as in (9.16). It follows that

$$\hat{\boldsymbol{\alpha}} = (\mathbf{X}^T \mathbf{W} \mathbf{X} + \tilde{\mathbf{G}})^{-1} \mathbf{X}^T \mathbf{W} \mathbf{y}.$$

To extract the estimators $\hat{\boldsymbol{\alpha}}_r, r = 0, 1, \cdots, d$, we use the matrices defined in (9.17) with $p_r = p, r = 0, 1, \cdots, d$ so that

$$\hat{\boldsymbol{\alpha}}_r = \mathbf{E}_r (\mathbf{X}^T \mathbf{W} \mathbf{X} + \tilde{\mathbf{G}})^{-1} \mathbf{X}^T \mathbf{W} \mathbf{y}, r = 0, 1, \cdots, d.$$

The P-spline estimators of $\beta_r(t), r = 0, 1, \cdots, d$ can then be expressed as

$$\hat{\beta}_r(t) = \boldsymbol{\Phi}_p(t)^T \mathbf{E}_r (\mathbf{X}^T \mathbf{W} \mathbf{X} + \tilde{\mathbf{G}})^{-1} \mathbf{X}^T \mathbf{W} \mathbf{y}, \ r = 0, 1, \cdots, d.$$

For a fixed truncated power basis $\boldsymbol{\Phi}_p(t)$ and a set of smoothing parameters $\lambda_r, r = 0, 1, \cdots, d$, it is easy to construct the smoother matrices and the standard deviation bands for the coefficient functions and the mean responses at all the design time points. The knots of $\boldsymbol{\Phi}_p(t)$ can be allocated using the methods described in Section 5.2.4 of Chapter 5. The number of knots in $\boldsymbol{\Phi}_p(t)$ can be selected roughly using the method proposed in Section 7.2.6 of Chapter 7. We shall delay discussing the choice of the smoothing parameters until Section 9.2.5.

9.2.4 Smoothing Spline Method

Hoover et al. (1998), Brumback and Rice (1998), Chiang, Rice and Wu (2000), and Eubank et al. (2004), among others, investigated the smoothing spline methods for fitting the TVC-NPM model (9.4). In the regression spline and P-spline methods described in the previous subsections, only part of the distinct design time points:

$$\tau_1, \tau_2, \cdots, \tau_M, \tag{9.23}$$

among all the design time points $\{t_{ij}, \ j = 1, 2, \cdots, n_i; \ i = 1, 2, \cdots, n\}$ are used as knots where $a = \tau_1, \tau_2 < \cdots < \tau_M = b$ with a and b specifying the data range of interest. For a smoothing spline method, all the distinct design time points (9.23) will be used as knots. Unlike the P-spline method where the penalties are assigned to the highest-order derivative jumps of the coefficient function regression splines, the smoothing spline method aims to penalize the roughness of the coefficient functions (Eubank 1988, Green and Silverman 1994). In this subsection, we first consider the cubic smoothing spline (CSS) method.

In the TVC-NPM model (9.4), assume that the coefficient functions $\beta_r(t), r = 0, 1, \cdots, d$ are twice continuously differentiable and their second derivatives $\beta_r''(t), r = 0, 1, \cdots, d$ are bounded and squared integrable. The CSS estimators $\hat{\beta}_r(t), r = 0, 1, \cdots, d$ are defined as the minimizers of the following PWLS criterion:

$$\sum_{i=1}^{n}\sum_{j=1}^{n_i} w_i \left[y_{ij} - \sum_{r=0}^{d} x_{ri}(t_{ij})\beta_r(t_{ij}) \right]^2 + \sum_{r=0}^{d} \lambda_r \int_a^b [\beta_r''(t)]^2 dt; \tag{9.24}$$

with respect to $\beta_r(t), r = 0, 1, \cdots, d$ where $w_i,\ i = 1, 2, \cdots, n$ are positive weights that can be specified using (9.8), and $\lambda_r, r = 0, 1, \cdots, d$ are the smoothing parameters. It can be shown that the CSS estimators $\hat{\beta}_r(t)$ are natural cubic splines with knots at all the distinct design time points (9.23).

Let $\boldsymbol{\beta}_r = [\beta_r(\tau_1), \cdots, \beta_r(\tau_M)]^T$ denote the values of $\beta_r(t)$ at all the distinct design time points (9.23). Let $\mathbf{h}_{ij}$ denote a M-dimensional unit vector whose l-th entry is 1 if $t_{ij} = \tau_l$ and 0 otherwise. Then we have

$$\beta_r(t_{ij}) = \mathbf{h}_{ij}^T \boldsymbol{\beta}_r,\ r = 0, 1, \cdots, d.$$

Moreover, by the properties of natural cubic splines and using (3.34), we have

$$\int_a^b [\beta_r''(t)]^2 dt = \boldsymbol{\beta}_r^T \mathbf{G} \boldsymbol{\beta}_r, r = 0, 1, \cdots, d,$$

where $\mathbf{G}$ is the cubic roughness matrix (3.33) with knots at (9.23). See Section 3.4.1 or Green and Silverman (1994) for a detailed scheme for computing $\mathbf{G}$. It follows that the PWLS criterion (9.24) can be further expressed as

$$\sum_{i=1}^{n}\sum_{j=1}^{n_i} w_i \left[y_{ij} - \sum_{r=0}^{d} x_{ri}(t_{ij}) \mathbf{h}_{ij}^T \boldsymbol{\beta}_r \right]^2 + \sum_{r=0}^{d} \lambda_r \boldsymbol{\beta}_r^T \mathbf{G} \boldsymbol{\beta}_r. \tag{9.25}$$

Let $\boldsymbol{\beta} = [\boldsymbol{\beta}_0^T, \boldsymbol{\beta}_1^T, \cdots, \boldsymbol{\beta}_d^T]^T$, $\mathbf{x}_{ij} = \mathbf{x}_i(t_{ij}) \otimes \mathbf{h}_{ij}$ and $\tilde{\mathbf{G}} = \text{diag}(\lambda_0 \mathbf{G}, \cdots, \lambda_d \mathbf{G})$. Then the PWLS criterion (9.25) can be further written as

$$(\mathbf{y} - \mathbf{X}\boldsymbol{\beta})^T \mathbf{W} (\mathbf{y} - \mathbf{X}\boldsymbol{\beta}) + \boldsymbol{\beta}^T \tilde{\mathbf{G}} \boldsymbol{\beta}, \tag{9.26}$$

where $\mathbf{y}, \mathbf{W}$ and $\mathbf{X}$ are similarly defined as in (9.16). It follows that

$$\hat{\boldsymbol{\beta}} = (\mathbf{X}^T \mathbf{W} \mathbf{X} + \tilde{\mathbf{G}})^{-1} \mathbf{X}^T \mathbf{W} \mathbf{y}. \tag{9.27}$$

To extract the estimators $\hat{\boldsymbol{\beta}}_r, r = 0, 1, \cdots, d$, we use the matrices defined in (9.17) with $p_r = M, r = 0, 1, \cdots, d$ so that

$$\hat{\boldsymbol{\beta}}_r = \mathbf{E}_r (\mathbf{X}^T \mathbf{W} \mathbf{X} + \tilde{\mathbf{G}})^{-1} \mathbf{X}^T \mathbf{W} \mathbf{y},\ r = 0, 1, \cdots, d. \tag{9.28}$$

The vector $\hat{\boldsymbol{\beta}}_r$ contains the values of $\hat{\beta}_r(t)$ at all the distinct design time points (9.23). For any design time point t_{ij}, we have $\hat{\beta}_r(t_{ij}) = \mathbf{h}_{ij}^T \hat{\boldsymbol{\beta}}_r,\ r = 0, 1, \cdots, d.$

For any other time point t, $\hat{\beta}_r(t)$ can be obtained using interpolation or using the definition of a cubic smoothing spline. In fact, a simple formula is given in Green and Silverman (1994). The choice of the smoothing parameters $\lambda_0, \lambda_1, \cdots, \lambda_d$ will not be discussed until Section 9.2.5.

Hoover et al. (1998) expressed $\beta_r(t)$ as $\mathbf{\Phi}_p(t)^T\boldsymbol{\alpha}_r$ where $\mathbf{\Phi}_p(t)$ is a B-spline basis (de Boor 1978). The technique is similar to the P-spline method discussed in the previous subsection except the roughness matrix $\mathbf{G}$ is computed using the B-spline basis as

$$\mathbf{G} = \int_a^b \mathbf{\Phi}_p''(t)\mathbf{\Phi}_p''(t)^T dt. \tag{9.29}$$

An advantage of the above approach is that it can be easily extended to minimize the following general PWLS criterion for the TVC-NPM model (9.4):

$$\sum_{i=1}^{n}\sum_{j=1}^{n_i} w_i \left[y_{ij} - \sum_{r=0}^{d} x_{ri}(t_{ij})\beta_r(t_{ij}) \right]^2 + \sum_{r=0}^{d} \lambda_r \int [\beta_r^{(k)}(t)]^2 dt, \tag{9.30}$$

over all the coefficient functions $\beta_r(t), r = 0, 1, \cdots, d$ having $(k-1)$-times absolutely continuous derivatives and squared integrable k times derivatives for $k \geq 1$, a specified integer. The minimizers $\hat{\beta}_r(t), r = 0, 1, \cdots, d$ are natural smoothing splines of degree $(2k-1)$ with knots at all the distinct design time points (9.23).

Following Hoover et al. (1998), we can express $\beta_r(t)$ as $\mathbf{\Phi}_p(t)^T\boldsymbol{\alpha}_r$ where $\mathbf{\Phi}_p(t)$ is a truncated power basis (8.12) of degree $(2k-1)$ with knots at (9.23). The resulting $(2k-1)$-st degree smoothing spline estimators $\hat{\beta}_r(t), r = 0, 1, \cdots, d$ can be obtained using the same technique as the P-spline method discussed in the previous subsection except that the roughness matrix $\mathbf{G}$ is now computed as

$$\mathbf{G} = \int_a^b \mathbf{\Phi}_p^{(k)}(t)\mathbf{\Phi}_p^{(k)}(t)^T dt. \tag{9.31}$$

Obviously, the roughness matrix (9.29) is a special case of the roughness matrix (9.31) when $\mathbf{\Phi}_p(t)$ in (9.29) is a cubic truncated power basis (8.12) with knots at (9.23).

Eubank et al. (2004) expressed the minimization problem (9.30) as a state-space form and solved it using the Kalman filter to achieve computational advantage. To describe their method briefly, one needs to re-write the TVC-NPM model (9.4) in terms of all the distinct design time points (9.23). That is,

$$y_{i_l j_l} = \sum_{r=0}^{d} x_{ri_l}(\tau_l)\beta_r(\tau_l) + e_{i_l j_l}, \tag{9.32}$$

where $i_l j_l$ denote those subscripts ij such that $t_{ij} = \tau_l$ and $e_{i_l j_l} \sim N(0, \sigma^2)$. It follows that the PWLS criterion (9.30) can be written as

$$\sum_{l=1}^{M}\sum_{i_l j_l} w_{i_l} \left[y_{i_l j_l} - \sum_{r=0}^{d} x_{ri_l}(\tau_l)\beta_r(\tau_l) \right]^2 + \sum_{r=0}^{d} \lambda_r \int_a^b [\beta_r^{(k)}(t)]^2 dt. \tag{9.33}$$

Since the minimizers $\hat{\beta}_r(t), r = 0, 1, \cdots, d$ of the above PWLS criterion are natural smoothing splines of degree $(2k-1)$ with knots at (9.23), as shown by Wahba (1978), they can be obtained through a signal extraction approach by introducing the following stochastic processes $P_r(t), r = 0, 1, \cdots, d$ for $\beta_r(t), r = 0, 1, \cdots, d$:

$$\begin{array}{rcl} P_r(t) & = & \sum_{l=0}^{k-1} \alpha_{rl} \frac{t^l}{l!} + \lambda_r^{-1/2} \sigma \int_a^t \frac{(t-h)^{k-1}}{(k-1)!} dW_r(h), \\ y_{i_l j_l} & = & \sum_{r=0}^{d} x_{r i_l}(\tau_l) P_r(\tau_l) + e_{i_l j_l}, \; r = 0, 1, \cdots, d, \end{array} \tag{9.34}$$

where $\boldsymbol{\alpha} = [\boldsymbol{\alpha}_0^T, \boldsymbol{\alpha}_1^T, \cdots, \boldsymbol{\alpha}_d^T]^T \sim N(0, \lambda_0^{-1} \mathbf{I}_{[(d+1)k]})$ with $\boldsymbol{\alpha}_r = [\alpha_{r0}, \cdots, \alpha_{r(k-1)}]^T$, and $W_r(h)$ are standard Wiener processes. Under some regularity conditions, we can show that the minimizers $\hat{\beta}_r(t), r = 0, 1, \cdots, d$ of the PWLS criterion (9.33) are the Bayesian estimates of $P_r(t), r = 0, 1, \cdots, d$ as $\lambda_0 \to 0$. That is,

$$\hat{\beta}_r(t) = \lim_{\lambda_0 \to 0} E_{\lambda_0} \left\{ P_r(t) | \mathbf{y} \right\}, \; r = 0, 1, \cdots, d.$$

To use the Kalman filter for achieving fast computation, we should re-write (9.34) in a state-space form. Let $\mathbf{z}_r(t) = [\beta_r(t), \beta_r'(t), \cdots, \beta_r^{(k-1)}(t)]^T$. Then (9.34) can be written as the following state-space form:

$$\begin{array}{rcl} \mathbf{z}_r(\tau_l) & = & \mathbf{U}_r(\tau_l - \tau_{l-1}) \mathbf{z}_r(\tau_{l-1}) + \lambda_r^{-1/2} \sigma \mathbf{u}_r(\tau_l - \tau_{l-1}), \\ y_{i_l j_l} & = & \sum_{r=0}^{d} x_{i_l}(\tau_l) \mathbf{e}_1^T \mathbf{z}_r(\tau_l) + e_{i_l j_l}, \; \text{for all } i_l j_l \;, \; r = 0, 1, \cdots, d, \end{array}$$

where $\mathbf{e}_1$ denotes a k-dimensional unit vector whose first entry is 1 and 0 otherwise; for each $r = 0, 1, \cdots, d$ and for some $\Delta \geq 0$,

$$\mathbf{U}_r(\Delta) = \begin{bmatrix} 1 & \Delta & \frac{\Delta^2}{2!} & \cdots & \frac{\Delta^{k-1}}{(k-1)!} \\ 0 & 1 & \Delta & \cdots & \frac{\Delta^{k-2}}{(k-2)!} \\ \vdots & \vdots & \vdots & \vdots & \vdots \\ 0 & 0 & 0 & \cdots & 1 \end{bmatrix},$$

and $\mathbf{u}_r(\Delta)$ is a k-dimensional vector, following $N(0, \mathbf{\Sigma}_r(\Delta))$, where $\mathbf{\Sigma}_r(\Delta)$ is a $k \times k$ matrix whose (s_1, s_2)-th entry is

$$\frac{\Delta^{(2k-1)-(s_1+s_2)}}{[(2k-1)-(s_1+s_2)](k-s_1)!(k-s_2)!}, \; s_1, s_2 = 1, 2, \cdots, k.$$

The above state-space equation system can be further combined as

$$\begin{array}{rcl} \mathbf{z}(\tau_l) & = & \mathbf{U}(\tau_l - \tau_{l-1}) \mathbf{z}(\tau_{l-1}) + \mathbf{u}(\tau_l - \tau_{l-1}), \\ y_{i_l j_l} & = & \mathbf{h}_{i_l j_l}^T \mathbf{z}(\tau_l) + e_{i_l j_l}, \; \text{for all } i_l j_l \;, \end{array}$$

where $\mathbf{h}_{i_l j_l} = \mathbf{x}_{i_l}(\tau_l) \otimes \mathbf{e}_1$, $\mathbf{z}(t) = [\mathbf{z}_0(t)^T, \mathbf{z}_1(t)^T, \cdots, \mathbf{z}_d(t)^T]^T$, and

$$\begin{array}{rcl} \mathbf{U}(\Delta) & = & \text{diag}(\mathbf{U}_0(\Delta), \mathbf{U}_1(\Delta), \cdots, \mathbf{U}_d(\Delta)), \\ \mathbf{u}(\Delta) & = & [\lambda_0^{-1/2} \sigma \mathbf{u}_0(\Delta)^T, \cdots, \lambda_d^{-1/2} \sigma \mathbf{u}_d(\Delta)^T]^T \sim N(0, \mathbf{\Sigma}(\Delta)), \end{array}$$

with $\mathbf{\Sigma}(\Delta) = \text{diag}(\lambda_0^{-1} \sigma^2 \mathbf{\Sigma}_0(\Delta), \cdots, \lambda_d^{-1} \sigma^2 \mathbf{\Sigma}_d(\Delta))$. The above state-space equation system can be solved efficiently using the Kalman filter with $O(N)$ operations as shown by Eubank et al. (2004).

9.2.5 Smoothing Parameter Selection

In the previous subsections, we described the methods for fitting the TVC-NPM model (9.4) with given smoothing parameters. In practice, we need to select these smoothing parameters to achieve the best performance of the resulting estimators. For the TVC-NPM model (9.4), the most popular smoothing parameter selector is the "leave-one-subject-out" cross-validation (SCV) rule first proposed by Rice and Silverman (1991), advocated by various authors including Zeger and Diggle (1994), Hoover et al. (1998), Fan and Zhang (1998, 2000), and Wu and Zhang (2002a), among others. This SCV rule has been described in various sections in the previous chapters.

For the TVC-NPM model (9.4), the SCV rule can be defined as

$$\mathrm{SCV}(\boldsymbol{\delta}) = \sum_{i=1}^{n}\sum_{j=1}^{n_i} w_i \left[y_{ij} - \mathbf{x}_i(t_{ij})^T \hat{\boldsymbol{\beta}}^{(-i)}(t_{ij})\right]^2, \tag{9.35}$$

where w_i are weights as specified in (9.8), $\hat{\boldsymbol{\beta}}^{(-i)}(t)$ is an estimator of $\boldsymbol{\beta}(t)$ using all the data except those data from the i-th subject and $\boldsymbol{\delta}$ denotes the smoothing parameter vector used in the estimator. Here the estimator $\hat{\boldsymbol{\beta}}(t)$ can be constructed using either the LPK, regression spline, P-spline or smoothing spline method described in the previous subsections. For the LPK method, $\boldsymbol{\delta} = h$, a single smoothing parameter, while for the other three methods $\boldsymbol{\delta}$ may be a $(d+1)$-dimensional vector. In fact, for the regression spline method, $\boldsymbol{\delta} = [p_0, p_1, \cdots, p_d]^T$ while for the P-spline and smoothing spline methods, $\boldsymbol{\delta} = [\lambda_0, \cdots, \lambda_d]^T$.

The advantages of using the SCV rule (9.35) include: (1) deletion of the entire data for one subject at a time preserves the within-subject correlation of the data; and (2) the rule does not require to model the within-subject correlation structure. However, when $\boldsymbol{\delta}$ is a $(d+1)$-dimensional vector, it is often challenging to minimize the SCV rule (9.35) especially when d or n is large. A strategy for partly reducing the burden of computational effort is suggested in Hoover et al. (1998) and Huang et al. (2002), among others.

Alternatively, Wang and Taylor (1995), and Eubank et al. (2004), among others, used the "leave-one-point-out" cross-validation (PCV) defined as

$$\mathrm{PCV}(\boldsymbol{\delta}) = \sum_{i=1}^{n}\sum_{j=1}^{n_i} w_i \left[y_{ij} - \mathbf{x}_i(t_{ij})^T \hat{\boldsymbol{\beta}}^{(-ij)}(t_{ij})\right]^2, \tag{9.36}$$

where $\hat{\boldsymbol{\beta}}^{(-ij)}(t)$ is an estimator of $\boldsymbol{\beta}(t)$ using all the data except the j-th measurement of the i-th subject. Let $\mathbf{A}$ be the smoother matrix for the responses at all the design time points such that $\hat{\mathbf{y}} = \mathbf{A}\mathbf{y}$. Then the PCV rule (9.36) can be expressed as

$$\mathrm{PCV}(\boldsymbol{\delta}) = (\mathbf{y} - \mathbf{A}\mathbf{y})^T \mathbf{W}(\mathbf{y} - \mathbf{A}\mathbf{y})/[1 - \mathrm{tr}(\mathbf{A})/N]^2, \tag{9.37}$$

where $N = \sum_{i=1}^{n} n_i$, $\mathbf{W} = \mathrm{diag}(w_1\mathbf{I}_{n_1}, \cdots, w_n\mathbf{I}_{n_n})$, and $\mathrm{tr}(\mathbf{A})$ denotes the trace of $\mathbf{A}$. The numerator of the right-hand side (9.37) is the weighted sum of the residuals, representing the goodness of fit and the denominator is a monotonic function of the number of degrees of freedom, $\text{tr}(\mathbf{A})$, representing the model complexity. Therefore, the PCV rule aims to select a good smoothing parameter vector $\boldsymbol{\delta}$ via trading-off the goodness-of-fit and the model complexity. The key advantage of using the PCV rule (9.37) is that it is less computationally intensive since there is no need to repeatly compute the estimators. A drawback of the PCV rule is that it does not account for the within-subject correlation effectively.

9.2.6 Backfitting Algorithm

To fit the TVC-NPM model (9.4), the LPK method uses only a single bandwidth h, while the regression spline, P-spline and smoothing spline methods use $(d+1)$ smoothing parameters. A single bandwidth in the LPK method is often not enough for fitting $(d+1)$ coefficient functions that admit quite different levels of smoothness (Hoover et al. 1998) while for the other three methods, it is often a challenge to select $(d+1)$ smoothing parameters (Fan and Zhang 2000).

These problems may be solved by using the backfitting algorithm proposed by Hastie and Tishirani (1990). Notice that when all the coefficient functions except a single coefficient function $\beta_r(t)$ are known, the TVC-NPM model (9.4) essentially reduces to the following special TVC-NPM model with only a single coefficient function:

$$\tilde{y}_{rij} = x_{ri}(t_{ij})\beta_r(t_{ij}) + e_{ij},\ j = 1, 2, \cdots, n_i;\ i = 1, 2, \cdots, n, \tag{9.38}$$

where $\tilde{y}_{rij} = y_{ij} - \sum_{l \neq r} x_{li}(t_{ij})\beta_l(t_{ij})$. The above TVC-NPM model can be solved using any one of the methods proposed in the previous subsections and most importantly, for each method, a single smoothing parameter is sufficient. The backfitting algorithm for fitting the TVC-NPM model (9.4) is given as follows:

(a) Get initial estimators $\hat{\beta}_l(t), l = 0, 1, \cdots, d$.

(b) Let $r = 0$.

(c) Based on the current coefficient estimators $\hat{\beta}_l(t), l \neq r$, calculate the residuals $\tilde{y}_{rij} = y_{ij} - \sum_{l \neq r} x_{li}(t_{ij})\hat{\beta}_l(t_{ij})$, fit the TVC-NPM model (9.38) and then update the estimator $\hat{\beta}_r(t)$.

(d) If $r < d$, set $r = r + 1$, go back to Step (c); else go to Step (e).

(e) Repeat Steps (b) to (d) until convergence.

The above backfitting algorithm allows one to use different smoothing parameters and even different smoothing techniques for different coefficient functions. Moreover, in each key step, only one smoothing parameter needs to be selected. This is in contrast to the LPK, regression spline, P-spline and smoothing spline methods that are directly applied to the general TVC-NPM model (9.4).

9.2.7 Two-Step Method

In the previous subsection, some problems encountered by the LPK, regression spline, P-spline and smoothing spline methods are overcomed by using a backfitting algorithm which is an iterative method. Alternatively, Fan and Zhang (2000) proposed a two-step method which involves no iterations. The two-step method is so named since it includes a "raw estimation step" and a "smoothing step". In the raw estimation step, one calculates the raw estimates of the coefficient functions via fitting a standard linear regression model and in the smoothing step, one smooths the raw estimates to obtain the smooth estimates of the coefficient functions via using one of existing smoothing techniques such as local polynomial smoothing, regression splines, penalized splines, smoothing splines or orthogonal series.

The raw estimation step can be described as follows. Let $\tau_1, \tau_2, \cdots, \tau_M$ be the distinct design time points among $\{t_{ij},\ j = 1, 2, \cdots, n_i;\ i = 1, 2, \cdots, n\}$. For each given time τ_l, let $\mathcal{D}_l$ be the collection of the subject indices of all y_{ij} observed at τ_l. Collect all $\mathbf{x}_i(t_{ij})$ and y_{ij} whose subject indices are in $\mathcal{D}_l$ and form the design matrix $\tilde{\mathbf{X}}_l$ and the response vector $\tilde{\mathbf{y}}_l$, respectively. Then from the TVC-NPM model (9.4), the data collected at time τ_l follow the standard linear regression model:

$$\tilde{\mathbf{y}}_l = \tilde{\mathbf{X}}_l \beta(\tau_l) + \tilde{\mathbf{e}}_l, \tag{9.39}$$

where $\tilde{\mathbf{e}}_l$ collects all the associated measurement errors at τ_l. It is easy to see that

$$\mathrm{E}\tilde{\mathbf{e}}_l = 0,\ \mathrm{Cov}(\tilde{\mathbf{e}}_l) = \sigma_l^2 I_{m_l},$$

where m_l denotes the number of subjects observed at time τ_l and σ_l^2 the variance of the measurement errors at τ_l.

Without loss of generality, one assumes that the design matrix $\tilde{\mathbf{X}}_l$ is of full rank; otherwise, one may collect more data within a small neighborhood of τ_l such that the associated $\tilde{\mathbf{X}}_l$ is of full rank. Then the OLS estimator of $\beta(\tau_l)$ is

$$\hat{\beta}_0(\tau_l) = (\tilde{\mathbf{X}}_l^T \tilde{\mathbf{X}}_l)^{-1} \tilde{\mathbf{X}}_l^T \tilde{\mathbf{y}}_l, \tag{9.40}$$

which is the raw estimator of $\beta(t)$ at τ_l. It is easy to notice that

$$\mathrm{E}\{\hat{\beta}_0(\tau_l)|\mathcal{D}\} = \beta(\tau_l),\ \mathrm{Cov}\{\hat{\beta}_0(\tau_l)|\mathcal{D}\} = \sigma_l^2 (\tilde{\mathbf{X}}_l^T \tilde{\mathbf{X}}_l)^{-1},$$

where here and throughout $\mathcal{D} = \{(\mathbf{x}_i(t_{ij}), t_{ij}),\ \ j = 1, 2, \cdots, n_i;\ i = 1, 2, \cdots, n\}$.

To extract the components of $\hat{\beta}_0(\tau_l)$, let $\mathbf{e}_r$ denote a $(d+1)$-dimensional unit vector whose r-th entry is 1 and 0 otherwise. Then the r-th component of $\hat{\beta}_0(\tau_l)$ is

$$\hat{\beta}_{0r}(\tau_l) = \mathbf{e}_r^T (\tilde{\mathbf{X}}_l^T \tilde{\mathbf{X}}_l)^{-1} \tilde{\mathbf{X}}_l^T \tilde{\mathbf{y}}_l,\ r = 0, 1, \cdots, d.$$

We have $\mathrm{E}\{\hat{\beta}_{0r}|\mathcal{D}\} = \beta_r(\tau_l)$ and

$$\mathrm{Cov}\{\hat{\beta}_{0r}(\tau_l), \hat{\beta}_{0r}(\tau_k)|\mathcal{D}\} = \mathbf{e}_r^T (\tilde{\mathbf{X}}_l^T \tilde{\mathbf{X}}_l)^{-1} \tilde{\mathbf{X}}_l^T \mathbf{M}_{lk} \tilde{\mathbf{X}}_k (\tilde{\mathbf{X}}_k^T \tilde{\mathbf{X}}_k)^{-1} \mathbf{e}_k,$$

where $\mathbf{M}_{lk} = \text{Cov}(\tilde{\mathbf{y}}_l, \tilde{\mathbf{y}}_k|\mathcal{D})$. In particular, $\mathbf{M}_{ll} = \text{Cov}(\tilde{\mathbf{y}}_l, \tilde{\mathbf{y}}_l|\mathcal{D}) = \sigma_l^2 \mathbf{I}_{m_l}$ so that the conditional variance of $\hat{\beta}_{0r}(\tau_l)$ is

$$\text{Var}\{\hat{\beta}_{0r}(\tau_l)|\mathcal{D}\} = \sigma_l^2 \mathbf{e}_r^T (\tilde{\mathbf{X}}_l^T \tilde{\mathbf{X}}_l)^{-1} \mathbf{e}_r. \tag{9.41}$$

Fan and Zhang (2000) proposed a simple method to estimate $\mathbf{M}_{lk}$ and σ_l^2. These quantities may be used in the smoothing step.

We now describe the smoothing step. For each $r = 0, 1, \cdots, d$, one can smooth the following pseudo-data:

$$\{\tau_l, \hat{\beta}_{0r}(\tau_l)\}, l = 1, 2, \cdots, M, \tag{9.42}$$

to obtain the refined estimator $\hat{\beta}_r(t), r = 0, 1, \cdots, d$, using any available smoothing techniques such as local polynomials, regression splines, penalized splines and smoothing splines. For some details of these techniques, see Chapter 3. Notice that by (9.41), unless all the m_l are the same and all the σ_l^2 are the same, the above smoothing step should take the heteroscedasticity of the pseudo-data (9.42) into account properly as done by Fan and Zhang (2000).

An advantage of the two-step method is that it is simple to understand and easy to implement. The method can flexibly use different smoothing methods and smoothing parameters. For a general TVC model, the two-step method not only allows different smoothing methods to be used for different coefficient functions, but also allows different smoothing parameters selected by different methods.

The two-step method can also be applied to fit the nested and crossed ANOVA models proposed and studied in Brumback and Rice (1998). For details, see Fan and Zhang (1998, 2000).

9.2.8 TVC-NPM Models with Time-Independent Covariates

In the previous subsections, we discussed how to fit the TVC-NPM model (9.4) where the covariates may depend on time. In many longitudinal studies, the covariates are time-independent so that the TVC-NPM model (9.4) can be further written as

$$y_{ij} = \mathbf{x}_i^T \beta(t_{ij}) + e_{ij}, \; j = 1, 2, \cdots, n_i; \; i = 1, 2, \cdots, n, \tag{9.43}$$

where the covariate vectors $\mathbf{x}_i$ are independent of time, but they are random across subjects. For the above special TVC-NPM model, componentwise methods are proposed in Chiang, Rice and Wu (2001) and Wu and Chiang (2000). The ANOVA model proposed and studied in Brumback and Rice (1998) and Fan and Zhang (1998, 2000) is a special case of the above TVC-NPM model (9.43).

To describe the componentwise methods, for simplicity, we start with the TVC-NPM model (9.3) with the time independent covariate vector $\mathbf{x}$, which can be written as

$$y(t) = \mathbf{x}^T \beta(t) + e(t). \tag{9.44}$$

Since $\mathbf{x}$ and $e(t)$ are uncorrelated or independent, we have $\text{E}\{\mathbf{x}e(t)\} = \{\text{E}\mathbf{x}\}\{\text{E}e(t)\} = 0$. Therefore,

$$\text{E}\{\mathbf{x}y(t)\} = \text{E}\{\mathbf{x}\mathbf{x}^T\}\beta(t).$$

Assume that $\mathbf{\Omega} = \mathrm{E}\{\mathbf{x}\mathbf{x}^T\}$ is invertible and hence positive definite. Then we can express $\beta(t)$ as $\beta(t) = \mathbf{\Omega}^{-1}\mathrm{E}\{\mathbf{x}y(t)\}$. Let $\mathbf{e}_{r+1}$ denote a $(d+1)$-dimensional unit vector whose $(r+1)$-th entry is 1 and 0 otherwise. Then $\beta_r(t), r = 0, 1, \cdots, d$ can be expressed as

$$\beta_r(t) = \mathbf{e}_{r+1}^T\mathbf{\Omega}^{-1}\mathrm{E}\left\{\mathbf{x}y(t)\right\} = \mathrm{E}\left\{\mathbf{e}_{r+1}^T\mathbf{\Omega}^{-1}\mathbf{x}y(t)\right\}, r = 0, 1, \cdots, d.$$

Since $\mathbf{x}_i,\ i = 1, 2, \cdots, n$ are observed and are i.i.d, we can estimate $\mathbf{\Omega}$ consistently as

$$\hat{\mathbf{\Omega}} = n^{-1}\sum_{i=1}^{n}\mathbf{x}_i\mathbf{x}_i^T,$$

with a parametric rate. It follows that $\left\{\mathbf{e}_{r+1}^T\mathbf{\Omega}\mathbf{x}\right\}y(t)$ at time t_{ij} can be estimated by $\tilde{y}_{rij} = \left\{\mathbf{e}_{r+1}^T\hat{\mathbf{\Omega}}^{-1}\mathbf{x}_i\right\}y_{ij}$. Therefore, for each $r = 0, 1, \cdots, d$, the componentwise estimator $\hat{\beta}_r(t)$ can be obtained via fitting the pseudo-data:

$$(t_{ij}, \tilde{y}_{rij}),\ j = 1, 2, \cdots, n_i;\ i = 1, 2, \cdots, n,$$

with the following nonparametric population mean (NPM) model:

$$\tilde{y}_{rij} = \beta_r(t_{ij}) + \epsilon_{ij},\ j = 1, 2, \cdots, n_i;\ i = 1, 2, \cdots, n, \tag{9.45}$$

where ϵ_{ij} denote the associated errors. The above model can be fitted by any one of the methods proposed in Chapters 4~7 for NPM models. Since $\mathbf{\Omega}$ is estimated with a parametric rate which is much faster than the usual nonparametric rate, it can be assumed to be known in fitting the NPM model (9.45). In fact, Wu and Chiang (2000) employed a kernel method to fit the NPM model (9.45). They derived the asymptotical biases, variances and optimal bandwidths of their componentwise estimators $\hat{\beta}_r(t), r = 0, 1, \cdots, d$, which are comparable to those obtained in Hoover et al. (1998). They also proposed a "resampling subjects" bootstrap procedure for constructing standard deviation bands for the estimators. Alternatively, Chiang, Rice and Wu (2001) used a smoothing spline method to fit the NPM model (9.45). They derived the asymptotical normality and risk representations of the associated componentwise estimators.

9.2.9 MACS Data

In Section 1.1.3 of Chapter 1, we introduced the MACS data where the response variable is the CD4 cell percentage and the three covariates are Smoking, Age, and PreCD4. In Section 8.2.7 of Chapter 8, we fit two semiparametric population mean (SPM) models (8.34) and (8.35) to the MACS data. In both SPM models, we assumed that the effects of the three covariates were polynomials of lower degrees. In many applications, this assumption may not be satisfied. In this subsection, as an illustration, we remove the polynomial model assumption, and instead assume that all the effects are nonparametric, resulting in the following TVC-NPM model:

$$\begin{aligned} y_{ij} = \beta_0(t_{ij}) + X_{1i}\beta_1(t_{ij}) + X_{2i}\beta_2(t_{ij}) + X_{3i}\beta_3(t_{ij}) + e_{ij}, \\ j = 1, 2, \cdots, n_i;\ i = 1, 2, \cdots, n, \end{aligned} \tag{9.46}$$

where X_{1i}, X_{2i} and X_{3i} denote the centralized values of the three covariates, Smoking, Age and PreCD4, for the i-th patient. These three covariates are time-independent, statisfying the conditions required by the componentwise methods of Wu and Chiang (2000) and Chiang, Rice and Wu (2001).

We fit the TVC-NPM model (9.46) to the MACS data using the P-spline method described in Section 9.2.3. We employed a quadratic truncated power basis (8.12) with 16 knots (about a quarter of all the distinct design time points). The knots were allocated using the "equally spaced sample quantitles as knots " method as described in Section 5.2.4 of Chapter 5. For simplicity, the smoothing parameters were selected using the PCV rule (9.37). The value of the selected smoothing parameters for the intercept, and the Age and PreCD4 coefficient functions was 3.1528×10^8, and the selected smoothing parameter value for the Smoking coefficient function was 849.36.

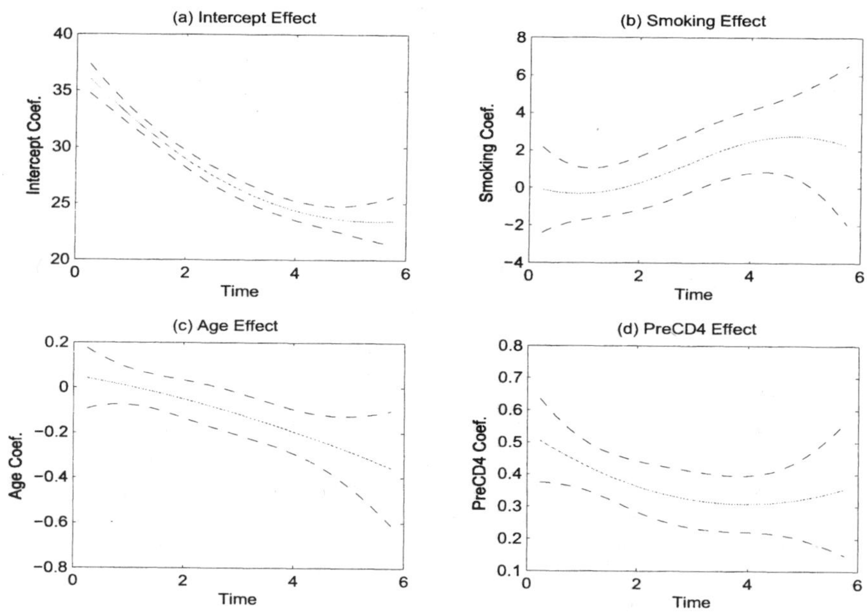

Fig. 9.1 TVC-NPM model (9.46) fit to the MACS data. Solid curves, estimated coefficient functions. Dashed curves, 95% pointwise SD bands.

Figure 9.1 shows the fitted coefficient functions (solid curves), together with the 95% pointwise SD bands (dashed curves). Comparing these estimated coefficient functions with those presented in Figure 8.2, we found that the estimated intercept, and the Smoking, Age and PreCD4 coefficient functions are quite similar, while the estimated 95% pointwise SD bands for the underlying Smoking coefficient function contain 0 all time, indicating that the effect of smoking status on the CD4 percentage of patients is not significant. The estimated coefficient functions presented in Figure 8.2 were obtained via fitting the SPM model (8.35) while assuming the Smoking, Age and PreCD4 coefficient functions to be polynomials with degrees less than 2, while in the TVC-NPM model (9.46), these assumptions were completely removed.

9.2.10 Progesterone Data

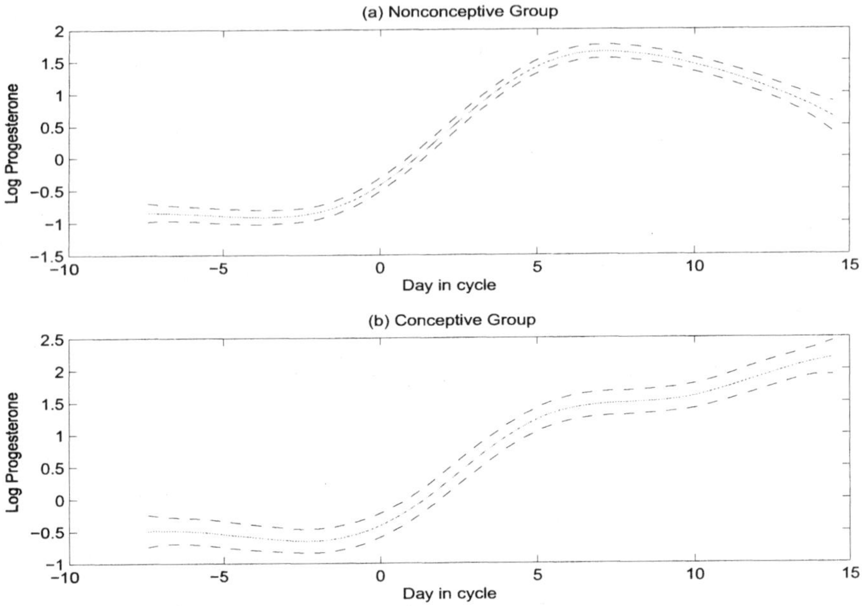

Fig. 9.2 TVC-NPM model (9.47) fit to the progesterone data. Solid curves, estimated coefficient functions. Dashed curves, 95% pointwise SD bands.

In Section 1.1.1 we introduced the progesterone data. The progesterone data consist of two groups: the nonconceptive progesterone group with 69 subjects, and the conceptive progesterone group with 22 subjects. Figures 1.1 and 1.2 show that the two groups have different group mean functions. To account for this feature, we introduce the following TVC-NPM model:

$$\begin{aligned} y_{ij} &= 1_{\{i\le 69\}}\alpha_1(t_{ij}) + 1_{\{i>69\}}\alpha_2(t_{ij}) + e_{ij}, \\ & \quad j = 1, 2, \cdots, n_i;\ i = 1, 2, \cdots, 71, \end{aligned} \tag{9.47}$$

where 1_A denotes the indicator of a set A, $\alpha_1(t)$ and $\alpha_2(t)$ are the group mean functions of the nonconceptive and conceptive progesterone groups, respectively. The above TVC-NPM model (9.47) is a special case of the general TVC-NPM model (9.4) with time independent covariates. When the componentwise methods of Wu and Chiang (2000) and Chiang, Rice and Wu (2001) are used, the resulting estimators of $\alpha_1(t)$ and $\alpha_2(t)$ are exactly the same as the estimators obtained by seperately fitting an nonparametric population mean (NPM) model of the form:

$$y_{ij} = \eta(t_{ij}) + e_{ij},\ \ j = 1, 2, \cdots, n_i;\ i = 1, 2, \cdots, n, \tag{9.48}$$

where the nonconceptive progesterone group, $\eta(t) = \alpha_1(t), n = 69$ and the conceptive group, $\eta(t) = \alpha_2(t); n = 22$.

We fit the TVC-NPM model (9.47) to the progesterone data using the P-spline method described in Section 9.2.3. We employed a quadratic truncated power basis (8.12) with 12 knots (half of all the distinct design time points). The knots were scattered using the "equally spaced sample quantitles as knots " method as described in Section 5.2.4 of Chapter 5. For simplicity, the smoothing parameters were selected using the PCV rule (9.37). The selected smoothing parameters for the two group mean functions were 957.08 and 225.25, respectively.

Figure 9.2 shows the fitted coefficient functions (solid curves), together with the 95% pointwise SD bands (dashed curves). The fitted group mean functions are comparable with those obtained Brumback and Rice (1998), and Fan and Zhang (1998, 2000), among others.

9.3 TIME-VARYING COEFFICIENT SPM MODEL

In the TVC-NPM model (9.3), we assume that all the coefficient functions are nonparametric and need to be estimated with some nonparametric techniques. Sun and Wu (2005) extended the model to a time-varying coefficient semiparametric population mean (TVC-SPM) model in which some of the coefficient functions are allowed to be parametric. That is, the parametric coefficient functions are known except an unknown parameter vector. The resultant TVC-SPM model can be written as

$$y_i(t) = \mathbf{c}_i(t)^T\boldsymbol{\alpha}(t;\boldsymbol{\theta}) + \mathbf{x}_i(t)^T\boldsymbol{\beta}(t) + e_i(t), \quad i = 1, 2, \cdots, n, \tag{9.49}$$

where the coefficient function vector $\boldsymbol{\alpha}(t;\boldsymbol{\theta}) = [\alpha_1(t;\boldsymbol{\theta}), \cdots, \alpha_p(t;\boldsymbol{\theta})]^T$ is known as long as the parameter vector $\boldsymbol{\theta} = [\theta_1, \cdots, \theta_k]^T$ is specified, $\mathbf{c}_i(t)$ denote the associated covariate vectors evaluated at time t, and other notations are the same as those in the TVC-NPM model (9.3), e.g., $\boldsymbol{\beta}(t) = [\beta_0(t), \cdots, \beta_d(t)]^T$. For example, suppose $\boldsymbol{\alpha}(t;\boldsymbol{\theta})$ has 3 components, being a constant θ_1, a linear function $\theta_2 + \theta_3 t$ and a quadratic function $\theta_4 + \theta_5 t + \theta_6 t^2$, respectively, then we have $\boldsymbol{\theta} = [\theta_1, \theta_2, \cdots, \theta_6]^T$ and

$$\begin{aligned} \alpha_1(t;\boldsymbol{\theta}) &= \theta_1, \\ \alpha_2(t;\boldsymbol{\theta}) &= \theta_2 + \theta_3 t, \\ \alpha_3(t;\boldsymbol{\theta}) &= \theta_4 + \theta_5 t + \theta_6 t^2. \end{aligned}$$

In the above example, $\boldsymbol{\alpha}(t;\boldsymbol{\theta})$ is a linear function of $\boldsymbol{\theta}$. In such a case, we can write $\boldsymbol{\alpha}(t;\boldsymbol{\theta}) = \mathbf{H}\boldsymbol{\theta}$ where $\mathbf{H}$ is a $p \times k$ design matrix. In this case, the TVC-SPM model (9.49) can be further written as

$$y_i(t) = \tilde{\mathbf{c}}_i(t)^T\boldsymbol{\theta} + \mathbf{x}_i(t)^T\boldsymbol{\beta}(t) + e_i(t), \quad i = 1, 2, \cdots, n, \tag{9.50}$$

where $\tilde{\mathbf{c}}_i(t) = \mathbf{H}^T\mathbf{c}_i(t), \ i = 1, 2, \cdots, n$. It is clear that the TVC-SPM model (9.50) is a special case of the TVC-SPM model (9.49). In particular, when the dimension of $\boldsymbol{\beta}(t)$ is 1, i.e., $d = 0$, the TVC-SPM model (9.50) reduces to the semiparametric population mean (SPM) model (8.1).

In Section 8.2 of Chapter 8, we introduced several methods for fitting the SPM model (8.1). These methods may be extended to fit the TVC-SPM model (9.50) and to fit the more general TVC-SPM model (9.49) as well. Sun and Wu (2005) provided an excellent example. In Section 8.2.6 of Chapter 8, we introduced the counting process method for fitting the SPM model (8.1). This method was proposed and studied by Martinussen and Scheike (1999, 2000, 2001), Lin and Ying (2001), and Fan and Li (2004), among others. The advantage of this method is that it involves no smoothing. Following these authors, Sun and Wu (2005) proposed a counting process method to fit the general TVC-SPM model (9.49). They developed weighted least squares type estimators for the parameter vector $\boldsymbol{\theta}$ and for the nonparametric coefficient functions $\beta_r(t), r = 0, 1, \cdots, d$ as well. They considered two kinds of the estimators: one uses the single nearest neighbor smoothing and the other uses the kernel smoothing. They pointed out that the procedure using the single nearest neighbor smoothing applies to a broader model for the sampling times than the proportional mean rate model as considered by Lin and Ying (2001), and that the estimators using the kernel smoothing avoid modeling the sampling times and are shown to be asymptotically more efficient.

An alternative example is provided as follows. In Section 8.2, we have introduced the backfitting algorithm of Hastie and Tishirani (1990) to fit the SPM model (8.1). We now extend it to fit the general TVC-SPM model (9.49). Notice that on the one hand, when $\boldsymbol{\theta}$ is known, the TVC-SPM model (9.49) essentially reduces to the TVC-NPM model (9.3). That is,

$$\tilde{y}_i(t) = \mathbf{x}_i(t)^T \boldsymbol{\beta}(t) + e_i(t), \; i = 1, 2, \cdots, n, \tag{9.51}$$

where $\tilde{y}_i(t) = y_i(t) - \mathbf{c}_i(t)^T \boldsymbol{\alpha}(t; \boldsymbol{\theta})$. The above TVC-NPM model may be solved using one of the methods proposed previously. On the other hand, when $\boldsymbol{\beta}(t)$ is known, the TVC-SPM model (9.49) reduces to the following parametric regression model:

$$\tilde{y}_i(t) = \mathbf{c}_i(t)^T \boldsymbol{\alpha}(t; \boldsymbol{\theta}) + e_i(t), \; i = 1, 2, \cdots, n, \tag{9.52}$$

where $\tilde{y}_i(t) = y_i(t) - \mathbf{x}_i(t)^T \boldsymbol{\beta}(t)$. The above parametric regression model can be fitted using the OLS or GLS method. However, if $\boldsymbol{\alpha}(t; \boldsymbol{\theta})$ is a nonlinear function of $\boldsymbol{\theta}$, a Newton-Raphson method or a quasi-Newton method is often needed. We then can state a backfitting algorithm for fitting the general TVC-SPM model (9.49) as follows:

(a) Get an initial estimate $\hat{\boldsymbol{\theta}}$.

(b) Calculate the residuals $\tilde{y}_i(t) = y_i(t) - \mathbf{c}_i(t)^T \boldsymbol{\alpha}(t; \hat{\boldsymbol{\theta}})$, fit the TVC-NPM model (9.51), and update $\hat{\boldsymbol{\beta}}(t)$.

(c) Calculate the residuals $\tilde{y}_i(t) = y_i(t) - \mathbf{x}_i(t)^T \hat{\boldsymbol{\beta}}(t)$, fit the parametric regression model (9.52), and update $\hat{\boldsymbol{\theta}}$.

(d) Repeat Steps (b) to (c) until convergence.

9.4 TIME-VARYING COEFFICIENT NPME MODEL

In the previous two sections, we described methods for fitting the TVC-NPM model (9.3) and the TVC-SPM model (9.49) where we made a working independence assumption for the measurement error processes $e_i(t),\ i = 1, 2, \cdots, n$. To take the within-subject correlation of the longitudinal data into account, one can model the within-subject correlation structure parametrically as done by many authors including Wang (1998a, b), Zhang et al. (1998), Lin and Zhang (1999), Lin and Carroll (2000, 2001a,b), among others. However, when a parametric model is not available for the within-subject correlation, a nonparametric technique has to be used. Examples can be found in Shi et al. (1996), Rice and Wu (2001), Wu and Zhang (2002a), Liang, Wu and Carroll (2003), and Liang and Wu (2004), among others. In this section, we generalize the TVC-NPM model (9.3) via modeling the within-subject correlation of $e_i(t)$ nonparametrically. That is, we divide the error term $e_i(t)$ into two parts, one part representing the subject-specific deviation from the population mean function, which may depend on a d^*-dimensional covariate vector, say $\mathbf{z}_i(t)$, and the other representing the measurement error so that $e_i(t) = \mathbf{z}_i(t)^T\mathbf{v}_i(t) + \epsilon_i(t)$. We then generalize the TVC-NPM model (9.3) to the following time-varying coefficient nonparametric mixed-effects (TVC-NPME) model:

$$\begin{array}{c} y_i(t) = \mathbf{x}_i(t)^T\boldsymbol{\beta}(t) + \mathbf{z}_i(t)^T\mathbf{v}_i(t) + \epsilon_i(t), \\ \mathbf{v}_i(t) \sim \text{GP}(0, \boldsymbol{\Gamma}),\ \ \epsilon_i(t) \sim \text{GP}(0, \gamma_\epsilon), \\ \mathbf{v}_i(t) \text{ and } \epsilon_i(t) \text{ are independent},\ \ i = 1, 2, \cdots, n, \end{array} \tag{9.53}$$

where $\mathbf{x}_i(t) = [x_{0i}(t), \cdots, x_{di}(t)]^T$ and $\boldsymbol{\beta}(t) = [\beta_0(t), \cdots, \beta_d(t)]^T$ are the covariates and coefficient functions of the fixed-effects component, and

$$\begin{aligned} \mathbf{z}_i(t) &= [z_{0i}(t), z_{1i}(t), \cdots, z_{d^*i}(t)]^T, \text{ and} \\ \mathbf{v}_i(t) &= [v_{0i}(t), v_{1i}(t), \cdots, v_{d^*i}(t)]^T, \end{aligned}$$

are the covariates and coefficient functions of the random-effects component. Notice that when we take $x_{0i}(t) \equiv 1$, we include a fixed-effect intercept function $\beta_0(t)$; when we take $z_{0i}(t) \equiv 1$, we include a random-effect intercept function $v_{0i}(t)$. The above TVC-NPME model (9.53) was first proposed by Guo (2002a) for analyzing functional data. He named this model as a "functional mixed-effects model" and fit it using a smoothing spline method. A wavelet-based method was proposed by Morris et al. (2003) and Morris and Carroll (2005).

When $d^* = 0$ and $z_{0i}(t) \equiv 1$, the TVC-NPME model (9.53) reduces to the following special TVC-NPME model:

$$\begin{aligned} y_i(t) &= \mathbf{x}_i(t)^T\boldsymbol{\beta}(t) + v_i(t) + \epsilon_i(t), \\ v_i(t) &\sim \text{GP}(0, \gamma), \epsilon_i(t) \sim \text{GP}(0, \gamma_\epsilon), \\ & \quad i = 1, 2, \cdots, n, \end{aligned} \tag{9.54}$$

where the random effect functions $v_i(t) \equiv v_{0i}(t)$. When $d = 0$, $x_{0i}(t) \equiv 1$, the above special TVC-NPME model (9.54) further reduces to the simple nonparametric mixed-effects (NPME) model discussed in Chapters 4~7.

When $d^* = d$ and the fixed-effects and random-effects covariates are the same, i.e., $\mathbf{x}_i(t) \equiv \mathbf{z}_i(t),\ i = 1, 2, \cdots, n$, the TVC-NPME model (9.53) reduces to the following random time-varying coefficient (RTVC) model:

$$\begin{aligned} y_i(t) &= \mathbf{x}_i(t)^T \boldsymbol{\beta}_i(t) + e_i(t), \\ \boldsymbol{\beta}_i(t) &= \boldsymbol{\beta}(t) + \mathbf{v}_i(t), \quad i = 1, 2, \cdots, n, \end{aligned} \tag{9.55}$$

where $\boldsymbol{\beta}_i(t)$ are the random coefficient function vectors, $\boldsymbol{\beta}(t) = \mathrm{E}[\boldsymbol{\beta}_i(t)]$ is the fixed-effect coefficient function vector and $\mathbf{v}_i(t) = \boldsymbol{\beta}_i(t) - \boldsymbol{\beta}(t)$ are the random-effects coefficient function vectors. When $d = 1$ and $x_{0i}(t) \equiv 1$, the above RTVC model reduces to the RTVC model proposed and studied by Liang, Wu and Carroll (2003) and Wu and Liang (2004). The former employed a regression spline method and the latter proposed a backfitting algorithm with a kernel method.

Given the design time points $\{t_{ij},\ j = 1, 2, \cdots, n_i;\ i = 1, 2, \cdots, n\}$, we can write the discrete version of the TVC-NPME model (9.53) as

$$\begin{gathered} y_{ij} = \mathbf{x}_i(t_{ij})^T \boldsymbol{\beta}(t_{ij}) + \mathbf{z}_i(t_{ij})^T \mathbf{v}_i(t_{ij}) + \epsilon_{ij}, \\ \mathbf{v}_i(t) \sim \mathrm{GP}(0, \boldsymbol{\Gamma}),\ \boldsymbol{\epsilon}_i = [\epsilon_{i1}, \cdots, \epsilon_{in_i}]^T \sim N(0, \mathbf{R}_i), \\ j = 1, 2, \cdots, n_i;\ i = 1, 2, \cdots, n, \end{gathered} \tag{9.56}$$

where $y_{ij} = y_i(t_{ij}), \epsilon_{ij} = \epsilon_i(t_{ij})$ and the (k, l)-th entry of $\mathbf{R}_i$ is $\gamma_\epsilon(t_{ik}, t_{il})$.

9.4.1 Local Polynomial Method

The LPK method for a NPME model was poineered by Wu and Zhang (2002a). Park and Wu (2005) considered a similar LPK method from a likelihood point of view. Wu and Liang (2004, 2005) generalized it to a special case of the TVC-NPME model (9.54). The key idea of the LPK method is to locally fit polynomials of some degree to t he fixed-effect coefficient functions $\beta_r(t), r = 0, 1, \cdots, d$ and the random-effects coefficient functions $v_{si}(t), s = 0, 1, \cdots, d^*$.

For an arbitrary fixed time point t, assume that $\beta_r(t), r = 0, 1, \cdots, d$ and $v_{si}(t), s = 0, 1, \cdots, d^*$ have up to $(p + 1)$-times continuous derivatives for some nonnegative integer p. Then by Taylor expansion, we have

$$\begin{gathered} \beta_r(t_{ij}) \approx \mathbf{h}_{ij}^T \boldsymbol{\alpha}_r,\ r = 0, 1, \cdots, d, \\ v_{si}(t_{ij}) \approx \mathbf{h}_{ij}^T \mathbf{b}_{si},\ s = 0, 1, \cdots, d^*, \\ j = 1, 2, \cdots, n_i;\ i = 1, 2, \cdots, n, \end{gathered} \tag{9.57}$$

where $\mathbf{h}_{ij} = [1, t_{ij} - t, \cdots, (t_{ij} - t)^p]^T, \boldsymbol{\alpha}_r = [\alpha_{r0}, \cdots, \alpha_{rp}]^T$ with $\alpha_{rl} = \beta_r^{(l)}(t)/l!$, and $\mathbf{b}_{si} = [b_{si0}, \cdots, b_{sip}]^T$ with $b_{sil} = v_{si}^{(l)}(t)/l!$. Let $\mathbf{b}_i = [\mathbf{b}_{0i}^T, \mathbf{b}_{1i}^T, \cdots, \mathbf{b}_{d^*i}^T]^T$. Obviously, $\mathbf{b}_i$ are i.i.d copies of a $[(p+1)(d^*+1)]$-dimensional normal distribution with mean 0 and some covariance matrix $\mathbf{D}$, i.e., $\mathbf{b}_i \sim N(0, \mathbf{D})$ where $\mathbf{D}$ is a $[(p+1)(d^*+1)] \times [(p+1)(d^*+1)]$ matrix depending on t. Notice that $\mathbf{h}_{ij}$ is common for all the fixed-effects and random-effect coefficient functions since the same degree of polynomial approximation is used. In principle, different polynomial approximations for different coefficient functions are allowed but we shall not pursue this for simplicity of presentation. Then, within a local neighborhood of t, the TVC-NPME model (9.53) can be approximately expressed as

$$\begin{aligned} y_{ij} &= \mathbf{x}_{ij}^T \boldsymbol{\alpha} + \mathbf{z}_{ij}^T \mathbf{b}_i + \epsilon_{ij}, \\ \mathbf{b}_i &\sim N(0, \mathbf{D}), \boldsymbol{\epsilon}_i \sim N(0, \mathbf{R}_i), \\ j &= 1, 2, \cdots, n_i;\ i = 1, 2, \cdots, n, \end{aligned} \tag{9.58}$$

where $\boldsymbol{\alpha} = [\boldsymbol{\alpha}_0^T, \boldsymbol{\alpha}_1^T, \cdots, \boldsymbol{\alpha}_d^T]^T$, $\mathbf{x}_{ij} = \mathbf{x}_i(t_{ij}) \otimes \mathbf{h}_{ij}$, and $\mathbf{z}_{ij} = \mathbf{z}_i(t_{ij}) \otimes \mathbf{h}_{ij}$.

Within a neighborhood of t, the TVC-NPME model (9.58) is a standard LME model. To incorporate the information of this locality, following Wu and Zhang (2002a), we can form the following localized generalized log-likelihood (LGLL) criterion:

$$\sum_{i=1}^{n} \left\{ (\mathbf{y}_i - \mathbf{X}_i \boldsymbol{\alpha} - \mathbf{Z}_i \mathbf{b}_i)^T \mathbf{K}_{ih}^{1/2} \mathbf{R}_i^{-1} \mathbf{K}_{ih}^{1/2} (\mathbf{y}_i - \mathbf{X}_i \boldsymbol{\alpha} - \mathbf{Z}_i \mathbf{b}_i) + \mathbf{b}_i^T \mathbf{D}^{-1} \mathbf{b}_i \right\}, \tag{9.59}$$

for estimating $\boldsymbol{\alpha}$ and $\mathbf{b}_i$, where

$$\begin{aligned} \mathbf{y}_i = [y_{i1}, \cdots, y_{in_i}]^T,\ \mathbf{K}_{ih} = \mathrm{diag}(K_h(t_{i1} - t), \cdots, K_h(t_{in_i} - t)), \\ \mathbf{X}_i = [\mathbf{x}_{i1}, \cdots, \mathbf{x}_{in_i}]^T,\ \mathbf{Z}_i = [\mathbf{z}_{i1}, \cdots, \mathbf{z}_{in_i}]^T, \end{aligned} \tag{9.60}$$

with $K_h(\cdot) = K(\cdot/h)/h$, K being a kernel function and h being a positive bandwidth. The above LGLL criterion is so called since when $\mathbf{D}$ and $\mathbf{R}_i,\ i = 1, 2, \cdots, n$ are given, it is proportional to the twice negative logarithm of the generalized likelihood of the localized fixed-effects and ranom-effects parameters $(\boldsymbol{\alpha}, \mathbf{b}_i,\ i = 1, 2, \cdots, n)$ (the joint probability density function of the localized responses $\mathbf{K}_{ih}^{1/2} \mathbf{y}_i,\ i = 1, 2, \cdots, n$ and the localized random-effects $\mathbf{b}_i,\ i = 1, 2, \cdots, n$). Except for the definitions of $\boldsymbol{\alpha}, \mathbf{b}_i, \mathbf{X}_i$ and $\mathbf{Z}_i$, the LGLL criterion (9.59) is almost the same as the criterion in Wu and Zhang (2002a) for a NPME model (see Chapter 4 for details). Therefore, it is expected that the minimizers to both criteria can be similarly obtained and in a similar form.

Denote further that

$$\begin{aligned} \mathbf{y} &= [\mathbf{y}_1^T, \cdots, \mathbf{y}_n^T]^T, & \mathbf{b} &= [\mathbf{b}_1^T, \cdots, \mathbf{b}_n^T]^T, \\ \mathbf{X} &= [\mathbf{X}_1^T, \cdots, \mathbf{X}_n^T]^T, & \mathbf{Z} &= \mathrm{diag}(\mathbf{Z}_1, \cdots, \mathbf{Z}_n), \\ \tilde{\mathbf{D}} &= \mathrm{diag}(\mathbf{D}, \cdots, \mathbf{D}), & \mathbf{R} &= \mathrm{diag}(\mathbf{R}_1, \cdots, \mathbf{R}_n), \\ \mathbf{K}_h &= \mathrm{diag}(\mathbf{K}_{1h}, \cdots, \mathbf{K}_{nh}). \end{aligned} \tag{9.61}$$

Then the LGLL criterion (9.59) can be rewritten as

$$(\mathbf{y} - \mathbf{X}\boldsymbol{\alpha} - \mathbf{Z}\mathbf{b})^T \mathbf{K}_h^{1/2} \mathbf{R}^{-1} \mathbf{K}_h^{1/2} (\mathbf{y} - \mathbf{X}\boldsymbol{\alpha} - \mathbf{Z}\mathbf{b}) + \mathbf{b}^T \tilde{\mathbf{D}}^{-1} \mathbf{b}. \tag{9.62}$$

Given the variance components $\tilde{\mathbf{D}}, \mathbf{R}$, the kernel K and the bandwidth h, minimizing (9.62) with respect to $\boldsymbol{\alpha}$ and $\mathbf{b}$ leads to

$$\begin{aligned} \hat{\boldsymbol{\alpha}} &= \left(\mathbf{X}^T \mathbf{K}_h^{1/2} \mathbf{V}^{-1} \mathbf{K}_h^{1/2} \mathbf{X} \right)^{-1} \mathbf{X}^T \mathbf{K}_h^{1/2} \mathbf{V}^{-1} \mathbf{K}_h^{1/2} \mathbf{y}, \\ \hat{\mathbf{b}} &= \tilde{\mathbf{D}} \mathbf{Z}^T \mathbf{K}_h^{1/2} \mathbf{V}^{-1} \mathbf{K}_h^{1/2} (\mathbf{y} - \mathbf{X}\hat{\boldsymbol{\alpha}}), \end{aligned} \tag{9.63}$$

so that $\hat{\mathbf{b}}_i = \mathbf{D}\mathbf{Z}_i\mathbf{K}_{ih}^{1/2}\mathbf{V}_i^{-1}\mathbf{K}_{ih}^{1/2}(\mathbf{y}_i - \mathbf{X}_i\hat{\boldsymbol{\alpha}}),\ i = 1, 2, \cdots, n$, where

$$\begin{aligned}\mathbf{V} &= \text{diag}(\mathbf{V}_1, \cdots, \mathbf{V}_n) = \mathbf{K}_h^{1/2}\mathbf{Z}\tilde{\mathbf{D}}\mathbf{Z}^T\mathbf{K}_h^{1/2} + \mathbf{R}, \\ \mathbf{V}_i &= \mathbf{K}_{ih}^{1/2}\mathbf{Z}_i\mathbf{D}\mathbf{Z}_i^T\mathbf{K}_{ih}^{1/2} + \mathbf{R}_i.\end{aligned}$$

The minimizers are indeed similar to those obtained in Wu and Zhang (2002a). See also Chapter 4.

The estimators $\hat{\boldsymbol{\alpha}}_r, r = 0, 1, \cdots, d$ and $\hat{\mathbf{b}}_{si}, s = 0, 1, \cdots, d^*$ can be easily extracted from $\hat{\boldsymbol{\alpha}}$ and $\hat{\mathbf{b}}_i,\ i = 1, 2, \cdots, n$, respectively. Let $\mathbf{e}_l$ denote a $(p+1)$-dimensional unit vector whose l-th entry is 1 and others are 0. Then for each $r = 0, 1, \cdots, d$ and $s = 0, 1, \cdots, d^*$, the estimators of the derivatives $\beta_r^{(l)}(t)$ and $v_{si}^{(l)}(t)$ for $l = 0, 1, \cdots, p$ are

$$\begin{aligned}\hat{\beta}_r^{(l)}(t) &= l!\mathbf{e}_{l+1}^T\hat{\boldsymbol{\alpha}}_r,\ r = 0, 1, \cdots, d, \\ \hat{v}_{si}^{(l)}(t) &= l!\mathbf{e}_{l+1}^T\hat{\mathbf{b}}_{si},\ s = 0, 1, \cdots, d^*.\end{aligned} \tag{9.64}$$

In particular, the LPK estimators of $\beta_r(t), r = 0, 1, \cdots, d$, and $v_{si}(t), s = 0, 1, \cdots, d^*$ are

$$\begin{aligned}\hat{\beta}_r(t) &= \mathbf{e}_1^T\hat{\boldsymbol{\alpha}}_r,\ r = 0, 1, \cdots, d, \\ \hat{v}_{si}(t) &= \mathbf{e}_1^T\hat{\mathbf{b}}_{si},\ s = 0, 1, \cdots, d^*.\end{aligned} \tag{9.65}$$

An advantage of the LPK method is that it can be easily implemented using existing software designed for standard LME models. In fact, for each given time point t, the estimators (9.63) can be obtained via operationally fitting the following standard LME model:

$$\begin{gathered}\tilde{\mathbf{y}} = \tilde{\mathbf{X}}\boldsymbol{\alpha} + \tilde{\mathbf{Z}}\mathbf{b} + \boldsymbol{\epsilon}, \\ \mathbf{b} \sim N(\mathbf{0}, \tilde{\mathbf{D}}),\ \boldsymbol{\epsilon} \sim N(\mathbf{0}, \mathbf{R}),\end{gathered} \tag{9.66}$$

where $\tilde{\mathbf{y}} = \mathbf{K}_h^{1/2}\mathbf{y}$ is the localized response vector, $\tilde{\mathbf{X}} = \mathbf{K}_h^{1/2}\mathbf{X}$ is the localized fixed-effects design matrix and $\tilde{\mathbf{Z}} = \mathbf{K}_h^{1/2}\mathbf{Z}$ the localized random-effects design matrix. The LPK estimators (9.63), their standard deviations, the variance components $\tilde{\mathbf{D}}$ and $\mathbf{R}$ can then be obtained via fitting (9.66) using the S-PLUS function *lme* or the SAS procedure *PROC MIXED*.

The bandwidth h for the fixed-effect coefficient functions $\hat{\beta}_r(t), r = 0, 1, \cdots, d$ and the random-effect functions $v_{si}(t), s = 0, 1, \cdots, d^*$ can be selected seperately using the "leave-one-subject-out" and "leave-one-point-out" cross-validation rules developed in Wu and Zhang (2002a), Liang and Wu (2004) and Park and Wu (2005), among others. See Chapter 4 for details.

9.4.2 Regression Spline Method

In the previous subsection, we fit the TVC-NPME model (9.56) using the LPK method. An advantage of this method is that it can be implemented using the existing software. A drawback is that it uses just one bandwidth for the fixed-effects and random-effect coefficient functions. In this subsection, we shall introduce a regression spline method to fit the TVC-NPME model (9.56). It has several advantages. First, different smoothing parameters can be used for different fixed-effects and random-effect coefficient functions. Second, it transforms the TVC-NPME model (9.56) into a standard LME model that can be fitted by standard statistical software such as SAS and S-PLUS. Regression spline methods for NPME models have been described in Section 5.4 of Chapter 5, and in Shi, Taylor, and Wang (1996), and Rice and Wu (2001). Wu and Zhang (2002b) proposed a regression spline method for a semiparametric nonlinear mixed-effects model. Liang, Wu and Carroll (2003) fit a special case of the TVC-NPME model (9.56) using a regression spline method.

The key idea of the regression spline method is to express the fixed-effect coefficient functions $\beta_r(t), r = 0, 1, \cdots, d$ and the random-effect coefficient functions $v_{si}(t), s = 0, 1, \cdots, d^*$ as regression splines using some regression spline bases. A regression spline basis can be a truncated power basis as defined in (8.12) or a B-spline basis (de Boor 1978). Assume that we have the following regression spline bases:

$$\begin{aligned} \mathbf{\Phi}_{rp_r}(t) &= [\phi_{r1}(t), \cdots, \phi_{rp_r}(t)]^T, \ r = 0, 1, \cdots, d, \\ \mathbf{\Psi}_{sq_s}(t) &= [\psi_{s1}(t), \cdots, \psi_{sq_s}(t)]^T, \ s = 0, 1, \cdots, d^*. \end{aligned}$$

Then we can express $\beta_r(t), r = 0, 1, \cdots, d$ and $v_{si}(t), s = 0, 1, \cdots, d^*$ as regression splines:

$$\begin{aligned} \beta_r(t) &= \mathbf{\Phi}_{rp_r}(t)^T \boldsymbol{\alpha}_r, \ r = 0, 1, \cdots, d, \\ v_{si}(t) &= \mathbf{\Psi}_{sq_s}(t)^T \mathbf{b}_{si}, \ s = 0, 1, \cdots, d^*, \end{aligned} \tag{9.67}$$

where $\boldsymbol{\alpha}_r = [\alpha_{r1}, \cdots, \alpha_{rp_r}]^T$ and $\mathbf{b}_{si} = [b_{si1}, \cdots, b_{siq_s}]^T$ are the associated coefficient vectors. Define

$$\begin{aligned} \mathbf{\Phi}(t) = \text{diag}[\mathbf{\Phi}_{0p_0}(t), \mathbf{\Phi}_{1p_1}(t), \cdots, \mathbf{\Phi}_{dp_d}(t)], \ \ \boldsymbol{\alpha} = [\boldsymbol{\alpha}_0^T, \boldsymbol{\alpha}_1^T, \cdots, \boldsymbol{\alpha}_d^T]^T, \\ \mathbf{\Psi}(t) = \text{diag}[\mathbf{\Psi}_{0q_0}(t), \mathbf{\Psi}_{1q_1}(t), \cdots, \mathbf{\Psi}_{d^*q_{d^*}}(t)], \ \ \mathbf{b}_i = [\mathbf{b}_{0i}^T, \mathbf{b}_{1i}^T, \cdots, \mathbf{b}_{d^*q_{d^*}}^T]^T. \end{aligned}$$

Then we can express the coefficient function vectors $\boldsymbol{\beta}(t)$ and $\mathbf{v}_i(t)$ as

$$\boldsymbol{\beta}(t) = \mathbf{\Phi}(t)^T \boldsymbol{\alpha}, \ \mathbf{v}_i(t) = \mathbf{\Psi}(t)^T \mathbf{b}_i, \ \ i = 1, 2, \cdots, n. \tag{9.68}$$

Notice that $\mathbf{b}_i, \ i = 1, 2, \cdots, n$ are i.i.d copies of a normal random variable with mean 0 and some covariance matrix $\mathbf{D}$. By (9.56) and (9.68), a direct connection between $\mathbf{\Gamma}$ and $\mathbf{D}$ is

$$\mathbf{\Gamma}(s, t) \approx \mathbf{\Psi}(s)^T \mathbf{D} \mathbf{\Psi}(t). \tag{9.69}$$

Thus, an estimator of $\mathbf{D}$ will lead to an estimator for $\mathbf{\Gamma}$.

The TVC-NPME model (9.56) can then be approximately expressed as

$$\begin{aligned} y_{ij} &= \mathbf{x}_{ij}^T \boldsymbol{\alpha} + \mathbf{z}_{ij}^T \mathbf{b}_i + \epsilon_{ij}, \\ \mathbf{b}_i &\sim N(0, \mathbf{D}), \epsilon_i \sim N(0, \mathbf{R}_i), \\ j &= 1, 2, \cdots, n_i; \ i = 1, 2, \cdots, n, \end{aligned} \tag{9.70}$$

where $\mathbf{x}_{ij} = \mathbf{\Phi}(t_{ij}) \mathbf{x}_i(t_{ij})$ and $\mathbf{z}_{ij} = \mathbf{\Psi}(t_{ij}) \mathbf{z}_i(t_{ij})$.

For given bases $\mathbf{\Phi}_{rp_r}(t), r = 0, 1, \cdots, d$ and $\mathbf{\Psi}_{sq_s}(t), s = 0, 1, \cdots, d^*$, the model (9.70) is a standard LME model that can be fitted using the S-PLUS function *lme* or the SAS procedure *PROC MIXED*. The numbers of basis functions $p_r, r = 0, 1, \cdots, d$ and $q_s, s = 0, 1, \cdots, d^*$ are the smoothing parameters. For the TVC-NPME model (9.56), the fixed-effects and random-effect coefficient functions $\beta_r(t), r = 0, 1, \cdots, d$ and $v_{si}(t), s = 0, 1, \cdots, d^*$ can admit different degrees of smoothness. It is more appropriate to use multiple smoothing parameters than to use a single smoothing parameter as in the LPK method described in the previous subsection.

To extract the estimators $\hat{\boldsymbol{\alpha}}_r, r = 0, 1, \cdots, d$, we use the matrices defined in (9.17) with $p_r, r = 0, 1, \cdots, d$ so that

$$\hat{\boldsymbol{\alpha}}_r = \mathbf{E}_r \hat{\boldsymbol{\alpha}}, r = 0, 1, \cdots, d.$$

Similarly, to extract the estimators $\hat{\mathbf{b}}_{si}, s = 0, 1, \cdots, d^*$, we can use the matrices defined in (9.17) via replacing $p_r, r = 0, 1, \cdots, d$ with $q_s, s = 0, 1, \cdots, d^*$. For convenience, we write these matrices as

$$\mathbf{F}_s = [\mathbf{0}_{q_s \times q_0}, \cdots, \mathbf{0}_{q_s \times q_{(s-1)}}, \mathbf{I}_{q_s}, \mathbf{0}_{q_s \times q_{(s+1)}}, \cdots, \mathbf{0}_{q_s \times q_{d^*}}]. \tag{9.71}$$

Then we have $\hat{\mathbf{b}}_{si} = \mathbf{F}_s \hat{\mathbf{b}}_i, s = 0, 1, \cdots, d^*$. It follows that $\hat{\beta}_r(t) = \mathbf{\Phi}_{rp_r}(t)^T \hat{\boldsymbol{\alpha}}_r, r = 0, 1, \cdots, d$ and $\hat{v}_{si}(t) = \mathbf{\Psi}_{sq_s}(t)^T \hat{\mathbf{b}}_{si},\ i = 1, 2, \cdots, n; s = 0, 1, \cdots, d^*$.

The smoothing parameters, $p_r, r = 0, 1, \cdots, d$, and $q_s, s = 0, 1, \cdots, d^*$ should be carefully chosen for good performance of the resultant regression spline estimators. Since the approximation LME model (9.70) is almost the same as the approximation LME model (5.38) for a NPME model, the AIC and BIC rules developed in Section 5.4.5 of Chapter 5 can be used to select these smoothing parameters.

9.4.3 Penalized Spline Method

The regression spline method in the previous subsection can use different bases for different fixed-effects and random-effect coefficient functions. The smoothness of a function is controlled by the number of the associated basis functions. It is often challenging to specify $[(d+1) + (d^*+1)]$ bases simultaneously especially when d and/or d^* is large. Alternatively, we can use a basis for all the fixed-effect coefficient functions and another basis for all the random-effect coefficient functions, while controlling the smoothness of a function via penalizing its roughness. The penalized spline (P-spline) method uses two truncated power bases of form (8.12) and penalizes the highest order derivative jumps of the associated regression splines.

Let $\mathbf{\Phi}_p(t)$ be a truncated power basis (8.12) of degree k with K knots $\tau_1, \cdots, \tau_K$ so that $p = K + k + 1$. Similarly let $\mathbf{\Psi}_q(t)$ be a truncated power basis (8.12) of degree k_v with K_v knots $\delta_1, \cdots, \delta_{K_v}$ so that $q = K_v + k_v + 1$. Then we can approximately express $\beta_r(t), r = 0, 1, \cdots, d$ and $v_{si}(t), s = 0, 1, \cdots, d^*$ as regression splines:

$$\begin{aligned} \beta_r(t) &= \mathbf{\Phi}_p(t)^T \boldsymbol{\alpha}_r,\ r = 0, 1, \cdots, d, \\ v_{si}(t) &= \mathbf{\Psi}_q(t)^T \mathbf{b}_{si},\ s = 0, 1, \cdots, d^*, \end{aligned} \tag{9.72}$$

where $\boldsymbol{\alpha}_r = [\alpha_{r1}, \cdots, \alpha_{rp}]^T$ and $\mathbf{b}_{si} = [b_{si1}, \cdots, b_{siq}]^T$. It is well known that the last K entries of $\boldsymbol{\alpha}_r$ are proportional to the k times derivative jumps of the regression spline $\beta_r(t)$ at the knots $\tau_1, \cdots, \tau_K$ and the last K_v entries of $\mathbf{b}_{si}$ are proportional to the k_v times derivative jumps of the regression spline $v_{si}(t)$ at the knots $\delta_1, \cdots, \delta_{K_v}$.

Let $\mathbf{\Phi}(t)$ and $\mathbf{\Psi}(t)$ be two block diagonal matrices with $(d+1)$ blocks of $\mathbf{\Phi}_p(t)$ and (d^*+1) blocks of $\mathbf{\Psi}_q(t)$, respectively. In addition, let $\boldsymbol{\alpha} = [\boldsymbol{\alpha}_0^T, \boldsymbol{\alpha}_1^T, \cdots, \boldsymbol{\alpha}_d^T]^T$ and $\mathbf{b}_i = [\mathbf{b}_{0i}^T, \mathbf{b}_{1i}^T, \cdots, \mathbf{b}_{d^*i}^T]^T$. Then we can express the coefficient function vectors $\boldsymbol{\beta}(t)$ and $\mathbf{v}_i(t)$ as

$$\boldsymbol{\beta}(t) = \mathbf{\Phi}(t)^T\boldsymbol{\alpha}, \ \mathbf{v}_i(t) = \mathbf{\Psi}(t)^T\mathbf{b}_i, \ \ i = 1, 2, \cdots, n. \tag{9.73}$$

It follows that $\mathbf{b}_i, \ i = 1, 2, \cdots, n$ are i.i.d copies of a normal random variable with mean 0 and some covariance matrix $\mathbf{D}$. By (9.56) and (9.73), a direct connection between $\mathbf{\Gamma}$ and $\mathbf{D}$ is $\mathbf{\Gamma}(s,t) \approx \mathbf{\Psi}(s)^T\mathbf{D}\mathbf{\Psi}(t)$.

The TVC-NPME model (9.56) can then be approximately expressed as

$$\begin{array}{c} y_{ij} = \mathbf{x}_{ij}^T\boldsymbol{\alpha} + \mathbf{z}_{ij}^T\mathbf{b}_i + \epsilon_{ij}, \\ \mathbf{b}_i \sim N(0, \mathbf{D}), \boldsymbol{\epsilon}_i \sim N(0, \mathbf{R}_i), \\ j = 1, 2, \cdots, n_i; \ i = 1, 2, \cdots, n, \end{array} \tag{9.74}$$

where $\mathbf{x}_{ij} = \mathbf{\Phi}(t_{ij})\mathbf{x}_i(t_{ij}) = \mathbf{x}_i(t_{ij}) \otimes \mathbf{\Phi}_p(t_{ij})$ and $\mathbf{z}_{ij} = \mathbf{\Psi}(t_{ij})\mathbf{z}_i(t_{ij}) = \mathbf{z}_i(t_{ij}) \otimes \mathbf{\Psi}_q(t_{ij})$. Using the vectors defined in (9.60), we can further write (9.74) as

$$\begin{array}{rcl} \mathbf{y}_i & = & \mathbf{X}_i\boldsymbol{\alpha} + \mathbf{Z}_i\mathbf{b}_i + \boldsymbol{\epsilon}_i, \\ \mathbf{b}_i & \sim & N(0, \mathbf{D}), \boldsymbol{\epsilon}_i \sim N(0, \mathbf{R}_i), \\ & & i = 1, 2, \cdots, n. \end{array} \tag{9.75}$$

To account for the roughness of the fixed-effect coefficient functions $\beta_r(t), r = 0, 1, \cdots, d$ and the random-effect coefficient functions $v_{si}(t), s = 0, 1, \cdots, d^*; \ i = 1, 2, \cdots, n$, let $\mathbf{G}$ and $\mathbf{G}_v$ be two P-spline roughness matrices associated with $\mathbf{\Phi}_p(t)$and $\mathbf{\Psi}_q(t)$, respectively. That is, $\mathbf{G} = \text{diag}(0, \cdots, 0, 1, \cdots, 1)$, a $p \times p$ diagonal matrix with the last K diagonal entries being 1, and $\mathbf{G}_v = \text{diag}(0, \cdots, 0, 1, \cdots, 1)$, a $q \times q$ diagonal matrix with the last K_v diagonal entries being 1. Then based on the model (9.75), the estimators of $\boldsymbol{\alpha}$ and $\mathbf{b}_i, \ i = 1, 2, \cdots, n$ can be obtained via minimizing the following penalized generalized log-likelihood (PGLL) criterion:

$$\begin{array}{c} \sum_{i=1}^n \left\{(\mathbf{y}_i - \mathbf{X}_i\boldsymbol{\alpha} - \mathbf{Z}_i\mathbf{b}_i)^T\mathbf{R}_i^{-1}(\mathbf{y}_i - \mathbf{X}_i\boldsymbol{\alpha} - \mathbf{Z}_i\mathbf{b}_i) + \mathbf{b}_i^T\mathbf{D}^{-1}\mathbf{b}_i\right\} \\ + \sum_{i=1}^n \sum_{s=0}^{d^*} \lambda_{sv}\mathbf{b}_{si}^T\mathbf{G}_v\mathbf{b}_{si} + \sum_{r=0}^d \lambda_r\boldsymbol{\alpha}_r^T\mathbf{G}\boldsymbol{\alpha}_r, \end{array} \tag{9.76}$$

where the first summation term is proportional to the twice negative logarithm of the generalized likelihood of the fixed-effects vector $\boldsymbol{\alpha}$ and the random-effects vectors $\mathbf{b}_i, \ i = 1, 2, \cdots, n$, representing the goodness of fit, the second term is the roughness of all the random-effects coefficient regression splines, and the third term is the roughness of all the fixed-effects coefficient regression splines. The smoothing parameters $\lambda_r, r = 0, 1, \cdots, d$ and $\lambda_{sv}, s = 0, 1, \cdots, d^*$ are used to trade-off the goodness of fit with the roughness of the regression splines.

Notice that the second term of (9.76) can be expressed as

$$\sum_{i=1}^n \mathbf{b}_i^T\tilde{\mathbf{G}}_v\mathbf{b}_i = \mathbf{b}^T\tilde{\mathbf{G}}_v^*\mathbf{b}^T,$$

where $\tilde{\mathbf{G}}_v = \text{diag}(\lambda_{0v}\mathbf{G}_v, \lambda_{1v}\mathbf{G}_v, \cdots, \lambda_{d^*v}\mathbf{G}_v)$ is a block diagonal matrix, $\tilde{\mathbf{G}}_v^*$ is a block diagonal matrix with n blocks of $\tilde{\mathbf{G}}_v$, and $\mathbf{b} = [\mathbf{b}_1^T, \cdots, \mathbf{b}_n^T]^T$. Notice also that the third term of (9.76) can be expressed as

$$\boldsymbol{\alpha}^T\tilde{\mathbf{G}}\boldsymbol{\alpha},$$

where $\tilde{\mathbf{G}} = \text{diag}(\lambda_0\mathbf{G}, \lambda_1\mathbf{G}, \cdots, \lambda_d\mathbf{G})$. Letting $\tilde{\mathbf{D}}$ denote a block diagonal matrix with n blocks of $\mathbf{D}$, we can further write the PGLL criterion (9.76) as

$$(\mathbf{y} - \mathbf{X}\boldsymbol{\alpha} - \mathbf{Z}\mathbf{b})^T\mathbf{R}^{-1}(\mathbf{y} - \mathbf{X}\boldsymbol{\alpha} - \mathbf{Z}\mathbf{b}) + \mathbf{b}^T\tilde{\mathbf{D}}^{-1}\mathbf{b} + \mathbf{b}^T\tilde{\mathbf{G}}_v^*\mathbf{b} + \boldsymbol{\alpha}^T\tilde{\mathbf{G}}\boldsymbol{\alpha}, \quad (9.77)$$

where $\mathbf{y}, \mathbf{X}, \mathbf{Z}, \mathbf{R}$ are similarly defined as in (9.61). Let $\tilde{\mathbf{D}}_v = (\tilde{\mathbf{D}}^{-1} + \tilde{\mathbf{G}}_v^*)^{-1} = \text{diag}(\mathbf{D}_v, \cdots, \mathbf{D}_v)$, a block diagonal matrix with n blocks of $\mathbf{D}_v = (\mathbf{D}^{-1} + \tilde{\mathbf{G}}_v)$. Then the PGLL criterion (9.77) can be further written as

$$(\mathbf{y} - \mathbf{X}\boldsymbol{\alpha} - \mathbf{Z}\mathbf{b})^T\mathbf{R}^{-1}(\mathbf{y} - \mathbf{X}\boldsymbol{\alpha} - \mathbf{Z}\mathbf{b}) + \mathbf{b}^T\tilde{\mathbf{D}}_v^{-1}\mathbf{b} + \boldsymbol{\alpha}^T\tilde{\mathbf{G}}\boldsymbol{\alpha}. \quad (9.78)$$

Let $\tilde{\mathbf{V}} = \text{diag}(\tilde{\mathbf{V}}_1, \cdots, \tilde{\mathbf{V}}_n)$ with $\tilde{\mathbf{V}}_i = \mathbf{Z}_i\mathbf{D}_v\mathbf{Z}_i^T + \mathbf{R}_i,\ i = 1, 2, \cdots, n$. Then the minimizers of the PGLL criterion (9.78) can be expressed as

$$\begin{aligned} \hat{\boldsymbol{\alpha}} &= \left(\mathbf{X}^T\tilde{\mathbf{V}}^{-1}\mathbf{X} + \tilde{\mathbf{G}}\right)^{-1}\mathbf{X}^T\tilde{\mathbf{V}}^{-1}\mathbf{y}, \\ \hat{\mathbf{b}} &= \tilde{\mathbf{D}}_v\mathbf{Z}^T\tilde{\mathbf{V}}^{-1}(\mathbf{y} - \mathbf{X}\hat{\boldsymbol{\alpha}}), \end{aligned} \quad (9.79)$$

so that $\hat{\mathbf{b}}_i = \mathbf{D}_v\mathbf{Z}_i^T\tilde{\mathbf{V}}_i^{-1}(\mathbf{y}_i - \mathbf{X}_i\hat{\boldsymbol{\alpha}}),\ i = 1, 2, \cdots, n$.

To extract the estimators $\hat{\boldsymbol{\alpha}}_r, r = 0, 1, \cdots, d$, we use the matrices defined in (9.17) with $p_r = p, r = 0, 1, \cdots, d$ so that

$$\hat{\boldsymbol{\alpha}}_r = \mathbf{E}_r(\mathbf{X}^T\tilde{\mathbf{V}}^{-1}\mathbf{X} + \tilde{\mathbf{G}})^{-1}\mathbf{X}^T\tilde{\mathbf{V}}^{-1}\mathbf{y}, r = 0, 1, \cdots, d.$$

Similarly, to extract the estimators $\hat{\mathbf{b}}_{si}, s = 0, 1, \cdots, d^*$, we can use the matrices defined in (9.71) with $q_s = q, s = 0, 1, \cdots, d^*$ so that

$$\hat{\mathbf{b}}_{si} = \mathbf{F}_s\mathbf{D}_v\mathbf{Z}_i^T\tilde{\mathbf{V}}_i^{-1}(\mathbf{y}_i - \mathbf{X}_i\hat{\boldsymbol{\alpha}}), \quad i = 1, 2, \cdots, n; \ s = 0, 1, \cdots, d^*.$$

The P-spline estimators of $\beta_r(t), r = 0, 1, \cdots, d$ and $v_{si}(t),\ i = 1, 2, \cdots, n; s = 0, 1, \cdots, d^*$ can then be expressed as

$$\begin{aligned} \hat{\beta}_r(t) &= \boldsymbol{\Phi}_p(t)^T\mathbf{E}_r(\mathbf{X}^T\tilde{\mathbf{V}}^{-1}\mathbf{X} + \tilde{\mathbf{G}})^{-1}\mathbf{X}^T\tilde{\mathbf{V}}^{-1}\mathbf{y},\ r = 0, 1, \cdots, d, \\ \hat{v}_{si}(t) &= \boldsymbol{\Psi}_q(t)^T\mathbf{F}_s\mathbf{D}_v\mathbf{Z}_i^T\tilde{\mathbf{V}}_i^{-1}(\mathbf{y}_i - \mathbf{X}_i\hat{\boldsymbol{\alpha}}),\ s = 0, 1, \cdots, d^*, \\ & \quad i = 1, 2, \cdots, n. \end{aligned} \quad (9.80)$$

For given truncated power bases $\boldsymbol{\Phi}_p(t)$, $\boldsymbol{\Psi}_q(t)$ and a set of smoothing parameters $\lambda_r, r = 0, 1, \cdots, d$, and $\lambda_{sv}, s = 0, 1, \cdots, d^*$, it is easy to construct the smoother matrices and the standard deviation bands for the fixed-effect coefficient functions and for the responses at all the design time points. The knots of $\boldsymbol{\Phi}_p(t)$ and $\boldsymbol{\Psi}_q(t)$ can be scattered using the methods described in Section 5.2.4 of Chapter 5. The numbers of knots in $\boldsymbol{\Phi}_p(t)$ and $\boldsymbol{\Psi}_q(t)$ can be selected roughly using the method discussed in Section 7.2.6 of Chapter 7. In principle, the smoothing parameters $\lambda_r, r = 0, 1, \cdots, d$ and $\lambda_{sv}, s = 0, 1, \cdots, d^*$ can be selected using the AIC and BIC rules developed in Section 7.5.6 of Chapter 7.

9.4.4 Smoothing Spline Method

Hoover et al. (1998), Brumback and Rice (1998), Chiang, Rice and Wu (2001), and Eubank et al. (2004), among others investigated smoothing spline methods for fitting the TVC-NPM model (9.4) but they do not account for the randomness of the subject specific deviation functions. Guo (2002a) only partially accounted for this randomness. In the regression spline and P-spline methods described in the previous subsections, only part of all the distinct design time points:

$$\tau_1, \tau_2, \cdots, \tau_M, \tag{9.81}$$

among the design time points $\{t_{ij},\ j = 1, 2, \cdots, n_i;\ i = 1, 2, \cdots, n\}$ are used as knots where $a < \tau_1 < \tau_2 < \cdots < \tau_M < b$ with $[a, b]$ denoting the range of the design time points. For a smoothing spline method, all the distinct design time points (9.81) will be used as knots. Unlike the P-spline method where the penalties are assigned to the highest order derivative jumps of the fixed-effect coefficient functions and the random-effect coefficient functions, the smoothing spline method aims to penalize the roughness of the resultant functions. Here we consider the cubic smoothing spline (CSS) method only. Extensions to a general degree smoothing spline method is straightforward.

Let $\boldsymbol{\beta}_r = [\beta_r(\tau_1), \cdots, \beta_r(\tau_M)]^T$ denote the values of $\beta_r(t)$ at all the distinct design time points (9.81). Similarly, let $\mathbf{v}_{si} = [v_{si}(\tau_1), \cdots, v_{si}(\tau_M)]^T$ denote the values of $v_{si}(t)$ at $\tau_1, \cdots, \tau_M$. Let $\mathbf{h}_{ij}$ denote a M-dimensional unit vector whose l-th entry is 1 if $t_{ij} = \tau_l$ and 0 otherwise. Then we have

$$\begin{aligned}\beta_r(t_{ij}) &= \mathbf{h}_{ij}^T\boldsymbol{\beta}_r,\ r = 0, 1, \cdots, d,\\ v_{si}(t_{ij}) &= \mathbf{h}_{ij}^T\mathbf{v}_{si},\ s = 0, 1, \cdots, d^*,\\ j &= 1, 2, \cdots, n_i;\ i = 1, 2, \cdots, n.\end{aligned}$$

Let $\mathbf{v}_i = [\mathbf{v}_{0i}^T, \mathbf{v}_{1i}^T, \cdots, \mathbf{v}_{d^*i}^T]^T$. Then $\mathbf{v}_i,\ i = 1, 2, \cdots, n$ are i.i.d, following a $[M(d^*+1)]$-dimensional normal distribution with mean 0 and some covariance matrix $\mathbf{D}$, i.e., $\mathbf{v}_i \sim N(0, \mathbf{D})$. The TVC-NPME model (9.56) can then be expressed as

$$\begin{aligned}y_{ij} &= \mathbf{x}_{ij}^T\boldsymbol{\beta} + \mathbf{z}_{ij}^T\mathbf{v}_i + \epsilon_{ij},\\ \mathbf{v}_i &\sim N(0, \mathbf{D}),\ \boldsymbol{\epsilon}_i \sim N(0, \mathbf{R}_i),\\ j &= 1, 2, \cdots, n_i;\ i = 1, 2, \cdots, n,\end{aligned}$$

where $\boldsymbol{\beta} = [\boldsymbol{\beta}_0^T, \boldsymbol{\beta}_1^T, \cdots, \boldsymbol{\beta}_d^T]^T$, $\mathbf{x}_{ij} = \mathbf{x}_i(t_{ij}) \otimes \mathbf{h}_{ij}$ and $\mathbf{z}_{ij} = \mathbf{z}_i(t_{ij}) \otimes \mathbf{h}_{ij}$. Using the vector notations defined in (9.60), the above model can be further written as

$$\begin{aligned}\mathbf{y}_i &= \mathbf{X}_i\boldsymbol{\beta} + \mathbf{Z}_i\mathbf{v}_i + \boldsymbol{\epsilon}_i,\\ \mathbf{v}_i &\sim N(0, \mathbf{D}),\ \boldsymbol{\epsilon}_i \sim N(0, \mathbf{R}_i),\\ i &= 1, 2, \cdots, n.\end{aligned} \tag{9.82}$$

For the TVC-NPME model (9.56), we assume that the fixed-effect coefficient functions $\beta_r(t), r = 0, 1, \cdots, d$ and the random-effect coefficient functions $v_{si}(t), s = 0, 1, \cdots, d^*;\ i = 1, 2, \cdots, n$ are twice continuously differentiable and their second derivatives $\beta_r''(t), r = 0, 1, \cdots, d$ and $v_{si}''(t), s = 0, 1, \cdots, d^*;\ i = 1, 2, \cdots, n$

are bounded and squared integrable. Then based on (9.82), the CSS estimators $\hat{\beta}_r(t), r = 0, 1, \cdots, d$ and $\hat{v}_{si}(t), s = 0, 1, \cdots, d^*;\ i = 1, 2, \cdots, n$ can be defined as the minimizers of the following penalized generalized log-likelihood (PGLL) criterion:

$$\begin{aligned} &\textstyle\sum_{i=1}^n \left[(\mathbf{y}_i - \mathbf{X}_i\beta - \mathbf{Z}_i\mathbf{v}_i)^T\mathbf{R}_i^{-1}(\mathbf{y}_i - \mathbf{X}_i\beta - \mathbf{Z}_i\mathbf{v}_i) + \mathbf{v}_i^T\mathbf{D}^{-1}\mathbf{v}_i\right] \\ &\quad + \textstyle\sum_{i=1}^n \sum_{s=0}^{d^*} \lambda_{sv} \int_a^b [v_{si}''(t)]^2 dt + \sum_{r=0}^d \lambda_r \int_a^b [\beta_r''(t)]^2 dt, \end{aligned} \tag{9.83}$$

with respect to $\beta_r(t), r = 0, 1, \cdots, d$ and $v_{si}(t), s = 0, 1, \cdots, d^*;\ i = 1, 2, \cdots, n$ where $\lambda_r, r = 0, 1, \cdots, d$ and $\lambda_{sv}, s = 0, 1, \cdots, d^*$ are the smoothing parameters. In the above PGLL criterion, the first summation term is proportional to the twice negative logarithm of the generalized likelihood of the fixed-effects vector β and the random-effects vectors $\mathbf{v}_i,\ i = 1, 2, \cdots, n$, representing the goodness of fit; the second term equals the weighted sum of the roughness of all the random-effect coefficient functions; and the third term equals the weighted sum of the roughness of all the fixed-effect coefficient functions. The smoothing parameters $\lambda_r, r = 0, 1, \cdots, d$ and $\lambda_{sv}, s = 0, 1, \cdots, d^*$ are used to trade off the goodness of fit with the roughness of the smoothing spline estimators.

It can be shown that the minimizers $\hat{\beta}_r(t)$ and $\hat{v}_{si}(t), s = 0, 1, \cdots, d^*;\ i = 1, 2, \cdots, n$ are natural cubic smoothing splines with knots at all the distinct design time points (9.81). Assuming that $\beta_r(t), r = 0, 1, \cdots, d$ and $v_{si}(t), s = 0, 1, \cdots, d^*;\ i = 1, 2, \cdots, n$ are natural cubic smoothing splines with the given knots, we have

$$\begin{aligned} \textstyle\int_a^b [\beta_r''(t)]^2 dt &= \boldsymbol{\beta}_r^T\mathbf{G}\boldsymbol{\beta}_r, r = 0, 1, \cdots, d, \\ \textstyle\int_a^b [v_{si}''(t)]^2 dt &= \mathbf{v}_{si}^T\mathbf{G}\mathbf{v}_{si}, s = 0, 1, \cdots, d^*, \quad i = 1, 2, \cdots, n, \end{aligned}$$

where $\mathbf{G}$ is the CSS roughness matrix (3.33) with knots at (9.81). See Section 3.4.1 or Green and Silverman (1994) for a detailed scheme for computing $\mathbf{G}$. By these equations, we can write the PGLL criterion (9.83) further as

$$\begin{aligned} &\textstyle\sum_{i=1}^n \left[(\mathbf{y}_i - \mathbf{X}_i\beta - \mathbf{Z}_i\mathbf{v}_i)^T\mathbf{R}_i^{-1}(\mathbf{y}_i - \mathbf{X}_i\beta - \mathbf{Z}_i\mathbf{v}_i) + \mathbf{v}_i^T\mathbf{D}^{-1}\mathbf{v}_i\right] \\ &\quad + \textstyle\sum_{i=1}^n \mathbf{v}_i^T\tilde{\mathbf{G}}_v\mathbf{v}_i + \beta^T\tilde{\mathbf{G}}\beta, \end{aligned} \tag{9.84}$$

where $\tilde{\mathbf{G}}_v = \text{diag}(\lambda_{0v}\mathbf{G}, \lambda_{1v}\mathbf{G}, \cdots, \lambda_{d^*v}\mathbf{G})$ and $\tilde{\mathbf{G}} = \text{diag}(\lambda_0\mathbf{G}, \lambda_1\mathbf{G}, \cdots, \lambda_d\mathbf{G})$ are two block diagonal matrices. Let $\tilde{\mathbf{D}}$ and $\tilde{\mathbf{G}}_v^*$ denote two block diagonal matrices with n blocks of $\mathbf{D}$ and $\tilde{\mathbf{G}}_v$, respectively. Then the PGLL criterion (9.84) can be further written as

$$(\mathbf{y} - \mathbf{X}\beta - \mathbf{Z}\mathbf{v})^T\mathbf{R}^{-1}(\mathbf{y} - \mathbf{X}\beta - \mathbf{Z}\mathbf{v}) + \mathbf{v}^T\tilde{\mathbf{D}}^{-1}\mathbf{v} + \mathbf{v}^T\tilde{\mathbf{G}}_v^*\mathbf{v} + \beta^T\tilde{\mathbf{G}}\beta, \tag{9.85}$$

where $\mathbf{y}, \mathbf{X}, \mathbf{Z}, \mathbf{R}$ are similarly defined as in (9.61) and $\mathbf{v} = [\mathbf{v}_1^T, \cdots, \mathbf{v}_n^T]^T$. Let $\tilde{\mathbf{D}}_v = (\tilde{\mathbf{D}}^{-1} + \tilde{\mathbf{G}}_v^*)^{-1} = \text{diag}(\mathbf{D}_v, \cdots, \mathbf{D}_v)$, a block diagonal matrix with n blocks, each being $\mathbf{D}_v = (\mathbf{D}^{-1} + \tilde{\mathbf{G}}_v)$. Then (9.85) can be further written as

$$(\mathbf{y} - \mathbf{X}\beta - \mathbf{Z}\mathbf{v})^T\mathbf{R}^{-1}(\mathbf{y} - \mathbf{X}\beta - \mathbf{Z}\mathbf{v}) + \mathbf{v}^T\tilde{\mathbf{D}}_v^{-1}\mathbf{v} + \beta^T\tilde{\mathbf{G}}\beta. \tag{9.86}$$

Let $\tilde{\mathbf{V}} = \text{diag}(\tilde{\mathbf{V}}_1, \cdots, \tilde{\mathbf{V}}_n)$ with $\tilde{\mathbf{V}}_i = \mathbf{Z}_i \mathbf{D}_v \mathbf{Z}_i^T + \mathbf{R}_i,\ i = 1, 2, \cdots, n$. Then the minimizers of the PGLL criterion (9.86) can be expressed as

$$\begin{aligned} \hat{\beta} &= \left(\mathbf{X}^T \tilde{\mathbf{V}}^{-1} \mathbf{X} + \tilde{\mathbf{G}}\right)^{-1} \mathbf{X}^T \tilde{\mathbf{V}}^{-1} \mathbf{y}, \\ \hat{\mathbf{v}} &= \tilde{\mathbf{D}}_v \mathbf{Z}^T \tilde{\mathbf{V}}^{-1} (\mathbf{y} - \mathbf{X}\hat{\beta}), \end{aligned} \tag{9.87}$$

so that $\hat{\mathbf{v}}_i = \mathbf{D}_v \mathbf{Z}_i^T \tilde{\mathbf{V}}_i^{-1} (\mathbf{y}_i - \mathbf{X}_i \hat{\beta}),\ i = 1, 2, \cdots, n$.

To extract the estimators $\hat{\beta}_r, r = 0, 1, \cdots, d$, we use the matrices defined in (9.17) with $p_r = M, r = 0, 1, \cdots, d$ so that

$$\hat{\beta}_r = \mathbf{E}_r \left(\mathbf{X}^T \tilde{\mathbf{V}}^{-1} \mathbf{X} + \tilde{\mathbf{G}}\right)^{-1} \mathbf{X}^T \tilde{\mathbf{V}}^{-1} \mathbf{y}, r = 0, 1, \cdots, d. \tag{9.88}$$

Similarly, to extract the estimators $\hat{\mathbf{v}}_{si}, s = 0, 1, \cdots, d^*$, we can use the matrices defined in (9.71) with $q_s = M, s = 0, 1, \cdots, d^*$ so that

$$\hat{\mathbf{v}}_{si} = \mathbf{F}_s \mathbf{D}_v \mathbf{Z}_i^T \tilde{\mathbf{V}}_i^{-1} (\mathbf{y}_i - \mathbf{X}_i \hat{\beta}),\ s = 0, 1, \cdots, d^*;\ i = 1, 2, \cdots, n.$$

For a given set of smoothing parameters $\lambda_r, r = 0, \cdots, d$ and $\lambda_{sv}, s = 0, \cdots, d^*$, it is easy to construct the smoother matrices and the standard deviation bands for the fixed-effect coefficient functions and for the subject-specific mean responses at all the design time points. In principle, the smoothing parameters $\lambda_r, r = 0, 1, \cdots, d$ and $\lambda_{sv}, s = 0, 1, \cdots, d^*$ can be selected using the AIC and BIC rules developed in Section 6.5.6 of Chapter 6.

For a special case of the TVC-NPME model (9.56), the smoothing spline method of Brumback and Rice (1998) only accounts for the roughness of the coefficient functions $\beta(t)$ and $\mathbf{v}_i(t),\ i = 1, 2, \cdots, n$ in the sense that the term $\sum_{i=1}^n \mathbf{v}_i^T \mathbf{D}^{-1} \mathbf{v}_i$ does not appear in their PGLL criterion. For the TVC-NPME model (9.56), the smoothing spline method of Guo (2002a) just partially accounts for the randomness of the coefficient functions $\mathbf{v}_i(t),\ i = 1, 2, \cdots, n$ in the sense that the covariance matrix $\mathbf{D}$ of $\mathbf{v}_i,\ i = 1, 2, \cdots, n$ in Guo (2002a) has a special structure. Guo (2002a) also applied the state-space representation of smoothing splines and used the Kalman filter to ease the computational burden.

9.4.5 Backfitting Algorithms

The LPK method in Section 9.4.1 uses only a single bandwidth h for all the fixed-effects and random-effect coefficient functions, while the regression spline, P-spline and smoothing spline methods in Sections 9.4.2~9.4.4 use $(d + d^* + 2)$ smoothing parameters. A single bandwidth in the LPK method is often not good enough for fitting all the fixed-effects and random-effect coefficient functions that admit quite different levels of smoothness while for the other three methods, it is often challenging to select $(d + d^* + 2)$ smoothing parameters.

These problems may be solved by extending the backfitting algorithm proposed by Hastie and Tishirani (1990). In fact, this has been done by Liang and Wu (2004) for a special case of the RTVC model (9.55) when $d = 1$ and $x_{0i}(t) \equiv 1$. Following Liang and Wu (2004), we can describe a backfitting algorithm for the general RTVC model (9.55). Let $\boldsymbol{\beta}_i(t) = [\beta_{0i}(t), \cdots, \beta_{di}(t)]^T$ where $\beta_{ri}(t) = \beta_r(t) + v_{ri}(t)$. Notice that when all the random coefficient functions except $\beta_{ri}(t)$ are known, the RTVC model (9.55) essentially reduces to the following special RTVC model with only a single random coefficient function:

$$\begin{aligned} \tilde{y}_{rij} &= x_{ri}(t_{ij})\beta_{ri}(t_{ij}) + \epsilon_{ij}, \quad \beta_{ri}(t) = \beta_r(t) + v_{ri}(t), \\ & j = 1, 2, \cdots, n_i; \; i = 1, 2, \cdots, n, \end{aligned} \tag{9.89}$$

where $\tilde{y}_{rij} = y_{ij} - \sum_{l \neq r} x_{li}(t_{ij})\beta_{li}(t_{ij})$ and $x_{li}(t)$ is the l-th component of $\mathbf{x}_i(t)$. The above RTVC model can be solved for $\hat{\beta}_r(t)$ and $\hat{v}_{ri}(t), \; i = 1, 2, \cdots, n$ using any of the methods proposed in the previous subsections and most importantly, for any method, at most only two smoothing parameters are needed. The backfitting algorithm for fitting the general RTVC model (9.55) can be written as

(a) Get initial estimators $\hat{\beta}_{ri}(t), r = 0, 1, \cdots, d; \; i = 1, 2, \cdots, n.$

(b) Let $r = 0$.

(c) Based on the current coefficient estimators $\hat{\beta}_{li}(t), l \neq r, \; i = 1, 2, \cdots, n$, calculate the residuals $\tilde{y}_{rij} = y_{ij} - \sum_{l \neq r} x_{li}(t_{ij})\hat{\beta}_{li}(t_{ij})$, fit the RTVC model (9.89) and then update the estimator $\hat{\beta}_{ri}(t), \; i = 1, 2, \cdots, n$.

(d) If $r < d$, set $r = r + 1$, go back to Step (c); else go to Step (e).

(e) Repeat Steps (b) to (d) until convergence.

(f) Record the estimators $\hat{\beta}_r(t), r = 0, 1, \cdots, d$ and $\hat{v}_{ri}(t), r = 0, 1, \cdots, d; \; i = 1, 2, \cdots, n$ and the associated variance components.

When $v_{ri}(t) \equiv 0, r = 0, 1, \cdots, d; \; i = 1, 2, \cdots, n$, the RTVC model (9.55) reduces to the standard TVC-NPM model (9.4). Therefore, we can use the estimators $\hat{\beta}_r(t), r = 0, 1, \cdots, d$ of the standard TVC-NPM model (9.4) as the initial estimators $\hat{\beta}_{ri}(t), r = 0, 1, \cdots, d; \; i = 1, 2, \cdots, n$ in Step (a).

The above backfitting algorithm can be extended to fit the general TVC-NPME model (9.56). Notice that for the general TVC-NPME model (9.56), when all the fixed-effect coefficient functions except the fixed-effect coefficient function $\beta_r(t)$ and all the random-effect coefficient functions $\mathbf{v}_i(t), \; i = 1, 2, \cdots, n$ are known, we can define the residual $\tilde{y}_{rij}$ as $\tilde{y}_{rij} = y_{ij} - \sum_{l \neq r} x_{li}(t_{ij})\beta_l(t_{ij}) - \mathbf{z}_i(t_{ij})^T\mathbf{v}_i(t_{ij})$ so that the model (9.56) can now be written as

$$\tilde{y}_{rij} = x_{ri}(t_{ij})\beta_r(t_{ij}) + \epsilon_{ij}, \; j = 1, 2, \cdots, n_i; \; i = 1, 2, \cdots, n. \tag{9.90}$$

On the other hand, when all the fixed-effect coefficient functions $\beta_r(t), r = 0, 1, \cdots, d$ and all the random-effect coefficient functions except the random-effect coefficient function $v_{si}(t)$ are known, we can fit the following special TVC-NPME model:

$$\tilde{y}_{svij} = z_{si}(t_{ij})v_{si}(t_{ij}) + \epsilon_{ij}, \; j = 1, 2, \cdots, n_i; \; i = 1, 2, \cdots, n, \tag{9.91}$$

where

$$\tilde{y}_{svij} = y_{ij} - \mathbf{x}_i(t_{ij})^T \boldsymbol{\beta}(t_{ij}) - \sum_{l \neq s} z_{li}(t_{ij}) v_{li}(t_{ij})$$

with $z_{li}(t)$ being the l-th component of $\mathbf{z}_i(t)$. The backfitting algorithm for the general TVC-NPME model (9.56) can then written as

(a) Get initial estimators $\hat{\boldsymbol{\beta}}(t)$ and $\hat{\mathbf{v}}_i(t),\ i = 1, 2, \cdots, n$.

(b) Let $r = 0$.

(c) Based on the current coefficient estimators $\hat{\beta}_l(t), l \neq r$ and $\hat{\mathbf{v}}_i(t),\ i = 1, 2, \cdots, n$, calculate the residuals $\tilde{y}_{rij} = y_{ij} - \sum_{l \neq r} x_{li}(t_{ij})\hat{\beta}_l(t_{ij}) - \mathbf{z}_i(t_{ij})^T \hat{\mathbf{v}}_i(t_{ij})$, fit the TVC-NPM model (9.90), and then update the estimator $\hat{\beta}_r(t)$.

(d) If $r < d$, set $r = r + 1$, go back to Step (c); else go to Step (e).

(e) Let $s = 0$.

(f) Based on the current coefficient estimators $\hat{\boldsymbol{\beta}}(t)$ and $\hat{v}_{li}(t), l \neq s,\ i = 1, 2, \cdots, n$, calculate the residuals $\tilde{y}_{svij} = y_{ij} - \mathbf{x}_i(t_{ij})^T \hat{\boldsymbol{\beta}}(t_{ij}) - \sum_{l \neq s} z_{li}(t_{ij})\hat{v}_{li}(t_{ij})$, fit the TVC-NPME model (9.91), and then update the estimator $\hat{v}_{si}(t),\ i = 1, 2, \cdots, n$.

(g) If $s < d^*$, set $s = s + 1$, go back to Step (e); else go to Step (h).

(h) Repeat Steps (b) to (g) until convergence.

Again, the initial estimators in Step (a) can be obtained by fitting a standard TVC-NPM model (9.4) and by taking $\hat{v}_{si}(t) \equiv 0, s = 0, 1, \cdots, d^*;\ i = 1, 2, \cdots, n$.

9.4.6 MACS Data Revisted

In Section 9.2.9, we fit the TVC-NPM model (9.46) to the MACS data introduced in Section 1.1.3 of Chapter 1. Our original analysis of these data in Section 9.2.9 did not take the within-subject correlations into account, potentially resulting in misleading conclusions about the covariate effects. In order to account for the within-subject correlations, in this subsection, we extend the TVC-NPM model (9.46) to the following TVC-NPME model:

$$\begin{aligned} y_{ij} = \beta_0(t_{ij}) + X_{1i}\beta_1(t_{ij}) + X_{2i}\beta_2(t_{ij}) + X_{3i}\beta_3(t_{ij}) + v_i(t_{ij}) + \epsilon_{ij}, \\ v_i(t) \sim \text{GP}(0, \gamma), \boldsymbol{\epsilon}_i = [\epsilon_{i1}, \cdots, \epsilon_{in_i}]^T \sim N(0, \mathbf{R}_i), \\ j = 1, 2, \cdots, n_i;\ i = 1, 2, \cdots, n, \end{aligned} \tag{9.92}$$

where $v_i(t),\ i = 1, 2, \cdots, n$ are used to model the subject-specific random-effects of the MACS data. The above TVC-NPME model can also be regarded as an extension of the semiparametric mixed-effects (SPME) model (8.80) where the fixed-effect coefficient functions $\beta_r(t), r = 0, 1, 2, 3$ are assumed to be polynomials with degrees less than 2.

We fit the TVC-NPME model (9.92) to the MACS data using the P-spline method described in Section 9.4.3. We used a quadratic truncated power basis for both the fixed-effect coefficient functions and the random-effect functions. The tunning parameters were selected in the same way as that for fitting the TVC-NPM model (9.46) except that the smoothing parameters were selected using the BIC rule proposed in Section 7.5.6 of Chapter 7. To save computation time, we used the same smoothing parameter for all the coefficient functions and all the random-effect functions. The selected smoothing parameter was 11.818.

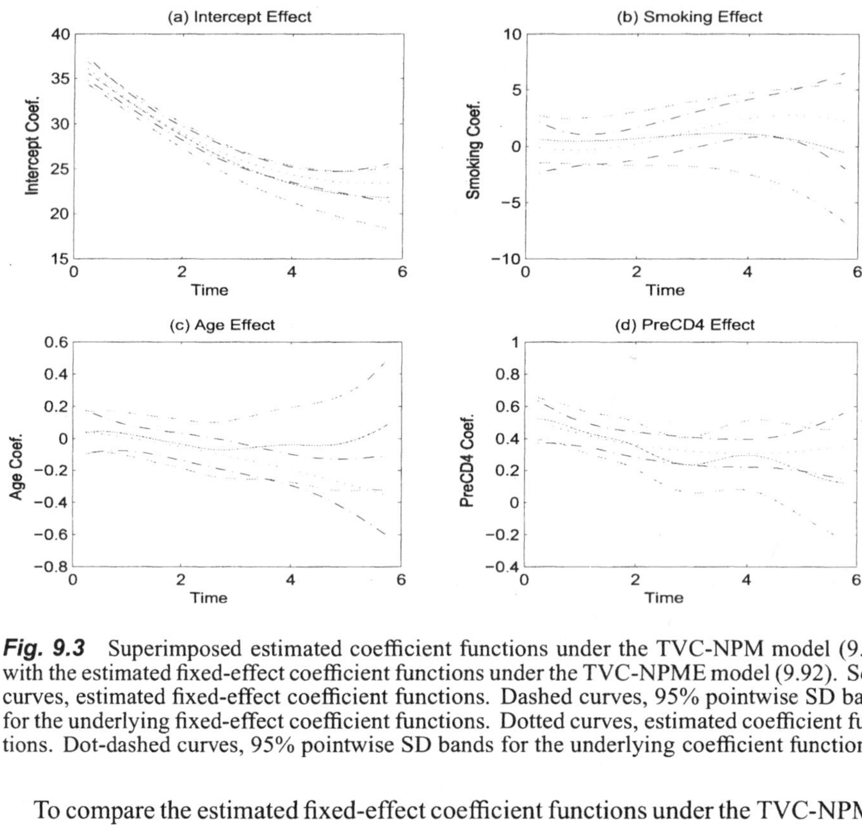

Fig. 9.3 Superimposed estimated coefficient functions under the TVC-NPM model (9.46) with the estimated fixed-effect coefficient functions under the TVC-NPME model (9.92). Solid curves, estimated fixed-effect coefficient functions. Dashed curves, 95% pointwise SD bands for the underlying fixed-effect coefficient functions. Dotted curves, estimated coefficient functions. Dot-dashed curves, 95% pointwise SD bands for the underlying coefficient functions.

To compare the estimated fixed-effect coefficient functions under the TVC-NPME model (9.92) with the estimated coefficient functions presented in Figure 9.46, we superimposed them in Figure 9.3 (solid and dotted curves, respectively), together with the associated 95% pointwise SD bands (dashed and dot-dashed curves, respectively). We can see that the 95% pointwise SD bands under the TVC-NPME model (9.92) are generally wider than, and almost cover completely those 95% pointwise SD bands under the TVC-NPM model (9.46). This is not surprising since the TVC-NPME model accounts for the within-subject correlations while the TVC-NPM model does not. Notice that under the TVC-NPME model, both the Smoking and Age effects are insignificant since the associated 95% pointwise SD bands contain 0 all the time. This conclusion is consistent with that obtained by fitting the SPME model (8.80) to

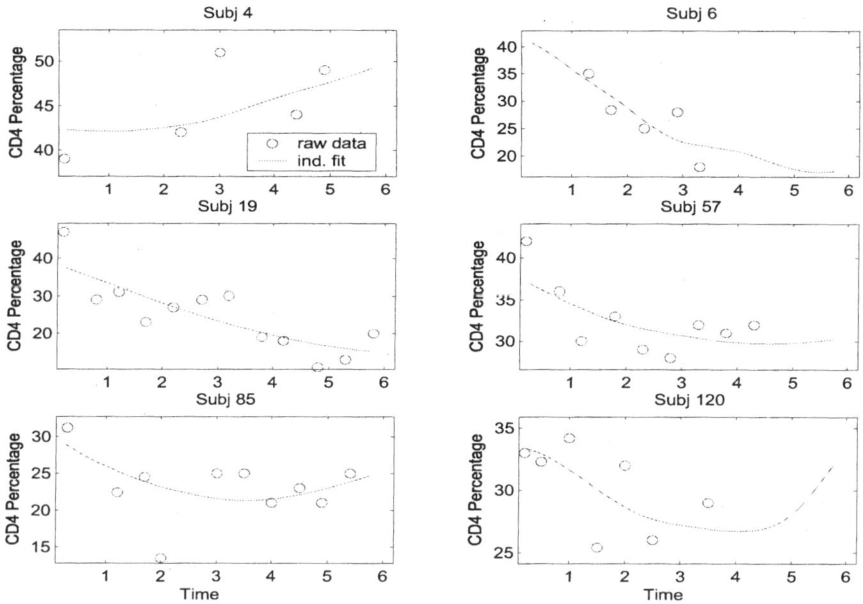

Fig. 9.4 TVC-NPM model (9.46) fit to the MACS data. Solid curves, estimated individual functions. Dots, observed data points.

the MACS data. However, under the TVC-NPM model, the Age effect is not always insignificant since the 95% pointwise SD band does not always contain 0. Therefore, ignoring the within-subject correlation may lead to misleading conclusions.

In Figure 8.9, we depicted the estimated individual functions of six subjects under the SPME model (8.80). For comparison, in Figure 9.4, we depict the estimated individual functions for the same six subjects under the TVC-NPME model (9.92). We can see that the individual function fits are comparable, and they fit the individual data points quite well, although the TVC-NPME model (9.92) does not require the assumption of parametric models for the fixed-effect coefficient functions.

9.4.7 Progesterone Data Revisted

Recall that the progesterone data consist of two groups: the nonconceptive progesterone group with 69 subjects, and the conceptive progesterone group with 22 subjects. In Section 9.2.10, we fit the special TVC-NPM model (9.47) to the progesterone data where the within-subject correlations were not accounted for. In order to account for the within-subject correlations and to predict the individual functions, we extend the TVC-NPM model (9.47) to the following TVC-NPME model:

$$\begin{gathered} y_{ij} = 1_{\{i \le 69\}}\alpha_1(t_{ij}) + 1_{\{i>69\}}\alpha_2(t_{ij}) + v_i(t_{ij}) + \epsilon_{ij}, \\ v_i(t) \sim \text{GP}(0, \gamma), \boldsymbol{\epsilon}_i = [\epsilon_{i1}, \cdots, \epsilon_{in_i}]^T \sim N(0, \mathbf{R}_i), \\ j = 1, 2, \cdots, n_i; \ i = 1, 2, \cdots, 71. \end{gathered} \tag{9.93}$$

We fit the above TVC-NPME model (9.93) to the progesterone data using the P-spline method described in Section 9.4.3. The quadratic truncated power basis (8.12) with 12 knots was used for the fixed-effect coefficient functions and the random-effect functions as well. The knots were scattered using the "equally spaced sample quantitles as knots" method. We selected the smoothing parameters using the BIC rule developed in Section 7.5.6 of Chapter 7. The smoothing parameters for the two fixed-effect coefficient (group mean) functions and the random-effect functions were 241.32, 257.80 and 8.6771×10^7, respectively. Figure 9.5 depicts the fitted

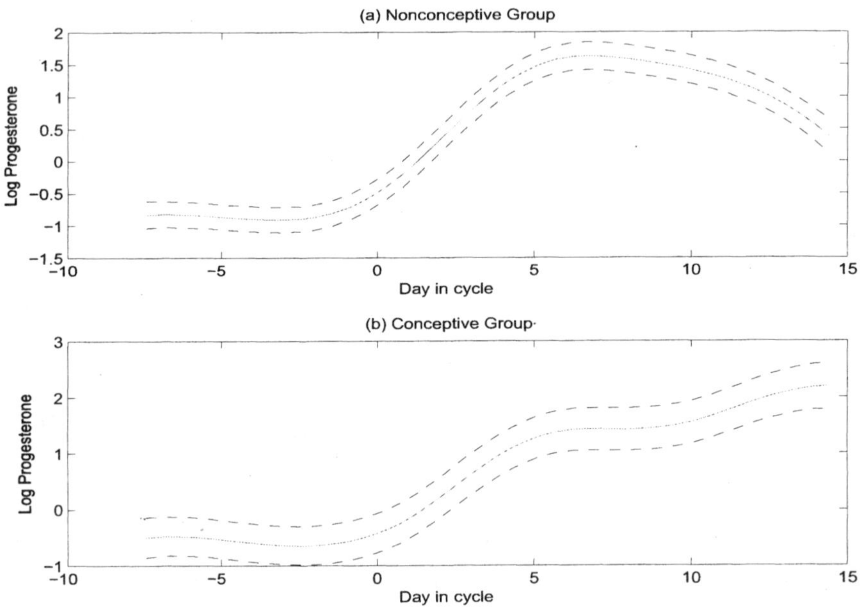

Fig. 9.5 TVC-NPME model (9.93) fit to the progesterone data. Solid curves, estimated fixed-effect coefficient functions. Dashed curves, 95% pointwise SD bands.

fixed-effect coefficient functions (solid curves), together with the associated 95% pointwise SD bands (dashed curves). These fitted fixed-effect coefficient functions are comparable to the fitted group mean functions (see Figure 9.2) obtained by fitting the TVC-NPM model (9.47) to the progesterone data but ignoring the within-subject correlations. It is seen, however, that the 95% pointwise SD bands under the TVC-NPME model are wider than those under the TVC-NPM model, as expected.

Figure 9.6 depicts the fitted individual functions (solid curves) of six subjects, superimposed with their group mean functions (dashed curves). It is seen that although the group mean functions are quite different, these individual functions fit the observed individual data points very well.

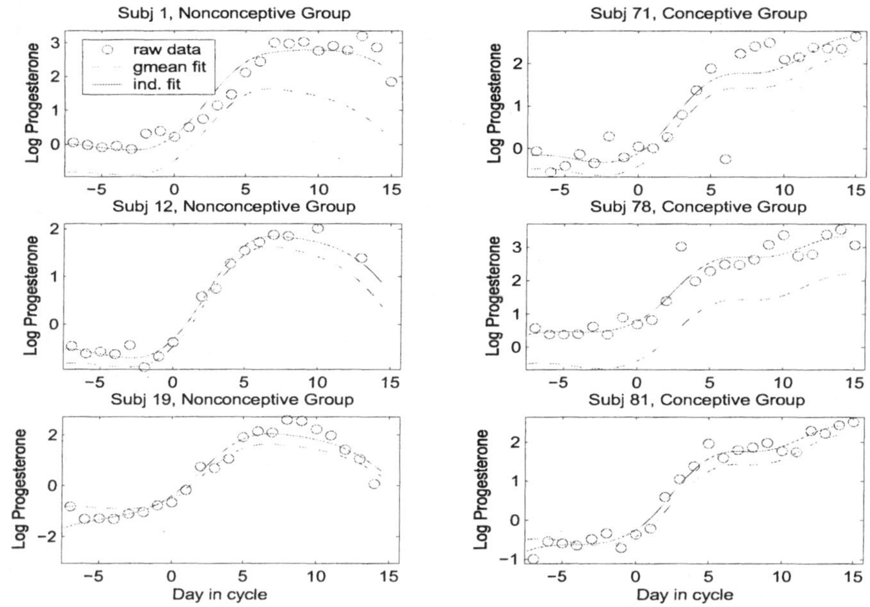

Fig. 9.6 TVC-NPME model (9.93) fit to the progesterone data. Solid curves, estimated individual functions. Dashed curves, estimated group mean functions. Dots, raw individual data points.

In Section 4.9 of Chapter 4, as an illustration, we fit the data from the nonconceptive progesterone group using the following nonparametric mixed-effects (NPME) model:

$$\begin{gathered} y_{ij} = \eta(t_{ij}) + v_i(t_{ij}) + \epsilon_{ij}, \\ v_i(t) \sim \text{GP}(0, \gamma), \boldsymbol{\epsilon}_i = [\epsilon_{i1}, \cdots, \epsilon_{in_i}]^T \sim N(0, \mathbf{R}_i), \\ j = 1, 2, \cdots, n_i;\ i = 1, 2, \cdots, n. \end{gathered} \tag{9.94}$$

We can also fit the data from the conceptive progesterone group using the above NPME model. If we know that the covariance structure of the nonconceptive progesterone group is significantly different from the conceptive progesterone group, we can fit the two groups seperately using the NPME model (9.94). However, if there is some evidence indicating that both groups have a similar covariance structure, then fitting the TVC-NPME model (9.93) to the progesterone data would be more efficient since it would use all the available data to estimate the common covariance structure of the progesterone data.

9.5 TIME-VARYING COEFFICIENT SPME MODEL

The TVC-NPME model (9.53) can be extended to a time-varying coefficient semi-parametric mixed-effects (TVC-SPME) model with the following form:

$$\begin{aligned} y_i(t) &= \mathbf{c}_i(t)^T\boldsymbol{\alpha} + \mathbf{x}_i(t)^T\boldsymbol{\beta}(t) + \mathbf{h}_i(t)^T\mathbf{a}_i + \mathbf{z}_i(t)^T\mathbf{v}_i(t) + \epsilon_i(t), \\ \mathbf{a}_i &\sim N(0, \mathbf{D}_a), \mathbf{v}_i(t) \sim \text{GP}(0, \boldsymbol{\Gamma}),\ \text{E}[\mathbf{a}_i\mathbf{v}_i(t)^T] = \boldsymbol{\Gamma}_a(t), \\ \boldsymbol{\epsilon}_i(t) &\sim \text{GP}(0, \gamma_\epsilon),\ i = 1, 2, \cdots, n, \end{aligned} \tag{9.95}$$

where $\mathbf{c}_i(t)$ and $\mathbf{h}_i(t)$ are observable parametric fixed-effects and random-effects covariate vectors, respectively, and $\boldsymbol{\alpha}$ and $\mathbf{a}_i$ are the associated parametric fixed-effects and random-effects vectors, and other notations are the same as in (9.53). When $\boldsymbol{\alpha}$ and $\mathbf{a}_i,\ i = 1, 2, \cdots, n$ are all zero, the above TVC-SPME model reduces to the TVC-NPME model (9.53). In principle, all the methods for fitting the TVC-NPME model (9.53) described in the previous section can be extended to fit the TVC-SPME model (9.95). In this section, as illustrations, we shall describe a backfitting algorithm and a regression spline method in the following two subsections.

9.5.1 Backfitting Algorithm

When $\boldsymbol{\alpha}$ and $\mathbf{a}_i,\ i = 1, 2, \cdots, n$ are known, the TVC-SPME model (9.95) essentially reduces to the TVC-NPME model (9.53) and can be written as

$$\tilde{y}_i(t) = \mathbf{x}_i(t)^T\boldsymbol{\beta}(t) + \mathbf{z}_i(t)^T\mathbf{v}_i(t) + \epsilon_i(t),\quad i = 1, 2, \cdots, n, \tag{9.96}$$

where $\tilde{y}_i(t) = y_i(t) - \mathbf{c}_i(t)^T\boldsymbol{\alpha} - \mathbf{h}_i(t)^T\mathbf{a}_i$. The above TVC-NPME model can be fitted using the methods proposed in the previous section. Similarly, when $\boldsymbol{\beta}(t)$ and $\mathbf{v}_i(t),\ i = 1, 2, \cdots, n$ are known, the TVC-SPME model (9.95) essentially reduces to the LME model discussed in Chapter 2. This LME model can be written as

$$\tilde{y}_i(t) = \mathbf{c}_i(t)^T\boldsymbol{\alpha} + \mathbf{h}_i(t)^T\mathbf{a}_i + \epsilon_i(t),\quad i = 1, 2, \cdots, n, \tag{9.97}$$

where $\tilde{y}_i(t) = y_i(t) - \mathbf{x}_i(t)^T\boldsymbol{\beta}(t) - \mathbf{z}_i(t)^T\mathbf{v}_i(t)$. Thus, we propose the following backfitting algorithm to fit the TVC-SPME model (9.95).

(a) Get initial estimates for $\boldsymbol{\alpha}$ and $\mathbf{a}_i,\ i = 1, 2, \cdots, n$, denoted as $\hat{\boldsymbol{\alpha}}$ and $\hat{\mathbf{a}}_i,\ i = 1, 2, \cdots, n$.

(b) Calculate the residuals $\tilde{y}_i(t) = y_i(t) - \mathbf{c}_i(t)^T\hat{\boldsymbol{\alpha}} - \mathbf{h}_i(t)^T\hat{\mathbf{a}}_i$, fit the TVC-NPME model (9.96), and update $\hat{\boldsymbol{\beta}}(t)$ and $\hat{\mathbf{v}}_i(t),\ i = 1, 2, \cdots, n$.

(c) Calculate the residuals $\tilde{y}_i(t) = y_i(t) - \mathbf{x}_i(t)^T\hat{\boldsymbol{\beta}}(t) - \mathbf{z}_i(t)^T\hat{\mathbf{v}}_i(t)$, fit the LME model (9.97), and update $\hat{\boldsymbol{\alpha}}$ and $\hat{\mathbf{a}}_i,\ i = 1, 2, \cdots, n$.

(d) Repeat Steps (b) to (c) until convergence.

9.5.2 Regression Spline Method

The regression spline method is straightforward and simple to fit the semiparametric TVC-NPME model (9.95). Assume that we have the following regression spline bases:

$$\begin{aligned}\mathbf{\Phi}_{rp_r}(t) &= [\phi_{r1}(t), \cdots, \phi_{rp_r}(t)]^T, \ r = 0, 1, \cdots, d,\\ \mathbf{\Psi}_{sq_s}(t) &= [\psi_{s1}(t), \cdots, \psi_{sq_s}(t)]^T, \ s = 0, 1, \cdots, d^*.\end{aligned}$$

Then we can express $\beta_r(t), r = 0, 1, \cdots, d$ and $v_{si}(t), s = 0, 1, \cdots, d^*$ as regression splines:

$$\begin{aligned}\beta_r(t) &= \mathbf{\Phi}_{rp_r}(t)^T \boldsymbol{\beta}_r, \ r = 0, 1, \cdots, d,\\ v_{si}(t) &= \mathbf{\Psi}_{sq_s}(t)^T \mathbf{b}_{si}, \ s = 0, 1, \cdots, d^*,\end{aligned}$$

where $\boldsymbol{\beta}_r = [\beta_{r1}, \cdots, \beta_{rp_r}]^T$ and $\mathbf{b}_{si} = [b_{si1}, \cdots, b_{siq_s}]^T$ are the associated coefficient vectors. Define

$$\begin{aligned}\mathbf{\Phi}(t) &= \text{diag}[\mathbf{\Phi}_{0p_0}(t), \mathbf{\Phi}_{1p_1}(t), \cdots, \mathbf{\Phi}_{dp_d}(t)], \ \boldsymbol{\beta} = [\boldsymbol{\beta}_0^T, \boldsymbol{\beta}_1^T, \cdots, \boldsymbol{\beta}_d^T]^T,\\ \mathbf{\Psi}(t) &= \text{diag}[\mathbf{\Psi}_{0q_0}(t), \mathbf{\Psi}_{1q_1}(t), \cdots, \mathbf{\Phi}_{d^*q_{d^*}}(t)], \ \mathbf{b}_i = [\mathbf{b}_{0i}^T, \mathbf{b}_{1i}^T, \cdots, \mathbf{b}_{d^*i}^T]^T.\end{aligned}$$

Then we can express the nonparametric coefficient function vectors $\boldsymbol{\beta}(t)$ and $\mathbf{v}_i(t)$ as

$$\boldsymbol{\beta}(t) = \mathbf{\Phi}(t)^T \boldsymbol{\beta}, \ \mathbf{v}_i(t) = \mathbf{\Psi}(t)^T \mathbf{b}_i, \ \ i = 1, 2, \cdots, n.$$

Notice that $\mathbf{b}_i, \ i = 1, 2, \cdots, n$ are i.i.d copies of a normal random variable with mean 0 and some covariance matrix $\mathbf{D}_b$. It follows that the TVC-SPME model (9.95) becomes

$$\begin{aligned} y_i(t) &= \mathbf{c}_i(t)^T \boldsymbol{\alpha} + \tilde{\mathbf{x}}_i(t)^T \boldsymbol{\beta} + \mathbf{h}_i(t)^T \mathbf{a}_i + \tilde{\mathbf{z}}_i(t)^T \mathbf{b}_i + \epsilon_i(t),\\ \mathbf{a}_i &\sim N(0, \mathbf{D}_a), \ \mathbf{b}_i \sim N(0, \mathbf{D}_b), \ \text{E}[\mathbf{a}_i \mathbf{b}_i^T] = \mathbf{D}_{ab},\\ \epsilon_i(t) &\sim \text{GP}(0, \gamma_\epsilon), \ \ i = 1, 2, \cdots, n,\end{aligned} \tag{9.98}$$

where $\tilde{\mathbf{x}}_i(t) = \mathbf{\Phi}(t)\mathbf{x}_i(t)$ and $\tilde{\mathbf{x}}_i(t) = \mathbf{\Psi}(t)\mathbf{z}_i(t)$. For the given bases, the model (9.98) is a standard LME model. Thus, using the regression spline method, the TVC-SPME model (9.95) is transformed into a standard LME model. This LME model can be fitted using the S-PLUS function *lme* or the SAS procedure *PROC MIXED*.

For the above regression spline method, the smoothing parameters $p_r, r = 0, \cdots, d$ and $q_s, s = 0, 1, \cdots, d^*$ need to be properly selected in order to optimally trade off the goodness-of-fit and smoothness of the fitted functions. We may use the AIC or BIC rules described in Chapter 5.

9.6 SUMMARY AND BIBLIOGRAPHICAL NOTES

In this chapter, we reviewed and proposed various methods for fitting various TVC models for longitudinal data analysis, including the TVC-NPM, TVC-SPM, TVC-NPME and TVC-SPME models.

West et al. (1985) were the first to introduce and study a "dynamic generalized linear" model which is also known as a TVC model. The TVC model is a special case of the general varying coefficient model, first proposed and studied by Hastie and Tibshirani (1993). For cross-sectional data and time series data, further work on TVC models and their applications can be found in Cai, Fan and Yao (2000), Fan and Zhang (1999, 2000), Cai, Fan and Li (2000), Fan, Zhang and Zhang (2001), among others.

Applications of the TVC models to longitudinal data analysis were poineered by Brumback and Rice (1998), Hoover et al. (1998), Wu, Chiang and Hoover (1998), Fan and Zhang (1998), among others. More work was done by Fan and Zhang (2000), Wu and Chiang (2000), Chiang, Rice and Wu (2001), Huang, Wu and Zhou (2002), and Eubank et al. (2004), among others. However, these authors did not incorporate the within-subject correlations into the construction of the estimators. When the longitudinal data are sparse, the resultant estimators work well.

However, when the longitudinal data are highly correlated or less sparse, properly modeling the within-subject correlations is often necessary. Wu and Zhang (2002a) first proposed and studied a nonparametric mixed-effects (NPME) model for longitudinal data analysis, which properly models the within-subject correlation using random-effect functions. Park and Wu (2005) considered a similar problem from the view of local likelihood. Liang et al. (2003) extended the NPME model to a simple TVC-NPME model. A regression spline method was used to fit the TVC-NPME model. Wu and Liang (2004) considered a RTVC model and a backfitting algorithm was proposed to fit the RTVC model. Further work in this area is still ongoing.

10
Discrete Longitudinal Data

10.1 INTRODUCTION

In Chapters 4~9, we presented the major nonparametric regression methods (including local polynomial, regression splines, smoothing splines and P-splines), as well as semiparametric models and time-varying coefficient models for longitudinal data analysis. The methodologies described in these chapters focus on continuous longitudinal data. However, many of these methods and ideas can be used to analyze discrete longitudinal data as well. In fact, the LPK-GEE methods for generalized nonparametric/semiparametric population mean models proposed by Lin and Carroll (2000, 2001a,b) are generally applicable to discrete longitudinal data as illustrated by the application examples in these articles. Wang (2003) and Wang, Carroll and Lin (2005) recently developed a more efficient LPK-GEE method for nonparametric/semiparametric population mean models to improve the methods by Lin and Carroll (2001a,b). In an unpublished manuscript by Cai, Li and Wu (2003), the local polynomial methods are proposed to fit generalized nonparametric mixed-effects models and generalized time-varying coefficient mixed-effects models. Lin and Zhang (1999) also proposed the inference methods for generalized semiparametric additive mixed-effects models for discrete longitudinal data analysis based on a smoothing spline approach. In this chapter we briefly review these models and methods for discrete longitudinal data analysis.

10.2 GENERALIZED NPM MODEL

Suppose we have a discrete longitudinal data set:

$$(y_{ij}, t_{ij}),\ j = 1, 2, \cdots, n_i;\ i = 1, 2, \cdots, n,$$

where y_{ij} denotes the outcome for the j-th measurement from the i-th subject (or cluster) at time point t_{ij}, n is the number of subjects and n_i is the number of measurements of the i-th subject. The population marginal mean and variance of the responses y_{ij} are specified as

$$\begin{array}{c} \mathrm{E}(y_{ij}|t_{ij}) = \mu_{ij},\ \mathrm{Var}(y_{ij}|t_{ij}) = \phi w_{ij}^{-1} V(\mu_{ij}), \\ j = 1, 2, \cdots, n_i;\ i = 1, 2, \cdots, n, \end{array} \tag{10.1}$$

where ϕ is a scale parameter, w_{ij} are the weights, and $V(\cdot)$ is a known variance function. The population marginal mean μ_{ij} is assumed to be related to the time points t_{ij} through a known differentiable and monotonic link function $g(\cdot)$:

$$g(\mu_{ij}) = \eta(t_{ij}),\ \ j = 1, 2, \cdots, n_i;\ i = 1, 2, \cdots, n, \tag{10.2}$$

where $\eta(\cdot)$ is an unknown smooth function that needs to be estimated nonparametrically. Since the model (10.2) aims to model the population marginal mean of the responses y_{ij}nonparametrically via a known link function, we call it a generalized nonparametric population mean (GNPM) model.

Lin and Carroll (2000) proposed a local polynomial kernel generalized estimating equation (LPK-GEE) method to fit the GNPM model (10.2). The key idea is to approximate the function $\eta(t)$ locally via a polynomial of some degree, say p. Assume that $\eta(t)$ has up to $(p+1)$-times continuous derivatives. Then by Taylor expansion, for a fixed time point t, we have

$$\eta(t_{ij}) \approx \eta(t) + \eta'(t)(t_{ij} - t) + \cdots + \eta^{(p)}(t)(t_{ij} - t)^p/p!. \tag{10.3}$$

Let $\mathbf{x}_{ij} = [1, t_{ij} - t, \cdots, (t_{ij} - t)^p]^T$ and $\boldsymbol{\beta} = [\beta_0, \cdots, \beta_p]^T$ with $\beta_l = \eta^{(l)}(t)/l!$. Then we have

$$g(\mu_{ij}) = \eta(t_{ij}) \approx \mathbf{x}_{ij}^T\boldsymbol{\beta},\ \ j = 1, 2, \cdots, n_i;\ i = 1, 2, \cdots, n. \tag{10.4}$$

The equation (10.4) is valid only in a local neighborhood of t. To reflect this feature, we use a kernel method. Let $K_h(\cdot) = K(\cdot/h)/h$ where K is a kernel function and h is a bandwidth that specifies the size of the local neighborhood $[t-h, t+h]$ at t. For simplicity, throughout this chapter, we denote $f(\mathbf{a}) = [f(a_1), \cdots, f(a_d)]^T$ for any scalar function f and any vector $\mathbf{a} = [a_1, \cdots, a_d]^T$. Moreover, let

$$\begin{array}{rcl} \mathbf{y}_i & = & [y_{i1}, \cdots, y_{in_i}]^T,\ \mathbf{X}_i = [\mathbf{x}_{i1}, \cdots, \mathbf{x}_{in_i}]^T, \\ \mathbf{K}_{ih} & = & \mathrm{diag}[K_h(t_{i1} - t), \cdots, K_h(t_{in_i} - t)], \end{array} \tag{10.5}$$

and

$$\tilde{\mu}_{ij} = g^{-1}(\mathbf{x}_{ij}^T\boldsymbol{\beta}),\ \ j = 1, 2, \cdots, n_i;\ i = 1, 2, \cdots, n, \tag{10.6}$$

where $g^{-1}(\cdot)$ denotes the inverse function of $g(\cdot)$ so that $g(\tilde{\mu}_{ij}) = \mathbf{x}_{ij}^T\boldsymbol{\beta}$. It follows that

$$\begin{aligned} \tilde{\boldsymbol{\mu}}_i = [\tilde{\mu}_{i1}, \cdots, \tilde{\mu}_{in_i}]^T = g^{-1}(\mathbf{X}_i\boldsymbol{\beta}), \\ g(\tilde{\boldsymbol{\mu}}_i) = \mathbf{X}_i\boldsymbol{\beta}, \quad i = 1, 2, \cdots, n. \end{aligned} \tag{10.7}$$

To estimate $\boldsymbol{\beta}$, Lin and Carroll (2000) considered the following two LPK-GEE equations:

$$\mathbf{X}_i^T \boldsymbol{\Delta}_i \mathbf{K}_{ih}^{1/2} \mathbf{V}_i^{-1} \mathbf{K}_{ih}^{1/2} \Big(\mathbf{y}_i - \tilde{\boldsymbol{\mu}}_i\Big) = 0, \quad i = 1, 2, \cdots, n, \tag{10.8}$$

and

$$\mathbf{X}_i^T \boldsymbol{\Delta}_i \mathbf{V}_i^{-1} \mathbf{K}_{ih} \Big(\mathbf{y}_i - \tilde{\boldsymbol{\mu}}_i\Big) = 0, \quad i = 1, 2, \cdots, n, \tag{10.9}$$

where $\boldsymbol{\Delta}_i = \Big[\text{diag}\{g'(\tilde{\boldsymbol{\mu}}_i)\}\Big]^{-1}$ with $g'(\cdot)$ denoting the first derivative of $g(\cdot)$, $\mathbf{V}_i = \mathbf{S}_i^{1/2}\mathbf{R}_i\mathbf{S}_i^{1/2}$ with $\mathbf{S}_i = \text{diag}\left\{\phi w_{ij}^{-1} V(\tilde{\mu}_{ij}), j = 1, 2, \cdots, n_i\right\}$, and $\mathbf{R}_i$ being a user-specified working correlation matrix for the i-th subject.

The solution to (10.8) or (10.9) can be obtained using the iterated re-weighted least squares method. To this end, define the following working responses:

$$\begin{aligned} \tilde{y}_{ij} = g(\tilde{\mu}_{ij}) + g'(\tilde{\mu}_{ij})(y_{ij} - \tilde{\mu}_{ij}), \\ j = 1, 2, \cdots, n_i; \; i = 1, 2, \cdots, n. \end{aligned} \tag{10.10}$$

The above working responses can be computed when the coefficient vector $\boldsymbol{\beta}$ is obtained. Using the above working responses and the expressions (10.7), we have

$$\mathbf{y}_i - \tilde{\boldsymbol{\mu}}_i = \boldsymbol{\Delta}_i[\tilde{\mathbf{y}}_i - g(\tilde{\boldsymbol{\mu}}_i)] = \boldsymbol{\Delta}_i(\tilde{\mathbf{y}}_i - \mathbf{X}_i\boldsymbol{\beta}), \quad i = 1, 2, \cdots, n. \tag{10.11}$$

Plugging the above equation into (10.8) and (10.9), we have

$$\mathbf{X}_i^T \boldsymbol{\Delta}_i \mathbf{K}_{ih}^{1/2} \mathbf{V}_i^{-1} \mathbf{K}_{ih}^{1/2} \boldsymbol{\Delta}_i \Big(\tilde{\mathbf{y}}_i - \mathbf{X}_i\boldsymbol{\beta}\Big) = 0, \quad i = 1, 2, \cdots, n, \tag{10.12}$$

and

$$\mathbf{X}_i^T \boldsymbol{\Delta}_i \mathbf{V}_i^{-1} \mathbf{K}_{ih} \boldsymbol{\Delta}_i \Big(\tilde{\mathbf{y}}_i - \mathbf{X}_i\boldsymbol{\beta}\Big) = 0, \quad i = 1, 2, \cdots, n. \tag{10.13}$$

Solving the above LPK-GEE equations, respectively, leads to

$$\hat{\boldsymbol{\beta}} = \Big(\sum_{i=1}^n \mathbf{X}_i^T \boldsymbol{\Phi}_i \mathbf{X}_i\Big)^{-1} \Big(\sum_{i=1}^n \mathbf{X}_i^T \boldsymbol{\Phi}_i \tilde{\mathbf{y}}_i\Big), \tag{10.14}$$

and

$$\hat{\boldsymbol{\beta}} = \Big(\sum_{i=1}^n \mathbf{X}_i^T \boldsymbol{\Psi}_i \mathbf{X}_i\Big)^{-1} \Big(\sum_{i=1}^n \mathbf{X}_i^T \boldsymbol{\Psi}_i \tilde{\mathbf{y}}_i\Big), \tag{10.15}$$

where $\boldsymbol{\Phi}_i = \boldsymbol{\Delta}_i \mathbf{K}_{ih}^{1/2} \mathbf{V}_i^{-1} \mathbf{K}_{ih}^{1/2} \boldsymbol{\Delta}_i$ and $\boldsymbol{\Psi}_i = \boldsymbol{\Delta}_i \mathbf{V}_i^{-1} \mathbf{K}_{ih} \boldsymbol{\Delta}_i$. For a given current estimate of $\boldsymbol{\beta}$, the working response vectors $\tilde{\mathbf{y}}_i$ and the matrices $\boldsymbol{\Phi}_i$ and $\boldsymbol{\Psi}_i$ are computable so that the updated estimate $\hat{\boldsymbol{\beta}}$ can be obtained using (10.14) or (10.15).

This process is repeated until convergence. Let $\hat{\beta}$ denote the resulting estimate of β at convergence. The estimator of $\eta(t)$ is then $\hat{\eta}(t) = \mathbf{e}_1^T\hat{\beta} = \hat{\beta}_0$, where $\mathbf{e}_1$ denotes a unit vector whose first entry is 1 and other entries are 0.

When $p = 0, 1$, Lin and Carroll (2000) showed that the two estimators of $\eta(t)$ obtained using (10.8) and (10.9), respectively, have different asymptotic properties: asymptotic properties of $\hat{\eta}(t)$ based on (10.9) are much harder to obtain. The most important result of Lin and Carroll (2000) is that, unlike the parametric GEE estimators of Liang and Zeger (1986), the asymptotically most efficient kernel ($p = 0$) estimator $\hat{\eta}(t)$ derived from (10.8) is obtained by entirely ignoring the within-subject correlation and pretending that the observations within the same subject were independent. That is, one makes a working independence assumption with $\mathbf{R}_i = \mathbf{I}_{n_i}$, $i = 1, 2, \cdots, n$. Correctly specifying the correlation matrix in fact has adverse effects and results in an asymptotically less efficient estimator of $\eta(t)$.

10.3 GENERALIZED SPM MODEL

In the previous section, we described the LPK-GEE method for fitting the GNPM model (10.2) where the only covariate is time t. Lin and Carroll (2001a) extended the GNPM model to a generalized semiparametric population mean (GSPM) model where the population marginal mean μ_{ij} depends on the time nonparametrically and other covariates linearly.

Suppose we have the following discrete longitudinal data set:

$$(y_{ij}, \mathbf{c}_{ij}^T, t_{ij}),\ j = 1, 2, \cdots, n_i;\ i = 1, 2, \cdots, n, \tag{10.16}$$

where y_{ij} denotes the j-th response for the i-th subject (or cluster) at time point t_{ij} and with the observed d-dimensional covariate vector $\mathbf{c}_{ij}$. The population marginal mean and variance of the responses y_{ij} are now specified as

$$\begin{gathered}\mathrm{E}(y_{ij}|\mathbf{c}_{ij}, t_{ij}) = \mu_{ij},\ \mathrm{Var}(y_{ij}|\mathbf{c}_{ij}, t_{ij}) = \phi w_{ij}^{-1} V(\mu_{ij}),\\ j = 1, 2, \cdots, n_i;\ i = 1, 2, \cdots, n,\end{gathered} \tag{10.17}$$

where again ϕ is a scale parameter, w_{ij} are the weights, and $V(\cdot)$ is a known varianc and the time points t_{ij} through a known differentiable and monotonic link function $g(\cdot)$:

$$g(\mu_{ij}) = \mathbf{c}_{ij}^T\boldsymbol{\alpha} + \eta(t_{ij}),\quad j = 1, 2, \cdots, n_i;\ i = 1, 2, \cdots, n, \tag{10.18}$$

where $\boldsymbol{\alpha}$ is an unknown d-dimensional parameter vector and $\eta(\cdot)$ is an unknown smooth function. Since the model (10.18) aims to model the population marginal mean of the responses y_{ij} semiparametrically via a known link function, we call it a GSPM model.

Note that the GSPM model (10.18) essentially reduces to the standard generalized linear mixed-effects model (Liang and Zeger 1986) when $\eta(t_{ij})$ is completely known, and it essentially reduces to the GNPM model (10.2) when the parameter vector $\boldsymbol{\alpha}$

is completely known. This motivated Lin and Carroll (2001a) to propose a profile LPK-GEE method for fitting the GSPM model (10.18).

As in Section 10.2, by assuming that $\eta(t)$ has up to $(p+1)$-times continuous derivatives, for a fixed time point t, one can obtain the Taylor expansion (10.3) for $\eta(t_{ij})$. Let $\mathbf{x}_{ij} = [1, t_{ij}-t, \cdots, (t_{ij}-t)^p]^T$ and $\boldsymbol{\beta} = [\beta_0, \cdots, \beta_p]^T$ with $\beta_l = \eta^{(l)}(t)/l!$. Then we have

$$g(\mu_{ij}) \approx \mathbf{c}_{ij}^T\boldsymbol{\alpha} + \mathbf{x}_{ij}^T\boldsymbol{\beta}, \quad j = 1, 2, \cdots, n_i; \; i = 1, 2, \cdots, n. \tag{10.19}$$

We now define

$$\tilde{\mu}_{ij} = g^{-1}(\mathbf{c}_{ij}^T\boldsymbol{\alpha} + \mathbf{x}_{ij}^T\boldsymbol{\beta}), \quad j = 1, 2, \cdots, n_i; \; i = 1, 2, \cdots, n, \tag{10.20}$$

so that $g(\tilde{\mu}_{ij}) = \mathbf{c}_{ij}^T\boldsymbol{\alpha} + \mathbf{x}_{ij}^T\boldsymbol{\beta}$. It follows that

$$\begin{aligned} \tilde{\boldsymbol{\mu}}_i = [\tilde{\mu}_{i1}, \cdots, \tilde{\mu}_{in_i}]^T &= g^{-1}(\mathbf{C}_i\boldsymbol{\alpha} + \mathbf{X}_i\boldsymbol{\beta}), \\ g(\tilde{\boldsymbol{\mu}}_i) &= \mathbf{C}_i\boldsymbol{\alpha} + \mathbf{X}_i\boldsymbol{\beta}, \quad i = 1, 2, \cdots, n, \end{aligned} \tag{10.21}$$

where $\mathbf{C}_i = [\mathbf{c}_{i1}, \cdots, \mathbf{c}_{in_i}]^T$ and $\mathbf{X}_i = [\mathbf{x}_{i1}, \cdots, \mathbf{x}_{in_i}]^T$.

Using the notations $\mathbf{y}_i, \mathbf{K}_{ih}$ defined in Section 10.2 and in (10.21), for a known parameter vector $\boldsymbol{\alpha}$, the $\boldsymbol{\beta}$ can be estimated using one of the following LPK-GEE equations:

$$\sum_{i=1}^n \mathbf{X}_i^T \boldsymbol{\Delta}_i \mathbf{K}_{ih}^{1/2} \mathbf{V}_{2i}^{-1} \mathbf{K}_{ih}^{1/2} (\mathbf{y}_i - \tilde{\boldsymbol{\mu}}_i) = 0, \tag{10.22}$$

and

$$\sum_{i=1}^n \mathbf{X}_i^T \boldsymbol{\Delta}_i \mathbf{V}_{2i}^{-1} \mathbf{K}_{ih} (\mathbf{y}_i - \tilde{\boldsymbol{\mu}}_i) = 0, \tag{10.23}$$

where $\boldsymbol{\Delta}_i = \left\{\text{diag}[g'(\tilde{\boldsymbol{\mu}}_i)]\right\}^{-1}$, $\mathbf{V}_{2i} = \mathbf{S}_i^{1/2}\mathbf{R}_{2i}\mathbf{S}_i^{1/2}$ with

$$\mathbf{S}_i = \text{diag}\left\{\phi w_{ij}^{-1} V(\tilde{\mu}_{ij}), j = 1, 2, \cdots, n_i\right\},$$

and $\mathbf{R}_{2i}$ being the working correlation matrix which may also contain unknown parameters. Then, for a given parameter vector $\boldsymbol{\alpha}$, the LPK-GEE estimator of $\eta(t)$ is $\hat{\eta}(t; \boldsymbol{\alpha}) = \hat{\beta}_0$.

The parameter vector $\boldsymbol{\alpha}$ can be estimated by solving the profile parametric estimating equations: ,

$$\sum_{i=1}^n \frac{\partial \tilde{\boldsymbol{\mu}}_i(\boldsymbol{\alpha})^T}{\partial \boldsymbol{\alpha}} \mathbf{V}_{1i}^{-1} \left[\mathbf{y}_i - \tilde{\boldsymbol{\mu}}_i(\boldsymbol{\alpha})\right] = 0, \tag{10.24}$$

where $\tilde{\boldsymbol{\mu}}_i(\boldsymbol{\alpha}) = [\tilde{\mu}_{i1}(\boldsymbol{\alpha}), \cdots, \tilde{\mu}_{in_i}(\boldsymbol{\alpha})]^T$ and $\mathbf{V}_i = \mathbf{S}_i^{1/2}\mathbf{R}_{1i}\mathbf{S}_i^{1/2}$ with

$$\begin{aligned} \tilde{\mu}_{ij}(\boldsymbol{\alpha}) &= g^{-1}[\mathbf{c}_{ij}^T\boldsymbol{\alpha} + \hat{\eta}(t_{ij}; \boldsymbol{\alpha})], \\ \mathbf{S}_i &= \text{diag}\left\{\phi w_{ij}^{-1} V\left[\tilde{\mu}_{ij}(\boldsymbol{\alpha})\right], j = 1, 2, \cdots, n_i\right\}, \end{aligned}$$

with $\mathbf{R}_{1i}$ being a working correlation matrix which may also contain unknown parameters. The unknown parameters in the correlation matrices can be estimated using the method of moments (Liang and Zeger 1986).

The asymptotic properties of the above profile LPK-GEE estimators have been studied by Lin and Carroll (2001a) when $p = 1$. They obtained the following interesting conclusions. (1) If standard local linear ($p = 1$) smoothing is used, then only when assuming working independence, $\hat{\alpha}$ is $\sqrt{n}$-consistent. (2) For other specifications of the working correlations $\{\mathbf{R}_{1i}, \mathbf{R}_{2i}\}$, except in special cases, $\hat{\boldsymbol{\alpha}}$ is $\sqrt{n}$-inconsistent unless $\eta(t)$ is undersmoothed. When $\eta(t)$ is undersmoothed and the true correlation matrix is assumed, the resulting profile LPK-GEE estimator $\hat{\boldsymbol{\alpha}}$ is not semiparametrically efficient. (3) The semiparametric efficient estimator of $\boldsymbol{\alpha}$ is complicated and difficult to obtain. Also note that conventional bandwidth selection techniques, such as cross-validation by deleting one subject or cluster data at a time, fail unless working independence is assumed. This is because the bandwidth h chosen by cross-validation satisfies $h = O(n^{-1/5})$, and $\hat{\boldsymbol{\alpha}}$ will be $\sqrt{n}$-inconsistent unless working independence is assumed.

Wang, Carroll and Lin (2005) extended the LPK-GEE method of Wang (2003) (see Section 4.2.2) for the GSPM model (10.18). The proposed method in Wang, Carroll and Lin (2005) differs from those proposed by Severini and Staniswalis (1994) and Lin and Carroll (2001a) in the way in which $\hat{\eta}(t;\boldsymbol{\alpha})$, the estimated $\eta(t)$ for a given $\boldsymbol{\alpha}$, is constructed based on the more efficient LPK-GEE method proposed by Wang (2003). Thus, the within-subject correlation can be accounted for in the estimator $\hat{\eta}(t;\boldsymbol{\alpha})$. We can write Wang, Carroll and Lin (2005)'s method in a general form.

Define $\mathbf{G}_{ij}$ to be an $n_i \times (p+1)$ matrix with the j-th row to be $\mathbf{x}_{ij}^T = [1, t_{ij} - t, \cdots, (t_{ij} - t)^p]$ and 0 otherwise. The new and efficient estimation procedure for the GSPM model (10.18) can be described as follows:

Step 1. Obtain an initial estimate of $\eta(t)$, e.g., the working independence estimator.

Step 2. Let $\tilde{\eta}(\cdot)$ be the current estimator of $\eta(\cdot)$. Given $\boldsymbol{\alpha}$, let $\hat{\boldsymbol{\beta}} = \hat{\boldsymbol{\beta}}(t;\boldsymbol{\alpha}) = [\hat{\beta}_0(t;\boldsymbol{\alpha}), \cdots, \hat{\beta}_p(t;\boldsymbol{\alpha})]^T$ be the solution to the LPK-GEE equations:

$$\sum_{i=1}^{n}\sum_{j=1}^{n_i} K_h(t_{ij} - t)\Delta_{ij}\mathbf{G}_{ij}^T\mathbf{V}_i^{-1}[\mathbf{y}_i - \boldsymbol{\mu}_{ij}^*(\boldsymbol{\beta})] = 0, \tag{10.25}$$

where $\Delta_{ij} = \left[g'(\tilde{\mu}_{ij})\right]^{-1}$ with $\tilde{\mu}_{ij} = g^{-1}(\mathbf{c}_{ij}^T\boldsymbol{\alpha} + \mathbf{x}_{ij}^T\boldsymbol{\beta})$, and the l-th entry of $\boldsymbol{\mu}_{ij}^*(\boldsymbol{\beta})$ is

$$g^{-1}\left\{\mathbf{c}_{il}^T\boldsymbol{\alpha} + 1_{\{l=j\}}\mathbf{x}_{ij}^T\boldsymbol{\beta} + 1_{\{l\neq j\}}\tilde{\eta}(t_{il};\boldsymbol{\alpha})\right\}.$$

The updated estimator of $\eta(t)$ is $\hat{\eta}(t;\boldsymbol{\alpha}) = \hat{\beta}_0(t;\boldsymbol{\alpha})$.

Step 3. Obtain $\hat{\boldsymbol{\alpha}}$ by solving the profile-type estimating equation:

$$\sum_{i=1}^{n} \frac{\partial\tilde{\boldsymbol{\mu}}_i(\boldsymbol{\alpha})^T}{\partial\boldsymbol{\alpha}}\mathbf{V}_i^{-1}[\mathbf{y}_i - \tilde{\boldsymbol{\mu}}_i(\boldsymbol{\alpha})] = 0, \tag{10.26}$$

where $\tilde{\boldsymbol{\mu}}_i(\boldsymbol{\alpha}) = [\tilde{\mu}_{i1}(\boldsymbol{\alpha}), \cdots, \tilde{\mu}_{in_i}(\boldsymbol{\alpha})]^T$ with $\tilde{\mu}_{ij}(\boldsymbol{\alpha}) = g^{-1}[\mathbf{c}_{ij}^T\boldsymbol{\alpha}+\hat{\eta}(t_{ij};\boldsymbol{\alpha})]$.

Step 4. Repeat Steps 2 and 3 until convergence.

The working covariance matrix $\mathbf{V}_i$ depends on a parameter vector $\boldsymbol{\tau}$ that can be estimated separately via the method of moments.

When $p = 1$, Wang, Carroll and Lin (2005) showed that, in the above estimation procedure, it is unnecessary to undersmooth $\eta(t)$ in order to obtain the semiparametric efficient ($\sqrt{n}$-consistent) estimator of $\boldsymbol{\alpha}$. This is in contrast with the working independence LPK-GEE estimation method whose estimating equations are given by (10.22) and (10.24) (Zeger and Diggle 1994, Lin and Carroll 2001a). In the new efficient estimation procedure, it is also not necessary to use different working covariances in the estimating equations of $\eta(t)$ and $\boldsymbol{\alpha}$. As with standard parametric GEEs (Liang and Zeger 1986), the estimator resulting from (10.25) and (10.26) is still consistent when the working covariance matrix is misspecified and is most efficient when it is correctly specified. The moment estimator $\hat{\boldsymbol{\tau}}$ in the working covariance matrix has no asymptotic effect on the estimators of $\eta(t)$ and $\boldsymbol{\alpha}$, once the moment estimator converges in probability to some $\boldsymbol{\tau}^*$ at a rate of $\sqrt{n}$.

10.4 GENERALIZED NPME MODEL

In this section, we propose a generalized nonparametric mixed-effects (GNPME) model which is an extension of the nonparametric mixed-effects (NPME) model in Section 4.3 of Chapter 4 in order to deal with discrete longitudinal data. This and next sections are based on an unpublished manuscript by Cai, Li and Wu (2003).

Suppose we have a discrete longitudinal data set:

$$(y_{ij}, t_{ij}),\ j = 1, 2, \cdots, n_i;\ i = 1, 2, \cdots, n, \tag{10.27}$$

where y_{ij} denotes the outcome for the j-th measurement from the i-th subject (or cluster) at time point t_{ij}, n is the number of subjects and n_i is the number of measurements of the i-th subject. Conditional to the given subject i, the subject-specific marginal mean and variance of the responses y_{ij} are specified as

$$\begin{aligned} \mathrm{E}(y_{ij}|t_{ij}) = \mu_{ij},\ \mathrm{Var}(y_{ij}|t_{ij}) = \phi w_{ij}^{-1} V(\mu_{ij}), \\ j = 1, 2, \cdots, n_i;\ i = 1, 2, \cdots, n, \end{aligned} \tag{10.28}$$

where ϕ is a scale parameter, w_{ij} are the weights, and $V(\cdot)$ is a known variance function. Notice that the subject-specific marginal mean and variance of y_{ij} are different from those population marginal mean and variance of y_{ij} given in (10.1).

The subject-specific marginal mean μ_{ij} is assumed to be related to the time points t_{ij} through a known differentiable and monotonic link function $g(\cdot)$:

$$g(\mu_{ij}) = \eta(t_{ij}) + v_i(t_{ij}),\ \ j = 1, 2, \cdots, n_i;\ i = 1, 2, \cdots, n, \tag{10.29}$$

where $\eta(t)$ is an unknown generalized population mean function, common for all the subjects, called a (generalized) fixed-effect function or population curve; and $v_i(t)$ is the subject-specific deviation function apart from the generalized population mean function $\eta(t)$, called the i-th (generalized) random-effect function or random curve. For simplicity, we shall drop the word "generalized" before the terms "fixed/random-effects". We call the model (10.29) a GNPME model. Cai, Li and Wu (2003) called it a generalized random curve (GRC) model.

When the link function g is an identity link, i.e., $g(t) = t$, the GNPME model (10.29) reduces to the NPME model proposed and studied by Wu and Zhang (2002a). They proposed a LPK method to fit the NPME model (4.4). We can extend their LPK method to fit the GNPME model (10.29). The key idea is to combine the local polynomial techniques and mixed-effects model approaches to develop the estimation procedures and model fitting algorithms. First, for any fixed time point t, the population curve and random-effects curves are approximated by a polynomial function around a neighborhood of t, and then the model (10.29) becomes a generalized linear mixed-effects (GLME) model. Combining the ideas of the local quasi-likelihood (Fan and Gijbels 1996) and the penalized quasi-likelihood proposed by Green (1987) and Breslow and Clayton (1993), we propose the penalized local quasi-likelihood approach to estimate the functionals and establish the asymptotic theories for the resulting estimators. To accommodate the feature that the population curve and random curves might possess different degrees of smoothness, similar to the bandwidth selection idea for the NPME model (4.19) in Section 4.5, we propose the leave-one-subject-out cross-validation for the population curve and the leave-one-point-out cross-validation for the random curve component, coupled with an iterative algorithm to implement these two criteria.

10.4.1 Penalized Local Polynomial Estimation

Assume that both $\eta(t)$ and $v_i(t)$ in the model (10.29) have up to $(p+1)$-times derivatives. For any fixed time point t, the population curve $\eta(t)$ and random-effects curves $v_i(t),\ i = 1, 2, \cdots, n$ at t_{ij} can be approximated by p-th degree polynomials within a neighborhood of t (different degrees may be used for $\eta(t)$ and $v_i(t)$, respectively). Then, we obtain the following Taylor expansions:

$$\begin{aligned}\eta(t_{ij}) &\approx \eta(t) + \eta'(t)(t_{ij} - t) + \cdots + \tfrac{\eta^{(p)}(t)}{p!}(t_{ij} - t)^p = \mathbf{x}_{ij}^T\beta,\\ v_i(t_{ij}) &\approx v_i(t) + v_i'(t)(t_{ij} - t) + \cdots + \tfrac{v_i^{(p)}(t)}{p!}(t_{ij} - t)^p = \mathbf{x}_{ij}^T\mathbf{b}_i,\end{aligned}$$

where, for $j = 1, 2, \cdots, n_i;\ i = 1, 2, \cdots, n$,

$$\mathbf{x}_{ij} = \begin{pmatrix} 1 \\ t_{ij} - t \\ \vdots \\ (t_{ij} - t)^p \end{pmatrix},\quad \beta = \begin{pmatrix} \eta(t) \\ \eta'(t) \\ \vdots \\ \eta^{(p)}(t)/p! \end{pmatrix},\ \text{and}\ \mathbf{b}_i = \begin{pmatrix} v_i(t) \\ v_i'(t) \\ \vdots \\ v_i^{(p)}(t)/p! \end{pmatrix}.$$

Thus, within a neighborhood of t, model (10.29) can be reasonably approximated by a GLME model (Zeger and Karim 1991, Breslow and Clayton 1993):

$$g(\mu_{ij}) \approx \mathbf{x}_{ij}^T\beta + \mathbf{x}_{ij}^T\mathbf{b}_i,\quad i = 1, 2, \cdots, n, \tag{10.30}$$

where $\mathbf{b}_i \sim (\mathbf{0}, \mathbf{D})$ is a vector of random-effects. Following convention, in the model (10.30) it is assumed that for a given $\mathbf{b}_i$, the repeated measurements $\{\mathbf{y}_i\}_{i=1}^n$ with $\mathbf{y}_i = [y_{i1}, \cdots, y_{in_i}]^T$ are independent and that $\mathbf{b}_i$ is normally distributed. See Diggle, Liang, and Zeger (1994) for details.

To estimate parameters β and $\mathbf{b}_i$ in a standard GLME model, Green (1987) and Breslow and Clayton (1993) proposed a penalized quasi-likelihood approach, which maximizes

$$-\frac{1}{2}\sum_{i=1}^{n}\left\{\sum_{j=1}^{n_i}\phi^{-1}w_{ij}d(y_{ij},\, \mu_{ij}) + \mathbf{b}_i^T\mathbf{D}^{-1}\mathbf{b}_i\right\} \tag{10.31}$$

with respect to β and $\mathbf{b}_i$, where the deviance is

$$d(y,\, \mu) = -2\int_y^{\mu}\frac{y-u}{V(u)}du.$$

The term $\mathbf{b}_i^T\mathbf{D}^{-1}\mathbf{b}_i$ is a penalty due to the random-effects. However, note that the model (10.30) is only approximately true within a neighborhood of t. Thus, to estimate $\eta(t)$ and $v_i(t)$ and their derivatives in the model (10.29), we propose a penalized local polynomial quasi-likelihood (PLPQL) method , which is a generalization of (10.31) by combining the local quasi-likelihood (see Section 5.4.3 of Fan and Gijbels 1996) and the penalized quasi-likelihood (Green 1987, Breslow and Clayton 1993). That is to maximize the following PLPQL:

$$-\frac{1}{2}\sum_{i=1}^{n}\left\{\sum_{j=1}^{n_i}K_h(t_{ij}-t)\phi^{-1}w_{ij}d(y_{ij},\, \tilde{\mu}_{ij}) + \mathbf{b}_i^T\,\mathbf{D}^{-1}\,\mathbf{b}_i\right\} \tag{10.32}$$

with respect to β and $\mathbf{b}_i$, where $K_h(\cdot) = K(\cdot/h)/h$ with $K(\cdot)$ being a kernel function and h being a bandwidth, and

$$\tilde{\mu}_{ij} = g^{-1}(\mathbf{x}_{ij}^T\beta + \mathbf{x}_{ij}^T\mathbf{b}_i), \;\; j = 1, 2, \cdots, n_i;\; i = 1, 2, \cdots, n. \tag{10.33}$$

Let $\tilde{\boldsymbol{\mu}}_i = [\tilde{\mu}_{i1}, \cdots, \tilde{\mu}_{in_i}]^T$. Then we have

$$\begin{aligned} g(\tilde{\mu}_{ij}) &= \mathbf{x}_{ij}^T\beta + \mathbf{x}_{ij}^T\mathbf{b}_i, \\ g(\tilde{\boldsymbol{\mu}}_i) &= \mathbf{X}_i\beta + \mathbf{X}_i\mathbf{b}_i, \;\; i = 1, 2, \cdots, n, \end{aligned} \tag{10.34}$$

where $\mathbf{X}_i = [\mathbf{x}_{i1}, \cdots, \mathbf{x}_{in_i}]^T$.

Differentiating (10.32) with respect to β and $\mathbf{b}_i$ leads to the following score equations:

$$\sum_{i=1}^{n}\sum_{j=1}^{n_i}K_h(t_{ij}-t)\frac{\mathbf{x}_{ij}(y_{ij}-\tilde{\mu}_{ij})}{\phi w_{ij}^{-1}V(\tilde{\mu}_{ij})g'(\tilde{\mu}_{ij})} = \mathbf{0}, \tag{10.35}$$

and, for each i,

$$\sum_{j=1}^{n_i}K_h(t_{ij}-t)\frac{\mathbf{x}_{ij}(y_{ij}-\tilde{\mu}_{ij})}{\phi w_{ij}^{-1}V(\tilde{\mu}_{ij})g'(\tilde{\mu}_{ij})} = \mathbf{D}^{-1}\,\mathbf{b}_i. \tag{10.36}$$

By solving the above equations, we obtain the p-th degree PLPQL estimates $\hat{\beta}$ and $\hat{\mathbf{b}}_i$. As a result, the first element of $\hat{\beta}$ gives the PLPQL estimate $\hat{\eta}(t)$, and $\hat{v}_i(t)$ is the first element of $\hat{\mathbf{b}}_i$. For practical applications, Fan and Gijbels (1996) recommended using the local linear fit ($p = 1$).

To express the above equations in matrix notation, we define the working responses as

$$\tilde{y}_{ij} = g(\tilde{\mu}_{ij}) + g'(\tilde{\mu}_{ij})(y_{ij} - \tilde{\mu}_{ij}), \;\; j = 1, 2, \cdots, n_i; \; i = 1, 2, \cdots, n. \quad (10.37)$$

Moreover, we define

$$\begin{aligned} \tilde{\mathbf{y}}_i = (\tilde{y}_{i1}, \cdots, \tilde{y}_{in_i})^T, \; \mathbf{K}_{ih} = \text{diag}[K_h(t_{i1} - t), \cdots, K_h(t_{in_i} - t)], \\ \mathbf{R}_i = \text{diag}(R_{i1}, \cdots, R_{in_i}), \; R_{ij} = \phi w_{ij}^{-1} V(\tilde{\mu}_{ij})[g'(\tilde{\mu}_{ij})]^2. \end{aligned} \quad (10.38)$$

Then, equations (10.35) and (10.36) become

$$\sum_{i=1}^{n} \mathbf{X}_i^T \mathbf{K}_{ih}^{1/2} \mathbf{R}_i^{-1} \mathbf{K}_{ih}^{1/2} \mathbf{X}_i \, (\beta + \mathbf{b}_i) = \sum_{i=1}^{n} \mathbf{X}_i^T \mathbf{K}_{ih}^{1/2} \mathbf{R}_i^{-1} \mathbf{K}_{ih}^{1/2} \tilde{\mathbf{y}}_i,$$

and, for each i,

$$\mathbf{X}_i^T \mathbf{K}_{ih}^{1/2} \mathbf{R}_i^{-1} \mathbf{K}_{ih}^{1/2} \mathbf{X}_i \, (\beta + \mathbf{b}_i) + \mathbf{D}^{-1} \mathbf{b}_i = \mathbf{X}_i^T \mathbf{K}_{ih}^{1/2} \mathbf{R}_i^{-1} \mathbf{K}_{ih}^{1/2} \tilde{\mathbf{y}}_i.$$

Define $\tilde{\mathbf{y}} = (\tilde{\mathbf{y}}_1^T, \cdots, \tilde{\mathbf{y}}_n^T)^T$, $\mathbf{X} = (\mathbf{X}_1^T, \cdots, \mathbf{X}_n^T)^T$, $\mathbf{Z} = \text{diag}(\mathbf{X}_1, \cdots, \mathbf{X}_n)$, $\mathbf{b} = (\mathbf{b}_1^T, \cdots, \mathbf{b}_n^T)^T$, and $\tilde{\mathbf{D}} = \text{diag}(\mathbf{D}, \cdots, \mathbf{D})$. Then, the solutions to (10.35) and (10.36) can be obtained by solving

$$\begin{pmatrix} \mathbf{X}^T \Theta_h \mathbf{X} & \mathbf{X}^T \Theta_h \mathbf{Z} \\ \mathbf{Z}^T \Theta_h \mathbf{X} & \mathbf{Z}^T \Theta_h \mathbf{Z} + \tilde{\mathbf{D}}^{-1} \end{pmatrix} \begin{pmatrix} \beta \\ \mathbf{b} \end{pmatrix} = \begin{pmatrix} \mathbf{X}^T \Theta_h \tilde{\mathbf{y}} \\ \mathbf{Z}^T \Theta_h \tilde{\mathbf{y}} \end{pmatrix}, \quad (10.39)$$

where $\Theta_h = \text{diag}(\Theta_{1h}, \cdots, \Theta_{nh})$ with $\Theta_{ih} = \mathbf{K}_{ih}^{1/2} \mathbf{R}_i^{-1} \mathbf{K}_{ih}^{1/2}$. This expression is the same as the iterative solution of the mixed model equations for the GLME model in Harville (1977) and Breslow and Clayton (1993) via the Fisher scoring algorithm (Green 1987).

To solve the maximization problem in (10.32) from a computational point of view, the Fisher scoring algorithm may be used, which is equivalent to an iterated reweighted least squares method . To show this point, by using (10.39), the maximization problem of (10.32) is approximately equivalent to iteratively fitting the following standard linear mixed-effects (LME) model:

$$\mathbf{K}_{ih}^{1/2} \tilde{\mathbf{y}}_i \approx \mathbf{K}_{ih}^{1/2} \mathbf{X}_i \, (\beta + \mathbf{b}_i) + \epsilon_i, \;\; i = 1, \cdots, n, \quad (10.40)$$

where $\epsilon_i \sim (\mathbf{0}, \mathbf{R}_i)$ and $\mathbf{b}_i \sim (\mathbf{0}, \mathbf{D})$. Therefore, a closed-form solution to equations (10.35) and (10.36) is approximately given by

$$\left\{ \begin{array}{lcl} \hat{\beta} & = & \left\{ \sum_{i=1}^n \mathbf{X}_i^T \mathbf{\Omega}_{ih} \mathbf{X}_i \right\}^{-1} \sum_{i=1}^n \mathbf{X}_i^T \mathbf{\Omega}_{ih} \tilde{\mathbf{y}}_i, \\ \hat{\mathbf{b}}_i & = & \mathbf{D} \mathbf{X}_i^T \mathbf{\Omega}_{ih} \, (\tilde{\mathbf{y}}_i - \mathbf{X}_i \hat{\beta}), \;\; i = 1, 2, \cdots, n, \end{array} \right. \quad (10.41)$$

where

$$\boldsymbol{\Omega}_{ih} = \mathbf{K}_{ih}^{1/2}\mathbf{V}_i^{-1}\mathbf{K}_{ih}^{1/2}, \text{ and } \mathbf{V}_i = \mathbf{K}_{ih}^{1/2}\mathbf{X}_i\mathbf{D}\mathbf{X}_i^T\mathbf{K}_{ih}^{1/2} + \mathbf{R}_i. \tag{10.42}$$

The estimator $\hat{\boldsymbol{\beta}}$ is a local quasi-likelihood estimate in spirit and $\hat{\mathbf{b}}_i$ are empirical Bayes estimates (Diggle, Liang, and Zeger 1994).

The above results are based on the assumption of known covariance matrices $\mathbf{D}$ and $\mathbf{R}_i$, which however, might not be true in reality. Therefore, $\mathbf{D}$ and $\mathbf{R}_i$ need to be estimated. One way to tackle this difficulty is to use the local marginal quasi-likelihood method in conjunction with the restricted maximum likelihood. For the detailed description of this methodology, we refer the reader to the paper by Lin and Zhang (1999). In fact, the estimation of the nonparametric functions in the GNPME model can be easily implemented by fitting the working GLM model (10.30) with an extra kernel weight $\mathbf{K}_{ih}$. Hence, some existing software packages, such as SAS macro *GLIMMIX* (Wolfinger 1996), SAS *PROC NLMIXED*, and S-PLUS *lme* function (Pinheiro and Bates 2000), can be used with some minor modifications.

When $\hat{\boldsymbol{\beta}}$ and $\hat{\mathbf{b}}_i$ are obtained by fitting the model (10.40), we can update the following quantities: $g(\hat{\mu}_{ij}) = \mathbf{x}_{ij}^T\hat{\boldsymbol{\beta}} + \mathbf{x}_{ij}^T\hat{\mathbf{b}}_i$, $\hat{\mu}_{ij} = g^{-1}(\mathbf{x}_{ij}^T\hat{\boldsymbol{\beta}} + \mathbf{x}_{ij}^T\hat{\mathbf{b}}_i)$, and $g'(\hat{\mu}_{ij})$. Then, the working response variable can be updated as $\tilde{y}_{ij} = g(\hat{\mu}_{ij}) + (y_{ij} - \hat{\mu}_{ij})\, g'(\hat{\mu}_{ij})$. Thus, we refit the kernel-weighted LME model (10.40) using the updated values and repeat this procedure until convergence (e.g., when the estimates of $\boldsymbol{\beta}$ and $\mathbf{b}_i$ or the deviances do not change significantly). See more details on the model fitting procedure below.

Once the final estimates $\hat{\boldsymbol{\beta}}$ and $\hat{\mathbf{b}}_i$ are obtained from the iterative procedure, one can easily get the PLPQL estimators of $\eta(t)$, $v_i(t)$, $\eta_i(t) = \eta(t) + v_i(t)$, and their q-th derivatives as

$$\hat{\eta}^{(q)}(t) = q!\mathbf{e}_{q+1}^T\hat{\boldsymbol{\beta}},\ \hat{v}_i^{(q)}(t) = q!\mathbf{e}_{q+1}^T\hat{\mathbf{b}}_i,\ \hat{\eta}_i^{(q)}(t) = \hat{\eta}^{(q)}(t) + \hat{v}_i^{(q)}(t),$$

for $i = 1, 2, \cdots, n$ and $q = 0, 1, \cdots, p$, where and throughout $\mathbf{e}_{q+1}$ is a $(p+1)$-dimensional vector with the $(q+1)$-th component being 1, and 0 otherwise. In particular, $\hat{\eta}(t) = \hat{\eta}^{(0)}(t)$ and $\hat{v}_i(t) = \hat{v}_i^{(0)}(t)$ are the estimators of $\eta(t)$ and $v_i(t)$.

10.4.2 Bandwidth Selection

For the model (10.29), our goal is to estimate both the population curve $\eta(t)$ and the random-effects curves $v_i(t)$ or individual curves $\eta_i(t) = \eta(t) + v_i(t)$. It is crucial to select the bandwidth to appropriately estimate both population and random components. Different characteristics of the population curve $\eta(t)$ and the random-effects curves $v_i(t)$ warrant different bandwidths for their estimations. The bandwidth selection may be affected by several factors which include the curvature (or smoothness) of curves, the kernel function, the number of measurements, and the allocation (distribution) of the design points. It is clear from (10.41) that the data from all subjects contribute to the estimate of the population curve $\eta(t)$, but only the data from an individual subject mainly contribute to the individual curve estimation if the population estimate is given. Thus, the data for estimating the random-effects curves $v_i(t)$

are much more sparse compared to those for estimating the population curve $\eta(t)$. This indicates that a larger bandwidth is needed for the estimation of random-effects curves $v_i(t)$, compared to that for the population curve estimation.

According to the above arguments, a bandwidth selection scheme based on the cross-validated deviance (CVD) is proposed as follows. First we introduce two CVD criteria for longitudinal data, leave-one-subject-out cross-validated deviance (SCVD) and leave-one-point-out cross-validated deviance (PCVD). The SCVD-score, targeting the population curve estimation, is defined as

$$\mathrm{SCVD}_s(h) = \sum_{i=1}^{n} \sum_{j=1}^{n_i} d\left(y_{ij}, \hat{\mu}_{ij}^{(-i)}\right) / (nn_i),$$

where $\hat{\mu}_{ij}^{(-i)}$ stands for the estimator of $\hat{\mu}_{ij}$ based on the data without the i-th subject, and the weights $1/(nn_i)$ take the number of measurements from the i-th subject into account. The optimal SCVD bandwidth h_s^* is the minimizer of $\mathrm{SCVD}_s(h)$. To reduce the computational burden, an approximation for $\hat{\mu}_{ij}^{(-i)}$ may be used. For a given bandwidth h, all data can be used to estimate $\mathbf{V}_i$, $\mathbf{\Omega}_{ih}$ and $\mathbf{D}$, and then $\hat{\beta}^{(-i)}$ is approximately obtained from the closed-form solution (10.41) for the estimate of β by deleting the term that involves the i-th subject. That is,

$$\begin{aligned} \hat{\beta}^{(-i)} &\approx \left\{ \sum_{k \neq i} \mathbf{X}_k^T \hat{\mathbf{\Omega}}_{kh} \mathbf{X}_k \right\}^{-1} \sum_{k \neq i} \mathbf{X}_k^T \hat{\mathbf{\Omega}}_{kh} \tilde{\mathbf{y}}_k, \\ \hat{\mathbf{b}}_i^{(-i)} &= \hat{\mathbf{D}} \mathbf{X}_i^T \hat{\mathbf{\Omega}}_{ih} \left(\tilde{\mathbf{y}}_i - \mathbf{X}_i \hat{\beta}^{(-i)} \right). \end{aligned}$$

The PCVD is designed for the estimation of the individual curve $s_i(t) = \eta(t) + v_i(t)$. Assume $t_1, \cdots, t_M$ to be all distinct design time points for the whole data set. For a given time t_J, we assume subjects $i_1, \cdots, i_{M_J}$ to have measurements at t_J, denoted by $y_{i_L}(t_J)$. Let $\hat{\mu}_{i_L}^{(-J)}(t_J)$ be the estimator of $\hat{\mu}_{i_L,J}$ when all data at the design point t_J are excluded. Then the PCVD-score is defined as

$$\mathrm{PCVD}_p(h) = \sum_{J=1}^{M} \sum_{L=1}^{M_J} d\left\{ y_{i_L}(t_J), \hat{\mu}_{i_L,J}^{(-J)} \right\} / (MM_J),$$

where the weights $1/(MM_J)$ take the number of measurements at t_J into account. The optimal PCVD bandwidth h_p^* is the minimizer of $\mathrm{PCVD}_p(h)$.

As pointed out by Diggle and Hutchinson (1989), Zeger and Diggle (1994), Rice and Silverman (1991), and Hart and Wehrly (1993), the use of the standard leave-one-point-out cross-validation bandwidth tracks the individual curves closely while the use of leave-one-subject-out cross-validation bandwidth tracks the population mean curve better in longitudinal data analysis. Therefore, to obtain a better estimate for both components, we propose using an iterative algorithm by combining the PCVD and SCVD in our estimation in the next subsection. By the same token, to reduce the computational cost, one might use the generalized cross-validation method (Hastie and Tibshirani 1990).

10.4.3 Implementation

A simple method for estimating $\eta(t)$ and $v_i(t)$ by different bandwidths is to fit the local GLME model (10.30) twice using the procedure proposed in the previous two subsections. First, the SCVD bandwidth is used to obtain the estimate of the population curve $\eta(t)$, denoted by $\hat{\eta}_{\text{subj}}(t)$, and then refit the model using the PCVD bandwidth to obtain the estimate of the random curve $v_i(t)$, denoted by $\hat{v}_{\text{pt},i}(t)$. The estimate of the individual curve is the summation of the two components, $\hat{\eta}_{\text{mix},i}(t) = \hat{\eta}_{\text{subj}}(t) + \hat{v}_{\text{pt},i}(t)$. As is expected, the population estimate $\hat{\eta}_{\text{subj}}(t)$ is better than that using the PCVD bandwidth, and the individual curve estimate $\hat{\eta}_{\text{mix},i}(t)$ is better than that either using the SCVD bandwidth or PCVD bandwidth alone. However, both population and individual curve estimates can be further improved by a backfitting algorithm. The cost is the substantial increase in computational effort, but the rapid development of computer power may attenuate this concern.

We propose the model fitting procedure to estimate $\eta(t)$ and $v_i(t)$ iteratively as follows (McCullagh and Nelder 1989):

Step 1. Get initial values for $\tilde{\mu}_{ij}$.

Step 2. Construct the working response variables:

$$\tilde{y}_{ij} = g(\tilde{\mu}_{ij}) + (y_{ij} - \tilde{\mu}_{ij})\, g'(\tilde{\mu}_{ij}).$$

Step 3. Fit a local LME model for a given t with a SCVD bandwidth, h_s,

$$\mathbf{K}_{ih}^{1/2}\mathbf{R}_i^{-1/2}\tilde{\mathbf{y}}_i = \mathbf{K}_{ih}^{1/2}\mathbf{R}_i^{-1/2}\mathbf{X}_i^T(\boldsymbol{\beta} + \mathbf{b}_i) + \boldsymbol{\epsilon}_i,\ i = 1, \cdots, n,$$

where $\tilde{\mathbf{y}}_i = [\tilde{y}_{i1}, \cdots, \tilde{y}_{in_i}]^T$, $\mathbf{R}_i = \text{diag}(R_{i1}, \cdots, R_{in_i})$ with

$$R_{ij} = [g'(\tilde{\mu}_{ij})]^2 \phi w_{ij}^{-1} V(\tilde{\mu}_{ij}),$$

$\boldsymbol{\epsilon}_i \sim N(\mathbf{0}, \sigma^2\mathbf{I})$, and $\mathbf{b}_i \sim N(\mathbf{0}, \mathbf{D})$. Then, we obtain the estimates of $\boldsymbol{\beta}$ and $\mathbf{b}_i$, denoted by $\hat{\boldsymbol{\beta}}_{\text{subj}}$ and $\hat{\mathbf{b}}_{\text{subj},i}$, respectively. The estimates of $\eta(t)$ and $v_i(t)$ can be obtained by the first element of $\hat{\boldsymbol{\beta}}_{\text{subj}}$ and $\hat{\mathbf{b}}_{\text{subj},i}$, respectively, denoted by $\hat{\eta}_{\text{subj}}(t)$ and $\hat{v}_{\text{subj},i}(t)$.

Step 4. Repeat Step 3 using a PCVD bandwidth, h_p, and obtain the estimates of $\boldsymbol{\beta}$ and $\mathbf{b}_i$, denoted by $\hat{\boldsymbol{\beta}}_{\text{pt}}$ and $\hat{\mathbf{b}}_{\text{pt},i}$, respectively. The corresponding estimates of $\eta(t)$ and $v_i(t)$ are denoted by $\hat{\eta}_{\text{pt}}(t)$ and $\hat{v}_{\text{pt},i}(t)$, respectively.

Step 5. Repeat Steps $2 sim 4$ until convergence, and collect the final estimates, $\hat{\eta}(t) = \hat{\eta}_{\text{subj}}(t)$ and $\hat{v}_i(t) = \hat{v}_{\text{pt},i}(t)$. In fact the algorithm converges very rapidly.

10.4.4 Asymptotic Theory

To study the asymptotic properties of $\hat{\eta}(t)$, first we introduce some notations. Define, for $j \geq 1$,

$$q_j(y, s) = -\frac{1}{2}\partial^j d\left(y, g^{-1}(s)\right)/\partial s^j.$$

Let $\mu_2(K) = \int u^2 K(u)du$ and $\nu_0 = \int K^2(u)du$. Note that $q_j(y, s)$ is linear in y for any fixed s such that $q_1(y, s) = [y - g^{-1}(s)]\,T(g^{-1}(s))$ where $T(\cdot) = [V(\cdot)g'(\cdot)]^{-1}$. Let $\eta_{ij} = \eta(t_{ij}) + v_i(t_{ij})$. Then $\mu_{ij} = g^{-1}(\eta_{ij})$ and

$$q_1(\mu_{ij}, \eta_{ij}) = 0, \quad \text{and} \quad q_2(\mu_{ij}, \eta_{ij}) = -R^{-1}(\mu_{ij}), \tag{10.43}$$

where $R(\cdot) = V(\cdot)[g'(\cdot)]^2$. For expositional purpose, we only consider the penalized local linear estimate ($p = 1$) although the theory is still true for general local polynomial fitting. Some regularity conditions are imposed as follows although some of them might not be the weakest.

Assumption A

A1. The function $q_2(y, s) < 0$ for $s \in \Re$ and y in the range of the response variable.

A2. Assume that $N\,h \to \infty$ and $h\,N_2/N \to \lambda$ with $0 \leq \lambda \leq \infty$, where $N = \sum_{i=1}^n n_i$ and $N_2 = \sum_{i=1}^n n_i^2$.

A3. The kernel function $K(\cdot)$ is a symmetric density function with a bounded support.

A4. Functions $g'''(\cdot)$ and $V''(\cdot)$ are continuous.

A5. Assume that for each i, the design points $\{t_{ij}\}$ follow a density function $f_t(\cdot)$, $\{t_{ij}\}$ are independent of the stochastic processes $\{v_i(t)\}$, and t is the interior point such that $f_t(t) > 0$.

A6. Assume that $v_i(t)$ is a random-effects curve with mean zero, $\{v_i(t)\}$ are iid for fixed t, and $v_i(t)$ has the same distribution as $v(t)$.

A7. Assume that $\gamma_y(s, t) = \text{Cov}\{[y(s) - \mu(s)]\,T[\mu(s)],\, [y(t) - \mu(t)]\,T[\mu(t)]\}$ is continuous for both s and t, where $\mu(t) = g^{-1}[\eta(t) + v(t)]$ and $v(t)$ is given in A6.

A8. $E|y(t)|^4 < \infty$.

Remark 10.1 *(Conditions) Condition A1 guarantees that the PLPQL function (10.32) is concave. It is satisfied for the canonical exponential family with a canonical link. In Condition A2, n_i can be either bounded or going to infinity. The bounded case is more realistic for many practical situations. For more discussions on this aspect, see Remarks 10.3 and 10.5 below. In Condition A3, the bounded support of $K(\cdot)$ is only a technical assumption that can be usually relaxed in real applications by using kernels with small tails; for example, the standard Gaussian density. Condition A4 implies that $q_j(\cdot,\cdot)$ is continuous. Condition A5 is commonly used in longitudinal studies (Hoover et al. 1998, Fan and Zhang 2000) to simplify some technical arguments, which might be modified if necessary. Finally, the covariance function $\gamma_y(\cdot,\cdot)$ in Condition A7 becomes $\gamma_\epsilon(\cdot,\cdot)$ for the model (4.19).*

Theorem 10.1 *Under Assumptions A1 - A8, if* $0 \le \lambda < \infty$, *we have*

$$\sqrt{N\,h}\left[\hat{\eta}(t) - \eta(t) - b_\eta(t) + o_p(h^2)\right] \to N(0, \sigma^2(t)),$$

where the asymptotic bias is

$$b_\eta(t) = \frac{h^2\mu_2(K)}{2}\left[\eta''(t) + \frac{E\left\{v''(t)\,R^{-1}(\mu(t))\right\}}{E\{R^{-1}(\mu(t))\}}\right], \tag{10.44}$$

and the asymptotic variance is

$$\sigma^2(t) = f_t^{-1}(t)\left[E\left\{R^{-1}(\mu(t))\right\}\right]^{-2}\left[\nu_0\,E\left\{R^{-1}(\mu(t))\right\} + \lambda\,f_t(t)\gamma_y(t,t)\right]. \tag{10.45}$$

Proof: see Appendix.

Remark 10.2 *(Asymptotic Bias) By a comparison of the result in Theorem 10.1 with that for the ordinary regression model for independent cross-sectional data (Fan and Gijbels 1996), it is interesting to note that the asymptotic bias depends on not only the approximation error but also on the variance function since there is an extra term (the second term in the right side of (10.44)). The extra term in the asymptotic bias vanishes when* $R(\cdot)$ *is constant, which is particularly true for the Gaussian case. It is also noted that the asymptotic bias is not affected by the within-subject correlation.*

Remark 10.3 *(Asymptotic Variance) Another implication of Theorem 10.1 is that the main influence of the within-subject correlation on the asymptotic distribution of* $\hat{\eta}(t)$ *appears at the second term on the right side of (10.45). When* $\lambda = 0$, *(10.45) is similar to the asymptotic variance of kernel regression estimators with independent cross-sectional data (Fan and Gijbels 1996). In the interesting case that all* n_i *are bounded, which is a case of practical interest, the probability that there are at least two data points from the same subject in a shrinking neighborhood tends to be zero, therefore, the effect of within-subject correlation on local quasi-likelihood fitting may not be significant when the ratio* n_i/n *is small.*

Remark 10.4 *(Asymptotic Mean Squared Error) One more direct implication of Theorem 10.1 is that the asymptotic mean squared error (AMSE) is*

$$\begin{aligned}\text{AMSE} &= bias^2 + variance \\ &= \tfrac{h^4}{4}\mu_2^2(K)\left[\eta''(t) + \tfrac{E\{v''(t)\,R^{-1}(\mu(t))\}}{E\{R^{-1}(\mu(t))\}}\right]^2 + \tfrac{\sigma^2(t)}{N\,h},\end{aligned} \tag{10.46}$$

which, by minimizing the AMSE, gives the optimal bandwidth h_{opt}:

$$h_{opt} = \left[\frac{\sigma^2(t)}{N\mu_2^2(K)}\right]^{1/5}\left[\eta''(t) + \frac{E\left\{v''(t)\,R^{-1}(\mu(t))\right\}}{E\left\{R^{-1}(\mu(t))\right\}}\right]^{-2/5},$$

so that the optimal bandwidth is of the order $N^{-1/5}$.

Remark 10.5 *(Sample Sizes) Finally, the asymptotic properties of* $\hat{\eta}(\cdot)$ *depend on the numbers of repeated measurements* $n_1, \cdots, n_n$*, and the sizes of* n*,* N*, and* N_2*. Further, as discussed in Remark 10.1, it is not required for* n_i *to go to infinity, however, it can go to infinity at a certain rate such as* $n_i = O(n^\tau)$ *with* $\tau \le 1/4$*. In such a case, the order of the optimal bandwidth is* $n^{-(1+\tau)/5}$*, which is smaller than* $n^{-1/5}$ *for independent cross-sectional data with* n *data points. If* n_i *goes to infinity too fast so that* $\lambda = \infty$*, then it can be shown with a slight modification of the method developed here that the convergence rate of* $\hat{\eta}(t)$ *is* $N_2^{1/2}/N$ *(see Theorem 10.2 below), which is slower than* $(N\,h)^{-1/2}$ *(see Theorem 10.1). In such a case, the asymptotic distribution of* $\hat{\eta}(\cdot)$ *can be derived in the same way. Now the asymptotic result is stated in the following theorem.*

Theorem 10.2 *Under Assumptions A1 - A8, if* $h\,N_2/N \to \infty$*,* $N/\sqrt{N_2} \to \infty$*, and* $\gamma_y(t,\,t) > 0$*, we have*

$$\frac{N}{\sqrt{N_2}}\left[\hat{\eta}(t) - \eta(t) - b_\eta(t) + o_p(h^2)\right] \to N(0,\, \sigma_\infty^2(t)),$$

where the asymptotic bias $b_\eta(t)$ *is given in (10.44) and the asymptotic variance is*

$$\sigma_\infty^2(t) = \gamma_y(t,\,t)\ \left[E\left\{R^{-1}(\mu(t))\right\}\right]^{-2}.$$

Proof: see Appendix.

Remark 10.6 *(Singularity) It is interesting and somewhat surprising to see from Theorem 10.2 that the asymptotic bias is the same as that in Theorem 10.1 but* $\sigma_\infty^2(t) < \sigma^2(t)$*. For the particular case that* $n_i = O(n^\tau)$ *for* $\tau > 1/4$*, the convergence rate of the estimator* $\hat{\eta}(t)$ *becomes* $n^{-1/2}$*. Note that* n *is the number of subjects and the rate* $n^{-1/2}$ *is slower than the standard nonparametric rate* $(N\,h)^{-1/2}$ *in this case, which indicates that the within-subject correlation slows down the convergence of* $\hat{\eta}(t)$*. By combining with Remark 10.5, it is easy to see that when* $n_i = O(n^\tau)$ *with* $\tau > 0$*, the bandwidth can be taken to be* $h = O(n^{-0.2\,(1+\tau\,I(\tau\le 1/4)+1/4\,I(\tau>1/4))})$ *and the convergence rate of* $\hat{\eta}(\cdot)$ *is* $n^{-0.4\,(1+\tau\,I(\tau\le 1/4)+1/4\,I(\tau>1/4))}$*. This implies that the order of the bandwidth taken in practice should be in between* $n^{-1/4}$ *and* $n^{-1/5}$*. It seems that these findings are novel and interesting.*

The asymptotic distribution of Theorem 10.1 provides a useful tool for making inferences for $\eta(t)$. In particular, an approximate $100(1-\alpha)\%$ pointwise confidence interval for $\eta(t)$ can be given by

$$\hat{\eta}(t) - \hat{b}_\eta(t) \pm z_{1-\alpha/2}\hat{\sigma}(t)/\sqrt{N\,h}, \tag{10.47}$$

where $0 < \alpha < 1$ and z_a is the α-th quantile of the standard normal distribution. In principle, it is possible to construct the consistent estimators $\hat{b}_\eta(t)$ and $\hat{\sigma}^2(t)$, and there are several methods available in the literature such as the bootstrap and plug-in method, see, for example, Hoover et al. (1998), Wu, Chiang and Hoover (1998), and Chiang, Rice and Wu (2001). However, these methods rely on cumbersome computations. In practice, a simple and quick way to estimate the asymptotic bias given in (10.44) and the variance given in (10.45) is desirable. Now, we propose the estimators of $b_\eta(t)$ and $\sigma^2(t)$ that can be easily obtained, as follows. First, we define

$$\hat{b}_\eta(t) = \sqrt{\frac{h}{N}} \sum_{i=1}^{n} \sum_{j=1}^{n_i} \left[g^{-1}(\hat{\eta}_{ij}) - g^{-1}(\hat{\tilde{\eta}}_{ij}) \right] T(g^{-1}(\hat{\tilde{\eta}}_{ij})) \, K_h(t_{ij} - t),$$

where $\hat{\eta}_{ij} = \hat{\eta}(t_{ij}) + \hat{v}_i(t_{ij})$, $\hat{\tilde{\eta}}_{ij} = \mathbf{x}_{ij}^T \hat{\boldsymbol{\beta}} + \mathbf{x}_{ij}^T \hat{\mathbf{b}}_i$. Also $\hat{\eta}(t_{ij})$, $\hat{v}_i(t_{ij})$, $\hat{\boldsymbol{\beta}}$, and $\hat{\mathbf{b}}_i$ can be obtained using (10.41). Then, by (10.75) and (10.76) in the Appendix in Section 10.8, $\hat{b}_\eta(t)$ can be used as a consistent estimator of the asymptotic bias $b_\eta(t)$ since $\hat{\eta}_{ij}$ is a consistent estimator of η_{ij} and $\hat{\tilde{\eta}}_{ij}$ is a consistent estimator of $\tilde{\eta}_{ij}$. To circumvent the difficulty of estimating $R(\mu(t))$ and $\gamma_y(t,\, t)$ in the expression of $\sigma^2(t)$, now, we propose an alternative way to estimate $\sigma^2(t)$. We define

$$\begin{aligned} \delta_{n1} &= \frac{1}{N} \sum_{i=1}^{n} \sum_{j=1}^{n_i} q_2(y_{ij}, \hat{\eta}_{ij}) \, K_h(t_{ij} - t), \quad \text{and} \\ \delta_{n2} &= \frac{h}{N} \sum_{i=1}^{n} \left\{ \sum_{j=1}^{n_i} q_1(y_{ij}, \hat{\eta}_{ij}) \, K_h(t_{ij} - t) \right\}^2 . \end{aligned}$$

By (10.79) in the Appendix, the "sandwich" method can be used to construct the estimate of $\sigma^2(t)$ as

$$\hat{\sigma}^2(t) = \delta_{n2} / \delta_{n1}^2 .$$

From the proof of Theorem 10.1 [see (10.72) and (10.78) in the Appendix], it can be seen (although some additional conditions might be needed) that in probability, as $n \to \infty$,

$$\delta_{n1} \to -f_t(t) \, E\left\{ R^{-1}(\mu(t)) \right\}, \quad \text{and} \quad \delta_{n2} \to \sigma_1^2(t),$$

where $\sigma_1^2(t)$ is defined in (10.78) in the Appendix. Therefore, by (10.79) in the Appendix, $\hat{\sigma}^2(t)$ is a consistent estimator of $\sigma^2(t)$. Note that the detailed proof of the consistency of $\hat{b}_\eta(t)$ and $\hat{\sigma}^2(t)$ is omitted since the proof is not very difficult. However, a practical procedure based on (10.47) requires further substantial development and the asymptotic confidence regions for the GNPME model (10.32) remain an open problem.

10.4.5 Application to an AIDS Clinical Study

We applied the proposed models and methods to an AIDS clinical study of HIV-infected patients receiving highly active antiretroviral therapy (HAART). See more discussions on this study (ACTG 388) and the CD4 percentage data in Section 1.1.3. Figure 10.1 shows the individual trajectories of plasma HIV RNA concentrations (viral load) after initiation of the antiviral treatments from one study arm. One objective of the antiviral treatment is to suppress the viral load below the limit of detection (50 copies per ml plasma). In this application, we are interested in modeling the dynamics of the binary response, with/without viral suppression, during nearly 2 years of treatment. Let $y_{ij} = 1$ if the viral load of the i-th subject is below the limit of detection (success) at treatment time t_{ij}; and $y_{ij} = 0$ otherwise. We applied the proposed GNPME model (10.29) to the data with a logistic link function for the longitudinal binary data and employed the above proposed methods to fit the model. Figure 10.2 shows the estimate of the population curve $\eta(t)$ with its 95% pointwise confidence band which indicates an increase in the likelihood of a positive outcome over the treatment time. This suggests that the antiviral treatment is successful overall in this patient population since the likelihood of viral suppression increases with time. However, when we look at individual responses, a different trend is observed for some patients. Figure 10.3 shows the individual curve estimates $\hat{\eta}_i(t) = \hat{\eta}(t) + \hat{v}_i(t)$ from four selected subjects (dotted lines). We can see that the likelihood of treatment success in some patients follows the pattern of the population curve, but not all. For example, Subject 81 was successfully treated initially (the likelihood of viral suppression increased), but later the treatment failed after week 57 (Figure 10.3).

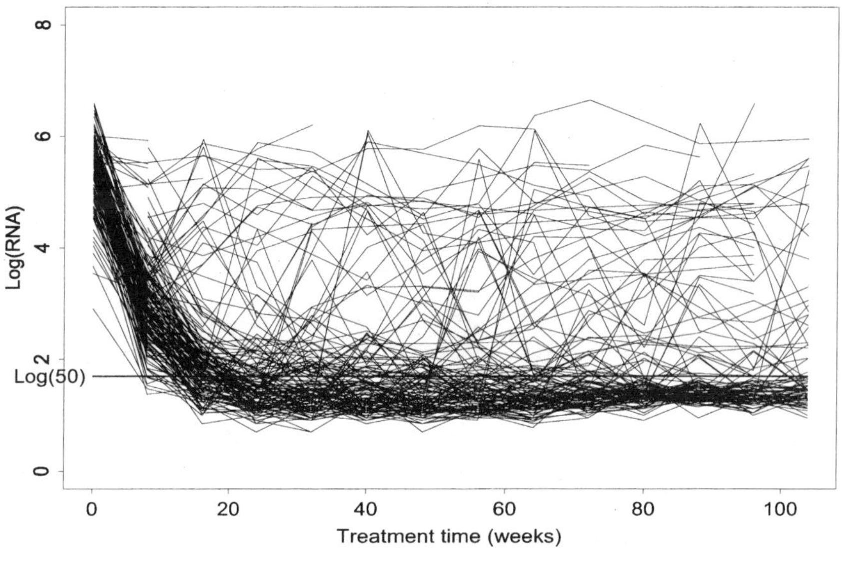

Fig. 10.1 Individual trajectories of plasma HIV RNA concentrations after initiation of antiviral treatment with a detection limit of 50 copies per mL plasma from patients in one study arm.

In summary, the proposed models and methods can be used to estimate both the population and the individual response curves. In fact, not all the individual curves follow the pattern of the population curve. Thus, it is important to estimate the population curve as well as individual profiles in biomedical applications since the individual estimates can be used by clinicians to individualize patient treatment and care.

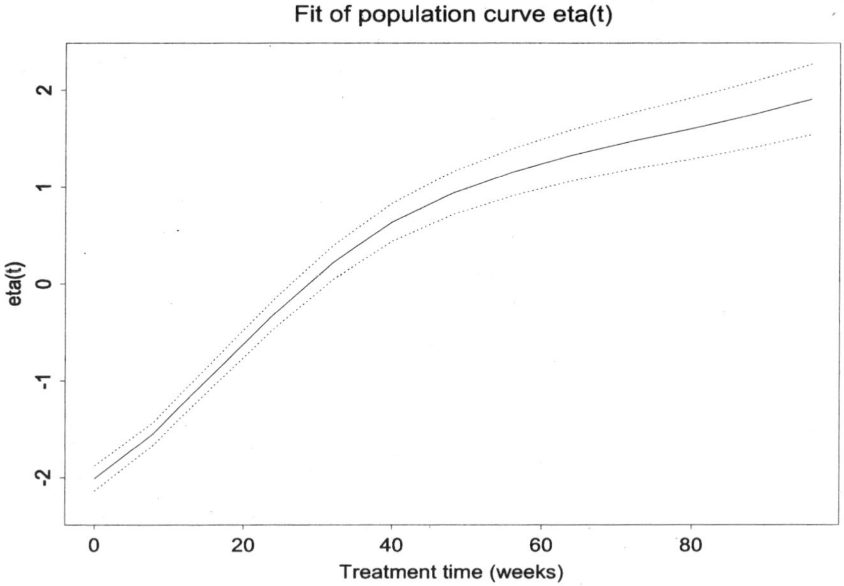

Fig. 10.2 Population curve fit $\hat{\eta}(t)$ with 95% pointwise confidence band.

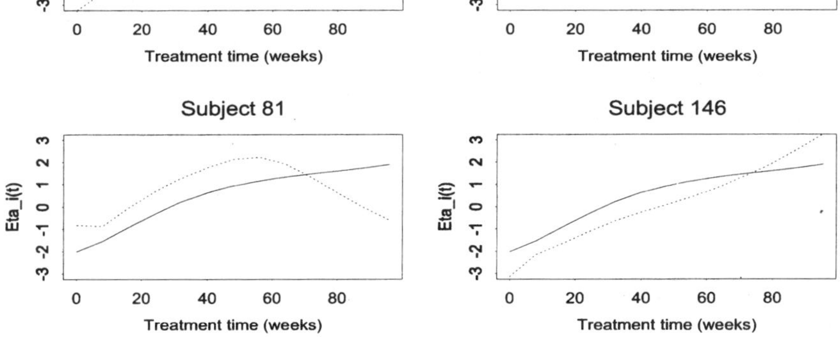

Fig. 10.3 Individual curve fits (dotted lines), $\hat{\eta}_i(t)$ from four selected subjects. Population curve fit (solid line) was also plotted for comparisons.

10.5 GENERALIZED TVC-NPME MODEL

In this section, we extend the GNPME model (10.29) to include covariates with time-varying coefficients. Suppose now we have the following discrete longitudinal data set:

$$(y_{ij}, \mathbf{c}_{ij}^T, t_{ij}),\ j = 1, 2, \cdots, n_i;\ i = 1, 2, \cdots, n,$$

where y_{ij} denotes the j-th response for the i-th subject (or cluster) at time point t_{ij} and with the observed d-dimensional covariate vector $\mathbf{c}_{ij}$. Conditional to the given subject i, the subject-specific marginal mean and variance of the responses y_{ij} are now specified as

$$\begin{array}{c} \mathrm{E}(y_{ij}|\mathbf{c}_{ij}, t_{ij}) = \mu_{ij},\ \mathrm{Var}(y_{ij}|\mathbf{c}_{ij}, t_{ij}) = \phi w_{ij}^{-1} V(\mu_{ij}), \\ j = 1, 2, \cdots, n_i;\ i = 1, 2, \cdots, n, \end{array} \tag{10.48}$$

where again ϕ is a scale parameter, w_{ij} are the weights, and $V(\cdot)$ is a known variance function. The subject-specific marginal mean μ_{ij} is assumed to be related to the covariate vectors $\mathbf{c}_{ij}$ and the time points t_{ij} through a known differentiable and monotonic link function $g(\cdot)$:

$$g(\mu_{ij}) = \mathbf{c}_{ij}^T \boldsymbol{\alpha}(t_{ij}) + v_i(t_{ij}),\ \ j = 1, 2, \cdots, n_i;\ i = 1, 2, \cdots, n, \tag{10.49}$$

where $\boldsymbol{\alpha}(t)$ is an unknown d-dimensional coefficient function vector and $v_i(t)$ is the i-th subject-specific generalized deviation function. We call the model (10.49) a generalized time-varying coefficient nonparametric mixed-effects (GTVC-NPME) model. The above GTVC-NPME model can be considered as an extension of the model proposed by Lin and Carroll (2001a) in the sense that we consider the time-varying coefficient as well as nonparametric random-effects.

Now we apply the penalized local quasi-likelihood technique previously described to estimate the time-varying coefficient function vector $\boldsymbol{\alpha}(t)$ and random curves $v_i(t)$ in the model (10.49). To this end, by assuming that $\boldsymbol{\alpha}(t)$ and $v_i(\cdot)$ have up to $(p+1)$-times derivatives, we apply a Taylor expansion to $\boldsymbol{\alpha}(t)$ and $v_i(t)$ to obtain

$$g(\mu_{ij}) \approx \mathbf{c}_{ij}^T(\mathbf{x}_{ij}^T \otimes \mathbf{I}_d)\boldsymbol{\beta} + \mathbf{x}_{ij}^T \mathbf{b}_i = \tilde{\mathbf{x}}_{ij}^T \boldsymbol{\beta} + \mathbf{x}_{ij}^T \mathbf{b}_i,$$

where

$$\boldsymbol{\beta} = \begin{pmatrix} \boldsymbol{\alpha}(t) \\ \boldsymbol{\alpha}'(t) \\ \vdots \\ \boldsymbol{\alpha}^{(p)}(t)/p! \end{pmatrix},\quad \tilde{\mathbf{x}}_{ij} = (\mathbf{x}_{ij} \otimes \mathbf{I}_d)\mathbf{c}_{ij},$$

$\mathbf{I}_d$ is the $d \times d$ identity matrix, and in what follows, for two matrixes $\mathbf{A}$ and $\mathbf{B}$, $\mathbf{A} \otimes \mathbf{B}$ denotes the Kronecker product $(a_{ij}\mathbf{B})$. The PLPQL for the model (10.49) is

$$-\frac{1}{2}\sum_{i=1}^{n}\left\{\sum_{j=1}^{n_i} K_h(t_{ij} - t)\phi^{-1} w_{ij} d(y_{ij}, \tilde{\mu}_{ij}) + \mathbf{b}_i^T \mathbf{D}^{-1} \mathbf{b}_i\right\}, \tag{10.50}$$

where $\tilde{\mu}_{ij} = g^{-1}(\tilde{\eta}_{ij})$ and $\tilde{\eta}_{ij} = \tilde{\mathbf{x}}_{ij}^T\boldsymbol{\beta}+\mathbf{x}_{ij}^T\mathbf{b}_i$. Differentiating (10.50) with respect to $\boldsymbol{\beta}$ and $\mathbf{b}_i$ leads to the equations similar to (10.35)-(10.39) and solving these equations yields the p-th degree PLPQL estimate $\hat{\boldsymbol{\beta}}$ and $\hat{\mathbf{b}}_i$ for the model (10.49). The first d components of $\hat{\boldsymbol{\beta}}$ give the PLPQL estimate of the coefficient functions $\boldsymbol{\alpha}(t)$, denoted by $\hat{\boldsymbol{\alpha}}(t)$. Similar to Section 10.4, we can obtain the estimate of the random-effects $\mathbf{b}_i$ and $v_i(t),\ \ i = 1, 2, \cdots, n$. Also, the bandwidth selection procedures and the model-fitting algorithm previously described can be applied here with a slight modification.

Similarly, we can obtain the asymptotic normality of $\hat{\boldsymbol{\alpha}}(t)$ in the model (10.49). To this end, more regularity conditions in addition to Conditions A1-A8 are needed and are listed as follows although some of them might not be the weakest. For simplicity, we only consider the local linear case ($p = 1$), which is recommended by Fan and Gijbels (1996) in practice. A special case of the model (10.49) that is of particular interest is the model (10.49) with time-independent covariates, $\mathbf{c}_{ij} \equiv \mathbf{c}_i$. In practice, time-independent covariates, such as treatment, dosage, and gender, are very common in longitudinal studies. Throughout this section, we assume that the covariates are time-independent but their coefficients are time-dependent. Note that this assumption is also imposed by Fan and Zhang (2000) and Chiang, Rice and Wu (2001) to simplify the technical proofs.

Assumption B

B1. The series $\{\mathbf{c}_i\}$ are iid with the same distribution as $\mathbf{c}$ and they are independent of both $\{t_{ij}\}$ and $\{v_i(\cdot)\}$.

B2. $E|\mathbf{c}|^4 < \infty$.

B3. $\Delta_1(t) = E\left\{R^{-1}(\mu(t,\mathbf{c}))\mathbf{c}\mathbf{c}^T\right\}$ is a positive definite and continuous function, where $\mu(t,\mathbf{c}) = g^{-1}[\mathbf{c}^T\boldsymbol{\alpha}(t) + v(t)]$.

Now, we state the asymptotic normality for $\hat{\boldsymbol{\alpha}}(t)$ in the following theorems.

Theorem 10.3 *Under Assumptions A1 - A8 and B1 - B3, if* $\lambda < \infty$, *we have*

$$\sqrt{N\,h}\left[\hat{\boldsymbol{\alpha}}(t) - \boldsymbol{\alpha}(t) - \mathbf{B}_{\boldsymbol{\alpha}}(t) + o_p(h^2)\right] \ \rightarrow\ N(\mathbf{0}, \boldsymbol{\Sigma}(t)),$$

where the asymptotic bias $\mathbf{B}_{\boldsymbol{\alpha}}(t)$ *is given by*

$$\mathbf{B}_{\boldsymbol{\alpha}}(t) = \frac{h^2\mu_2(K)}{2}\left[\boldsymbol{\alpha}''(t) + \Delta_1^{-1}(t)\,E\left\{v''(t)\,R^{-1}(\mu(t,\mathbf{c}))\mathbf{c}\right\}\right], \tag{10.51}$$

and the asymptotic variance $\boldsymbol{\Sigma}(t)$ *is given by*

$$\boldsymbol{\Sigma}(t) = f_t(t)^{-1}\,\Delta_1^{-1}(t)\left[\nu_0\Delta_1(t) + \lambda\,f_t(t)\,\gamma_y(t,\,t)\,E\left\{\mathbf{c}\mathbf{c}^T\right\}\right]\Delta_1^{-1}(t). \tag{10.52}$$

Proof: see Appendix.

Theorem 10.4 *Under Assumptions A1 - A8 and B1 - B3, if* $h\,N_2/N \rightarrow \infty$, $N/\sqrt{N_2} \rightarrow \infty$, *and* $\gamma_y(t,\,t) > 0$, *we have*

$$\frac{N}{\sqrt{N_2}}\left[\hat{\boldsymbol{\alpha}}(t) - \boldsymbol{\alpha}(t) - \mathbf{B}_{\boldsymbol{\alpha}}(t) + o_p(h^2)\right] \ \rightarrow\ N(\mathbf{0}, \boldsymbol{\Sigma}_\infty(t)),$$

where the asymptotic variance $\mathbf{\Sigma}_\infty(t)$ is given by

$$\mathbf{\Sigma}_\infty(t) = \gamma_y(t,\, t)\Delta_1^{-1}(t)\, E\left\{\mathbf{c}\mathbf{c}^T\right\}\, \Delta_1^{-1}(t).$$

Proof: see Appendix.

Remark 10.7 *(Comparisons) By comparing the results in Theorem 10.3 with those in Cai, Fan and Li (2000) for independent cross-sectional data, it is interesting to note that there is an extra term in both asymptotic bias and variance (the second term in (10.51) and (10.52)) due to the random-effects and repeated measurements. By comparing the results in Theorem 10.3 with those in Wu, Chiang and Hoover (1998) for the Gaussian case, it is not surprising to see that the asymptotic variance is the same but there is an extra term (the second term in the right side of (10.51)) in the asymptotic bias due to the structure of variance function.*

Remark 10.8 *Remarks 10.2-10.6 on Theorems 10.1 and 10.2 are still applicable to the GTVC-NPME models with covariates. The ad hoc method proposed to estimate the asymptotic bias and variance in the previous section is also applicable.*

10.6 GENERALIZED SAME MODEL

Lin and Zhang (1999) proposed a generalized semiparametric additive mixed-effects (GSAME) model for discrete longitudinal data analysis. They employed a smoothing spline approach to make inferences for the proposed model. We review the main results of Lin and Zhang (1999) in this section. Karcher and Wang (2001) extended Lin and Zhang (1999)'s model into a general generalized semiparametric mixed-effects model.

Suppose now we have the following discrete longitudinal data set:

$$(y_{ij}, \mathbf{x}_{ij}^T, \mathbf{z}_{ij}^T, t_{ij}),\ j = 1, 2, \cdots, n_i;\ i = 1, 2, \cdots, n, \tag{10.53}$$

where y_{ij} denotes the j-th response for the i-th subject (or cluster) at time point t_{ij} and with the observed $(p+1)$-dimensional fixed-effects covariate vector $\mathbf{x}_{ij} = [1, x_{1ij}, \cdots, x_{pij}]^T$ and the observed q-dimensional random-effects covariate vector $\mathbf{z}_{ij} = [z_{1ij}, \cdots, z_{qij}]^T$. Conditional to the given subject i, the subject-specific marginal mean and variance of the responses y_{ij} are now specified as

$$\begin{gathered}\mathrm{E}(y_{ij}|\mathbf{x}_{ij}, \mathbf{z}_{ij}, t_{ij}) = \mu_{ij},\ \mathrm{Var}(y_{ij}|\mathbf{x}_{ij}, \mathbf{z}_{ij}, t_{ij}) = \phi w_{ij}^{-1} V(\mu_{ij}),\\ j = 1, 2, \cdots, n_i;\ i = 1, 2, \cdots, n,\end{gathered}$$

where again ϕ is a scale parameter, w_{ij} are the weights, and $V(\cdot)$ is a known variance function. The subject-specific marginal mean μ_{ij} is assumed to be related to the covariate vectors $\mathbf{x}_{ij}, \mathbf{z}_{ij}$ and the time points t_{ij} through a known differentiable and monotonic link function $g(\cdot)$:

$$\begin{aligned} g(\mu_{ij}) &= \beta_0 + f_1(x_{1ij}) + \cdots + f_p(x_{pij}) + \mathbf{z}_{ij}^T\mathbf{b}_i,\\ \mathbf{b}_i &\sim N(0, \mathbf{D}),\ \ j = 1, 2, \cdots, n_i;\ i = 1, 2, \cdots, n,\end{aligned} \tag{10.54}$$

where $f_r(\cdot), r = 1, \cdots, p$ are unknown centered twice-differentiable smooth functions, and $\mathbf{D}$ is a $q \times q$ covariance matrix, depending on a parameter vector $\boldsymbol{\theta}$.

The GSAME model (10.54) is a hybrid extension of generalized linear mixed-effects (GLME) models (Breslow and Clayton 1993) and generalized additive models (Hastie and Tibshirani 1990). If $f_r(\cdot), r = 1, \cdots, p$ all are linear functions, the GSAME model (10.54) reduces to the GLME model. The semiparametric mixed-effects models presented in Chapter 8 are special cases of the GSAME models. The GSAME models use additive nonparametric functions to model covariate effects while accounting for over-dispersion and data correlation by adding random-effects to the additive predictor.

The integrated log-quasi-likelihood of $\{\beta_0, f_1, \cdots, f_p, \boldsymbol{\theta}\}$ for the GSAME model (10.54) can be written as

$$\begin{array}{c} \exp[l\{\beta_0, f_1, \cdots, f_p, \boldsymbol{\theta}|\mathbf{y}\}] \propto |\mathbf{D}|^{-n/2} \\ \times \int \exp\left\{-\frac{1}{2}\sum_{i=1}^{n}[\sum_{j=1}^{n_i} \phi^{-1} w_{ij} d(y_{ij}; \mu_{ij}) + \mathbf{b}_i^T \mathbf{D}^{-1}\mathbf{b}_i]\right\} d\mathbf{b}, \end{array} \tag{10.55}$$

where $\mathbf{y} = [\mathbf{y}_1^T, \cdots, \mathbf{y}_n^T]^T$ with $\mathbf{y}_i = [y_{i1}, \cdots, y_{in_i}]^T$, and

$$d(y, \mu) = -2\int_y^{\mu} (y - u)/V(u) du,$$

defines the deviance function.

Lin and Zhang (1999) proposed using natural cubic smoothing splines (NCSS) method to estimate the nonparametric functions $f_r(\cdot)$. That is, for given smoothing parameters $\lambda_r, r = 1, \cdots, p$ and covariance parameters $\boldsymbol{\theta}$, the NCSS estimators of the $f_r(\cdot)$ maximize the penalized log-quasi-likelihood

$$\begin{array}{rl} & l\{\beta_0, f_1, \cdots, f_p, \boldsymbol{\theta}|\mathbf{y}\} - \frac{1}{2}\sum_{r=1}^{p} \lambda_r \int_{a_r}^{b_r} f_r''(t)^2 dt \\ = & l(\beta_0, f_1, \cdots, f_p, \boldsymbol{\theta}|\mathbf{y}) - \frac{1}{2}\sum_{r=1}^{p} \lambda_r \mathbf{f}_r^T \mathbf{G}_r \mathbf{f}_r, \end{array} \tag{10.56}$$

where (a_r, b_r) defines the range of the r-th fixed-effects covariate x_{rij} and the smoothing parameters $\lambda_1, \cdots, \lambda_p$ control the trade-off between the goodness of fit and the smoothness of the estimated functions. For $r = 1, \cdots, p$, let

$$\tau_{r1}, \tau_{r2}, \cdots, \tau_{rM_r}, \tag{10.57}$$

be all the ordered distinct values (in an increasing order) among the r-th observed fixed-effects covariate values x_{rij}, $j = 1, 2, \cdots, n_i$; $i = 1, 2, \cdots, n$. Then $\mathbf{f}_r$ is a M_r-dimensional unknown vector of the values of $f_r(\cdot)$ evaluated at (10.57), $\mathbf{G}_r$ is the corresponding cubic smoothing spline roughness matrix with knots at (10.57). See Section 3.4.1 or Green and Silverman (1994) for a detailed scheme for computing $\mathbf{G}_r$.

In vector notation, with $\boldsymbol{\mu}_i = [\mu_{i1}, \cdots, \mu_{in_i}]^T$, $g(\boldsymbol{\mu}_i) = [g(\mu_{i1}), \cdots, g(\mu_{in_i})]^T$ and $\mathbf{Z}_i = [\mathbf{z}_{i1}, \cdots, \mathbf{z}_{in_i}]^T$, the GSAME (10.54) can be written as

$$g(\boldsymbol{\mu}_i) = \mathbf{1}_{n_i}\beta_0 + \mathbf{N}_{i1}\mathbf{f}_1 + \cdots + \mathbf{N}_{ip}\mathbf{f}_p + \mathbf{Z}_i\mathbf{b}_i, , \; i = 1, 2, \cdots, n, \tag{10.58}$$

where $\mathbf{1}_{n_i}$ is an n_i-dimensional vector of 1s and $\mathbf{N}_{ir}$ is an $n_i \times M_r$ incidence matrix with entries (j,l) being 1 if $x_{rij} = \tau_{rl}$ and 0 otherwise. Then, the j-th component of $\mathbf{N}_{ir}\mathbf{f}_r$ is $f_r(x_{rij})$.

Since maximization of expression (10.56) is often intractable, the Laplace approximation was proposed to tackle this difficulty (Breslow and Clayton 1993). It can be shown that the approximate NCSS estimators $(\hat{\beta}_0, \hat{\mathbf{f}}_1, \cdots, \hat{\mathbf{f}}_p)$ can be obtained by maximizing the double penalized quasi-likelihood (DPQL) with respect to $(\beta_0, \mathbf{f}_1, \cdots, \mathbf{f}_p)$ and $\mathbf{b}$:

$$-\frac{1}{2}\sum_{i=1}^{n}\Big\{\sum_{j=1}^{n_i}\phi^{-1}w_{ij}d(y_{ij},\mu_{ij}) + \mathbf{b}_i^T\mathbf{D}^{-1}\mathbf{b}_i\Big\} - \frac{1}{2}\sum_{r=1}^{p}\lambda_r\mathbf{f}_r^T\mathbf{G}_r\mathbf{f}_r. \tag{10.59}$$

The penalty term $\frac{1}{2}\sum_{i=1}^{n}\mathbf{b}_i^T\mathbf{D}^{-1}\mathbf{b}_i$ is due to the random-effects and the Laplace approximation, whereas the penalty term $\frac{1}{2}\sum_{r=1}^{p}\lambda_r\mathbf{f}_r^T\mathbf{G}_r\mathbf{f}_r$ is due to the roughness penalty of the NCSS estimators. To make $f_r(t)$ identifiable, $\mathbf{f}_r$ needs to satisfy $\mathbf{f}_r^T\mathbf{1} = 0$ so that $f_r(t)$ is approximately centered.

Differentiating expression (10.59) with respect to $(\beta_0, \mathbf{f}_1, \cdots, \mathbf{f}_p)$ and $\mathbf{b}$ yields the estimating equations:

$$\begin{aligned}
\mathbf{1}_N^T\mathbf{W}\boldsymbol{\Delta}(\mathbf{y}-\boldsymbol{\mu}) &= 0,\\
\mathbf{N}_r^T\mathbf{W}\Delta(\mathbf{y}-\boldsymbol{\mu}) - \lambda_r\mathbf{G}_r\mathbf{f}_r &= 0,\ r = 1,\cdots,p,\\
\mathbf{Z}^T\mathbf{W}\Delta(\mathbf{y}-\boldsymbol{\mu}) - \tilde{\mathbf{D}}^{-1}\mathbf{b} &= 0,\ \ i = 1,2,\cdots,n,
\end{aligned} \tag{10.60}$$

where $N = \sum_{i=1}^{n} n_i$, and

$$\begin{aligned}
\mathbf{y} &= [\mathbf{y}_1^T,\cdots,\mathbf{y}_n^T]^T, & \boldsymbol{\mu} &= [\boldsymbol{\mu}_1^T,\cdots,\boldsymbol{\mu}_n^T]^T,\\
\mathbf{W} &= \text{diag}(\mathbf{W}_1,\cdots,\mathbf{W}_n), & \boldsymbol{\Delta} &= \text{diag}(\boldsymbol{\Delta}_1,\cdots,\boldsymbol{\Delta}_n),\\
\mathbf{N}_r &= [\mathbf{N}_{1r}^T,\cdots,\mathbf{N}_{nr}^T]^T, & \mathbf{Z} &= \text{diag}(\mathbf{Z}_1,\cdots,\mathbf{Z}_n),\\
\tilde{\mathbf{D}} &= \text{diag}(\mathbf{D},\cdots,\mathbf{D}), & \mathbf{b} &= [\mathbf{b}_1^T,\cdots,\mathbf{b}_n^T]^T,
\end{aligned}$$

with $\Delta_i = \text{diag}\{g'(\boldsymbol{\mu}_i)\}^{-1}$, and $\mathbf{W}_i = \text{diag}\{[\phi w_{ij}^{-1}V(\mu_{ij})g'(\mu_{ij})]^{-1}\}$ for $i = 1,2,\cdots,n$.

The above equations can be solved by using the Fisher scoring algorithm, i.e., to solve the following equation system:

$$\begin{pmatrix}
\mathbf{1}_N & \mathbf{S}_0\mathbf{N}_1 & \cdots & \mathbf{S}_0\mathbf{N}_p & \mathbf{S}_0\mathbf{Z}\\
\mathbf{S}_1\mathbf{1}_N & \mathbf{I}_N & \cdots & \mathbf{S}_1\mathbf{N}_p & \mathbf{S}_1\mathbf{Z}\\
\vdots & \vdots & \vdots & \vdots & \vdots\\
\mathbf{S}_p\mathbf{1}_N & \mathbf{S}_p\mathbf{N}_1 & \cdots & \mathbf{I}_N & \mathbf{S}_p\mathbf{Z}\\
\mathbf{S}_b\mathbf{1}_N & \mathbf{S}_b\mathbf{N}_1 & \cdots & \mathbf{S}_b\mathbf{N}_p & \mathbf{I}_N
\end{pmatrix}
\begin{pmatrix}\beta_0\\ \mathbf{f}_1\\ \vdots\\ \mathbf{f}_p\\ \mathbf{b}\end{pmatrix}
=
\begin{pmatrix}\mathbf{S}_0\tilde{\mathbf{y}}\\ \mathbf{S}_1\tilde{\mathbf{y}}\\ \vdots\\ \mathbf{S}_p\tilde{\mathbf{y}}\\ \mathbf{S}_b\tilde{\mathbf{y}}\end{pmatrix}, \tag{10.61}$$

where

$$\tilde{\mathbf{y}} = \beta_0\mathbf{1}_N + \sum_{r=1}^{p}\mathbf{N}_r\mathbf{f}_r + \mathbf{Z}\mathbf{b} + \Delta(\mathbf{y}-\boldsymbol{\mu}) \tag{10.62}$$

is a modified GSAME working vector, $\mathbf{S}_r$ is a centered smoother for $\mathbf{f}_r$ and satisfied $\mathbf{S}_r^T\mathbf{1}_N = 0$, and $\mathbf{S}_r$ and $\mathbf{S}_b$ are defined as

$$
\begin{aligned}
\mathbf{S}_0 &= (\mathbf{1}_N^T\mathbf{W}\mathbf{1}_N)^{-1}\mathbf{1}_N^T\mathbf{W}, \\
\mathbf{S}_r &= \left(\mathbf{I}_N - \frac{\mathbf{1}_N\mathbf{1}_N^T}{M_r}\right)(\mathbf{N}_r^T\mathbf{W}\mathbf{N}_r + \lambda_r\mathbf{G}_r)^{-1}\mathbf{N}_r^T\mathbf{W}, \quad r = 1, \cdots, p, \\
\mathbf{S}_b &= (\mathbf{Z}^T\mathbf{W}\mathbf{Z} + \tilde{\mathbf{D}}^{-1})^{-1}\mathbf{Z}^T\mathbf{W}.
\end{aligned}
$$

The resultant estimators $\hat{f}_r(t)$ are also centered.

If the coefficient matrix in the left-hand side of (10.61) is of full rank (no concurvity), we can use the results below to show that the solution to equation (10.61) is unique and the following backfitting algorithm converges to this unique solution:

$$
\begin{aligned}
\beta_0 &= \mathbf{S}_0\left(\tilde{\mathbf{y}} - \sum_{r=1}^{p}\mathbf{N}_r\mathbf{f}_r - \mathbf{Z}\mathbf{b}\right), \\
\mathbf{f}_r &= \mathbf{S}_r\left(\tilde{\mathbf{y}} - \beta_0\mathbf{1}_N - \sum_{l\neq r}\mathbf{N}_l\mathbf{f}_l - \mathbf{Z}\mathbf{b}\right), \quad r = 1, \cdots, p, \\
\mathbf{b} &= \mathbf{S}_b\left(\tilde{\mathbf{y}} - \beta_0\mathbf{1}_N - \sum_{r=1}^{p}\mathbf{N}_r\mathbf{f}_r\right).
\end{aligned}
$$

The estimate of $\mathbf{b}$ can be viewed as an empirical Bayes estimator.

The above DPQL estimators $\hat{\mathbf{f}}_r$ can be easily obtained by fitting a GLME model using existing statistical software if we reparameterize $\mathbf{f}_r, r = 1, 2, \cdots, p$ in a form of a LME model. Following Green (1987), Zhang et al. (1998), and Wang (1998a,b), we can reparameterize $\mathbf{f}_r$ (note that $\mathbf{f}_r$ is a centered parameter vector) as

$$
\mathbf{f}_r = \mathbf{h}_r\beta_r + \mathbf{B}_r\mathbf{a}_r, \tag{10.63}
$$

where β_r is a scalar and $\mathbf{a}_r$ is a $(M_r - 2) \times 1$ vector, $\mathbf{h}_r = [\tau_{r1}, \cdots, \tau_{rM_r}]^T$ is a $M_r \times 1$ vector, and $\mathbf{B}_r = \mathbf{L}_r(\mathbf{L}_r^T\mathbf{L}_r)^{-1}$ and $\mathbf{L}_r$ is a $M_r \times (M_r - 2)$ full rank matrix satisfying $\mathbf{G}_r = \mathbf{L}_r\mathbf{L}_r^T$ and $\mathbf{L}_r^T\mathbf{h}_r = 0$. Using the identity $\mathbf{f}_r^T\mathbf{G}_r\mathbf{f}_r = \mathbf{a}_r^T\mathbf{a}_r$, the DPQL (10.59) criterion becomes

$$
-\frac{1}{2}\sum_{i=1}^{n}\sum_{j=1}^{n_i}\phi^{-1}w_{ij}d(y_{ij}, \mu_{ij}) - \frac{1}{2}\mathbf{b}^T\tilde{\mathbf{D}}^{-1}\mathbf{b} - \frac{1}{2}\mathbf{a}^T\mathbf{\Lambda}^{-1}\mathbf{a}, \tag{10.64}
$$

where $\mathbf{a} = [\mathbf{a}_1^T, \cdots, \mathbf{a}_p^T]^T$ and $\mathbf{\Lambda} = \text{diag}(\tau_1\mathbf{I}_{M_1}, \cdots, \tau_p\mathbf{I}_{M_p})$ with $\tau_r = 1/\lambda_r$.

Combining equation (10.63) and equation (10.58), we can show that maximizing expression (10.64) for estimation of $\mathbf{f}_r$ is equivalent to fitting the following GLME model:

$$
g(\boldsymbol{\mu}) = \mathbf{X}\boldsymbol{\beta} + \mathbf{B}\mathbf{a} + \mathbf{Z}\mathbf{b}, \tag{10.65}
$$

using the penalized quasi-likelihood approach, where $\mathbf{X} = [\mathbf{1}_N, \mathbf{N}_1\mathbf{h}_1, \cdots, \mathbf{N}_p\mathbf{h}_p]$, $\mathbf{B} = [\mathbf{N}_1\mathbf{B}_1, \cdots, \mathbf{N}_p\mathbf{B}_p]$, and $\beta = [\beta_0, \beta_1, \cdots, \beta_p]^T$. Moreover, $\mathbf{a}$ and $\mathbf{b}$ are independent random-effects with distributions $\mathbf{a} \sim N(0, \Lambda)$ and $\mathbf{b} \sim N(\mathbf{0}, \tilde{\mathbf{D}})$. The DPQL estimators are $\hat{\mathbf{f}}_r = \mathbf{h}_r\hat{\beta}_r + \mathbf{B}_r\hat{\mathbf{a}}_r, r = 1, \cdots, p$.

Similarly, the maximization of the DPQL (10.64) criterion with respect to $(\beta, \mathbf{a}, \mathbf{b})$ can proceed by using the Fisher scoring algorithm to solve

$$\begin{pmatrix} \mathbf{X}^T\mathbf{W}\mathbf{X} & \mathbf{X}^T\mathbf{W}\mathbf{B} & \mathbf{X}^T\mathbf{W}\mathbf{Z} \\ \mathbf{B}^T\mathbf{W}\mathbf{X} & \mathbf{B}^T\mathbf{W}\mathbf{B} + \Lambda^{-1} & \mathbf{B}^T\mathbf{W}\mathbf{Z} \\ \mathbf{Z}^T\mathbf{W}\mathbf{X} & \mathbf{Z}^T\mathbf{W}\mathbf{B} & \mathbf{Z}^T\mathbf{W}\mathbf{Z} + \tilde{\mathbf{D}}^{-1} \end{pmatrix} \begin{pmatrix} \beta \\ \mathbf{a} \\ \mathbf{b} \end{pmatrix} = \begin{pmatrix} \mathbf{X}^T\mathbf{W}\tilde{\mathbf{y}} \\ \mathbf{B}^T\mathbf{W}\tilde{\mathbf{y}} \\ \mathbf{Z}^T\mathbf{W}\tilde{\mathbf{y}} \end{pmatrix}, \tag{10.66}$$

where $\tilde{\mathbf{y}}$ is the GSAME working vector defined in (10.62). One can easily show that equation (10.66) has a unique solution $\hat{\mathbf{f}}_r = \mathbf{h}_r\hat{\beta}_r + \mathbf{B}_r\hat{\mathbf{a}}_r \;\; r = 1, \cdots, p$, if $\mathbf{X}$ is of full rank, and the $\hat{\mathbf{f}}_r, r = 1, \cdots, p$ obtained from equation (10.66) are identical with those obtained from equation (10.61).

We notice that the estimating equation (10.66) corresponds to the normal equation of the LME model:

$$\tilde{\mathbf{y}} = \mathbf{X}\beta + \mathbf{B}\mathbf{a} + \mathbf{Z}\mathbf{b} + \epsilon, \tag{10.67}$$

where $\mathbf{a}$ and $\mathbf{b}$ are independent random-effects with $\mathbf{a} \sim N(0, \Lambda)$ and $\mathbf{b} \sim N(\mathbf{0}, \tilde{\mathbf{D}})$, and $\epsilon \sim N(0, \mathbf{W}^{-1})$. Thus, the DPQL estimators $\hat{\mathbf{f}}_r$ and the random-effect estimators $\hat{\mathbf{b}}$ can be easily obtained by iteratively fitting model (10.67) to the GSAME working vector $\tilde{\mathbf{y}}$.

To find Cov$(\hat{f}_r), r = 1, \cdots, p$, we may rewrite the estimating equation for $[\beta^T, \mathbf{a}^T]^T$ as

$$\begin{pmatrix} \mathbf{X}^T\mathbf{R}^{-1}\mathbf{X} & \mathbf{X}^T\mathbf{R}^{-1}\mathbf{B} \\ \mathbf{B}^T\mathbf{R}^{-1}\mathbf{X} & \mathbf{B}^T\mathbf{R}^{-1}\mathbf{B} + \Lambda^{-1} \end{pmatrix} \begin{pmatrix} \beta \\ \mathbf{a} \end{pmatrix} = \begin{pmatrix} \mathbf{X}^T\mathbf{R}^{-1}\tilde{\mathbf{y}} \\ \mathbf{B}^T\mathbf{R}^{-1}\tilde{\mathbf{y}} \end{pmatrix}, \tag{10.68}$$

where $\mathbf{R}^{-1} = \mathbf{W}^{-1} + \mathbf{Z}\tilde{\mathbf{D}}\mathbf{Z}^T$. Denoting by $\mathbf{H}$ the coefficient matrix on the left-hand side of equation (10.68) and $\mathbf{H}_0 = [\mathbf{X}, \mathbf{B}]^T\mathbf{R}^{-1}[\mathbf{X}, \mathbf{B}]$, the approximate covariance matrix of $\hat{\beta}$ and $\hat{\mathbf{a}}$ is

$$\text{Cov}\left([\hat{\beta}^T, \hat{\mathbf{a}}^T]^T\right) = \mathbf{H}^{-1}\mathbf{H}_0\mathbf{H}^{-1}.$$

Similarly we can derive the NCSS estimators of the $\mathbf{f}_r$ from the Bayesian perspective and the corresponding inference can be established under the Bayesian framework (Wahba 1983, Zhang et al. 1998, Wang 1998a, b, Lin and Zhang 1999).

We assumed that the smoothing parameters $\lambda_r, r = 1, \cdots, p$ and the variance-covariance parameter vector $\boldsymbol{\theta}$ are known when we derive the estimators of $\mathbf{f}_r$ previously. However, they often need to be estimated from the data as well. The extended restricted maximum likelihood (REML) approach can be employed to estimate $\boldsymbol{\lambda} = [\lambda_1, \cdots, \lambda_p]^T$ and $\boldsymbol{\theta}$. In fact, the marginal quasi-likelihood of $\boldsymbol{\tau} = [1/\lambda_1, \cdots, 1/\lambda_p]^T$ and $\boldsymbol{\theta}$ can be constructed under the GSAME (10.54) by assuming that $\mathbf{f}_r = \mathbf{h}_r\beta_r + \mathbf{B}_r\mathbf{a}_r$ with $\mathbf{a}_r \sim N(0, \tau_r\mathbf{I}_{M_r}), \;\; r = 1, \cdots, p$ and a flat prior for

β, and integrating the $\mathbf{a}_r$ and β out as follows:

$$\exp\{l_M(\boldsymbol{\tau},\boldsymbol{\theta}|\tilde{\mathbf{y}})\} \propto |\Lambda|^{-1/2} \int \exp\left\{l(\beta,\mathbf{a},\boldsymbol{\theta}|\tilde{\mathbf{y}}) - \tfrac{1}{2}\mathbf{a}^T\Lambda^{-1}\mathbf{a}\right\} d\mathbf{a}d\beta$$
$$\propto (|\tilde{\mathbf{D}}||\Lambda|)^{-1/2} \int \exp\left\{-\tfrac{1}{2}\left[\sum_{i,j} \tfrac{d(y_{ij},\mu_{ij})}{\phi w_{ij}^{-1}} + \mathbf{b}^T\tilde{\mathbf{D}}^{-1}\mathbf{b} + \mathbf{a}^T\Lambda^{-1}\mathbf{a}\right]\right\} d\mathbf{b}d\mathbf{a}d\beta,$$

where $l(\beta,\mathbf{a},\boldsymbol{\theta}|\tilde{\mathbf{y}}) = l(\beta_0,\mathbf{f}_1,\cdots,\mathbf{f}_p,\boldsymbol{\theta}|\tilde{\mathbf{y}})$ was defined in expression (10.55). Note that evaluation of the marginal quasi-likelihood for non-Gaussian responses is often difficult if not intractable since the evaluation of the marginal quasi-likelihood $l_M(\tau,\theta|\tilde{\mathbf{y}})$ often involves high-dimensional integration. The Laplace method may be used to approximate $l_M(\boldsymbol{\tau},\boldsymbol{\theta}|\tilde{\mathbf{y}})$ as

$$l_M(\boldsymbol{\tau},\boldsymbol{\theta}|\tilde{\mathbf{y}}) \approx -\frac{1}{2}\left\{\log|\mathbf{V}| + \log|\mathbf{X}^T\mathbf{V}^{-1}\mathbf{X}| + (\tilde{\mathbf{y}}-\mathbf{X}\hat{\beta})^T\mathbf{V}^{-1}(\tilde{\mathbf{y}}-\mathbf{X}\hat{\beta})\right\}, \tag{10.69}$$

where $\mathbf{V} = \mathbf{B}\Lambda\mathbf{B}^T + \mathbf{Z}\mathbf{D}\mathbf{Z}^T + \mathbf{W}^{-1}$. It can be shown that the approximate marginal log-quasi-likelihood (10.69) corresponds to the restricted log-likelihood of the working vector $\tilde{\mathbf{y}}$ under the LME model (10.67) with both $\mathbf{a}$ and $\mathbf{b}$ as random-effects and $\boldsymbol{\tau}$ and $\boldsymbol{\theta}$ as variance components. Thus, we can easily estimate $\boldsymbol{\tau}$ and $\boldsymbol{\theta}$ by iteratively fitting model (10.67) using REML. The Fisher information matrix of the approximate marginal quasi-likelihood estimators can be used to obtain the standard errors of these estimators. However, when the data are sparse (e.g. binary data), the estimators of the variance components $\boldsymbol{\theta}$ are subject to considerable bias. The GLME model bias correction procedure (Lin and Breslow 1996) may be applied to GSAME models with some modifications to obtain better estimates of the variance components (Lin and Zhang 1999).

In summary, inference on all model components in GSAME models (10.54), including $(\mathbf{f},\boldsymbol{\theta},\boldsymbol{\tau})$, can be easily implemented by fitting the working GLME model (10.65) using the penalized quasi-likelihood approach (Breslow and Clayton 1993, Lee and Nelder 1996). Equivalently, we only need to fit the working LME model (10.67) iteratively to the working vector $\tilde{\mathbf{y}}$, and to use the BLUP estimators of β_r and $\mathbf{a}_r$ to construct approximate NCSS estimators $\hat{\mathbf{f}}_r$ and to estimate $\boldsymbol{\theta}$ and $\boldsymbol{\tau}$ by using REML. Hence the existing statistical software SAS macro *GLIMMIX* (Wolfinger 1996), can be used.

10.7 SUMMARY AND BIBLIOGRAPHICAL NOTES

In this chapter, we reviewed the generalized mixed-effects modeling approaches for discrete longitudinal data. Generalized nonparametric/semiparametric population mean models were proposed and studied by Lin and Carroll (2000, 2001a, b) using LPK-GEE methods. Generalized nonparametric/time-varying coefficient mixed-effects models were discussed in an unpublished manuscript by Cai, Li and Wu (2003). Generalized semiparametric additive mixed-effects models were investigated by Lin and Zhang (1999) and Karcher and Wang (2001) using smoothing spline approaches. Other smoothing methods and models described in Chapters 4∼9 can be extended to make inferences for discrete longitudinal data. However, the literature in semiparametric/nonparametric methods and models for discrete longitudinal data is sparse. More work in this direction is warranted.

10.8 APPENDIX: PROOFS

Proof of Theorem 10.1 Let

$$\boldsymbol{\beta}^* = \gamma_n^{-1}\begin{pmatrix}\beta_1 - \eta(t)\\ h\,[\beta_2 - \eta'(t)]\end{pmatrix}, \qquad \text{and} \qquad \mathbf{U}_{ij} = \begin{pmatrix}1\\ (t_{ij}-t)/h\end{pmatrix},$$

where $\gamma_n = (N\,h)^{-1/2}$. Then,

$$\tilde{\eta}_{ij} = \mathbf{x}_{ij}^T\boldsymbol{\beta} + \mathbf{x}_{ij}^T\mathbf{b}_i = \overline{\eta}_{ij} + \gamma_n \mathbf{U}_{ij}^T\boldsymbol{\beta}^*,$$

where $\overline{\eta}_{ij} = \eta(t) + \eta'(t)\,(t_{ij}-t) + \mathbf{x}_{ij}^T\mathbf{b}_i$. Thus, the PLPQL defined in (10.32) becomes

$$\ell_n(\boldsymbol{\beta}) = -\frac{1}{2}\sum_{i=1}^{n}\sum_{j=1}^{n_i} K_h(t_{ij}-t)d\left[y_{ij},\, g^{-1}\left\{\overline{\eta}_{ij} + \gamma_n\mathbf{U}_{ij}^T\boldsymbol{\beta}^*\right\}\right] - \frac{1}{2}\sum_{i=1}^{n}\mathbf{b}_i^T\mathbf{D}^{-1}\mathbf{b}_i,$$

which is a function of $\boldsymbol{\beta}^*$, denoted by $\ell_n(\boldsymbol{\beta}^*)$. Let

$$\hat{\boldsymbol{\beta}}^* = \gamma_n^{-1}\begin{pmatrix}\hat{\beta}_1 - \eta(t)\\ h\,(\hat{\beta}_2 - \eta'(t))\end{pmatrix}.$$

Then, $\hat{\boldsymbol{\beta}}^*$ is the solution of $\ell_n'(\boldsymbol{\beta}^*) = \mathbf{0}$, because $\hat{\boldsymbol{\beta}}$ is the solution of $\ell_n'(\boldsymbol{\beta}) = \mathbf{0}$. Equivalently, $\hat{\boldsymbol{\beta}}^*$ maximizes $\ell_n^*(\boldsymbol{\beta}^*)$ since $\ell_n^*(\boldsymbol{\beta}^*)$ is concave by Condition A1, where

$$\begin{aligned}\ell_n^*(\boldsymbol{\beta}^*) \;&=\; -\tfrac{1}{2}\textstyle\sum_{i=1}^{n}\sum_{j=1}^{n_i}\Big[d\left\{y_{ij},\, g^{-1}\left(\overline{\eta}_{ij} + \gamma_n\mathbf{U}_{ij}^T\boldsymbol{\beta}^*\right)\right\}\\ &\quad - d\left\{y_{ij},\, g^{-1}(\overline{\eta}_{ij})\right\}\Big] K((t_{ij}-t)/h).\end{aligned}$$

By a Taylor expansion of $d(y,\, g^{-1}(\cdot))$, we have

$$\ell_n^*(\boldsymbol{\beta}^*) = \mathbf{W}_n^T\boldsymbol{\beta}^* - \frac{1}{2}\boldsymbol{\beta}^{*T}\Delta_n\boldsymbol{\beta}^* + R_n, \tag{10.70}$$

where

$$\mathbf{W}_n = \gamma_n\sum_{i=1}^{n}\sum_{j=1}^{n_i} q_1(y_{ij}, \overline{\eta}_{ij})\mathbf{U}_{ij}\, K((t_{ij}-t)/h),$$

$$\Delta_n = \gamma_n^2\sum_{i=1}^{n}\sum_{j=1}^{n_i} q_2(y_{ij},\, \overline{\eta}_{ij})\mathbf{U}_{ij}\mathbf{U}_{ij}^T\, K((t_{ij}-t)/h),$$

$$R_n = \frac{\gamma_n^3}{6} \sum_{i=1}^{n} \sum_{j=1}^{n_i} q_3(y_{ij}, \eta_{ij}^*) \, (\mathbf{U}_{ij}^T \beta^*)^3 \, K((t_{ij} - t)/h),$$

and η_{ij}^* is between $\overline{\eta}_{ij}$ and $\overline{\eta}_{ij} + \gamma_n \mathbf{U}_{ij}^T \beta^*$. According to the definition of $\overline{\eta}_{ij}$, it is easy to see that

$$\eta_{ij} - \overline{\eta}_{ij} = \frac{1}{2} \eta''(t) \, (t_{ij} - t)^2 + \frac{1}{2} \, v_i''(t) \, (t_{ij} - t)^2 + o_p(h^2), \qquad (10.71)$$

so that, by (10.43) and Conditions A2, A3, and A5,

$$\begin{aligned} E(\Delta_n) &= \gamma_n^2 \, h \sum_{i=1}^{n} \sum_{j=1}^{n_i} E\{q_2(\mu_{ij}, \overline{\eta}_{ij}) \mathbf{U}_{ij} \mathbf{U}_{ij}^T \, K_h(t_{ij} - t)\} \\ &\approx -\gamma_n^2 \, h \sum_{i=1}^{n} \sum_{j=1}^{n_i} E\left\{\mathbf{U}_{ij} \mathbf{U}_{ij}^T \, K_h(t_{ij} - t) R^{-1}(\mu_{ij})\right\} \\ &\approx -\gamma_n^2 \, N \, h \, \Delta = -\Delta, \end{aligned}$$

where

$$\Delta = f_t(t) \, E\left\{R^{-1}(\mu(t))\right\} \begin{pmatrix} 1 & 0 \\ 0 & \mu_2(K) \end{pmatrix}.$$

A similar argument shows that $\text{Var}(\Delta_n) = O(\gamma_n^2)$. Therefore,

$$\Delta_n = -\Delta + o_p(1). \qquad (10.72)$$

It is easy to see that (10.71) implies that $\eta_{ij} - \eta_{ij}^* = o_p(1)$. Then,

$$E|R_n| = O\left\{N \gamma_n^3 \, E\left(|q_3(y_{ij}, \eta_{ij})| \, K((t_{ij} - t)/h)\right)\right\} = O(\gamma_n) = o(1), \qquad (10.73)$$

since $K(\cdot)$ has a bounded support, $q_3(\cdot, \cdot)$ is linear in y_{ij}, and $E(|y_{ij}| \, |\mathbf{b}_i) < \infty$. This, in conjunction with (10.70) and (10.72), gives

$$\ell_n^*(\beta^*) = \mathbf{W}_n^T \beta^* - \frac{1}{2} \, \beta^{*T} \Delta \beta^* + o_p(1).$$

By the quadratic approximation lemma (see Fan and Gijbels 1996), we have

$$\beta^* = \Delta^{-1} \mathbf{W}_n + o_p(1), \qquad (10.74)$$

if $\mathbf{W}_n$ is a sequence of stochastically bounded random vectors. Clearly, the asymptotic normality of β^* follows from that of $\mathbf{W}_n$. Hence, what remains is to establish the asymptotic normality of $\mathbf{W}_n$.

Now, $\mathbf{W}_n$ is re-expressed as

$$\begin{aligned} \mathbf{W}_n &= \gamma_n \sum_{i=1}^{n} \sum_{j=1}^{n_i} [y_{ij} - g^{-1}(\overline{\eta}_{ij})] \, T(g^{-1}(\overline{\eta}_{ij})) \mathbf{U}_{ij} \, K((t_{ij} - t)/h) \\ &\equiv \mathbf{W}_{n1} + \mathbf{W}_{n2}, \end{aligned} \qquad (10.75)$$

where

$$\mathbf{W}_{n1} = \gamma_n \sum_{i=1}^{n} \sum_{j=1}^{n_i} [y_{ij} - \mu_{ij}]\, T(g^{-1}(\overline{\eta}_{ij}))\mathbf{U}_{ij}\, K((t_{ij} - t)/h),$$

and

$$\mathbf{W}_{n2} = \gamma_n \sum_{i=1}^{n} \sum_{j=1}^{n_i} [g^{-1}(\eta_{ij}) - g^{-1}(\overline{\eta}_{ij})]\, T(g^{-1}(\overline{\eta}_{ij}))\mathbf{U}_{ij}\, K((t_{ij} - t)/h).$$

First, we consider $\mathbf{W}_{n2}$. To this end, by (10.71),

$$\begin{aligned} & g^{-1}(\eta_{ij}) - g^{-1}(\overline{\eta}_{ij}) \\ = \;& \tfrac{1}{2}\left[\eta''(t)\,(t_{ij} - t)^2 + v_i''(t)\,(t_{ij} - t)^2\right]\,[g'(\mu_{ij})]^{-1} + o_p(h^2). \end{aligned}$$

Now, define

$$\mathbf{W}_{n21} \equiv \gamma_n \sum_{i=1}^{n} \sum_{j=1}^{n_i} \left[\frac{1}{2}\eta''(t)\,(t_{ij} - t)^2\right] R^{-1}(\mu_{ij})\mathbf{U}_{ij}\, K((t_{ij} - t)/h),$$

and

$$\mathbf{W}_{n22} \equiv \gamma_n \sum_{i=1}^{n} \sum_{j=1}^{n_i} \left[\frac{1}{2}\, v_i''(t)\,(t_{ij} - t)^2\right] R^{-1}(\mu_{ij})\mathbf{U}_{ij}\, K((t_{ij} - t)/h).$$

Then, it is easy to see that

$$E(\mathbf{W}_{n21}) = \frac{h^2}{2}\eta''(t)\gamma_n^{-1}\, f_t(t)\, E\left\{R^{-1}(\mu(t))\right\} \begin{pmatrix} \mu_2(K) \\ 0 \end{pmatrix} \{1 + o(1)\},$$

and

$$E(\mathbf{W}_{n22}) = \frac{h^2}{2}\gamma_n^{-1}\, f_t(t)\, E\left\{v''(t)\, R^{-1}(\mu(t))\right\} \begin{pmatrix} \mu_2(K) \\ 0 \end{pmatrix} \{1 + o(1)\}.$$

Similarly, $\mathrm{Var}(\mathbf{W}_{n21}) = o(h^4)$ and $\mathrm{Var}(\mathbf{W}_{n22}) = o(h^4)$. Therefore,

$$\begin{aligned} \mathbf{W}_{n2} \;=\; & \frac{h^2}{2}\gamma_n^{-1}\, f_t(t)\Big[\eta''(t)\, E\left\{R^{-1}(\mu(t))\right\} \\ & + E\left\{v''(t)\, R^{-1}(\mu(t))\right\}\Big] \begin{pmatrix} \mu_2(K) \\ 0 \end{pmatrix} \{1 + o_p(1)\}. \end{aligned} \quad (10.76)$$

It is easy to see, by the definition of Δ and (10.74) and (10.75), that

$$\begin{aligned} & \sqrt{N\,h}\,\left[\hat{\eta}(t) - \eta(t) - b_\eta(t) + o_p(h^2)\right] \\ = \;& \left[f_t(t)\, E\left\{R^{-1}(\mu(t))\right\}\right]^{-1}\, W_n^*\{1 + o_p(1)\}, \end{aligned} \quad (10.77)$$

where

$$W_n^* = \gamma_n \sum_{i=1}^{n} \sum_{j=1}^{n_i} \xi_{ij}, \text{ and } \xi_{ij} = (y_{ij} - \mu_{ij})\, T(\mu_{ij})\, K((t_{ij} - t)/h).$$

Clearly, $E(W_n^*) = 0$ and

$$\text{Var}(W_n^*) = \gamma_n^2 \sum_{i=1}^{n} \text{Var}\left(\sum_{j=1}^{n_i} \xi_{ij}\right) \equiv I_1 + I_2,$$

where

$$I_1 = \gamma_n^2 \sum_{i=1}^{n} \sum_{j=1}^{n_i} \text{Var}(\xi_{ij}), \text{ and } I_2 = \gamma_n^2 \sum_{i=1}^{n} \sum_{1 \le j \ne k \le n_i} \text{Cov}(\xi_{ij}, \xi_{ik}).$$

It is easy to see that

$$\begin{aligned} h^{-1}\text{Var}(\xi_{ij}) &= h^{-1} E\left[V(\mu_{ij})\, T^2(\mu_{ij})\, K^2((t_{ij} - t)/h)\right] \\ &= \nu_0\, f_t(t)\, E\left[V(\mu(t))\, T^2(\mu(t))\right] + o(1), \end{aligned}$$

which implies that

$$I_1 = \nu_0\, f_t(t)\, E\left\{R^{-1}(\mu(t))\right\} + o(1).$$

Next, for $j \ne k$, by Condition A7,

$$\begin{aligned} \text{Cov}(\xi_{ij}, \xi_{ik}) &= E\{\gamma_y(t_{ij}, t_{ik})\, K((t_{ij} - t)/h)\, K((t_{ik} - t)/h)\} \\ &= h^2\, f_t^2(t)\gamma_y(t, t)\{1 + o(1)\}, \end{aligned}$$

which, in conjunction with Condition A2, implies that

$$I_2 = f_t^2(t)\gamma_y(t, t)\gamma_n^2\, h^2 \sum_{i=1}^{n} n_i(n_i - 1)\{1 + o(1)\} = \lambda\, f_t^2(t)\gamma_y(t, t)\{1 + o(1)\}.$$

Therefore,

$$\begin{aligned} \text{Var}(W_n^*) &= I_1 + I_2 \\ &= \nu_0\, f_t(t)\, E\left\{R^{-1}(\mu(t))\right\} + \lambda\, f_t^2(t)\gamma_y(t, t) + o(1) \\ &\equiv \sigma_1^2(t) + o(1). \end{aligned} \quad (10.78)$$

Hence,

$$\sigma^2(t) = \left[f_t(t)\, E\left\{R^{-1}(\mu(t))\right\}\right]^{-2} \sigma_1^2(t). \quad (10.79)$$

By (10.77), to accomplish the proof of the theorem, it suffices to show that $W_n^* \to N(0, \sigma_1^2(t))$, which can be verified by expressing W_n^* as a sum of independent random variables $W_n^* = \gamma_n \sum_{i=1}^{n} \xi_i$, where $\xi_i = \sum_{j=1}^{n_i} \xi_{ij}$, and by checking Lyapounov's condition. That is to show that $\gamma_n^4 \sum_{i=1}^n E|\xi_i|^4 \to 0$ as $n \to \infty$, which can be shown to be true by using the same arguments as those used in the proof of (10.72). This completes the proof of Theorem 10.1.

Sketch Proof of Theorem 10.2 Since the proof of Theorem 10.2 is similar to that for Theorem 10.1, we only present a brief outline here. γ_n now becomes $\gamma_n = \sqrt{N_2}/N$. Let $\lambda_n = h\, N_2/N$. Then, $\gamma_n \to 0$ and $\lambda_n \to \infty$ by the assumptions in Theorem 10.2. Divide $\ell_n^*(\beta^*)$ by λ_n. Then, (10.70) becomes

$$\ell_n^*(\beta^*)/\lambda_n = \mathbf{W}_n^T/\lambda_n\beta^* + \frac{1}{2}\beta^{*T}\Delta_n/\lambda_n\beta^* + R_n/\lambda_n.$$

Similar to (10.72) and (10.73), one can show that

$$\Delta_n/\lambda_n = -\Delta + o_p(1), \text{ and } E|R_n|/\lambda_n = O(N\gamma_n^3\, h/\lambda_n) = O(\gamma_n) = o(1).$$

Therefore, (10.74) becomes

$$\beta^* = \Delta^{-1}\mathbf{W}_n/\lambda_n + o_p(1),$$

and (10.76) holds. Thus, (10.77) becomes

$$\begin{aligned} & \tfrac{N}{\sqrt{N_2}}\left[\hat{\eta}(t) - \eta(t) - b_\eta(t) + o_p(h^2)\right] \\ = \; & \left[f_t(t)\, E\left\{R^{-1}(\mu(t))\right\}\right]^{-1} W_n^*/\lambda_n\{1 + o_p(1)\}. \end{aligned}$$

It is easy to see that $I_1/\lambda_n^2 = O(\lambda_n^{-1}) = o(1)$ and

$$\begin{aligned} I_2/\lambda_n^2 &= f_t^2(t)\gamma_y(t,\, t)\, \tfrac{\gamma_n^2\, h}{\lambda_n^2} \textstyle\sum_{i=1}^n n_i(n_i - 1)\{1 + o(1)\} \\ &= f_t^2(t)\gamma_y(t,\, t)\{1 + o(1)\}, \end{aligned}$$

so that $\text{Var}(W_n^*/\lambda_n) = f_t^2(t)\gamma_y(t,\, t)\{1 + o(1)\}$. The rest of proof is same as that for Theorem 10.1, and omitted.

Sketch Proofs of Theorems 10.3 and 10.4 Note that $\mathbf{U}_{ij}$ in the proof of Theorem 10.1 now becomes

$$\mathbf{U}_{ij} = \begin{pmatrix} \mathbf{X}_i \\ \mathbf{X}_i\,(t_{ij} - t)/h \end{pmatrix}.$$

Then, the rest of proofs are same as those for Theorems 10.1 and 10.2, and are omitted.

References

1. Akaike, H. (1973). Information theory and an extension of the entropy maximization principle. *2nd International Symposium on Information Theory,* eds. B.N. Petrov and F. Csak, Kiado: Akademia. pp. 267-281.

2. Altman, N.S. (1991). Kernel smoothing of data with correlated errors. *Journal of American Statistical Association*, **85**, 749-759.

3. Anderson, T.W. (1984). *An Introduction to Multivariate Statistical Analysis.* Wiley, New York.

4. Beal, S.L. and Sheiner, L.B. (1982). Estimating population kinetics. *CRC Critical Reviews in Biomedical Engineering*, **8**, 195-222.

5. Beal, S.L. and Sheiner, L.B. (1992). *NONMEM User's Guides.* NONMEM Project Group, University of California, San Francisco.

6. Besse, P., Cardot, H. and Ferraty, F. (1997). Simultaneous nonparametric regressions of unbalanced longitudinal data. *Computational Statistics and Data Analysis*, **24**, 255-270.

7. Breslow, N.E. and Clayton, D.G. (1993). Approximate inference in generalized linear mixed models. *Journal of American Statistical Association*, **88**, 9-25.

8. Breslow, N.E. and Lin, X. (1995). Bias correction in generalized linear mixed models with a single component of dispersion. *Biometrika*, **82**, 81-91.

9. Brumback, B. and Rice, J.A. (1998). Smoothing spline models for the analysis of nested and crossed samples of curves. *Journal of American Statistical Association*, **93**, 961-994.

10. Buja, A., Hastie, T. and Tibshirani, R. (1989). Linear smoothers and additive models. *Annals of Statistics*, **17**, 453-510.

11. Cai, Z., Fan, J. and Li, R. (2000). Efficient estimation and inferences for varying-coefficient models. *Journal of American Statistical Association*, **95**, 888-902.

12. Cai, Z., Fan, J. and Yao, Q. (2000). Functional-coefficient regression models for nonlinear time series. *Journal of American Statistical Association*, **95**, 941-956.

13. Cai, Z., Li, H. and Wu, H. (2003). Generalized random curve models for longitudinal data. Manuscript.

14. Carroll, R. J., Fan, J., Gijbels, I. and Wand, M. P. (1997). Generalized partially linear single-index models. *Journal of American Statistical Association*, **92**, 477-489.

15. Chen, K. and Jin, Z. (2005). Local polynomial regression analysis of clustered data. *Biometrika*, **92**, 59-74.

16. Cheng, M. Y., Fan, J. and Marron, J. S. (1997). On automatic boundary corrections. *Annals of Statistics*, **25**, 1691-1708.

17. Cheng, S.C. and Wei, L. J. (2000). Inferences for semiparametric model with panel data. *Biometrika*, **87**, 89-97.

18. Chiang, C.T., Rice, J.A. and Wu, C.O. (2001). Smoothing spline estimation for varying coefficient models with repeatedly measured dependent variables. *Journal of American Statistical Association*, **96**, 605-619.

19. Cleveland, W. (1979). Robust locally weighted regression and smoothing scatterplots. *Journal of American Statistical Association*, **83**, 596-610.

20. Cox, D. R. and Solomon, P. J. (2003). *Components of Variance*. Chapman and Hall, London.

21. Craven, P. and Wahba, G. (1979). Smoothing noisy data with spline functions. *Numerische Mathematik*, **31**, 377-390.

22. Crouch, A.C. and Spiegelman, E. (1990). The evaluation of integrals of the form $\int f(t)\exp(-t^2)dt$: application to logistic-normal models. *Journal of American Statistical Association*, **85**, 464-469.

23. Davidian, M. and Giltinan, D. M. (1993). Some general estimation methods for nonlinear mixed effects models. *Journal of Biopharmaceutical Statistics*, **3**, 23-55.

24. Davidian, M. and Giltinan, D. M. (1995). *Nonlinear Models for Repeated Measurement Data*. Chapman and Hall, London.

25. de Boor, C. (1978). *A Practical Guide to Splines*. Springer-Verlag, New York.

26. Demidenko, E. (2004). *Mixed Models: Theory and Applications*. Wiley, New York.

27. Dempster, A.P., Rubin, D.B. and Tsutakawa, R.K. (1981). Estimation in covariance components models. *Journal of American Statistical Association*, **76**, 341-353.

28. Diggle, P. J. (1988). An approach to the analysis of repeated measurements. *Biometrics*, **44**, 959-971.

29. Diggle, P. J., Liang, K. Y. and Zeger, S. L. (1994). *Analysis of Longitudinal Data*. Oxford University Press, Oxford, U.K.

30. Diggle, P. J., Heagerty, P., Liang, K. Y. and Zeger, S. L. (2002). *Analysis of Longitudinal Data (2nd ed.)*. Oxford University Press, Oxford, U.K.

31. Diggle, P.J. and Hutchinson, M.F. (1989). On spline smoothing with autocorrelated errors. *Australian Journal of Statistics*, **31**, 166-168.

32. Donnelly, C.A., Laird, N. M. and Ware, J. H. (1995). Prediction and creation of smooth curves for temporally correlated longitudinal data. *Journal of American Statistical Association*, **90**, 984-989.

33. Donoho, D. L. and Johnstone, I. M. (1994). Ideal spatial adaption by wavelet shrinkage. *Biometrika*, **81**, 425-455.

34. Durleman, S. and Simon, R. (1989). Flexible regression models with cubic splines. *Statistics in Medicine*, **8**, 551-561.

35. Durban, M., Harezlak, J., Wand, M. P. and Carroll, L. J. (2005). Simple fitting of subject-specific curves for longitudinal data. *Statistics in Medicine*, **24**, 1153-1167.

36. Draper, N. R. and Smith, H. (1998). *Applied Regression Analysis*. 3rd ed., Wiley, New York.

37. Efromovich, S. (1999). *Nonparametric Curve Estimation*. Springer-Verlag, New York.

38. Eilers, P. H. C. and Marx, B. D. (1996). Flexible smoothing with B-splines and penalties. *Statistics Science*, **11**, 89-102.

39. Engle, R. F., Granger, C. W. J., Rice, J. and Weiss, A. (1986). Semiparametric estimates of the relation between weather and electricity slaes. *Journal of American Statistical Association*, **81**, 310-320.

40. Eubank, R.L. (1988). *Spline Smoothing and Nonparametric Regression*. Marcel Dekker, New York.

41. Eubank, R.L. (1999). *Nonparametric Regression and Spline Smoothing*. Marcel Dekker, New York.

42. Eubank, R.L., Huang, C., Maldonado, Y., Wang, N., Wang, S. and Buchanan, R. J. (2004). Smoothing spline estimation in varying-coefficient models. *Journal of Royal Statistical Society, Series B*, **66**, 653-667.

43. Fan, J. (1992). Design-adaptive nonparametric regression. *Journal of American Statistical Association*, **87**, 998-1004.

44. Fan, J. (1993). Local linear regression smoothers and their minimax efficiency. *Annals of Statistics*, **21**, 196-216.

45. Fan, J. (1996). Test of significance based on wavelet thresholding and Neyman's truncation. *Journal of American Statistical Association*, **91**, 674-688.

46. Fan, J., Farmen, M. and Gijbels, I. (1998). Local maximum likelihood estimation and inference. *Journal of Royal Statistical Society, Series B*, **60**, 591-608.

47. Fan, J. and Huang, L. (2001). Goodness-of fit test for parametric regression models. *Journal of American Statistical Association*, **96**, 640-652.

48. Fan, J. and Gijbels, I. (1992). Variable bandwidth and local linear regression smoothers. *Annals of Statistics*, **20**, 2008-36.

49. Fan, J. and Gijbels, I. (1996). *Local Polynomial Modelling and its Applications*. Chapman and Hall, London.

50. Fan, J. and Li, R. (2001). Variable selection via nonconcave penalized likelihood and its oracle properties. *Journal of American Statistical Association*, **96**, 1348-1360.

51. Fan, J. and Li, R. (2004). New estimation and model selection procedures for semiparametric modeling in longitudinal data analysis. *Journal of American Statistical Association*, **99**, 710-723.

52. Fan, J. and Marron, J. S. (1994). Fast implementations of nonparametric curve estimators. *Journal of Computational and Graphical Statistics*, **3**, 35-56.

53. Fan, J., Zhang, C. and Zhang, J. (2001). Generalized likelihood ratio statistics and wilks phenomenon. *Annals of Statistics*, **29**, 153-193.

54. Fan, J. and Zhang, J.-T. (1998). Comments on " Smoothing spline models for the analysis of nested and crossed samples of curves" by Brumback and Rice (1998). *Journal of American Statistical Association*, **93**, 980-983.

55. Fan, J. and Zhang, J.-T. (2000). Two-step estimation of functional linear models with applications to longitudinal data. *Journal of Royal Statistical Society, Series B*, **62**, 303-322.

56. Fan, J. and Zhang, W. (1999). Statistical estimation in varying coefficient models. *Annals of Statistics*, **27**, 1491-1518.

57. Fan, J. and Zhang, W. (2000). Simultaneous confidence bands and hypothesis testing in varying coefficient models. *Scandinavian Journal of Statistics*, **27**, 715-731.

58. Fischl, M.A., Ribaudo, H.J., Collier, A.C., et al. (2003). A randomized trial of 2 different 4-drug antiretroviral regimens versus a 3-drug regimen, in advanced human immunodeficiency virus disease. *Journal of Infectious Diseases*, **188**, 625-634.

59. Friedman, J. H. and Silverman, B. W. (1989). Flexible parsimonious smoothing and additive modeling (with discussion). *Technometrics*, **31**, 3-39.

60. Friedman, J. H. (1991). Multivariate adaptive regression splines. *Annals of Statistics*, **19**, 1-68.

61. Gasser, T. and Müller, H. G. (1979). Kernel estimation of regression functions. In *Smoothing Techniques for Curve Estimation* (eds. T. Gasser and M, Rosenblatt). Springer-Verlag, Heidelberg, pp. 23-68.

62. Gasser, T., Müller, H. G. and Mammitzsch, V. (1985). Kernels for nonparametric curve estimation. *Journal of Royal Statistical Society, Series B*, **86**, 665-672.

63. Gu, C. (1989). RKPACK and its application: fitting smoothing spline models. In *Proceedings of the Statistica; Computing Section, American Statistical Association*. pp. 42-51.

64. Gu, C. (2002). *Smoothing Spline ANOVA Models*. Springer-Verlag, New York.

65. Guo, W. (2002a). Functional mixed-effects models. *Biometrics*, **58**, 121-128.

66. Guo, W. (2002b). Inference in smoothing spline analysis of variance. *Journal of Royal Statistical Society, Series B*, **64**, 887-889.

67. Guo, W., Wang, Y. and Brown, M. B. (1999). A signal extraction approach to modeling hormone time series with pulses and a changing baseline. *Journal of American Statistical Association*, **94**, 746-756.

68. Green, P. (1987). Penalized likelihood for general semiparametric regression models. *International Statistical Review*, **55**, 245-259.

69. Green, P., Jennsion, C. and Seheult, A. (1985). Analysis of field experiments by least squares smoothing. *Journal of Royal Statistical Society, Series B*, **47**, 299-315.

70. Green, P. and Silverman, B.W. (1994). *Nonparametric Regression and Generalized Linear Models*. Chapman and Hall, London.

71. Härdle, W. (1990). *Applied Nonparametric Regression*. Cambridge University Press, Boston.

72. Härdle, W., Liang, H. and Gao, J. (2000). *Partially Linear Models*. Springer-Verlag, New York.

73. Hart, J.D. (1991). Kernel regression estimation with time series errors. *Journal of Royal Statistical Society, Series B*, **53**, 173-187.

74. Hart, J. D. (1997). *Nonparametric Smoothing and Lack-of-Fit Tests*. Springer-Verlag, New York.

75. Hart, J.D. and Wehrly, T.E. (1986). Kernel regression estimation using repeated measurements data. *Journal of American Statistical Association*, **81**, 1080-1088.

76. Hart, J.D. and Wehrly, T.E. (1993). Consistency of cross-validation when the data are curves. *Stochastic Processes and their Applications*, **45**, 351-361.

77. Harville, D.A. (1976). Extension of the Gauss-Markov theorem to include the estimation of random effects. *Annals of Statistics*, **4**, 384-395.

78. Harville, D.A. (1977). Maximum likelihood approaches to variance component estimation and to related problems. *Journal of American Statistical Association*, **72**, 320-340.

79. Hastie, T.J. and Loader, C. (1993). Local regression: automatic kernel carpentry(with discussion). *Statistics Science*, **8**, 120-143.

80. Hastie, T. and Tibshirani, R. (1990). *Generalized Additive Models*. Chapman and Hall, London.

81. Hastie, T. and Tibshirani, R. (1993). Varying-coefficients models. *Journal of Royal Statistical Society, Series B*, **55**, 757-796.

82. He, X., Zhu, Z.-Y. and Fung, W.-K. (2002). Estimation in a semiparametric model for longitudinal data with unspecified dependence structure. *Biometrika*, **89**, 579-590.

83. Heckman, N. (1986). Spline smoothing in a partly linear model. *Journal of Royal Statistical Society, Series B*, **48**, 244-248.

84. Heckman, N. and Ramsay, J. O. (2000). Penalized regression with model-based penalties. *Canadian Journal of Statistics*, **28**, 241-248.

85. Heuer, C. (1997). Modeling of time trends and interactions in vital rates using restricted regression splines. *Biometrics*, **53**, 161-177.

86. Hoover, D. R., Rice, J.A., Wu, C.O., and Yang, L. P. (1998). Nonparametric smoothing estimates of time-varying coefficient models with longitudinal data. *Biometrika*, **85**, 809-822.

87. Hu, Z., Wang, N. and Carroll, J. R. (2004). Profile-kernel versus backfitting in the partially linear models for longitudinal/clustered data. *Biometrika*, **91**, 251-262.

88. Huang, J.Z., Wu, C.O. and Zhou, L. (2002). Varying-coefficient models and basis function approximations for the analysis of repeated measurements. *Biometrika*, **89**, 111-128.

89. Huggins, R. M. and Loesch, D. Z. (1998). On the analysis of mixed longitudinal growth data. *Biometrics*, **54**, 583-595.

90. Jacqmin-Gadda, H., Joly, P., Commenges, D., Binquet, C. and Chene, G. (2002). Penalized likelihood approach to estimate a smooth mean curve on longitudinal data. *Statistics in Medicine*, **21**, 2391-2402.

91. Jennrich, R.I. and Schluchter, M.D. (1986). Unbalanced repeated measures models with structured covariance matrices. *Biometrics*, **42**, 805-820.

92. Jones, B. and Kenward, M.G. (1987). Modelling binary data from a three-point cross-over trial. *Statistics in Medicine*, **6**, 555-564.

93. Jones, R. H. (1993). *Longitudinal Data with Serial Correlation: A State-space Approach*. Chapman & Hall, London.

94. Karcher, P. and Wang, Y. (2001). Generalized nonparametric mixed-effects models. *Journal of Computational and Graphical Statistics*, **10**, 641-655.

95. Kaslow, R. A., Ostrow, D. G., Detels., R., Phair, J.P., Polk, B. F. and Rinaldo, C. R. (1978). The multicenter AIDS cohort study: rationale, organization and selected characteristics of the participants. *American Journal of Epidemiology*, **126**, 310-318.

96. Ke, C. and Wang, Y. (2001). Semiparametric nonlinear mixed-effects models and their applications (with discussions). *Journal of American Statistical Association*, **96**, 1272-1298.

97. Kohn, R., Ansley, C. F. and Tharm, D. (1991). The performance of cross-validation and maximum likelihood estimators of spline smoothing parameters. *Journal of American Statistical Association*, **86**, 1042-1050.

98. Ladd, W. M, Lindstrom, M. J. (2000). Self modeling for two-dimensional response curves. *Biometrics*, **56**, 89-97.

99. Laird, N. M. and Ware, J. H. (1982). Random effects models for longitudinal data. *Biometrics*, **38**, 963-974.

100. Laird, N. M., Lange, N. and Stram, D. (1987). Maximum likelihood computations with repeated measures: Application of the EM algorithm. *Journal of American Statistical Association*, **82**, 97-105.

101. Lange, N., Carlin, B. P. and Gelfand, A. E. (1992). Hierarchical bayes models for progression of HIV infection using longitudinal CD4 T-cell numbers. *Journal of American Statistical Association*, **87**, 615-626.

102. Lee, Y. and Nelder, J.A. (1996). Hierarchical generalized linear models (with discussion). *Journal of Royal Statistical Society, Series B*, **58**, 619-678.

103. Liang, H., Wu, H. and Carroll, R.J. (2003). The relationship between virologic and immunologic responses in AIDS clinical research using mixed-effects varying-coefficient semiparametric models with measurement error. *Biostatistics*, **4**, 297-312.

104. Liang, H. and Wu, H. (2004). An efficient kernel linear mixed-effects model estimator in nonparametric kernel regression for longitudinal data. Manuscript.

105. Liang, K. Y. and Zeger, S. L. (1986). Longitudinal data analysis using generalized linear models. *Biometrika*, **73**, 13-22.

106. Lin, D. Y. and Ying, Z. (2001). Semiparametric and nonparametric regression analysis of longitudinal data (with discussions). *Journal of American Statistical Association*, **96**, 103-113.

107. Lin, X. and Breslow, N.E. (1996). Bias correction in generalized linear mixed models with multiple components of dispersion. *Journal of American Statistical Association*, **91**, 1007-1016.

108. Lin, X. and Carroll, R. J. (2000). Nonparametric function estimation for clustered data when the predictor is measured without/with error. *Journal of American Statistical Association*, **95**, 520-534.

109. Lin, X. and Carroll, R. J. (2001a). Semiparametric regression for clustered data using generalized estimating equations. *Journal of American Statistical Association*, **96**, 1045-1056.

110. Lin, X. and Carroll, R. J. (2001b). Semiparametric regression for clustered data. *Biometrika*, **88**, 1179-1185.

111. Lin, X. and Carroll, R. J. (2005). Semiparametric estimation in general repeated measures problems. *Harvard University Biostatistics Working Paper Series*.

112. Lin, X. and Zhang, D. (1999). Inference in generalized additive mixed models by using smoothing splines. *Journal of Royal Statistical Society, Series B*, **61**, 381-400.

113. Lin, X., Wang, N., Welsh, A. and Carroll, R. J. (2004). Equivalent kernels of smoothing splines in nonparametric regression for clustered data. *Biometrika*, **91**, 177-193.

114. Lindsey, J. K. (1993). *Models for Repeated Measurements.* Oxford University Press, Oxford, U.K.

115. Lindstrom, M. J. and Bates, D. M. (1990). Nonlinear mixed-effects models for repeated measures. *Biometrics*, **46**, 673-687.

116. Longford, N.T. (1993). *Random Coefficient Models.* Oxford University Press, Oxford, UK.

117. Marron, J. S. and Nolan, D. (1989). Canonical kernels for density estimation. *Statistics and Probability Letters*, **7**, 195-9.

118. Martinussen, T. and Scheike, T. H. (1999). A semiparametric additive regression model for longitudinal data. *Biometrika*, **86**, 691-702.

119. Martinussen, T. and Scheike, T. H. (2000). A nonparametric dynamic additive regression model for longitudinal data. *Annals of Statistics*, **28**, 1000-1025.

120. Martinussen, T. and Scheike, T. H. (2001). Sampling-adjusted analysis of dynamic additive regression models for longitudinal data. *Scandinavian Journal of Statistics*, **28**, 303-323.

121. McCullagh, P. and Nelder, J.A. (1989). *Generalized Linear Models.* Chapman and Hall, New York.

122. Morris, J.S., Vannucci, M., Brown, P.J. and Carroll, R.J. (2003). Wavelet-based nonparametric modeling of hierarchical functions in colon carcinogenesis (with discussions). *Journal of American Statistical Association*, **98**, 573-597.

123. Morris, J.S. and Carroll, R.J. (2005). Wavelet-based functional mixed models. Manuscript.

124. Morton, R. (1987). A generalized linear model with nested strat of extra-Poisson variation. *Biometrika*, **74**, 247-257.

125. Moyeed, R. A. and Diggle, P. J. (1994). Rates of convergence in semiparametric modeling of longitudinal data. *Australian Journal of Statistics*, **36**, 75-93.

126. Müller, H.G. (1988). *Nonparametric Regression Analysis of Longitudinal Data.* Lecture Notes in Statistics. Springer-Verlag, New York.

127. Nadaraya, E. A. (1964). On estimating regression. *Theory of Probability and its Applications*, **9**, 141-42.

128. Nelder, J.A. and Wedderburn, R.W.M. (1972). Generalized linear models. *Journal of Royal Statistical Society, Series A*, **135**, 370-384.

129. Opsomer, J., Wang, Y. and Yang, Y. (2001). Nonparametric regression with correlated errors. *Statistics Science*, **16**, 134-153.

130. O'Sullivan, F. (1986). A statistical perspective on ill-posed in verse problems (with discussion). *Statistics Science*, **1**, 505-527.

131. O'Sullivan, F. (1988). Fast computation of fully automated log-density and log-hazard estimators. *SIAM Journal on Scientific and Statistical Computing*, **9**, 363-379.

132. Park, J.G. and Wu, H. (2005). Backfitting and local likelihood methods for nonparametric mixed-effects models with longitudinal data. Manuscript.

133. Pinheiro, J. and Bates, D. (2000). *Mixed-effects Models in S and S-plus*. Springer-Verlag, New York.

134. Racine-Poon, A. (1985). A Bayesian approach to nonlinear random effects models. *Biometrics*, **41**, 1015-1023.

135. Ramsay, J.O. and Silverman, B.W. (1997). *Functional Data Analysis*. Springer-Verlag, New York.

136. Ramsay, J.O. and Silverman, B.W. (2002). *Applied Functional Data Analysis*. Springer-Verlag, New York.

137. Rice, J. A. (1986). Convergence rates for partially splined models. *Statistics and Probability Letters*, **4**, 203-208.

138. Rice, J.A. and Silverman, B.W. (1991). Estimating the mean and covariance structure nonparametrically when the data are curves. *Journal of Royal Statistical Society, Series B*, **53**, 233-243.

139. Rice, J. A. and Wu, C.O. (2001). Nonparametric mixed effects models for unequally sampled noisy curves. *Biometrics*, **57**, 253-259.

140. Robinson, G. K. (1991). That BLUP is a good thing: the estimation of random effects (with discussions). *Statistics Science*, **6**, 15-32.

141. Rosner, B. (1984). Multivariate methods in ophthalmology with applications to other paired-data situations. *Biometrics*, **40**, 1025-1035.

142. Rotnitzky, A. and Jewell, N. P. (1990). Hypothesis testing of regression parameters in semiparametric generalized linear models for cluster correlated data. *Biometrika*, **77**, 485-497.

143. Ruppert, D. (2002). Selecting the number of knots for penalized splines. *Journal of Computational and Graphical Statistics*, **11**, 735-757.

144. Ruppert, D. and Carroll, R. J. (2000). Spatially-adaptive penalties for spline fitting. *Australian and New Zealand Journal of Statistics*, **42**, 205-224.

145. Ruppert, D., Sheather, S.J. and Wand, M.P. (1995). An effective bandwidth selector for local least squares regression. *Journal of American Statistical Association*, **90**, 1257-1270.

146. Ruppert, D., Wand, M. P. and Carroll, R. J. (2003). *Semiparametric Regression*. Cambridge University Press.

147. Schall, R. (1991). Estimation in generalized linear models with random effects. *Biometrika*, **78**, 719-727.

148. Schwarz, G. (1978). Estimating the dimension of a model. *Annals of Statistics*, **6**, 461-464.

149. Searle, S.R., Casella, G. and McCulloch, C.E. (1992). *Variance Components*. Wiley, New York.

150. Severini, T. A. and Staniswalis, J. G. (1994). Quasi-likelihood estimation in semiparametric models. *Journal of American Statistical Association*, **89**, 501-11.

151. Sheiner, L.B., Rosenberg, B. and Melmon, K.L. (1972). Modeling of individual pharmacokinetics for computer-aided drug dosage. *Computers and Biomedical Research*, **5**, 441-459.

152. Sheiner, L.B., Rosenberg, B. and Marathe, V.V. (1977). Estimation of population characteristics of pharmacokinetic parameters from routine clinical data. *Journal of Pharmacokinetics and Biopharmaceutics*, **5**, 445-479.

153. Sheiner, L.B. and Beal, S.L. (1980). Evaluation of methods for estimating population pharmacokinetic parameters. I. Michaelis-Menten model: routine pharmacokinetic data. *Journal of Pharmacokinetics and Biopharmaceutics*, **8**, 553-571.

154. Sheiner, L.B. and Beal, S.L. (1981). Evaluation of methods for estimating population pharmacokinetic parameters. II. Biexponential model: experimental pharmacokinetic data. *Journal of Pharmacokinetics and Biopharmaceutics*, **9**, 635-651.

155. Shi, M., Weiss, R.E. and Taylor, J.M. (1996). An analysis of paediatric CD4 counts for acquired immune deficiency syndrome using flexible random curves. *Applied Statistics*, **45**, 151-163.

156. Silverman, B. W. (1984). Spline smoothing: the equivalent variable kernel method. *Annals of Statistics*, **12**, 898-916.

157. Simonoff, J. S. (1996). *Smoothing Methods in Statistics*. Springer-Verlag, New York.

158. Smith, P. L. (1979). Splines as a useful and convenient statistical tool. *American Statistician*, **33**, 57-62.

159. Smith, M. and Kohn, R. (1996). Nonparametric regression via Bayesian variable selection. *Journal of Econometrics*, **75**, 317-344.

160. Speed, T. P. (1991). Discussion of "That BLUP is a good thing: The estimation of random effects" by Robinson. *Statistics Science*, **6**, 42-44.

161. Speckman, P. (1988). Kernel smoothing in partial linear models. *Journal of Royal Statistical Society, Series B*, **50**, 413-436.

162. Staniswalis, J. G. (1989). The kernel estimate of a regression function in likelihood-based models. *Journal of American Statistical Association*, **84**, 276-283.

163. Steimer, J. L., Mallet, A., Golmard, J.L. and Bisvieux, J.F. (1984). Alternative approaches to estimation of population pharmacokineti parameters: comparison with the nonlinear mixed effect mode. *Drug Metabolism Review*, **15**, 265-292.

164. Stiratelli, R., Laird, N.M. and Ware, J. (1984). Random effects models for serial observations with binary responses. *Biometrics*, **40**, 961-971.

165. Stone, C. J. (1984). An asymptotically optimal window selection rule for kernel density estimation. *Annals of Statistics*, **12**, 1285-97.

166. Stone, C. J., Hansen, M.H., Kooperberg, C. and Truong, Y. K. (1997). Polynomial splines and their tensor products in extended linear modeling. *Annals of Statistics*, **25**, 1371-1425.

167. Sun, Y. and Wu, H. (2005). Semiparametric time-varying regression model for longitudinal data. *Scandinavian Journal of Statistics*, **32**, 21-47.

168. Tao, H., Palta, M., Yandell, B.S. and Newton, M. A. (2005). An estimation method for the semiparametric mixed-effects model. *Biometrics*, **55**, 102-110.

169. Tibshirani, R. and Hastie, T. (1987). Local likelihood estimation. *Journal of American Statistical Association*, **82**, 559-567.

170. Verbeke, G. and Molenberghs, G. (2000). *Linear Mixed Models for Longitudinal Data.* Springer-Verlag, New York.

171. Verbyla, A.P., Cullis, B. R., Kenward, M. G. and Welham, S. J. (1999). The analysis of designed experiments and longitudinal data by using smoothing splines. *Applied Statistics*, **48**, 269-311.

172. Vonesh, E. F. and Carter, R.L. (1992). Mixed-effects nonlinear regression for unbalanced repeated measures. *Biometrics*, **48**, 1-18.

173. Vonesh, E. F. and Chinchilli, V. M. (1996). *Linear and Nonlinear Models for the Analysis of Repeated Measurements.* Marcel Dekker, New York.

174. Wahba, G. (1977). A survey of some smoothing problems and the method of generalized cross-validation for solving them. In *Applications of Statistics*, (P. R. Krisnaiah, ed.), 507-523. North Holland, Amsterdam.

175. Wahba, G. (1978). Improper priors, spline smoothing and the problem of guarding against model errors in regression. *Journal of Royal Statistical Society, Series B*, **40**, 364-372.

176. Wahba, G. (1983). Bayesian "confidence intervals" for the cross-validated smoothing spline. *Journal of Royal Statistical Society, Series B*, **45**, 133-150.

177. Wahba, G. (1985). A comparison of GCV and GML for choosing the smoothing parameter in the generalized spline smoothing problem. *Annals of Statistics*, **13**, 1378-1402.

178. Wahba, G. (1990). *Spline Models for Observational Data*. SIAM, Philadelphia. CBMS-NSF Regional Conference Series in Applied Mathematics, Vol. 59.

179. Wand, M.P. and Jones, M.C. (1995). *Kernel Smoothing*. Chapman and Hall, London.

180. Wang, N. (2003). Marginal nonparametric kernel regression accounting for within-subject correlation. *Biometrika*, **90**, 43-52.

181. Wang, N., Carroll, R. J. and Lin X. (2005) . Efficient semiparametric marginal estimation for longitudinal /clustered data. *Journal of American Statistical Association*, **100**, 147-157.

182. Wang, Y. (1998a). Mixed-effects smoothing spline ANOVA. *Journal of Royal Statistical Society, Series B*, **60**, 159-174.

183. Wang, Y. (1998b). Smoothing spline models with correlated random errors. *Journal of American Statistical Association*, **93**, 341-348.

184. Wang, Y.X. and Taylor, J.M. (1995). Inference for smooth curves in longitudinal data with application to an AIDS clinical trial. *Statistics in Medicine*, **14**, 1205-1218.

185. Watson, G. S. (1964). Smooth regression analysis. *Sankhya*, **26**, 101-116.

186. Wecker, W. E. and Ansley, C. F. (1983). The signal extraction approach to nonlinear regression and spline smoothing. *Journal of American Statistical Association*, **78**, 81-89.

187. Wedderburn, R.W.M. (1974). Quasilikelihood functions, generalized linear models and the Gauss-Newton method. *Biometrika*, **61**, 439-447.

188. Wegman, E. J. and Wright, I. W. (1983). Splines in statistics. *Journal of American Statistical Association*, **78**, 351-265.

189. Welsh, A.H., Lin, X. and Carroll, R.J. (2002). Marginal longitudinal nonparametric regression: locality and efficiency of spline and kernel methods. *Journal of American Statistical Association*, **97**, 482-493.

190. West, M., Harrison, P. J. and Migon, H.S. (1985). Dynamic generalized linear models and Bayesian forecasting. *Journal of American Statistical Association*, **80**, 73-83.

191. Wold, S. (1974). Spline functions in data analysis. *Technometrics*, **16**, 1-11.

192. Wolfinger, R. (1996). *The GLIMMIX SAS Macro*. SAS Institute, Cary, NC, USA.

193. Wu, C. O. and Chiang, C. T. (2000). Kernel smoothing on varying coefficient models with longitudinal dependent variable. *Statistica Sinica*, **10**, 433-456.

194. Wu, C. O., Chiang, C. T. and Hoover, D.R. (1998). Asymptotic confidence regions for kernel smoothing of a varying-coefficient model with longitudinal data. *Journal of American Statistical Association*, **93**, 1388-1402.

195. Wu, H. and Liang, H. (2004). Backfitting random varying-coefficient models with time-dependent smoothing covariates. *Scandinavian Journal of Statistics*, **31**, 3-19.

196. Wu, H. and Liang, H. (2005). Random varying-coefficient models with smoothing covariates, applications to an AIDS clinical study. Manuscript.

197. Wu, H. and Zhang, J.-T. (2002a). Local polynomial mixed-effects models for longitudinal data. *Journal of American Statistical Association*, **97**, 883-897.

198. Wu, H. and Zhang, J.-T. (2002b). The study of long-term HIV dynamics using semiparametric nonlinear mixed-effects models. *Statistics in Medicine*, **21**, 3655-3675.

199. Yen, S. and Jaffe, R. B. (1991). *Reproductive Endocrinology: Physiology, Pathophysiology, and Clinical Management*. W. B. Saunders, Philadephia.

200. Zeger, S.L. and Diggle, P.J. (1994). Semiparametric models for longitudinal data with application to CD4 cell numbers in HIV seroconverters. *Biometrics*, **50**, 689-699.

201. Zeger, S.L. and Karim, M.R. (1991). Generalized linear models with random effects: a Gibbs sampling approach. *Journal of American Statistical Association*, **86**, 79-86.

202. Zeger, S.L., Liang, K. and Albert, P. (1988). Models for longitudinal data: A generalized estimating equation approach. *Biometrics*, **44**, 1049-1060.

203. Zellner, A. (1962). An efficient method of estimating seemingly unrelated regressions and tests of aggregation bias. *Journal of American Statistical Association*, **57**, 348-368.

204. Zhang, D., Lin, X., Raz, J. and Sowers, M. (1998). Semiparametric stochastic mixed models for longitudinal data. *Journal of American Statistical Association*, **93**, 710-719.

205. Zhang, H.P. (1994). Maximal correlation and adaptive splines . *Technometrics*, **36**, 196-201.

206. Zhang, H.P. (1997). Multivariate adaptive splines for the analysis of longitudinal data. *Journal of Computational and Graphical Statistics*, **6**, 74-91.

207. Zhang, H.P. (1999). Analysis of infant growth curves using multivariate adaptive splines. *Biometrics*, **55**, 452-459.

208. Zhang, J.-T. and Fan, J. (2000). Minimax kernels for nonparametric curve estimation. *Journal of Nonparametric Statistics*, **12**, 417-445.

Index

WILEY SERIES IN PROBABILITY AND STATISTICS

ESTABLISHED BY WALTER A. SHEWHART AND SAMUEL S. WILKS

The ***Wiley Series in Probability and Statistics*** is well established and authoritative. It covers many topics of current research interest in both pure and applied statistics and probability theory. Written by leading statisticians and institutions, the titles span both state-of-the-art developments in the field and classical methods.

Reflecting the wide range of current research in statistics, the series encompasses applied, methodological and theoretical statistics, ranging from applications and new techniques made possible by advances in computerized practice to rigorous treatment of theoretical approaches.

This series provides essential and invaluable reading for all statisticians, whether in academia, industry, government, or research.

† ABRAHAM and LEDOLTER · Statistical Methods for Forecasting
AGRESTI · Analysis of Ordinal Categorical Data
AGRESTI · An Introduction to Categorical Data Analysis
AGRESTI · Categorical Data Analysis, *Second Edition*
ALTMAN, GILL, and McDONALD · Numerical Issues in Statistical Computing for the Social Scientist
AMARATUNGA and CABRERA · Exploration and Analysis of DNA Microarray and Protein Array Data
ANDĚL · Mathematics of Chance
ANDERSON · An Introduction to Multivariate Statistical Analysis, *Third Edition*
* ANDERSON · The Statistical Analysis of Time Series
ANDERSON, AUQUIER, HAUCK, OAKES, VANDAELE, and WEISBERG · Statistical Methods for Comparative Studies
ANDERSON and LOYNES · The Teaching of Practical Statistics
ARMITAGE and DAVID (editors) · Advances in Biometry
ARNOLD, BALAKRISHNAN, and NAGARAJA · Records
* ARTHANARI and DODGE · Mathematical Programming in Statistics
* BAILEY · The Elements of Stochastic Processes with Applications to the Natural Sciences
BALAKRISHNAN and KOUTRAS · Runs and Scans with Applications
BARNETT · Comparative Statistical Inference, *Third Edition*
BARNETT and LEWIS · Outliers in Statistical Data, *Third Edition*
BARTOSZYNSKI and NIEWIADOMSKA-BUGAJ · Probability and Statistical Inference
BASILEVSKY · Statistical Factor Analysis and Related Methods: Theory and Applications
BASU and RIGDON · Statistical Methods for the Reliability of Repairable Systems
BATES and WATTS · Nonlinear Regression Analysis and Its Applications
BECHHOFER, SANTNER, and GOLDSMAN · Design and Analysis of Experiments for Statistical Selection, Screening, and Multiple Comparisons
BELSLEY · Conditioning Diagnostics: Collinearity and Weak Data in Regression

*Now available in a lower priced paperback edition in the Wiley Classics Library.
†Now available in a lower priced paperback edition in the Wiley–Interscience Paperback Series.

† BELSLEY, KUH, and WELSCH · Regression Diagnostics: Identifying Influential Data and Sources of Collinearity
BENDAT and PIERSOL · Random Data: Analysis and Measurement Procedures, *Third Edition*
BERRY, CHALONER, and GEWEKE · Bayesian Analysis in Statistics and Econometrics: Essays in Honor of Arnold Zellner
BERNARDO and SMITH · Bayesian Theory
BHAT and MILLER · Elements of Applied Stochastic Processes, *Third Edition*
BHATTACHARYA and WAYMIRE · Stochastic Processes with Applications
† BIEMER, GROVES, LYBERG, MATHIOWETZ, and SUDMAN · Measurement Errors in Surveys
BILLINGSLEY · Convergence of Probability Measures, *Second Edition*
BILLINGSLEY · Probability and Measure, *Third Edition*
BIRKES and DODGE · Alternative Methods of Regression
BLISCHKE AND MURTHY (editors) · Case Studies in Reliability and Maintenance
BLISCHKE AND MURTHY · Reliability: Modeling, Prediction, and Optimization
BLOOMFIELD · Fourier Analysis of Time Series: An Introduction, *Second Edition*
BOLLEN · Structural Equations with Latent Variables
BOLLEN and CURRAN · Latent Curve Models: A Structural Equation Perspective
BOROVKOV · Ergodicity and Stability of Stochastic Processes
BOULEAU · Numerical Methods for Stochastic Processes
BOX · Bayesian Inference in Statistical Analysis
BOX · R. A. Fisher, the Life of a Scientist
BOX and DRAPER · Empirical Model-Building and Response Surfaces
* BOX and DRAPER · Evolutionary Operation: A Statistical Method for Process Improvement
BOX, HUNTER, and HUNTER · Statistics for Experimenters: Design, Innovation, and Discovery, *Second Editon*
BOX and LUCEÑO · Statistical Control by Monitoring and Feedback Adjustment
BRANDIMARTE · Numerical Methods in Finance: A MATLAB-Based Introduction
BROWN and HOLLANDER · Statistics: A Biomedical Introduction
BRUNNER, DOMHOF, and LANGER · Nonparametric Analysis of Longitudinal Data in Factorial Experiments
BUCKLEW · Large Deviation Techniques in Decision, Simulation, and Estimation
CAIROLI and DALANG · Sequential Stochastic Optimization
CASTILLO, HADI, BALAKRISHNAN, and SARABIA · Extreme Value and Related Models with Applications in Engineering and Science
CHAN · Time Series: Applications to Finance
CHARALAMBIDES · Combinatorial Methods in Discrete Distributions
CHATTERJEE and HADI · Sensitivity Analysis in Linear Regression
CHATTERJEE and PRICE · Regression Analysis by Example, *Third Edition*
CHERNICK · Bootstrap Methods: A Practitioner's Guide
CHERNICK and FRIIS · Introductory Biostatistics for the Health Sciences
CHILÈS and DELFINER · Geostatistics: Modeling Spatial Uncertainty
CHOW and LIU · Design and Analysis of Clinical Trials: Concepts and Methodologies, *Second Edition*
CLARKE and DISNEY · Probability and Random Processes: A First Course with Applications, *Second Edition*
* COCHRAN and COX · Experimental Designs, *Second Edition*
CONGDON · Applied Bayesian Modelling
CONGDON · Bayesian Statistical Modelling
CONOVER · Practical Nonparametric Statistics, *Third Edition*
COOK · Regression Graphics

*Now available in a lower priced paperback edition in the Wiley Classics Library.
†Now available in a lower priced paperback edition in the Wiley–Interscience Paperback Series.

COOK and WEISBERG · Applied Regression Including Computing and Graphics
COOK and WEISBERG · An Introduction to Regression Graphics
CORNELL · Experiments with Mixtures, Designs, Models, and the Analysis of Mixture Data, *Third Edition*
COVER and THOMAS · Elements of Information Theory
COX · A Handbook of Introductory Statistical Methods
* COX · Planning of Experiments
CRESSIE · Statistics for Spatial Data, *Revised Edition*
CSÖRGŐ and HORVÁTH · Limit Theorems in Change Point Analysis
DANIEL · Applications of Statistics to Industrial Experimentation
DANIEL · Biostatistics: A Foundation for Analysis in the Health Sciences, *Eighth Edition*
* DANIEL · Fitting Equations to Data: Computer Analysis of Multifactor Data, *Second Edition*
DASU and JOHNSON · Exploratory Data Mining and Data Cleaning
DAVID and NAGARAJA · Order Statistics, *Third Edition*
* DEGROOT, FIENBERG, and KADANE · Statistics and the Law
DEL CASTILLO · Statistical Process Adjustment for Quality Control
DeMARIS · Regression with Social Data: Modeling Continuous and Limited Response Variables
DEMIDENKO · Mixed Models: Theory and Applications
DENISON, HOLMES, MALLICK and SMITH · Bayesian Methods for Nonlinear Classification and Regression
DETTE and STUDDEN · The Theory of Canonical Moments with Applications in Statistics, Probability, and Analysis
DEY and MUKERJEE · Fractional Factorial Plans
DILLON and GOLDSTEIN · Multivariate Analysis: Methods and Applications
DODGE · Alternative Methods of Regression
* DODGE and ROMIG · Sampling Inspection Tables, *Second Edition*
* DOOB · Stochastic Processes
DOWDY, WEARDEN, and CHILKO · Statistics for Research, *Third Edition*
DRAPER and SMITH · Applied Regression Analysis, *Third Edition*
DRYDEN and MARDIA · Statistical Shape Analysis
DUDEWICZ and MISHRA · Modern Mathematical Statistics
DUNN and CLARK · Basic Statistics: A Primer for the Biomedical Sciences, *Third Edition*
DUPUIS and ELLIS · A Weak Convergence Approach to the Theory of Large Deviations
* ELANDT-JOHNSON and JOHNSON · Survival Models and Data Analysis
ENDERS · Applied Econometric Time Series
† ETHIER and KURTZ · Markov Processes: Characterization and Convergence
EVANS, HASTINGS, and PEACOCK · Statistical Distributions, *Third Edition*
FELLER · An Introduction to Probability Theory and Its Applications, Volume I, *Third Edition,* Revised; Volume II, *Second Edition*
FISHER and VAN BELLE · Biostatistics: A Methodology for the Health Sciences
FITZMAURICE, LAIRD, and WARE · Applied Longitudinal Analysis
* FLEISS · The Design and Analysis of Clinical Experiments
FLEISS · Statistical Methods for Rates and Proportions, *Third Edition*
† FLEMING and HARRINGTON · Counting Processes and Survival Analysis
FULLER · Introduction to Statistical Time Series, *Second Edition*
FULLER · Measurement Error Models
GALLANT · Nonlinear Statistical Models
GEISSER · Modes of Parametric Statistical Inference
GEWEKE · Contemporary Bayesian Econometrics and Statistics

*Now available in a lower priced paperback edition in the Wiley Classics Library.
†Now available in a lower priced paperback edition in the Wiley–Interscience Paperback Series.

GHOSH, MUKHOPADHYAY, and SEN · Sequential Estimation
GIESBRECHT and GUMPERTZ · Planning, Construction, and Statistical Analysis of Comparative Experiments
GIFI · Nonlinear Multivariate Analysis
GIVENS and HOETING · Computational Statistics
GLASSERMAN and YAO · Monotone Structure in Discrete-Event Systems
GNANADESIKAN · Methods for Statistical Data Analysis of Multivariate Observations, *Second Edition*
GOLDSTEIN and LEWIS · Assessment: Problems, Development, and Statistical Issues
GREENWOOD and NIKULIN · A Guide to Chi-Squared Testing
GROSS and HARRIS · Fundamentals of Queueing Theory, *Third Edition*
† GROVES · Survey Errors and Survey Costs
* HAHN and SHAPIRO · Statistical Models in Engineering
HAHN and MEEKER · Statistical Intervals: A Guide for Practitioners
HALD · A History of Probability and Statistics and their Applications Before 1750
HALD · A History of Mathematical Statistics from 1750 to 1930
† HAMPEL · Robust Statistics: The Approach Based on Influence Functions
HANNAN and DEISTLER · The Statistical Theory of Linear Systems
HEIBERGER · Computation for the Analysis of Designed Experiments
HEDAYAT and SINHA · Design and Inference in Finite Population Sampling
HELLER · MACSYMA for Statisticians
HINKELMANN and KEMPTHORNE · Design and Analysis of Experiments, Volume 1: Introduction to Experimental Design
HINKELMANN and KEMPTHORNE · Design and Analysis of Experiments, Volume 2: Advanced Experimental Design
HOAGLIN, MOSTELLER, and TUKEY · Exploratory Approach to Analysis of Variance
HOAGLIN, MOSTELLER, and TUKEY · Exploring Data Tables, Trends and Shapes
* HOAGLIN, MOSTELLER, and TUKEY · Understanding Robust and Exploratory Data Analysis
HOCHBERG and TAMHANE · Multiple Comparison Procedures
HOCKING · Methods and Applications of Linear Models: Regression and the Analysis of Variance, *Second Edition*
HOEL · Introduction to Mathematical Statistics, *Fifth Edition*
HOGG and KLUGMAN · Loss Distributions
HOLLANDER and WOLFE · Nonparametric Statistical Methods, *Second Edition*
HOSMER and LEMESHOW · Applied Logistic Regression, *Second Edition*
HOSMER and LEMESHOW · Applied Survival Analysis: Regression Modeling of Time to Event Data
† HUBER · Robust Statistics
HUBERTY · Applied Discriminant Analysis
HUNT and KENNEDY · Financial Derivatives in Theory and Practice
HUSKOVA, BERAN, and DUPAC · Collected Works of Jaroslav Hajek—with Commentary
HUZURBAZAR · Flowgraph Models for Multistate Time-to-Event Data
IMAN and CONOVER · A Modern Approach to Statistics
† JACKSON · A User's Guide to Principle Components
JOHN · Statistical Methods in Engineering and Quality Assurance
JOHNSON · Multivariate Statistical Simulation
JOHNSON and BALAKRISHNAN · Advances in the Theory and Practice of Statistics: A Volume in Honor of Samuel Kotz
JOHNSON and BHATTACHARYYA · Statistics: Principles and Methods, *Fifth Edition*

*Now available in a lower priced paperback edition in the Wiley Classics Library.
†Now available in a lower priced paperback edition in the Wiley–Interscience Paperback Series.

JOHNSON and KOTZ · Distributions in Statistics
JOHNSON and KOTZ (editors) · Leading Personalities in Statistical Sciences: From the Seventeenth Century to the Present
JOHNSON, KOTZ, and BALAKRISHNAN · Continuous Univariate Distributions, Volume 1, *Second Edition*
JOHNSON, KOTZ, and BALAKRISHNAN · Continuous Univariate Distributions, Volume 2, *Second Edition*
JOHNSON, KOTZ, and BALAKRISHNAN · Discrete Multivariate Distributions
JOHNSON, KEMP, and KOTZ · Univariate Discrete Distributions, *Third Edition*
JUDGE, GRIFFITHS, HILL, LÜTKEPOHL, and LEE · The Theory and Practice of Econometrics, *Second Edition*
JUREČKOVÁ and SEN · Robust Statistical Procedures: Aymptotics and Interrelations
JUREK and MASON · Operator-Limit Distributions in Probability Theory
KADANE · Bayesian Methods and Ethics in a Clinical Trial Design
KADANE AND SCHUM · A Probabilistic Analysis of the Sacco and Vanzetti Evidence
KALBFLEISCH and PRENTICE · The Statistical Analysis of Failure Time Data, *Second Edition*
KASS and VOS · Geometrical Foundations of Asymptotic Inference
† KAUFMAN and ROUSSEEUW · Finding Groups in Data: An Introduction to Cluster Analysis
KEDEM and FOKIANOS · Regression Models for Time Series Analysis
KENDALL, BARDEN, CARNE, and LE · Shape and Shape Theory
KHURI · Advanced Calculus with Applications in Statistics, *Second Edition*
KHURI, MATHEW, and SINHA · Statistical Tests for Mixed Linear Models
* KISH · Statistical Design for Research
KLEIBER and KOTZ · Statistical Size Distributions in Economics and Actuarial Sciences
KLUGMAN, PANJER, and WILLMOT · Loss Models: From Data to Decisions, *Second Edition*
KLUGMAN, PANJER, and WILLMOT · Solutions Manual to Accompany Loss Models: From Data to Decisions, *Second Edition*
KOTZ, BALAKRISHNAN, and JOHNSON · Continuous Multivariate Distributions, Volume 1, *Second Edition*
KOTZ and JOHNSON (editors) · Encyclopedia of Statistical Sciences: Volumes 1 to 9 with Index
KOTZ and JOHNSON (editors) · Encyclopedia of Statistical Sciences: Supplement Volume
KOTZ, READ, and BANKS (editors) · Encyclopedia of Statistical Sciences: Update Volume 1
KOTZ, READ, and BANKS (editors) · Encyclopedia of Statistical Sciences: Update Volume 2
KOVALENKO, KUZNETZOV, and PEGG · Mathematical Theory of Reliability of Time-Dependent Systems with Practical Applications
LACHIN · Biostatistical Methods: The Assessment of Relative Risks
LAD · Operational Subjective Statistical Methods: A Mathematical, Philosophical, and Historical Introduction
LAMPERTI · Probability: A Survey of the Mathematical Theory, *Second Edition*
LANGE, RYAN, BILLARD, BRILLINGER, CONQUEST, and GREENHOUSE · Case Studies in Biometry
LARSON · Introduction to Probability Theory and Statistical Inference, *Third Edition*
LAWLESS · Statistical Models and Methods for Lifetime Data, *Second Edition*
LAWSON · Statistical Methods in Spatial Epidemiology
LE · Applied Categorical Data Analysis
LE · Applied Survival Analysis
LEE and WANG · Statistical Methods for Survival Data Analysis, *Third Edition*

*Now available in a lower priced paperback edition in the Wiley Classics Library.
†Now available in a lower priced paperback edition in the Wiley–Interscience Paperback Series.

LePAGE and BILLARD · Exploring the Limits of Bootstrap
LEYLAND and GOLDSTEIN (editors) · Multilevel Modelling of Health Statistics
LIAO · Statistical Group Comparison
LINDVALL · Lectures on the Coupling Method
LIN · Introductory Stochastic Analysis for Finance and Insurance
LINHART and ZUCCHINI · Model Selection
LITTLE and RUBIN · Statistical Analysis with Missing Data, *Second Edition*
LLOYD · The Statistical Analysis of Categorical Data
LOWEN and TEICH · Fractal-Based Point Processes
MAGNUS and NEUDECKER · Matrix Differential Calculus with Applications in Statistics and Econometrics, *Revised Edition*
MALLER and ZHOU · Survival Analysis with Long Term Survivors
MALLOWS · Design, Data, and Analysis by Some Friends of Cuthbert Daniel
MANN, SCHAFER, and SINGPURWALLA · Methods for Statistical Analysis of Reliability and Life Data
MANTON, WOODBURY, and TOLLEY · Statistical Applications Using Fuzzy Sets
MARCHETTE · Random Graphs for Statistical Pattern Recognition
MARDIA and JUPP · Directional Statistics
MASON, GUNST, and HESS · Statistical Design and Analysis of Experiments with Applications to Engineering and Science, *Second Edition*
McCULLOCH and SEARLE · Generalized, Linear, and Mixed Models
McFADDEN · Management of Data in Clinical Trials
* McLACHLAN · Discriminant Analysis and Statistical Pattern Recognition
McLACHLAN, DO, and AMBROISE · Analyzing Microarray Gene Expression Data
McLACHLAN and KRISHNAN · The EM Algorithm and Extensions
McLACHLAN and PEEL · Finite Mixture Models
McNEIL · Epidemiological Research Methods
MEEKER and ESCOBAR · Statistical Methods for Reliability Data
MEERSCHAERT and SCHEFFLER · Limit Distributions for Sums of Independent Random Vectors: Heavy Tails in Theory and Practice
MICKEY, DUNN, and CLARK · Applied Statistics: Analysis of Variance and Regression, *Third Edition*
* MILLER · Survival Analysis, *Second Edition*
MONTGOMERY, PECK, and VINING · Introduction to Linear Regression Analysis, *Third Edition*
MORGENTHALER and TUKEY · Configural Polysampling: A Route to Practical Robustness
MUIRHEAD · Aspects of Multivariate Statistical Theory
MULLER and STOYAN · Comparison Methods for Stochastic Models and Risks
MURRAY · X-STAT 2.0 Statistical Experimentation, Design Data Analysis, and Nonlinear Optimization
MURTHY, XIE, and JIANG · Weibull Models
MYERS and MONTGOMERY · Response Surface Methodology: Process and Product Optimization Using Designed Experiments, *Second Edition*
MYERS, MONTGOMERY, and VINING · Generalized Linear Models. With Applications in Engineering and the Sciences
† NELSON · Accelerated Testing, Statistical Models, Test Plans, and Data Analyses
† NELSON · Applied Life Data Analysis
NEWMAN · Biostatistical Methods in Epidemiology
OCHI · Applied Probability and Stochastic Processes in Engineering and Physical Sciences
OKABE, BOOTS, SUGIHARA, and CHIU · Spatial Tesselations: Concepts and Applications of Voronoi Diagrams, *Second Edition*
OLIVER and SMITH · Influence Diagrams, Belief Nets and Decision Analysis

*Now available in a lower priced paperback edition in the Wiley Classics Library.
†Now available in a lower priced paperback edition in the Wiley–Interscience Paperback Series.

PALTA · Quantitative Methods in Population Health: Extensions of Ordinary Regressions
PANKRATZ · Forecasting with Dynamic Regression Models
PANKRATZ · Forecasting with Univariate Box-Jenkins Models: Concepts and Cases
* PARZEN · Modern Probability Theory and Its Applications
PEÑA, TIAO, and TSAY · A Course in Time Series Analysis
PIANTADOSI · Clinical Trials: A Methodologic Perspective
PORT · Theoretical Probability for Applications
POURAHMADI · Foundations of Time Series Analysis and Prediction Theory
PRESS · Bayesian Statistics: Principles, Models, and Applications
PRESS · Subjective and Objective Bayesian Statistics, *Second Edition*
PRESS and TANUR · The Subjectivity of Scientists and the Bayesian Approach
PUKELSHEIM · Optimal Experimental Design
PURI, VILAPLANA, and WERTZ · New Perspectives in Theoretical and Applied Statistics
† PUTERMAN · Markov Decision Processes: Discrete Stochastic Dynamic Programming
QIU · Image Processing and Jump Regression Analysis
* RAO · Linear Statistical Inference and Its Applications, *Second Edition*
RAUSAND and HØYLAND · System Reliability Theory: Models, Statistical Methods, and Applications, *Second Edition*
RENCHER · Linear Models in Statistics
RENCHER · Methods of Multivariate Analysis, *Second Edition*
RENCHER · Multivariate Statistical Inference with Applications
* RIPLEY · Spatial Statistics
RIPLEY · Stochastic Simulation
ROBINSON · Practical Strategies for Experimenting
ROHATGI and SALEH · An Introduction to Probability and Statistics, *Second Edition*
ROLSKI, SCHMIDLI, SCHMIDT, and TEUGELS · Stochastic Processes for Insurance and Finance
ROSENBERGER and LACHIN · Randomization in Clinical Trials: Theory and Practice
ROSS · Introduction to Probability and Statistics for Engineers and Scientists
† ROUSSEEUW and LEROY · Robust Regression and Outlier Detection
* RUBIN · Multiple Imputation for Nonresponse in Surveys
RUBINSTEIN · Simulation and the Monte Carlo Method
RUBINSTEIN and MELAMED · Modern Simulation and Modeling
RYAN · Modern Regression Methods
RYAN · Statistical Methods for Quality Improvement, *Second Edition*
SALEH · Theory of Preliminary Test and Stein-Type Estimation with Applications
* SCHEFFE · The Analysis of Variance
SCHIMEK · Smoothing and Regression: Approaches, Computation, and Application
SCHOTT · Matrix Analysis for Statistics, *Second Edition*
SCHOUTENS · Levy Processes in Finance: Pricing Financial Derivatives
SCHUSS · Theory and Applications of Stochastic Differential Equations
SCOTT · Multivariate Density Estimation: Theory, Practice, and Visualization
SEARLE · Linear Models for Unbalanced Data
SEARLE · Matrix Algebra Useful for Statistics
SEARLE, CASELLA, and McCULLOCH · Variance Components
SEARLE and WILLETT · Matrix Algebra for Applied Economics
SEBER and LEE · Linear Regression Analysis, *Second Edition*
† SEBER · Multivariate Observations
† SEBER and WILD · Nonlinear Regression
SENNOTT · Stochastic Dynamic Programming and the Control of Queueing Systems
* SERFLING · Approximation Theorems of Mathematical Statistics
SHAFER and VOVK · Probability and Finance: It's Only a Game!

*Now available in a lower priced paperback edition in the Wiley Classics Library.
†Now available in a lower priced paperback edition in the Wiley–Interscience Paperback Series.

SILVAPULLE and SEN · Constrained Statistical Inference: Inequality, Order, and Shape Restrictions
SMALL and McLEISH · Hilbert Space Methods in Probability and Statistical Inference
SRIVASTAVA · Methods of Multivariate Statistics
STAPLETON · Linear Statistical Models
STAUDTE and SHEATHER · Robust Estimation and Testing
STOYAN, KENDALL, and MECKE · Stochastic Geometry and Its Applications, *Second Edition*
STOYAN and STOYAN · Fractals, Random Shapes and Point Fields: Methods of Geometrical Statistics
STYAN · The Collected Papers of T. W. Anderson: 1943–1985
SUTTON, ABRAMS, JONES, SHELDON, and SONG · Methods for Meta-Analysis in Medical Research
TAKEZAWA · Introduction to Nonparametric Regression
TANAKA · Time Series Analysis: Nonstationary and Noninvertible Distribution Theory
THOMPSON · Empirical Model Building
THOMPSON · Sampling, *Second Edition*
THOMPSON · Simulation: A Modeler's Approach
THOMPSON and SEBER · Adaptive Sampling
THOMPSON, WILLIAMS, and FINDLAY · Models for Investors in Real World Markets
TIAO, BISGAARD, HILL, PEÑA, and STIGLER (editors) · Box on Quality and Discovery: with Design, Control, and Robustness
TIERNEY · LISP-STAT: An Object-Oriented Environment for Statistical Computing and Dynamic Graphics
TSAY · Analysis of Financial Time Series, *Second Edition*
UPTON and FINGLETON · Spatial Data Analysis by Example, Volume II: Categorical and Directional Data
VAN BELLE · Statistical Rules of Thumb
VAN BELLE, FISHER, HEAGERTY, and LUMLEY · Biostatistics: A Methodology for the Health Sciences, *Second Edition*
VESTRUP · The Theory of Measures and Integration
VIDAKOVIC · Statistical Modeling by Wavelets
VINOD and REAGLE · Preparing for the Worst: Incorporating Downside Risk in Stock Market Investments
WALLER and GOTWAY · Applied Spatial Statistics for Public Health Data
WEERAHANDI · Generalized Inference in Repeated Measures: Exact Methods in MANOVA and Mixed Models
WEISBERG · Applied Linear Regression, *Third Edition*
WELSH · Aspects of Statistical Inference
WESTFALL and YOUNG · Resampling-Based Multiple Testing: Examples and Methods for p-Value Adjustment
WHITTAKER · Graphical Models in Applied Multivariate Statistics
WINKER · Optimization Heuristics in Economics: Applications of Threshold Accepting
WONNACOTT and WONNACOTT · Econometrics, *Second Edition*
WOODING · Planning Pharmaceutical Clinical Trials: Basic Statistical Principles
WOODWORTH · Biostatistics: A Bayesian Introduction
WOOLSON and CLARKE · Statistical Methods for the Analysis of Biomedical Data, *Second Edition*
WU and HAMADA · Experiments: Planning, Analysis, and Parameter Design Optimization
WU and ZHANG · Nonparametric Regression Methods for Longitudinal Data Analysis
YANG · The Construction Theory of Denumerable Markov Processes
* ZELLNER · An Introduction to Bayesian Inference in Econometrics
ZHOU, OBUCHOWSKI, and McCLISH · Statistical Methods in Diagnostic Medicine

*Now available in a lower priced paperback edition in the Wiley Classics Library.
†Now available in a lower priced paperback edition in the Wiley–Interscience Paperback Series.